Extreme Values in Finance, Telecommunications, and the Environment

MONOGRAPHS ON STATISTICS AND APPLIED PROBABILITY

General Editors

V. Isham, N. Keiding, T. Louis, N. Reid, R. Tibshirani, and H. Tong

1 Stochastic Population Models in Ecology and Epidemiology *M.S. Barlett* (1960)
2 Queues *D.R. Cox and W.L. Smith* (1961)
3 Monte Carlo Methods *J.M. Hammersley and D.C. Handscomb* (1964)
4 The Statistical Analysis of Series of Events *D.R. Cox and P.A.W. Lewis* (1966)
5 Population Genetics *W.J. Ewens* (1969)
6 Probability, Statistics and Time *M.S. Barlett* (1975)
7 Statistical Inference *S.D. Silvey* (1975)
8 The Analysis of Contingency Tables *B.S. Everitt* (1977)
9 Multivariate Analysis in Behavioural Research *A.E. Maxwell* (1977)
10 Stochastic Abundance Models *S. Engen* (1978)
11 Some Basic Theory for Statistical Inference *E.J.G. Pitman* (1979)
12 Point Processes *D.R. Cox and V. Isham* (1980)
13 Identification of Outliers *D.M. Hawkins* (1980)
14 Optimal Design *S.D. Silvey* (1980)
15 Finite Mixture Distributions *B.S. Everitt and D.J. Hand* (1981)
16 Classification *A.D. Gordon* (1981)
17 Distribution-Free Statistical Methods, 2nd edition *J.S. Maritz* (1995)
18 Residuals and Influence in Regression *R.D. Cook and S. Weisberg* (1982)
19 Applications of Queueing Theory, 2nd edition *G.F. Newell* (1982)
20 Risk Theory, 3rd edition *R.E. Beard, T. Pentikäinen and E. Pesonen* (1984)
21 Analysis of Survival Data *D.R. Cox and D. Oakes* (1984)
22 An Introduction to Latent Variable Models *B.S. Everitt* (1984)
23 Bandit Problems *D.A. Berry and B. Fristedt* (1985)
24 Stochastic Modelling and Control *M.H.A. Davis and R. Vinter* (1985)
25 The Statistical Analysis of Composition Data *J. Aitchison* (1986)
26 Density Estimation for Statistics and Data Analysis *B.W. Silverman* (1986)
27 Regression Analysis with Applications *G.B. Wetherill* (1986)
28 Sequential Methods in Statistics, 3rd edition
G.B. Wetherill and K.D. Glazebrook (1986)
29 Tensor Methods in Statistics *P. McCullagh* (1987)
30 Transformation and Weighting in Regression
R.J. Carroll and D. Ruppert (1988)
31 Asymptotic Techniques for Use in Statistics
O.E. Bandorff-Nielsen and D.R. Cox (1989)
32 Analysis of Binary Data, 2nd edition *D.R. Cox and E.J. Snell* (1989)
33 Analysis of Infectious Disease Data *N.G. Becker* (1989)
34 Design and Analysis of Cross-Over Trials *B. Jones and M.G. Kenward* (1989)
35 Empirical Bayes Methods, 2nd edition *J.S. Maritz and T. Lwin* (1989)

36 Symmetric Multivariate and Related Distributions
K.T. Fang, S. Kotz and K.W. Ng (1990)

37 Generalized Linear Models, 2nd edition *P. McCullagh and J.A. Nelder* (1989)

38 Cyclic and Computer Generated Designs, 2nd edition
J.A. John and E.R. Williams (1995)

39 Analog Estimation Methods in Econometrics *C.F. Manski* (1988)

40 Subset Selection in Regression *A.J. Miller* (1990)

41 Analysis of Repeated Measures *M.J. Crowder and D.J. Hand* (1990)

42 Statistical Reasoning with Imprecise Probabilities *P. Walley* (1991)

43 Generalized Additive Models *T.J. Hastie and R.J. Tibshirani* (1990)

44 Inspection Errors for Attributes in Quality Control
N.L. Johnson, S. Kotz and X. Wu (1991)

45 The Analysis of Contingency Tables, 2nd edition *B.S. Everitt* (1992)

46 The Analysis of Quantal Response Data *B.J.T. Morgan* (1992)

47 Longitudinal Data with Serial Correlation—A State-Space Approach
R.H. Jones (1993)

48 Differential Geometry and Statistics *M.K. Murray and J.W. Rice* (1993)

49 Markov Models and Optimization *M.H.A. Davis* (1993)

50 Networks and Chaos—Statistical and Probabilistic Aspects
O.E. Barndorff-Nielsen, J.L. Jensen and W.S. Kendall (1993)

51 Number-Theoretic Methods in Statistics *K.-T. Fang and Y. Wang* (1994)

52 Inference and Asymptotics *O.E. Barndorff-Nielsen and D.R. Cox* (1994)

53 Practical Risk Theory for Actuaries
C.D. Daykin, T. Pentikäinen and M. Pesonen (1994)

54 Biplots *J.C. Gower and D.J. Hand* (1996)

55 Predictive Inference—An Introduction *S. Geisser* (1993)

56 Model-Free Curve Estimation *M.E. Tarter and M.D. Lock* (1993)

57 An Introduction to the Bootstrap *B. Efron and R.J. Tibshirani* (1993)

58 Nonparametric Regression and Generalized Linear Models
P.J. Green and B.W. Silverman (1994)

59 Multidimensional Scaling *T.F. Cox and M.A.A. Cox* (1994)

60 Kernel Smoothing *M.P. Wand and M.C. Jones* (1995)

61 Statistics for Long Memory Processes *J. Beran* (1995)

62 Nonlinear Models for Repeated Measurement Data
M. Davidian and D.M. Giltinan (1995)

63 Measurement Error in Nonlinear Models
R.J. Carroll, D. Rupert and L.A. Stefanski (1995)

64 Analyzing and Modeling Rank Data *J.J. Marden* (1995)

65 Time Series Models—In Econometrics, Finance and Other Fields
D.R. Cox, D.V. Hinkley and O.E. Barndorff-Nielsen (1996)

66 Local Polynomial Modeling and its Applications *J. Fan and I. Gijbels* (1996)

67 Multivariate Dependencies—Models, Analysis and Interpretation
D.R. Cox and N. Wermuth (1996)

68 Statistical Inference—Based on the Likelihood *A. Azzalini* (1996)

69 Bayes and Empirical Bayes Methods for Data Analysis
B.P. Carlin and T.A Louis (1996)

70 Hidden Markov and Other Models for Discrete-Valued Time Series
I.L. Macdonald and W. Zucchini (1997)

71 Statistical Evidence—A Likelihood Paradigm *R. Royall* (1997)

72 Analysis of Incomplete Multivariate Data *J.L. Schafer* (1997)

73 Multivariate Models and Dependence Concepts *H. Joe* (1997)

74 Theory of Sample Surveys *M.E. Thompson* (1997)

75 Retrial Queues *G. Falin and J.G.C. Templeton* (1997)

76 Theory of Dispersion Models *B. Jørgensen* (1997)

77 Mixed Poisson Processes *J. Grandell* (1997)

78 Variance Components Estimation—Mixed Models, Methodologies and Applications
P.S.R.S. Rao (1997)

79 Bayesian Methods for Finite Population Sampling
G. Meeden and M. Ghosh (1997)

80 Stochastic Geometry—Likelihood and computation
O.E. Barndorff-Nielsen, W.S. Kendall and M.N.M. van Lieshout (1998)

81 Computer-Assisted Analysis of Mixtures and Applications—
Meta-analysis, Disease Mapping and Others *D. Böhning* (1999)

82 Classification, 2nd edition *A.D. Gordon* (1999)

83 Semimartingales and their Statistical Inference *B.L.S. Prakasa Rao* (1999)

84 Statistical Aspects of BSE and vCJD—Models for Epidemics
C.A. Donnelly and N.M. Ferguson (1999)

85 Set-Indexed Martingales *G. Ivanoff and E. Merzbach* (2000)

86 The Theory of the Design of Experiments *D.R. Cox and N. Reid* (2000)

87 Complex Stochastic Systems
O.E. Barndorff-Nielsen, D.R. Cox and C. Klüppelberg (2001)

88 Multidimensional Scaling, 2nd edition *T.F. Cox and M.A.A. Cox* (2001)

89 Algebraic Statistics—Computational Commutative Algebra in Statistics
G. Pistone, E. Riccomagno and H.P. Wynn (2001)

90 Analysis of Time Series Structure—SSA and Related Techniques
N. Golyandina, V. Nekrutkin and A.A. Zhigljavsky (2001)

91 Subjective Probability Models for Lifetimes
Fabio Spizzichino (2001)

92 Empirical Likelihood *Art B. Owen* (2001)

93 Statistics in the 21st Century
Adrian E. Raftery, Martin A. Tanner, and Martin T. Wells (2001)

94 Accelerated Life Models: Modeling and Statistical Analysis
Vilijandas Bagdonavičius and Mikhail Nikulin (2001)

95 Subset Selection in Regression, Second Edition *Alan Miller* (2002)

96 Topics in Modelling of Clustered Data
Marc Aerts, Helena Geys, Geert Molenberghs, and Louise M. Ryan (2002)

97 Components of Variance *D.R. Cox and P.J. Solomon* (2002)

98 Design and Analysis of Cross-Over Trials, 2nd Edition
Byron Jones and Michael G. Kenward (2003)

99 Extreme Values in Finance, Telecommunications, and the Environment
Bärbel Finkenstädt and Holger Rootzén (2003)

Extreme Values in Finance, Telecommunications, and the Environment

Edited by
Bärbel Finkenstädt
and
Holger Rootzén

A CRC Press Company
Boca Raton London New York Washington, D.C.

Library of Congress Cataloging-in-Publication Data

Séminaire européen de statistique (5th : 2001 : Gothenburg, Sweden)
Extreme values in finance, telecommunications, and the environment / edited by Barbel Finkenstadt, Holger Rootzen.
p. cm. — (Monographs on statistics and applied probability ; 99)
Includes bibliographical references and index.
ISBN 1-58488-411-8 (alk. paper)
1. Extreme value theory—Congresses. I. Finkenstädt, Bärbel. II. Rootzén, Holger. III. Title. IV. Series.

QA273.6.S45 2001
519.2′4—dc21 2003051602

Direct all inquiries to CRC Press LLC, 2000 N.W. Corporate Blvd., Boca Raton, Florida 33431.

International Standard Book Number 1-58488-411-8
Library of Congress Card Number 2003051602
Printed in the United States of America 1 2 3 4 5 6 7 8 9 0
Printed on acid-free paper

Contents

Contributors

Stuart Coles
Department of Mathematics
University of Bristol
Bristol, United Kingdom
Stuart.Coles@bristol.ac.uk, http://www.stats.bris.ac.uk/masgc/

Paul Embrechts
Department of Mathematics
Eidgenössische Technische Hochschule Zürich
Zürich, Switzerland
embrechts@math.ethz.ch, http://www.math.ethz.ch/~embrechts

Anne-Laure Fougères
Laboratoire de Statistique et Probabilités
Institut National des Sciences Appliquées de Toulouse — Université Paul Sabatier
Dépt. GMM, INSA
Toulouse, France
fougeres@gmm.insa-tlse.fr, http://www.gmm.insa-tlse.fr/fougeres/englishal.html

Claudia Klüppelberg
Center of Mathematical Sciences
Munich University of Technology
Munich, Germany
cklu@mathematik.tu-muenchen.de, http://www.ma.tum.de/stat/

Thomas Mikosch
Laboratory of Actuarial Mathematics
University of Copenhagen
Copenhagen, Denmark
mikosch@math.ku.dk, http://www.math.ku.dk/~mikosch

Sidney Resnick
School of Operations Research and Industrial Engineering
Cornell University
Ithaca, New York
sid@orie.cornell.edu, http://www.orie.cornell.edu/~sid

Richard L. Smith
Department of Statistics
University of North Carolina
Chapel Hill, North Carolina
rls@email.unc.edu, http://www.unc.edu/depts/statistics/faculty/rsmith.html

Participants

Andriy Adreev, Helsinki (Finland), andriy.andreev@shh.fi
A note on histogram approximation in Bayesian density estimation.

Jenny Andersson, Gothenburg (Sweden), jennya@math.chalmers.se
Analysis of corrosion on aluminium and magnesium by statistics of extremes.

Bojan Basrak, Eindhoven (Netherlands), basrak@eurandom.tue.nl
On multivariate regular variation and some time-series models.

Nathanaël Benjamin, Oxford (United Kingdom), nathanael.benjamin@centraliens.net
Bound on an approximation for the distribution of the extreme fluctuations of exchange rates.

Paola Bortot, Bologna (Italy), bortot@stat.unibo.it
Extremes of volatile Markov chains.

Natalia Botchkina, Bristol (United Kingdom), natasha.botchkina@mail.com
Wavelets and extreme value theory.

Leonardo Bottolo, Pavia (Italy), lbottolo@eco.unipv.it
Mixture models in Bayesian risk analysis.

Boris Buchmann, Munich (Germany), bbuch@mathematik.tu-muenchen.de
Decompounding: an estimation problem for the compound Poisson distribution.

Adam Butler, Lancaster (United Kingdom), a.butler@lancaster.ac.uk
The impact of climate change upon extreme sea levels.

Ana Cebrian, Louvain-La-Neuve (Belgium), cebrian@stat.ucl.ac.be
Analysis of bivariate extreme dependence using copulas with applications to insurance.

Ana Ferreira, Eindhoven (Netherlands), ferreira@eurandom.tue.nl
Confidence intervals for the tail index.

Christopher Ferro, Lancaster (United Kingdom), c.ferro@lancaster.ac.uk
Aspects of modelling extremal temporal dependence.

John Greenhough, Warwick (United Kingdom), greenh@astro.warwick.ac.uk
Characterizing anomalous transport in accretion disks from X-ray observations.

Viviane Grunert da Fonseca, Faro (Portugal), vgrunert@ualg.pt
Stochastic multiobjective optimization and the attainment function.

Janet Heffernan, Lancaster (United Kingdom), j.heffernan@lancaster.ac.uk
A conditional approach for multivariate extreme values.

Rachel Hilliam, Birmingham (United Kingdom), rmh@for.mat.bham.ac.uk
Statistical aspects of chaos-based communications modelling.

Daniel Hlubinka, Prague (Czech Republic), hlubinka@karlin.mff.cuni.cz
Stereology of extremes: shape factor.

Marian Hristache, Bruz (France), marian.hristache@ensai.fr
Structure adaptive approach for dimension reduction.

Pär Johannesson, Gothenburg (Sweden), par.johannesson@fcc.chalmers.se
Crossings of intervals in fatigue of materials.

Joachim Johansson, Gothenburg (Sweden), joachimj@math.chalmers.se
A semi-parametric estimator of the mean of heavy-tailed distributions.

Elisabeth Joossens, Leuven (Belgium), bettie.joossens@wis.kuleuven.ac.be
On the estimation of the largest inclusions in a piece of steel using extreme value analysis.

Vadim Kuzmin, St. Petersburg (Russia), kuzmin@rw.ru
Stochastic forecasting of extreme flood transformation.

Fabrizio Laurini, Padova (Italy), flaurini@stat.unipd.it
Estimating the extremal index in financial time series.

Tao Lin, Rotterdam (Netherlands), lin@few.eur.nl
Statistics of extremes in C[0,1].

Alexander Lindner, Munich (Germany), lindner@ma.tum.de
Angles and linear reconstruction of missing data.

Owen Lyne, Nottingham (United Kingdom), owen.lyne@nottingham.ac.uk
Statistical inference for multitype households SIR epidemics.

Hans Malmsten, Stockholm (Sweden), hans.malmsten@hhs.se
Moment structure of a family of first-order exponential GARCH models.

Alex Morton, Warwick (United Kingdom), a.morton@warwick.ac.uk
A new class of models for irregularly sampled time series.

Natalie Neumeyer, Bochum (Germany), natalie.neumeyer@ruhr-uni-bochum.de
Nonparametric comparison of regression functions—an empirical process approach.

Paul Northrop, Oxford (United Kingdom), northrop@stats.ox.ac.uk
An empirical Bayes approach to flood estimation.

Grégory Nuel, Evry (France), gnuel@maths.univ-evry.fr
Unusual word frequencies in Markov chains: the large deviations approach.

Fehmi Özkan, Freiburg im Breisgau (Germany), oezkan@fdm.uni-freiburg.de
The defaultable Lévy term structure: ratings and restructuring.

Francesco Pauli, Trieste (Italy), francescopauli@interfree.it
A multivariate model for extremes.

Olivier Perrin, Toulouse (France), perrin@cict.fr
On a time deformation reducing stochastic processes to local stationarity.

Martin Schlather, Bayreuth (Germany), martin.schlather@uni-bayreuth.de
A dependence measure for extreme values.

Manuel Scotto, Algueirão (Portugal), arima@mail.telepac.pt
Extremal behaviour of certain transformations of time series.

Scott Sisson, Bristol (United Kingdom), scott.sisson@bristol.ac.uk
An application involving uncertain asymptotic temporal dependence in the extremes of time series.

Alwin Stegeman, Groningen (Netherlands), stegeman@math.rug.nl
Long-range dependence in computer network traffic: theory and practice.

Scherbakov Vadim, Glasgow (United Kingdom), vadim@stats.gla.ac.uk
Voter model with mean-field interaction.

Yingcun Xia, Cambridge (United Kingdom), ycxia@zoo.cam.ac.uk
A childhood epidemic model with birthrate-dependent transmission.

Preface

The chapters in this volume are the invited papers presented at the fifth Séminaire Européen de Statistique (SemStat) on extreme value theory and applications, held under the auspices of Chalmers and Gothenburg University at the Nordic Folk Academy in Gothenburg, 10–16 December, 2001.

The volume is thus the most recent in a sequence of conference volumes that have appeared as a result of each Séminaire Européen de Statistique. The first of these workshops took place in 1992 at Sandbjerg Manor in the southern part of Denmark. The topic was statistical aspects of chaos and neural networks. A second meeting on time series models in econometrics, finance, and other fields was held in Oxford in December, 1994. The third meeting on stochastic geometry: likelihood and computation took place in Toulouse, 1996, and a fourth meeting, on complex stochastic systems, was held at EURANDOM, Eindhoven, 1999. Since August, 1996, SemStat has been under the auspices of the European Regional Committee of the Bernoulli Society for Mathematical Statistics and Probability.

The aim of the Séminaire Européen de Statistique is to provide young scientists with an opportunity to get quickly to the forefront of knowledge and research in areas of statistical science which are of current major interest. About 40 young researchers from various European countries participated in the 2001 séminaire. Each of them presented his or her work either by giving a seminar talk or contributing to a poster session. A list of the invited contributors and the young attendants of the séminaire, along with the titles of their presentations, can be found on the preceding pages.

The central paradigm of extreme value theory is semiparametric: you cannot trust standard statistical modeling by normal, lognormal, Weibull, or other distributions all the way out into extreme tails and maxima. On the other hand, nonparametric methods cannot be used either, because interest centers on more extreme events than those one already has encountered. The solution to this dilemma is semiparametric models which only specify the distributional shapes of maxima, as the extreme value distributions, or of extreme tails, as the generalized Pareto distributions. The rationales for these models are very basic limit and stability arguments.

The first chapter, written by Richard Smith, gives a survey of how this paradigm answers a variety of questions of interest to an applied scientist in climatology, insurance, and finance. The chapter also reviews parts of univariate extreme value theory and discusses estimation, diagnostics, multivariate extremes, and max-stable processes.

In the second chapter, Stuart Coles focuses on the particularly extreme event of the 1999 rainfall in Venezuela that caused widespread distruction and loss of life. He demonstrates that the probability for such an event would have been miscalculated even by the standard extreme value models, and discusses the use of various options available for extension in order to achieve a more satisfactory analysis.

The next three chapters consider applications of extreme value theory to risk management in finance and economics. First, in Chapter 3, Claudia Klüppelberg reviews aspects of Value-at-Risk (VaR) and its estimation based on extreme value theory. She presents results of a comprehensive investigation of the extremal behavior of some of the most important continuous and discrete time series models that are of current interest in finance. Her discussions are followed by an historic overview of financial risk management given by Paul Embrechts in Chapter 4. In Chapter 5, Thomas Mikosch introduces the stylized facts of financial time series, in particular the heavy tails exhibited by log-returns. He studies, in depth, their connection with standard econometric models such as the GARCH and stochastic volatility processes. The reader is also introduced to the mathematical concept of regular variation.

Another important area where extreme value theory plays a significant role is data network modelling. In Chapter 6, Sidney Resnick reviews issues to consider for data network modelling, some of the basic models and statistical techniques for fitting these models.

Finally, in Chapter 7, Anne-Laure Fougères gives an overview of multivariate extreme value distributions and the problem of measuring extremal dependence.

The order in which the chapters are compiled approximately follows the order in which they were presented at the conference. Naturally it is not possible to cover all aspects of this interesting and exciting research area in a single conference volume. The most important omission may be the extensive use of extreme value theory in reliability theory. This includes modelling of extreme wind and wave loads on structures, of strength of materials, and of metal corrosion and fatigue. In addition to methods discussed in this volume, these areas use the deep and interesting theory of extremes of Gaussian processes. Nevertheless it is our hope that the coverage provided by this volume will help the readers to acquaint themselves speedily with current research issues and techniques in extreme value theory.

The scientific programme of the fifth Séminaire Européen de Statistique was organized by the steering group, which, at the time of the conference, consisted of O.E. Barndorff-Nielsen (Aarhus University), B. Finkenstädt (University of Warwick), W.S. Kendall (University of Warwick), C. Klüppelberg (Munich University of Technology), D. Picard (Paris VII), H. Rootzén (Chalmers University Gothenburg), and A. van der Vaart (Free University Amsterdam). The local organization of the séminaire was in the hands of H. Rootzén and the smooth running was to a large part due to Johan Segers (Tilburg University), Jenny Andersson, and Jacques de Maré (both at Chalmers University Gothenburg).

The fifth Séminaire Européen de Statistique was supported by the TMR-network in statistical and computational methods for the analysis of spatial data, the Stochastic Centre in Gothenburg, the Swedish Institute of Applied Mathematics, the Swedish

Technical Sciences Research Council, the Swedish Natural Sciences Research Council, and the Knut and Alice Wallenberg Foundation. We are grateful for this support, without which the séminaire could not have taken place.

On behalf of the SemStat steering group
B. Finkenstädt and H. Rootzén
Warwick, Gothenburg

CHAPTER 1

Statistics of Extremes, with Applications in Environment, Insurance, and Finance

Richard L. Smith
University of North Carolina

Contents

1-58488-411-8/04/$0.00+$.50

1.1 Motivating examples

Extreme value theory is concerned with probabilistic and statistical questions related to very high or very low values in sequences of random variables and in stochastic processes. The subject has a rich mathematical theory and also a long tradition of applications in a variety of areas. Among many excellent books on the subject, Embrechts et al. (1997) give a comprehensive survey of the mathematical theory with an orientation toward applications in insurance and finance, while the recent book by Coles (2001) concentrates on data analysis and statistical inference for extremes.

The present survey is primarily concerned with statistical applications, and especially with how the mathematical theory can be extended to answer a variety of questions of interest to an applied scientist. Traditionally, extreme value theory has been employed to answer questions relating to the distribution of extremes (e.g., what is the probability that a windspeed over a given level will occur in a given location during a given year?) or the inverse problem of return levels (e.g., what height of a river will be exceeded with probability 1/100 in a given year? — this quantity is often called the 100-year return level). During the last 30 years, many new techniques have been developed concerned with exceedances over high thresholds, the dependence among extreme events in various types of stochastic processes, and multivariate extremes.

These new techniques make it possible to answer much more complex questions than simple distributions of extremes. Among those considered in the present review are whether probabilities of extreme events are changing with time or corresponding to other measured covariates (e.g., Section 1.5.1 through Section 1.5.3 and Section 1.6.4), the simultaneous fitting of extreme value distributions to several related time series (Section 1.6.1 through Section 1.6.3), the spatial dependence of extreme value distributions (Section 1.6.4) and the rather complex forms of extreme value calculations that arise in connection with financial time series (Section 1.8). Along the way, we shall also review relevant parts of the mathematical theory for univariate extremes (Section 1.2 through Section 1.4) and one recent approach (among several that are available) to the characterization of multivariate extreme value distributions (Section 1.7).

For the rest of this section, we give some specific examples of data-oriented questions which will serve to motivate the rest of the chapter.

1.1.1 Snowfall in North Carolina

On January 25, 2000, a snowfall of 20.3 inches was recorded at Raleigh-Durham airport in North Carolina. This is an exceptionally high snowfall for this part of the U.S. and caused widespread disruption to travel, power supplies, and the local school system. Various estimates that appeared in the press at the time indicated that such an event could be expected to occur once every 100 to 200 years. The question we consider here is how well one can estimate the probability of such an event based on data available prior to the actual event. Associated with this is the whole question of what is the uncertainty of such an assessment of an extreme value probability.

To simplify the question and to avoid having to consider time-of-year effects, we shall confine our discussion to the month of January, implicitly assuming that an extreme snowfall event is equally likely to occur at any time during the month. Thus the question we are trying to answer is, for any large value of x, "What is the probability that a snowfall exceeding x inches occurs at Raleigh-Durham airport, sometime during the month of January, in any given year?"

A representative data set was compiled from the publicly available data base of the U.S. National Climatic Data Center. Table 1.1 lists all the January snow events (i.e., daily totals where a nonzero snowfall was recorded) at Raleigh-Durham airport, for the period 1948 to 1998. We shall take this as a data base from which we try to answer the question just quoted, with $x = 20.3$, for some arbitrary year after 1998. It can be seen that no snowfall anywhere close to 20.3 inches occurs in the given data set, the largest being 9.0 inches on January 19, 1955. There are earlier records of daily snowfall events over 20 inches in this region, but these were prior to the establishment of a regular series of daily measurements, and we shall not take them into account.

In Section 1.3, we shall return to this example and show how a simple threshold-based analysis may be used to answer this question, but with particular attention to the sensitivity to the chosen threshold and to the contrast between maximum likelihood and Bayesian approaches.

1.1.2 Insurance risk of a large company

This example is based on Smith and Goodman (2000). A data set was compiled consisting of insurance claims made by an international oil company over a 15-year period. In the data set originally received from the company, 425 claims were recorded over a nominal threshold level, expressed in U.S. dollars and adjusted for inflation to 1994 cost equivalents. As a preliminary to the detailed analysis, two further preprocessing steps were performed: (i) the data were multiplied by a common but unspecified scaling factor — this has the effect of concealing the precise sums of money involved, without in any other way changing the characteristics of the data set, and (ii) simultaneous claims of the same type arising on the same day were aggregated

Table 1.1 *January snow events at Raleigh-Durham Airport, 1948–1998.*

Year	Day	Amount	Year	Day	Amount	Year	Day	Amount
1948	24	1.0	1965	15	0.8	1977	7	0.3
1948	31	2.5	1965	16	3.7	1977	24	1.8
1954	11	1.2	1965	17	1.3	1979	31	0.4
1954	22	1.2	1965	30	3.8	1980	30	1.0
1954	23	4.1	1965	31	0.1	1980	31	1.2
1955	19	9.0	1966	16	0.1	1981	30	2.6
1955	23	3.0	1966	22	0.2	1982	13	1.0
1955	24	1.0	1966	25	2.0	1982	14	5.0
1955	27	1.4	1966	26	7.6	1985	20	1.7
1956	23	2.0	1966	27	0.1	1985	28	2.4
1958	7	3.0	1966	29	1.8	1987	25	0.1
1959	8	1.7	1966	30	0.5	1987	26	0.5
1959	16	1.2	1967	19	0.5	1988	7	7.1
1961	21	1.2	1968	10	0.5	1988	8	0.2
1961	26	1.1	1968	11	1.1	1995	23	0.7
1962	1	1.5	1968	25	1.4	1995	30	0.1
1962	10	5.0	1970	12	1.0	1996	6	2.7
1962	19	1.6	1970	23	1.0	1996	7	2.9
1962	28	2.0	1973	7	0.7	1997	11	0.4
1963	26	0.1	1973	8	5.7	1998	19	2.0
1964	13	0.4	1976	17	0.4			

Table 1.2 *The seven types of insurance claims, with the total number of claims and the mean size of claim for each type.*

Type	Description	Number	Mean
1	Fire	175	11.1
2	Liability	17	12.2
3	Offshore	40	9.4
4	Cargo	30	3.9
5	Hull	85	2.6
6	Onshore	44	2.7
7	Aviation	2	1.6

into a single total claim for that day — the motivation for this was to avoid possible clustering effects due to claims arising from the same cause, though it is likely that this effect is minimal for the data set under consideration. With these two changes to the original data set, the analysed data consisted of 393 claims over a nominal threshold of 0.5, grouped into seven "types" as shown in Table 1.2.

The total of all 393 claims was 2989.6, and the ten largest claims, in order, were 776.2, 268.0, 142.0, 131.0, 95.8, 56.8, 46.2, 45.2, 40.4, and 30.7. These figures give some indication of the type of data we are talking about: the total loss to the company is dominated by the value of a few very large claims, with the largest claim itself

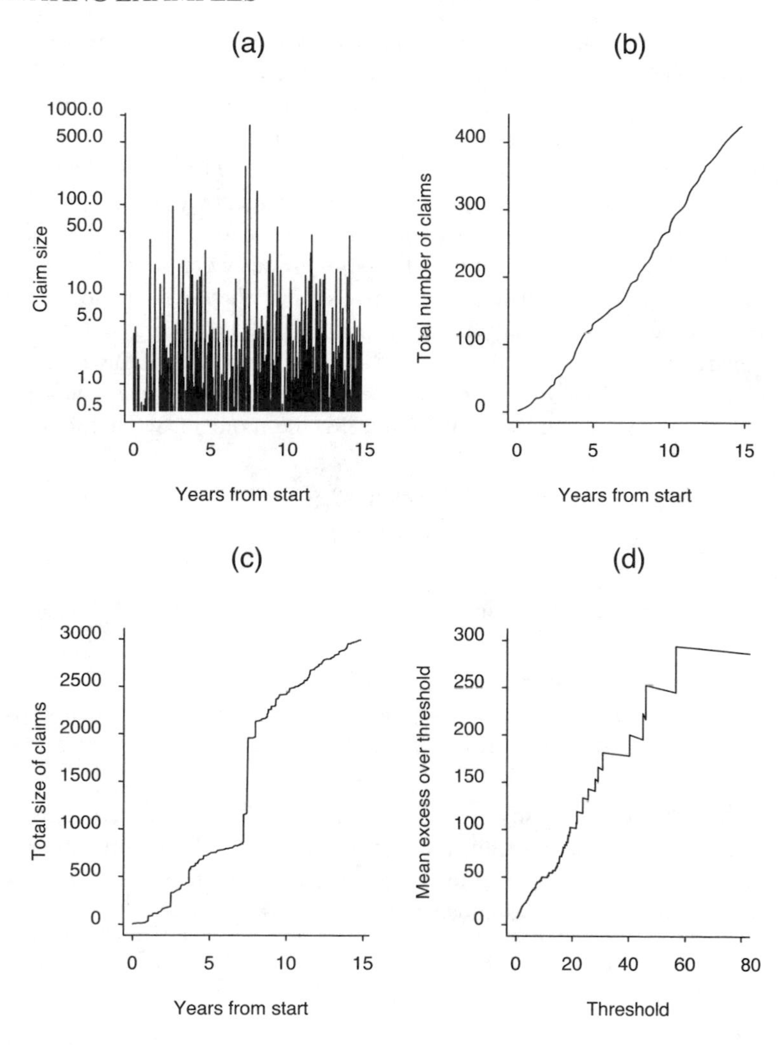

Figure 1.1 *Insurance data: (a) plot of raw data, (b) cumulative number of claims vs. time, (c) cumulative claim amount vs. time, and (d) mean excess plot.*

accounting for 26% of the total. In statistical terms, the data clearly represent a very skewed, long-tailed distribution, though these features are entirely typical of insurance data.

Further information about the data can be gained from Figure 1.1, which shows (a) a scatterplot of the individual claims against time — note that claims are drawn on a logarithmic scale; (b) cumulative number of claims against time — this serves as a visual indicator of whether there are trends in the frequency of claims; (c) cumulative claim amounts against time, as an indicator of trends in the total amounts of claims; and (d) the so-called mean excess plot, in which for a variety of possible thresholds, the mean excess over the threshold was computed for all claims that were above that threshold, and plotted against the threshold itself. As will be seen later (Section 1.4), this is a useful diagnostic of the generalized Pareto distribution (GPD) which is widely used as a probability distribution for excesses over thresholds — in this case, the fact that the plot in Figure 1.1(d) is close to a straight line over most of its range is an indicator that the GPD fits the data reasonably well. Of the other plots in Figure 1.1, plot (b) shows no visual evidence of a trend in the frequency of claims, while in (c), there is a sharp rise in the cumulative total of claims during year 7, but this arises largely because the two largest claims in the whole series were both in the same year, which raises the question of whether these two claims should be treated as outliers, and therefore analyzed separately from the rest of the data. The case for doing this is strengthened by the fact that these were the only two claims in the entire data set that resulted from the total loss of a facility. We shall return to these issues when the data are analysed in detail in Section 1.6, but for the moment, we list four possible questions for discussion:

1. What is the distribution of very large claims?
2. Is there any evidence of a change of the distribution of claim sizes and frequencies over time?
3. What is the influence of the different types of claims on the distribution of total claim size?
4. How should one characterize the risk to the company? More precisely, what probability distribution can one put on the amount of money that the company will have to pay out in settlement of large insurance claims over a future time period of, say, one year?

Published statistical analyses of insurance data often concentrate exclusively on question 1, but it is arguable that the other three questions are all more important and relevant than a simple characterisation of the probability distribution of claims, for a company planning its future insurance policies.

1.1.3 Value at risk in finance

Much of the recent research in extreme value theory has been stimulated by the possibility of large losses in the financial markets, which has resulted in a large amount of literature on "value at risk" and other measures of financial vulnerability. As an example of the types of data analysed and the kinds of questions asked, Figure 1.2 shows negative daily returns from closing prices of 1982 to 2001 stock prices in three companies, Pfizer, General Electric, and Citibank. If X_t is the closing price of a stock or financial index on day t, then the daily return (in effect, the percentage loss or gain on the day) is defined either by

$$Y_t = 100\left(\frac{X_t}{X_{t-1}} - 1\right) \tag{1.1}$$

or, more conveniently for the present discussion, by

$$Y_t = 100 \log \frac{X_t}{X_{t-1}}. \tag{1.2}$$

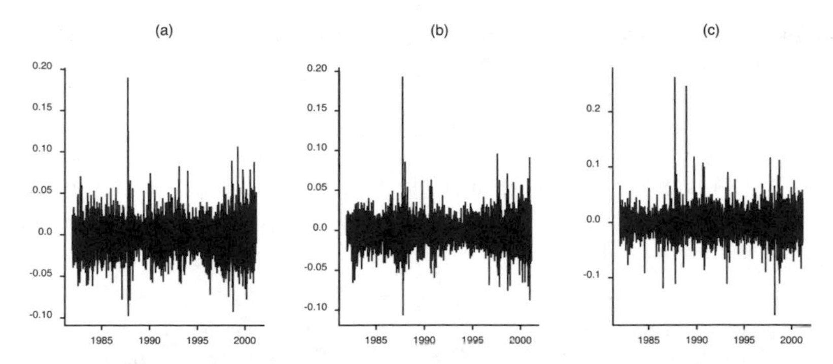

Figure 1.2 *Negative daily returns, defined by (1.3), for three stocks, 1982 to 2001, (a) Pfizer, (b) General Electric, and (c) Citibank.*

We are mainly interested in the possibility of large losses rather than large gains, so we rewrite (1.2) in terms of negative returns,

$$Y_t = 100 \log \frac{X_{t-1}}{X_t}, \tag{1.3}$$

which is the quantity actually plotted in Figure 1.2.

Typical problems here are:

1. Calculating the value at risk, i.e., the amount which might be lost in a portfolio of assets over a specified time period with a specified small probability;
2. Describing dependence among the extremes of different series, and using this description in the problem of managing a portfolio of investments; and
3. Modeling extremes in the presence of volatility — like all financial time series, those in Figure 1.2 show periods when the variability or volatility in the series is high, and others where it is much lower, but simple theories of extreme values in independent and identically distributed (i.i.d.) random variables or simple stationary time series do not account for such behaviour.

In Section 1.8, we shall return to this example and suggest some possible approaches to answering these questions.

1.2 Univariate extreme value theory

1.2.1 The extreme value distributions

In this section, we outline the basic theory that applies to univariate sequences of i.i.d. random variables. This theory is by now very well established and is the starting point for all the extreme value methods we shall discuss.

Suppose we have an i.i.d. sequence of random variables, $X_1, X_2, \ldots$, whose common cumulative distribution function is F, i.e.,

$$F(x) = \Pr\{X_i \le x\}.$$

Also let $M_n = \max(X_1, \ldots, X_n)$ denote the nth sample maximum of the process. Then

$$\Pr\{M_n \le x\} = F(x)^n. \tag{1.4}$$

Result (1.4) is of no immediate interest, since it simply says that for any fixed x for which $F(x) < 1$, we have $\Pr\{M_n \le x\} \to 0$. For nontrivial limit results, we must renormalize: find $a_n > 0$, b_n such that

$$\Pr\left\{\frac{M_n - b_n}{a_n} \le x\right\} = F(a_n x + b_n)^n \to H(x). \tag{1.5}$$

The Three Types Theorem, originally stated without detailed mathematical proof by Fisher and Tippett (1928), and later derived rigorously by Gnedenko (1943), asserts that if a nondegenerate H exists (i.e., a distribution function which does not put all its mass at a single point), it must be one of three types:

$$H(x) = \exp(-e^{-x}), \text{ all } x, \tag{1.6}$$

$$H(x) = \begin{cases} 0, & x < 0, \\ \exp(-x^{-\alpha}), & x > 0, \end{cases} \tag{1.7}$$

$$H(x) = \begin{cases} \exp(-|x|^{\alpha}), & x < 0, \\ 1, & x > 0. \end{cases} \tag{1.8}$$

Here two distribution functions H_1 and H_2 are said to be of the same type if one can be derived from the other through a simple location-scale transformation,

$$H_1(x) = H_2(Ax + B), \quad A > 0.$$

Very often, (1.6) is called the Gumbel type, (1.7) the Fréchet type, and (1.8) the Weibull type. In (1.7) and (1.8), $\alpha > 0$.

The three types may be combined into a single generalized extreme value (GEV) distribution:

$$H(x) = \exp\left\{-\left(1 + \xi\frac{x - \mu}{\psi}\right)_+^{-1/\xi}\right\}, \tag{1.9}$$

($y_+ = \max(y, 0)$) where μ is a location parameter, $\psi > 0$ is a scale parameter, and ξ is a shape parameter. The limit $\xi \to 0$ corresponds to the Gumbel distribution, $\xi > 0$ to the Fréchet distribution with $\alpha = 1/\xi$, and $\xi < 0$ to the Weibull distribution with $\alpha = -1/\xi$.

In more informal language, the case $\xi > 0$ is the "long-tailed" case for which $1 - H(x) \propto x^{-1/\xi}$, $\xi = 0$ is the "medium-tailed" case for which $1 - H(x)$ decreases exponentially for large x, and $\xi < 0$ is the "short-tailed" case, in which the distribution has a finite endpoint (the minimum value of x for which $H(x) = 1$) at $x = \mu - \psi/\xi$.

1.2.2 Exceedances over thresholds

Consider the distribution of X conditionally on exceeding some high threshold u (so $Y = X - u > 0$):

$$F_u(y) = \Pr\{Y \leq y \mid Y > 0\} = \frac{F(u+y) - F(u)}{1 - F(u)}.$$

As $u \to \omega_F = \sup\{x : F(x) < 1\}$, we often find a limit

$$F_u(y) \approx G(y; \sigma_u, \xi), \tag{1.10}$$

where G is generalized Pareto distribution (GPD)

$$G(y; \sigma, \xi) = 1 - \left(1 + \xi \frac{y}{\sigma}\right)_+^{-1/\xi}. \tag{1.11}$$

Although the Pareto and similar distributions have long been used as models for long-tailed processes, the rigorous connection with classical extreme value theory was established by Pickands (1975). In effect, Pickands showed that for any given F, a GPD approximation arises from (1.10) if and only there exist normalizing constants and a limiting H such that the classical extreme value limit result (1.5) holds; in that case, if H is written in GEV form (1.9), then the shape parameter ξ is the same as the corresponding GPD parameter in (1.11). Thus there is a close parallel between limit results for sample maxima and limit results for exceedances over thresholds, which is quite extensively exploited in modern statistical methods for extremes.

In the GPD, the case $\xi > 0$ is long-tailed, for which $1 - G(x)$ decays at the same rate as $x^{-1/\xi}$ for large x. This is reminiscent of the usual Pareto distribution, $G(x) = 1 - cx^{-\alpha}$, with $\xi = 1/\alpha$. For $\xi = 0$, we may take the limit as $\xi \to 0$ to get

$$G(y; \sigma, 0) = 1 - \exp\left(-\frac{y}{\sigma}\right),$$

i.e., exponential distribution with mean σ. For $\xi < 0$, the distribution has finite upper endpoint at $-\sigma/\xi$. Some other elementary results about the GPD are

$$\begin{aligned} \mathrm{E}(Y) &= \frac{\sigma}{1-\xi}, \quad (\xi < 1), \\ \mathrm{Var}(Y) &= \frac{\sigma^2}{(1-\xi)^2(1-2\xi)}, \quad \left(\xi < \frac{1}{2}\right), \\ \mathrm{E}(Y - y \mid Y > y > 0) &= \frac{\sigma + \xi y}{1-\xi}, \quad (\xi < 1). \end{aligned} \tag{1.12}$$

Poisson-GPD model for exceedances

Suppose we observe i.i.d. random variables $X_1, \ldots, X_n$, and observe the indices i for which $X_i > u$. If these indices are rescaled to points i/n, they can be viewed as a point process of rescaled exceedance times on [0, 1]. If $n \to \infty$ and $1 - F(u) \to 0$ such that $n(1 - F(u)) \to \lambda$ $(0 < \lambda < \infty)$, the process converges weakly to a homogeneous Poisson process on [0, 1], of intensity λ.

Motivated by this, we can imagine a limiting form of the joint point process of exceedance times and excesses over the threshold, of the following form:

1. The number, N, of exceedances of the level u in any one year has a Poisson distribution with mean λ; and
2. Conditionally on $N \geq 1$, the excess values $Y_1, \ldots, Y_N$ are i.i.d. from the GPD.

We call this the Poisson–GPD model.

Of course, there is nothing special here about one year as the unit of time — we could just as well use any other time unit — but for environmental processes in particular, a year is often the most convenient reference time period.

The Poisson–GPD process is closely related to the GEV distribution for annual maxima. Suppose $x > u$. The probability that the annual maximum of the Poisson–GPD process is less than x is

$$\begin{aligned}
\Pr\{\max_{1 \leq i \leq N} Y_i \leq x\} &= \Pr\{N = 0\} + \sum_{n=1}^{\infty} \Pr\{N = n,\ Y_1 \leq x, \ldots Y_n \leq x\} \\
&= e^{-\lambda} + \sum_{n=1}^{\infty} \frac{\lambda^n e^{-\lambda}}{n!} \cdot \left\{1 - \left(1 + \xi \frac{x - u}{\sigma}\right)_+^{-1/\xi}\right\}^n \\
&= \exp\left\{-\lambda \left(1 + \xi \frac{x - u}{\sigma}\right)_+^{-1/\xi}\right\}. \qquad (1.13)
\end{aligned}$$

If we substitute

$$\sigma = \psi + \xi(u - \mu), \qquad \lambda = \left(1 + \xi \frac{u - \mu}{\psi}\right)^{-1/\xi}, \qquad (1.14)$$

(1.13) reduces to the GEV form (1.9). Thus the GEV and GPD models are entirely consistent with one another above the threshold u, and (1.14) gives an explicit relationship between the two sets of parameters.

The Poisson–GPD model is closely related to the peaks over threshold (POT) model originally developed by hydrologists. In cases with high serial correlation, the threshold exceedances do not occur singly but in clusters, and, in that case, the method is most directly applied to the peak values within each cluster. For more detailed discussion, see Davison and Smith (1990).

Another issue is seasonal dependence. For environmental processes in particular, it is rarely the case that the probability of an extreme event is independent of the time of year, so we need some extension of the model to account for seasonality. Possible strategies include:

1. Remove seasonal trend before applying the threshold approach.
2. Apply the Poisson–GPD model separately to each season.
3. Expand the Poisson–GPD model to include covariates.

All three approaches have been extensively applied in past discussions of threshold methods. In the present chapter, we focus primarily on method 3. (e.g., Section 1.5 and Section 1.6.4), though only after first rewriting the Poisson-GPD model in a different form (Section 1.2.5).

1.2.3 Examples

In this section, we present four examples to illustrate how the extreme value and GPD limiting distributions work in practice, given various assumptions on the distribution function F from which the random variables are drawn. From a mathematical viewpoint, these examples are all special cases of the domain of attraction problem, which has been dealt with extensively in texts on extreme value theory, e.g., Leadbetter et al. (1983) or Resnick (1987). Here we make no attempt to present the general theory, but the examples serve to illustrate the concepts in some of the most typical cases.

The exponential distribution

Suppose $F(x) = 1 - e^{-x}$. Let $a_n = 1$ and $b_n = \log n$. Then

$$F^n(a_n x + b_n) = (1 - e^{-x - \log n})^n = \left(1 - \frac{e^{-x}}{n}\right)^n \to \exp(-e^{-x}),$$

in other words, the limiting distribution of sample extremes in this case is the Gumbel distribution.

For the threshold version of the result, set $\sigma_u = 1$. Then

$$F_u(\sigma_u z) = \frac{F(u+z) - F(u)}{1 - F(u)} = \frac{e^{-u} - e^{-u-z}}{e^{-u}} = 1 - e^{-z}$$

so the exponential distribution of mean 1 is the exact distribution for exceedances in this case.

Pareto-type tail

Suppose $1 - F(x) \sim cx^{-\alpha}$ as $x \to \infty$, with $c > 0$ and $\alpha > 0$. Let $b_n = 0,\ a_n = (nc)^{1/\alpha}$. Then for $x > 0$,

$$F^n(a_n x) \approx \{1 - c(a_n x)^{-\alpha}\}^n = \left(1 - \frac{x^{-\alpha}}{n}\right)^n \to \exp(-x^{-\alpha}),$$

which is the Fréchet limit.

For the threshold result, let $\sigma_u = ub$ for some $b > 0$. Then

$$\begin{aligned} F_u(\sigma_u z) &= \frac{F(u+ubz) - F(u)}{1 - F(u)} \approx \frac{cu^{-\alpha} - c(u+ubz)^{-\alpha}}{cu^{-\alpha}} \\ &= 1 - (1 + bz)^{-\alpha}. \end{aligned}$$

Set $\xi = \frac{1}{\alpha}$ and $b = \xi$ to get the result in GPD form.

Finite upper endpoint

Suppose $\omega_F = \omega < \infty$ and $1 - F(\omega - y) \sim cy^\alpha$ as $y \downarrow 0$ for $c > 0, \alpha > 0$. Set $b_n = \omega$, $a_n = (nc)^{-1/\alpha}$. Then for $x < 0$,

$$F^n(a_n x + b_n) = F^n(\omega + a_n x) \approx \{1 - c(-a_n x)^\alpha\}^n$$
$$= \left\{1 - \frac{(-x)^\alpha}{n}\right\}^n \to \exp\{-(-x)^\alpha\},$$

which is of Weibull type.

For the threshold version, let u be very close to ω and consider $\sigma_u = b(\omega - u)$ for $b > 0$ to be determined. Then for $0 < z < \frac{1}{b}$,

$$F_u(\sigma_u z) = \frac{F(u + \sigma_u z) - F(u)}{1 - F(u)}$$
$$\approx \frac{c(\omega - u)^\alpha - c(\omega - u - \sigma_u z)^\alpha}{c(\omega - u)^\alpha} = 1 - (1 - bz)^\alpha.$$

This is of GPD form with $\xi = -\frac{1}{\alpha}$ and $b = -\xi$.

Normal extremes

Let $\Phi(x) = \frac{1}{\sqrt{2\pi}} \int_{-\infty}^x e^{-y^2/2} dy$. By Feller (1968), page 193,

$$1 - \Phi(x) \sim \frac{1}{x\sqrt{2\pi}} e^{-x^2/2} \quad \text{as } x \to \infty.$$

Then

$$\lim_{u\to\infty} \frac{1 - \Phi(u + z/u)}{1 - \Phi(u)} = \lim_{u\to\infty} \left[\left(1 + \frac{z}{u^2}\right)^{-1} \cdot \exp\left\{-\frac{1}{2}\left(u + \frac{z}{u}\right)^2 + \frac{1}{2}u^2\right\}\right]$$
$$= e^{-z}.$$

For a first application, let $\sigma_u = 1/u$, then

$$\frac{\Phi(u + \sigma_u z) - \Phi(u)}{1 - \Phi(u)} \to 1 - e^{-z} \quad \text{as } u \to \infty,$$

so the limiting distribution of exceedances over thresholds is exponential.

For a second application, define b_n by $\Phi(b_n) = 1 - 1/n$, $a_n = 1/b_n$. Then

$$n\{1 - \Phi(a_n x + b_n)\} = \frac{1 - \Phi(a_n x + b_n)}{1 - \Phi(b_n)} \to e^{-x}$$

and hence

$$\lim_{n\to\infty} \Phi^n(a_n x + b_n) = \lim_{n\to\infty} \left(1 - \frac{e^{-x}}{n}\right)^n = \exp(-e^{-x}),$$

establishing convergence to Gumbel limit.

In practice, although the Gumbel and exponential distributions are the correct limits for sample maxima and threshold exceedances respectively, better approximations are obtained using the GEV and GPD, allowing $\xi \neq 0$. This is known as the penultimate approximation and was investigated in detail by Cohen (1982a, 1982b). The practical implication of this is that it is generally better to use the GEV/GPD distributions even when we suspect that Gumbel/exponential are the correct limits.

1.2.4 The r largest order statistics model

An extension of the annual maximum approach is to use the r largest observations in each fixed time period (say, one year), where $r > 1$. The mathematical result on which this relies is that (1.5) is easily extended to the joint distribution of the r largest order statistics, as $n \to \infty$ for a fixed $r > 1$, and this may therefore be used as a basis for statistical inference. A practical caution is that the r-largest result is more vulnerable to departures from the i.i.d. assumption (say, if there is seasonal variation in the distribution of observations, or if observations are dependent) than the classical results about extremes.

The main result is as follows: if $Y_{n,1} \geq Y_{n,2} \geq \cdots \geq Y_{n,r}$ are r largest order statistics of i.i.d. sample of size n, and a_n and b_n are the normalising constants in (1.5), then

$$\left(\frac{Y_{n,1} - b_n}{a_n}, \ldots, \frac{Y_{n,r} - b_n}{a_n}\right)$$

converges in distribution to a limiting random vector $(X_1, \ldots, X_r)$, whose density is

$$h(x_1, \ldots, x_r) = \psi^{-r} \exp\left\{ - \left(1 + \xi \frac{x_r - \mu}{\psi}\right)^{-1/\xi} - \left(1 + \frac{1}{\xi}\right) \sum_{j=1}^{r} \log\left(1 + \xi \frac{x_j - \mu}{\psi}\right) \right\}. \quad (1.15)$$

Some examples using this approach are the papers of Smith (1986) and Tawn (1988) on hydrological extremes, and Robinson and Tawn (1995) and Smith (1997) for a novel application to the analysis of athletic records. The latter application is discussed in Section 1.3.3.

1.2.5 Point process approach

This was introduced as a statistical approach by Smith (1989), though the basic probability theory from which it derives had been developed by a number of earlier authors. In particular, the books by Leadbetter et al. (1983) and Resnick (1987) contain much information on point-process viewpoints of extreme value theory.

In this approach, instead of considering the times at which high-threshold exceedances occur and the excess values over the threshold as two separate processes, they are combined into one process based on a two-dimensional plot of exceedance times and exceedance values. The asymptotic theory of threshold exceedances shows that under suitable normalisation, this process behaves like a nonhomogeneous Poisson process.

In general, a nonhomogeneous Poisson process on a domain $\mathcal{D}$ is defined by an intensity $\lambda(x),\ x \in \mathcal{D}$, such that if A is a measurable subset of $\mathcal{D}$ and $N(A)$ denotes the number of points in A, then $N(A)$ has a Poisson distribution with mean

$$\Lambda(A) = \int_A \lambda(x)dx.$$

If $A_1, A_2, \ldots,$ are *disjoint* subsets of $\mathcal{D}$, then $N(A_1), N(A_2), \ldots$ are independent Poisson random variables.

For the present application, we assume x is two-dimensional and identified with (t, y) where t is time, and $y \geq u$ is the value of the process, $\mathcal{D} = [0, T] \times [u, \infty)$, and we write

$$\lambda(t, y) = \frac{1}{\psi}\left(1 + \xi\frac{y - \mu}{\psi}\right)^{-1/\xi - 1}, \tag{1.16}$$

defined wherever $\{1 + \xi(y - \mu)/\psi\} > 0$ (elsewhere $\lambda(t, y) = 0$). If A is a set of the form $[t_1, t_2] \times [y, \infty)$ (see Figure 1.3), then

$$\Lambda(A) = (t_2 - t_1)\left(1 + \xi\frac{y - \mu}{\psi}\right)^{-1/\xi}$$
$$\text{provided } y \geq u, \quad 1 + \xi(y - \mu)/\psi > 0. \tag{1.17}$$

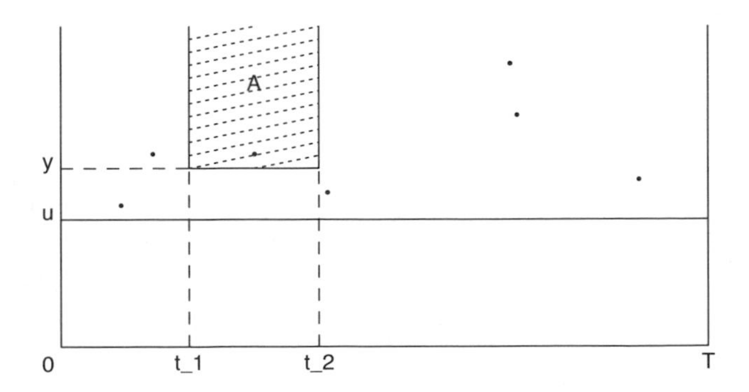

Figure 1.3 *Illustration of point process approach. Assume the process is observed over a time interval* $[0, T]$, *and that all observations above a threshold level u are recorded. These points are marked on a two-dimensional scatterplot as shown in the diagram. For a set A of the form shown in the figure, the count N(A) of observations in the set A is assumed to be Poisson with mean of the form given by (1.17).*

The mathematical justification for this approach lies in limit theorems as $T \to \infty$ and $1 - F(u) \to 1$, which we shall not go into here. To fit the model, we note that if a nonhomogeneous process of intensity $\lambda(t, y)$ is observed on a domain $\mathcal{D}$, and if $(T_1, Y_1), \ldots, (T_N, Y_N)$ are the N observed points of the process, then the joint density is

$$\prod_{i=1}^{N} \lambda(T_i, Y_i) \cdot \exp\left\{ -\int_{\mathcal{D}} \lambda(t, y)\, dt\, dy \right\}, \tag{1.18}$$

so (1.18) may be treated as a likelihood function and maximized with respect to the unknown parameters of the process. In practice, the integral in (1.18) is approximated by a sum, e.g., over all days if the observations are recorded daily.

An extension of this approach allows for nonstationary processes in which the parameters μ, ψ, and ξ are all allowed to be time-dependent, denoted μ_t, ψ_t, and ξ_t. Thus, (1.16) is replaced by

$$\lambda(t, y) = \frac{1}{\psi_t} \left(1 + \xi_t \frac{y - \mu_t}{\psi_t} \right)^{-1/\xi_t - 1}. \tag{1.19}$$

In the homogeneous case where μ, ψ, and ξ are constants, the model is mathematically equivalent to the Poisson-GPD model discussed above, though with a different parameterization. The extension (1.19) is particularly valuable in connection with extreme value regression problems, which are extensively discussed later (Section 1.5 and Section 1.6).

As an illustration of how the point process viewpoint may be used as a practical guide to visualising extreme value data, Figure 1.4 presents two plots derived from a 35-year series of the River Nidd in northern England (Davison and Smith 1990).

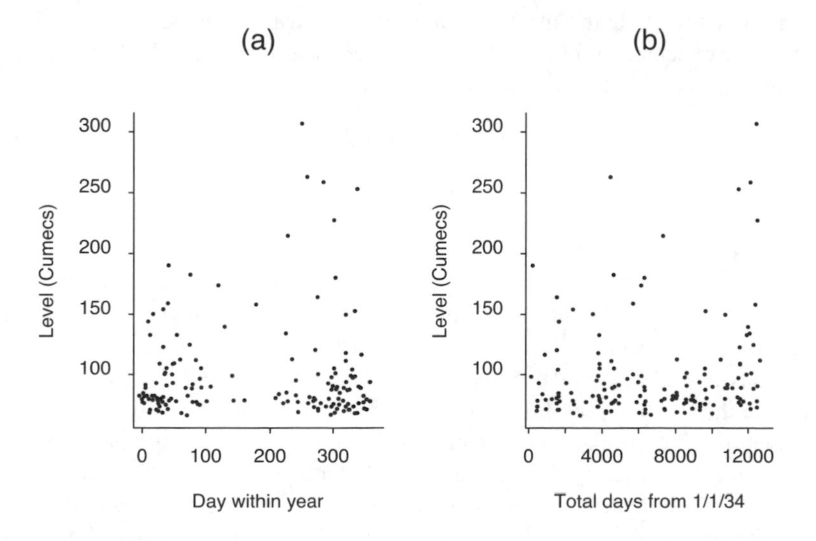

Figure 1.4 *Plots of exceedances of River Nidd, (a) against day within year, and (b) against total days from January 1, 1934. Adapted from Davison and Smith (1990).*

The data in this case consist of daily river flows above the level of 65 cumecs/second, and have been crudely declustered to remove successive high values that are part of the same extreme event. Figure 1.4(a) shows the data plotted against time measured as day within the year (1 to 366). Figure 1.4(b) shows a corresponding plot where time is the total cumulative number of days since the start of the series in 1934. Plot (a) is a visual diagnostic for seasonality in the series and shows, not surprisingly, that there are very few exceedances during the summer months. Plot (b) may be used as a diagnostic for overall trends in the series; in this case, there are three large values at the right-hand end of the series which could possibly indicate a trend in the extreme values of the series.

1.3 Estimation

1.3.1 Maximum likelihood estimation

Suppose we have observations $Y_1, \ldots, Y_N$ which are data for which the GEV distribution (1.9) is appropriate. For example, perhaps we take one year as the unit of time, and $Y_1, \ldots, Y_N$ represent annual maximum values for each of N years. The corresponding log likelihood is

$$\ell_Y(\mu, \psi, \xi) = -N \log \psi - \left(\frac{1}{\xi} + 1\right) \sum_i \log\left(1 + \xi \frac{Y_i - \mu}{\psi}\right) - \sum_i \left(1 + \xi \frac{Y_i - \mu}{\psi}\right)^{-1/\xi} \tag{1.20}$$

provided $1 + \xi(Y_i - \mu)/\psi > 0$ for each i.

For the Poisson-GPD model discussed above, suppose we have a total of N observations above a threshold in time T, and the excesses over the threshold are $Y_1, \ldots, Y_N$. Suppose the expected number of Poisson exceedances is λT, and the GPD parameters are σ and ξ, as in (1.11). Then the log likelihood is

$$\ell_{N,Y}(\lambda, \sigma, \xi) = N \log \lambda - \lambda T - N \log \sigma - \left(1 + \frac{1}{\xi}\right) \sum_{i=1}^{N} \log\left(1 + \xi \frac{Y_i}{\sigma}\right) \tag{1.21}$$

provided $1 + \xi Y_i/\sigma > 0$ for all i. Similar log likelihoods may be constructed from the joint densities (1.15) and (1.18) for the r largest order statistics approach and the point process approach.

The maximum likelihood estimators are the values of the unknown parameters that maximize the log likelihood. In practice these are local maxima found by nonlinear optimization. The standard asymptotic results of consistency, asymptotic efficiency, and asymptotic normality hold for these distributions if $\xi > -\frac{1}{2}$ (Smith 1985). In particular, the elements of the Hessian matrix of $-\ell$ (the matrix of second-order partial derivatives, evaluated at the maximum likelihood estimators) are known as the observed information matrix, and the inverse of this matrix is a widely used approximation for the variance-covariance matrix of the maximum likelihood estimators.

The square roots of the diagonal entries of this inverse matrix are estimates of the standard deviations of the three parameter estimates, widely known as the standard errors of those estimates. All these results are asymptotic approximations valid for large sample sizes, but in practice they are widely used even when the sample sizes are fairly small.

1.3.2 Profile likelihoods for quantiles

Suppose we are interested in the n-year return level y_n, i.e., the $(1 - 1/n)$-quantile of the annual maximum distribution. This is given by solving the equation

$$\exp\left\{-\left(1+\xi\frac{y_n-\mu}{\psi}\right)^{-1/\xi}\right\} = 1 - \frac{1}{n}. \tag{1.22}$$

Exploiting the approximation $1 - \frac{1}{n} \approx \exp\left(-\frac{1}{n}\right)$, this simplifies to

$$\left(1+\xi\frac{y_n-\mu}{\psi}\right)^{-1/\xi} = \frac{1}{n},$$

and hence

$$y_n = \mu + \psi\frac{n^\xi - 1}{\xi}. \tag{1.23}$$

One approach to the estimation of y_n is simply to substitute the maximum likelihood estimates $\hat{\mu}, \hat{\psi}, \hat{\xi}$ for the unknown parameters μ, ψ, ξ, thus creating a maximum likelihood estimator $\hat{y}_n$. The variance of $\hat{y}_n$ may be estimated through a standard delta function approximation, i.e., if we define a vector of partial derivatives

$$g(\mu, \psi, \xi) = \left(\frac{\partial y_n}{\partial \mu}, \frac{\partial y_n}{\partial \psi}, \frac{\partial y_n}{\partial \xi}\right)$$

and write $\hat{g}$ for $g(\hat{\mu}, \hat{\psi}, \hat{\xi})$, and also write H for the observed information matrix for $(\hat{\mu}, \hat{\psi}, \hat{\xi})$, then the variance of $\hat{y}_n$ is approximately

$$\hat{g} \cdot H^{-1} \cdot \hat{g}^T, \tag{1.24}$$

and the square root of (1.24) is an approximate standard error. In practice, this often gives a rather poor approximation which does not account for the skewness of the distribution of $\hat{y}_n$, especially when n is large.

An alternative approach is via a profile likelihood. Equation (1.23) shows how y_n (for a given value of n) may be expressed as a function of (μ, ψ, ξ). Suppose we rewrite this as

$$\mu = y_n - \psi\frac{n^\xi - 1}{\xi}$$

and substitute in (1.20) so that the log likelihood ℓ_Y is written as a function of new parameters (y_n, ψ, ξ). If for any given value of y_n we maximise this function with respect to ψ and ξ, we obtain a function of y_n alone, say $\ell_Y^*(y_n)$, which is known as the profile log likelihood function for y_n. This function may be plotted to determine the relative plausibility of different values of y_n.

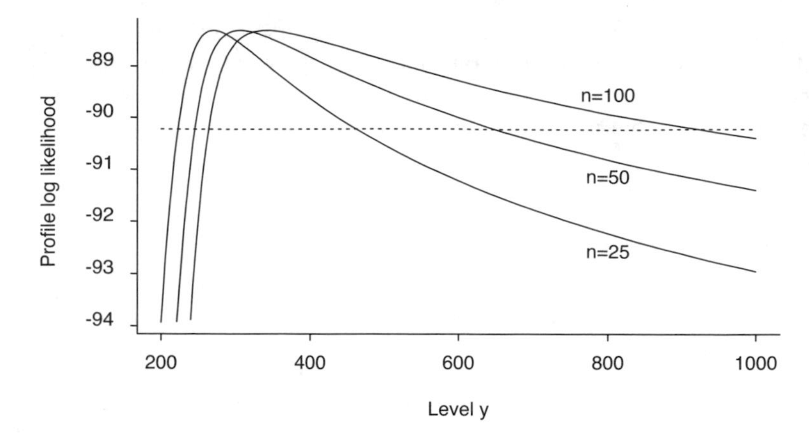

Figure 1.5 *Profile log likelihood plots for the n-year return value y_n for the River Nidd, for $n = 25$, 50, and 100. The horizontal dotted line is at a level 1.92 below the common maximum of the three curves; for each n, a 95% confidence interval for y_n consists of all values for which the profile log likelihood is above the dotted line.*

A confidence interval for y_n may be constructed by exploiting the following property: under standard regularity conditions for maximum likelihood (which, as noted already, are valid provided $\xi > -\frac{1}{2}$), under the null hypothesis that y_n is the true value,

$$2\{\log \ell_Y^*(\hat{y}_n) - \log \ell_Y^*(y_n)\} \sim \chi_1^2 \text{ (approximately).}$$

Therefore, an approximate $100(1 - \alpha)\%$ confidence interval for y_n consists of all values for which

$$\log \ell_Y^*(\hat{y}_n) - \log \ell_Y^*(y_n) \leq \frac{1}{2}\chi_{1;1-\alpha}^2 \tag{1.25}$$

where $\chi_{1;1-\alpha}^2$ is the $(1 - \alpha)$-quantile of the χ_1^2 distribution. For example, in the case $\alpha = .05$, the right-hand side of (1.25) is 1.92.

The same concept may be used in connection with (1.22) or any other model for which the standard regularity conditions for maximum likelihood hold. For example, Figure 1.5 (adapted from Davison and Smith, 1990) shows a profile likelihood plot for y_n, for each of the values $n = 25, 50$, and 100, constructed from the River Nidd threshold exceedances mentioned already in Section 1.2.5. The key point here is that the curves are highly skewed to the right, and correspondingly, so are the confidence intervals — in sharp contrast to the confidence intervals derived from the delta method that are always symmetric about the maximum likelihood estimator. This in turn reflects that there is much less information in the data about the behavior of the process at very high threshold levels (i.e., above 400) compared with lower levels where there is much more data. Although there is no proof that the confidence intervals derived by the profile likelihood method necessarily have better coverage probabilities than those derived by the delta method, simulations and practical experience suggest that they do.

1.3.3 Bayesian approaches

Bayesian methods of statistics are based on specifying a density function for the unknown parameters, known as the prior density, and then computing a posterior density for the parameters given the observations. In practice, such computations are nearly always carried out using some form of Markov chain Monte Carlo (MCMC) sampling, which we shall not describe here, as a number of excellent texts on the subject are available, e.g., Gamerman (1997) or Robert and Casella (2000). In the present discussion, we shall not dwell on the philosophical differences between Bayesian and frequentist approaches to statistics, but concentrate on two features that may be said to give Bayesian methods a practical advantage: their effectiveness in handling models with very large numbers of parameters (in particular, hierarchical models, which we shall see in an extreme values context in Section 1.6.3); and their usefulness in predictive inference, where the ultimate objective is not so much to learn the values of unknown parameters, but to establish a meaningful probability distribution for future unobserved random quantities.

For the rest of the present section, we focus on a specific example, first given by Smith (1997), that brings out the contrast between maximum likelihood inference about parameters and Bayesian predictive inference in a particularly striking way. Further instances of Bayesian predictive inference applied to extremes will be found in several subsequent sections, e.g., Section 1.3.4 and Section 1.6.2.

The example concerns the remarkable series of track performances achieved during 1993 by the Chinese athlete Wang Junxia, including new world records at 3000 and 10,000 meters, which were such an improvement on previous performances that there were immediate suspicions that they were drug assisted. However, although other Chinese athletes have tested positive for drugs, Wang herself never did, and her records still stand. The question considered here, and in an earlier paper by Robinson and Tawn (1995), is to assess just how much of an outlier the performance really was, in comparison with previous performances. The detailed discussion is confined to the 3000-meter event.

Figure 1.6(a) shows the five best running times by different athletes in the women's 3000-meter track event for each year from 1972 to 1992, along with Wang Junxia's world record from 1993. The first step is to fit a probability model to the data up to 1992.

Recall from (1.15) that there is an asymptotic distribution for the joint distribution of the r largest order statistics in a random sample, in terms of the usual GEV parameters (μ, ψ, ξ). Recall also that the upper endpoint of the distribution is at $\mu - \psi/\xi$ when $\xi < 0$. In this example, this is applied with $r = 5$, the observations (running times) are negated to convert minima into maxima, and the endpoint parameter is denoted x_{ult}, the nominal "ultimate running performance." A profile likelihood for x_{ult} may therefore be constructed by the same method as in Section 1.3.2. For the present study, the analysis is confined to the data from 1980 onwards, for which there is no visible evidence of a time trend.

The profile log likelihood constructed by this process is shown in Figure 1.6(b). A 95% confidence interval for x_{ult} is again calculated as the set of values for which the profile log likelihood is above the dashed line, and leads to an approximate confidence

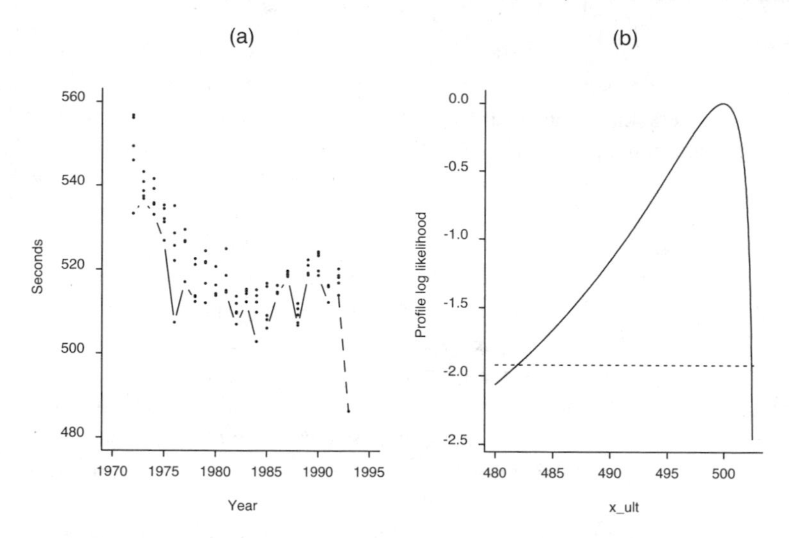

Figure 1.6 *(a) Five best performances by different athletes in the women's 3000-meter event, for each year from 1972 to 1992, together with Wang Junxia's record from 1993, and (b) profile log likelihood for* x_{ult}*, the ultimate endpoint parameter, based on the joint density (1.15) fitted to the data (multiplied by* -1 *to convert minima into maxima) from 1980 to 1992.*

interval (481 to 502). Wang's 1993 record — 486.1 seconds — lies within this confidence interval, so on the basis of the analysis so far, there is no clear-cut basis on which to say her record was anomalous. Robinson and Tawn (1995) considered a number of other models for the data, for example, allowing for various forms of time trend from 1972 onwards, but their main conclusions were consistent with this, i.e., that likelihood ratio confidence intervals for x_{ult} were wide and typically included Wang's record.

The alternative Bayesian analysis introduced by Smith (1997) was to consider the problem, not as one about estimating an ultimate limit parameter, but a more specific problem of *predicting* the best performance of 1993 given the preceding results for 1972 to 1992. The idea underlying this is that a prediction interval should give more precise information about what is likely to happen in a given year than the estimation of a parameter such as x_{ult}.

More precisely, Smith considered the conditional probability of a record equal to or better than the one actually achieved by Wang, given the event that the previous world record was broken. The conditioning was meant to provide some correction for the obvious selection effect, i.e., we would not even be considering these questions unless we had already observed some remarkable performance, such as a new world record. This conditional probability may be expressed as a specific analytic function of the three parameters μ, ψ, and ξ, say $\phi(\mu, \psi, \xi)$, and hence estimated through the Bayesian formula

$$\int\int\int \phi(\mu, \psi, \xi)\pi(\mu, \psi, \xi | Y) d\mu d\psi d\xi \tag{1.26}$$

where $\pi(\ldots|Y)$ denotes the posterior density given past data Y. Once again, the analysis was confined to the years 1980 to 1992 and did not take account of any time trend. A diffuse but proper prior distribution was assumed. The result, in this case, was .0006 (a modification of the result .00047 that was actually quoted in the paper by Smith (1997)). Such a small estimated probability provides strong evidence that Wang's performance represented an actual change in the distribution of running times. It does not, of course, provide any direct evidence that drugs were involved.

In this case, the sharp contrast between the maximum likelihood and Bayesian approaches is not a consequence of the prior distribution, nor of the MCMC method of computation, though the precise numerical result is sensitive to these. The main reason for the contrasting results lies in the change of emphasis from estimating a parameter of the model — for which the information in the data is rather diffuse, resulting in wide confidence intervals — to predicting a specific quantity, for which much more precise information is available. Note that the alternative "plug-in" approach to (1.26), in which $\phi(\mu, \psi, \xi)$ is estimated by $\phi(\hat{\mu}, \hat{\psi}, \hat{\xi})$, where $\hat{\mu}$, $\hat{\psi}$, and $\hat{\xi}$ are the maximum likelihood estimators, would result in a predicted probability (of a performance as good as Wang's) of 0. This is a consequence of the fact that the maximum likelihood estimate of x_{ult} is greater than 486.1. However, this result could not be accepted as a realistic estimate of the probability, because it takes no account whatsoever of the uncertainty in estimating the model parameters.

1.3.4 Raleigh snowfall example

The example of Section 1.3.3 is unusual in providing such a sharp contrast between the maximum likelihood parameter estimation approach and the Bayesian predictive approach, but the general implications of this example are relevant to a variety of problems connected with extreme values. Many extreme value problems arising in applied science are really concerned with estimating probabilities of specific outcomes rather than estimating model parameters, but until recently, this distinction was usually ignored.

As a further example, we consider again the Raleigh snowfall data set of Section 1.1.1. Specifically, we return to the data set of Table 1.1 and ask what is the probability, in a single January, of a snowfall equal to or greater than 20.3 inches. Assuming either the Poisson-GPD model, above, with parameters (λ, σ, ξ), or the equivalent point process approach (Section 1.2.5) with parameters (μ, ψ, ξ), we can estimate this probability assuming either the maximum likelihood plug-in approach or the Bayesian approach. In either case, it is necessary to choose a specific threshold, confining the estimation to those observations that are above the given threshold.

Figure 1.7 shows the predictive probability computed both by the Bayesian formula (1.26) (denoted by B on the figure) or by the maximum likelihood plug-in approach (denoted by M). Both quantities are in turn computed for a variety of different thresholds. For ease of plotting and annotation, the quantity actually plotted is N, where $1/N$ is the predictive probability.

In this case we can see, with both the maximum likelihood and the Bayesian results, that there is a huge dependence on the threshold, but the Bayesian results are all below

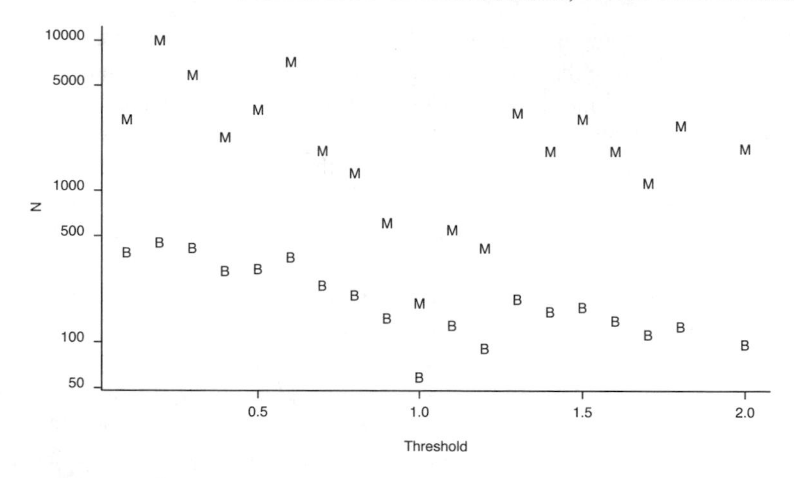

Figure 1.7 *Maximum likelihood (M) and Bayesian (B) estimates of predictive probability* $1/N$ *for different thresholds.*

the corresponding maximum likelihood results (i.e., the Bayesian predictive probability of an extreme event is larger than the maximum likelihood probability) and also arguably more consistent across different thresholds. This does not, of course, prove that the Bayesian estimates perform better than the maximum likelihood estimates, but it does serve to illustrate the contrasts between the two approaches. Most approaches to extreme value analysis keep separate the procedures of estimating unknown parameters and calculating probabilities of extreme events, and therefore by default use the plug-in approach.

The fact that the predictive probabilities, whether Bayesian or maximum likelihood, vary considerably with the somewhat arbitrary choice of a threshold, is still of concern. However, it can be put in some perspective when the variability between predictive probabilities for different thresholds is compared with the inherent uncertainty of those estimates. Confidence intervals for the predictive probability computed by the delta method (not shown) are typically very wide, while Figure 1.8 shows the posterior density of the predictive probability $1/N$ for two of the thresholds, 0.5 and 1, for which the point predictive probabilities are at opposite ends of the spectrum. The substantial overlap between these two posterior densities underlines the inherent variability of the procedure.

In summary, the main messages of this example are:

1. The point estimates (maximum likelihood or Bayes) are quite sensitive to the chosen threshold, and in the absence of a generally agreed criterion for choosing the threshold, this is an admitted difficulty of the approach.
2. The Bayesian estimates of N are nearly always smaller than the maximum likelihood estimates — in other words, Bayesian methods tend to lead to a larger (more conservative) estimate of the probability of an extreme event.

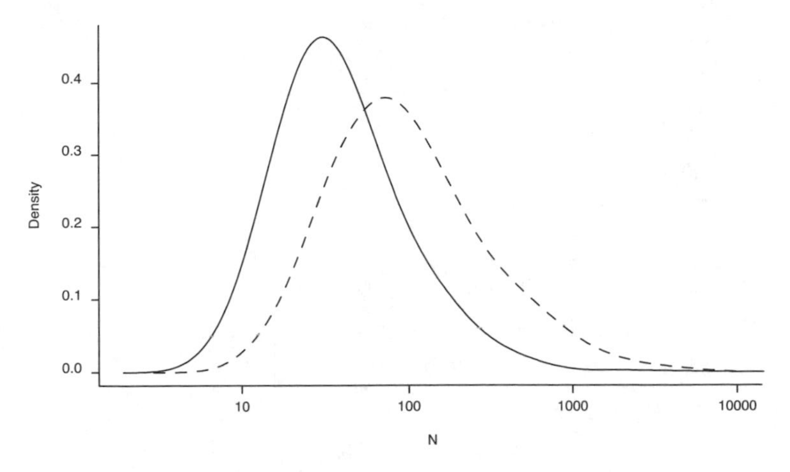

Figure 1.8 *Posterior densities of* $1/N$ *based on thresholds 1 (solid) and 0.5 (dashed).*

3. The variability among point estimates for different thresholds is less important than the inherent variability of the procedure, based on the standard error in the maximum likelihood case or the posterior density in the Bayesian case.

This example is somewhat unusual in that it represents a very considerable extrapolation beyond the range of the observed data, so it should not be surprising that all the estimates have very high variability. However, the Bayesian results are generally consistent with a return period of between 100 and 200 years, which in turn seems to be consistent with the judgement of most meteorologists based on newspaper reports at the time this event occurred.

1.4 Diagnostics

The example of Section 1.3.4 has highlighted one difficulty in applying threshold-based methods: the lack of a clear-cut criterion for choosing the threshold. If the threshold is chosen too high, then there are not enough exceedances over the threshold to obtain good estimates of the extreme value parameters, and consequently, the variances of the estimators are high. Conversely, if the threshold is too low, the GPD may not be a good fit to the excesses over the threshold and consequently there will be a bias in the estimates. There is an extensive literature on the attempt to choose an optimal threshold by, for example, a minimum mean squared error criterion, but it is questionable whether these techniques are preferable in practice to more ad hoc criteria, based on the fit of the model to the data. In any case, it is clearly desirable to have some diagnostic procedures to decide how well the models fit the data, and we consider some of these here. The emphasis is on graphical procedures.

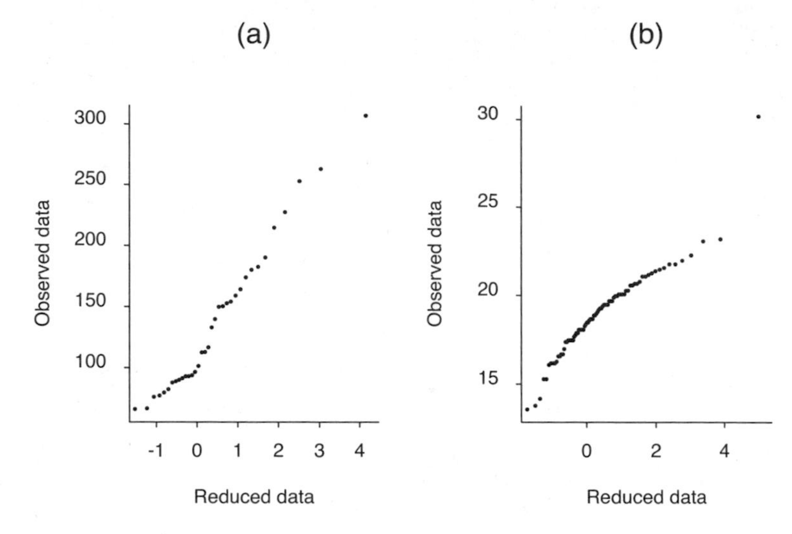

Figure 1.9 *Gumbel plots, (a) annual maxima for River Nidd river flow series, and (b) annual maximum temperatures in Ivigtut, Iceland.*

1.4.1 Gumbel plots

This is the oldest method, appropriate for examining the fit of annual maxima data (or maxima over some other time period) to a Gumbel distribution. Suppose the annual maxima over N years are $Y_1, \ldots, Y_N$, ordered as $Y_{1:N} \leq \cdots \leq Y_{N:N}$; then $Y_{i:N}$, for $1 \leq i \leq N$, is plotted against the *reduced value* $x_{i:N}$, where

$$x_{i:N} = -\log(-\log p_{i:N}),$$

$p_{i:N}$ being the ith *plotting position*, usually taken to be $(i - \frac{1}{2})/N$.

A straight line indicates good fit to the Gumbel distribution. Curvature upwards or downwards may indicate, respectively, a Fréchet or Weibull distribution. The method is also a useful way to detect outliers.

Examples. Figure 1.9(a) is a Gumbel plot based on the annual maxima of the River Nidd river flow series. This is a fairly typical example of a Gumbel plot in practice: although it is not a perfect straight line, there is no systematic evidence of curvature upwards or downwards, nor do there seem to be any outliers. On this basis we conclude that the Gumbel distribution would be a reasonable fit to the data.

Figure 1.9(b), taken from Smith (1990), is another example based on annual maximum temperatures (in °C) at Ivigtut, Iceland. Two features stand out: (a) the largest observation seems to be a clear outlier relative to the rest of the data, and (b) when this observation is ignored, the rest of the plot shows a clear downward curvature, indicating the Weibull form of extreme value distribution and a finite upper endpoint.

Plots of this nature were very widely used in the early days of the subject when, before automatic methods such as maximum likelihood became established, they were widely used for estimation as well as model checking (Gumbel 1958). This aspect is now not important, but the use of Gumbel plots as a diagnostic device is still useful.

1.4.2 *QQ plots*

A second type of probability plot is drawn *after* fitting the model. Suppose $Y_1, \ldots, Y_N$ are i.i.d. observations whose common distribution function is $G(y;\theta)$ depending on parameter vector θ. Suppose θ has been estimated by $\hat{\theta}$, and let $G^{-1}(p;\theta)$ denote the inverse distribution function of G, written as a function of θ. A QQ (quantile-quantile) plot consists of first ordering the observations $Y_{1:N} \leq \cdots \leq Y_{N:N}$, and then plotting $Y_{i:N}$ against the reduced value

$$x_{i:N} = G^{-1}(p_{i:N};\hat{\theta}),$$

where $p_{i:N}$ may be again taken as $(i-\frac{1}{2})/N$. If the model is a good fit, the plot should be roughly a straight line of unit slope through the origin.

Figure 1.10 illustrates this idea for the Ivigtut data of Figure 1.9(b). In Figure 1.10(a), the GEV distribution is fitted by maximum likelihood to the whole data set, and a QQ plot is drawn. The shape of the plot — with several points below the straight line at the right-hand end of the plot, except for the final data point which is well above — supports the treatment of the final data point as an outlier. In Figure 1.10(b), the same points are plotted (including the final data point), but for the purpose of estimating the parameters, the final observation was omitted. In this case, the plot seems to stick very closely to the straight line, except for the final data point, which is an even more obvious outlier than in Figure 1.10(a). Taken together, the two plots show that the largest data point is not only an outlier but also an influential data point, i.e., the fitted model is substantially different when the data point is included from when it is not. On the other hand the plot also confirms that if this suspect data point is omitted, the GEV indeed fits the rest well.

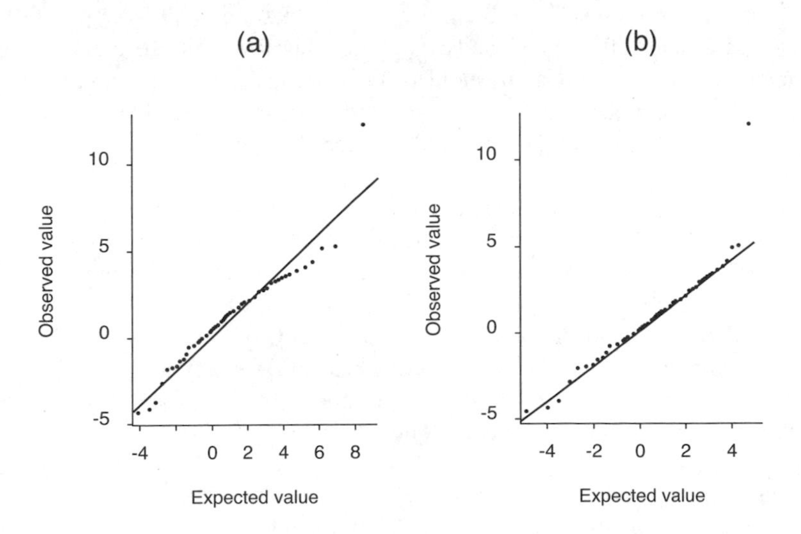

Figure 1.10 *GEV model to Ivigtut data, (a) without adjustment, and (b) excluding largest value from model fit but including it in the plot.*

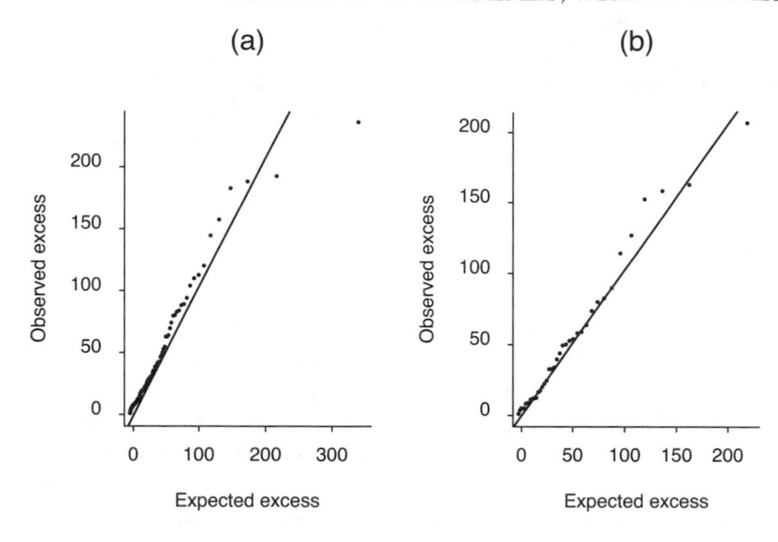

Figure 1.11 *QQ plots for GPD, Nidd data: (a)* $u = 70$*, and (b)* $u = 100$.

Figure 1.11, from Davison and Smith (1990), shows QQ plots for the Nidd data introduced in Section 1.2.5. In this case, $Y_1, \ldots, Y_N$ are the excess values over a high threshold, to which the GPD is fitted. The entire calculation (model fit followed by QQ plot) is carried out for two thresholds, 70 in plot (a) and 100 in plot (b). Plot (a) shows some strange behavior, the final two data points below the straight line but before them, a sequence of plotted points above the straight line. No single observation appears as an outlier, but the plot suggests that the GPD does not fit the data very well. Plot (b) shows no such problem, suggesting that the GPD is a good fit to the data over threshold 100. Davison and Smith cited this along with several other pieces of evidence to support using a threshold 100 in their analysis. This example serves to illustrate a general procedure, that the suitability of different possible thresholds may be assessed based on their QQ plots, and this can be used as a guide to selecting the threshold.

Figure 1.12 shows two examples from insurance data sets. Plot (a) again shows a scatterplot of the oil company data set from Section 1.1.2, and (b) a QQ plot based on the GPD fitted to all exceedances above threshold 5. The main point here is that, although there are reasons for treating the largest two observations as outliers, they are not in fact very far from the straight line — in other words, the data are in fact consistent with the fitted GPD, which in this case is extremely long-tailed (see Section 1.6 for further discussion). If this interpretation is accepted, there is no reason to treat those observations as outliers. In contrast, plots (c) and (d), taken from another, unpublished, study of oil industry insurance claims, in this case spanning several companies and a worldwide data base, show the largest observation as an enormous outlier; several alternative analyses were tried, including varying the threshold and performing regression analysis on various covariates (including the size of the spill in barrels, the x coordinate of plot (c)), but none succeeded in explaining this outlier.

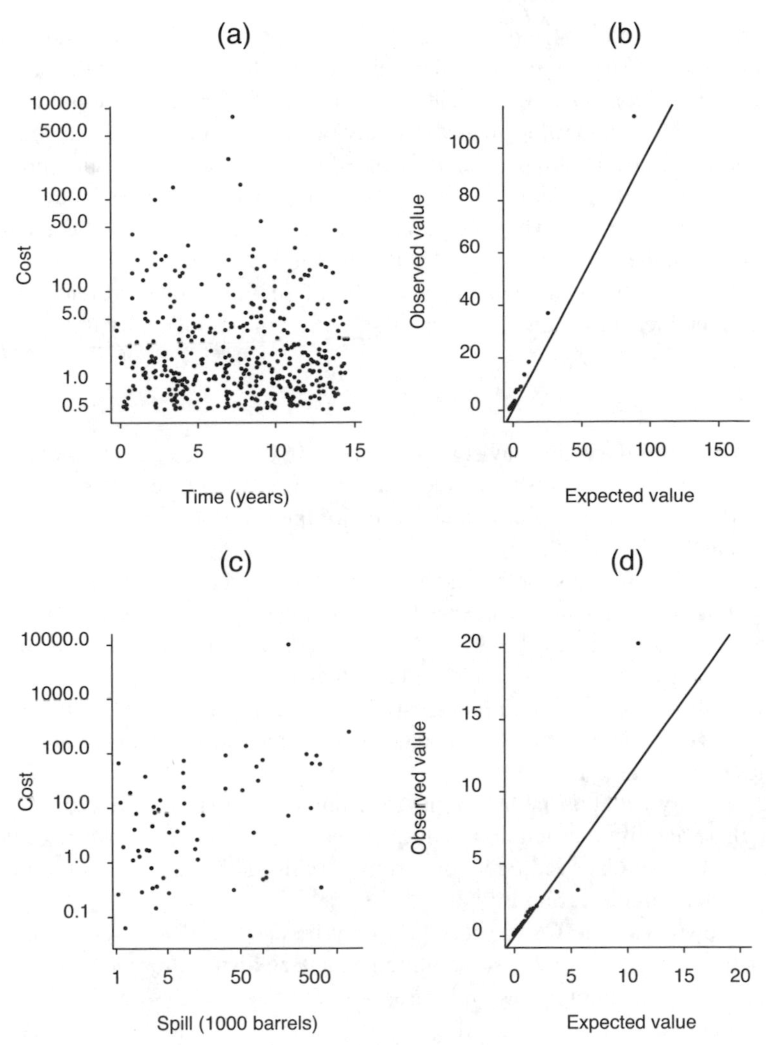

Figure 1.12 *Insurance claims data set, (a) scatterplot of insurance claims data, (b) QQ plot from GPD fit to insurance claims data, (c) scatterplot of costs of oil spills vs. size of spill, and (d) QQ plot for oil spills data based on GPD fit with regressors.*

The outlier is the Exxon Valdez oil spill in Alaska, and the largest component of the cost was the punitive damages assessed in litigation (9 billion dollars at the time the analysis was conducted, though the amount was recently reduced to 3 billion dollars on appeal).

The implications of these examples are that observations which first appear to be outliers (as in the first oil company data set) may not in fact be inconsistent with the rest of the data if they come from a long-tailed distribution, whereas in other cases (such as the Ivigtut temperature series and the second oil company example), no amount of fitting different models will make the outlier go away. This is a somewhat different interpretation of outliers from that usually given in statistics; in most statistical applications, the primary interest is in the center of the distribution, not the tails, so the main concern with outliers is to identify them so that they may be eliminated from the study. In an extreme value analysis, it may be important to determine whether the largest observations are simply the anticipated outcome of a long-tailed process, or are truly anomalous. QQ plots are one device to try to make that distinction.

QQ plots can be extended beyond the case of i.i.d. data, for example, to a regression model in which the Y_is are residuals, but it still requires that the residuals have a common distribution. On the other hand, the W plot of Section 1.4.4 represents a different variant on the idea, in which there is no assumption of homogeneity in the data.

1.4.3 The mean excess plot

This idea was introduced by Davison and Smith (1990), and is something of an analog of the Gumbel plot for threshold-exceedance data, in the sense that it is a diagnostic plot drawn before fitting any model and can therefore give guidance about what threshold to use.

The mathematical basis for this method is equation (1.12), the key feature of which is that if Y is GPD, then the mean excess over a threshold y, for any $y > 0$, is a linear function of y, with slope $\xi/(1-\xi)$. Thus, we can draw a plot in which the abscissa is the threshold, and the ordinate is the sample mean of all excesses over that threshold. The slope of the plot leads to a quick estimate of ξ: in particular, an increasing plot indicates $\xi > 0$, a decreasing plot indicates $\xi < 0$, and one of roughly constant slope indicates that ξ is near 0.

One difficulty with this method is that the sample mean excess plot typically shows very high variability, particularly at high thresholds. This can make it difficult to decide whether an observed departure from linearity is in fact due to failure of the GPD or is just sample variability.

The following Monte Carlo procedure may be used to give a rough confidence band on the plot. Suppose, for some finite u, the true distribution of excesses over u is exactly GPD with parameters (σ, ξ). Suppose the estimates of σ and ξ, based on all exceedances over u, are $\hat{\sigma}$ and $\hat{\xi}$. Also let $\mu(y) = \{\sigma + \xi(y-u)\}/(1-\xi)$ be the theoretical mean excess over threshold for $y > u$, and let $\hat{\mu}(y)$ be the sample mean excess.

A natural test statistic for the GPD assumption is

$$\hat{\mu}(y) - \frac{\hat{\sigma} + \hat{\xi}(y-u)}{1-\hat{\xi}} \tag{1.27}$$

for any given y: this represents the estimated difference between the empirical and theoretical mean excesses at that y.

We can simulate the distribution of (1.27), as follows. For $j = 1, 2, \ldots, 99$, generate a random sample from the GPD over threshold u, of the same size as the original sample, based on parameters $\hat{\xi}, \hat{\sigma}$. For each j, calculate new MLEs $\hat{\xi}^{(j)}, \hat{\sigma}^{(j)}$ and also the sample mean excess function $\hat{\mu}^{(j)}(y)$. For each u, compute the fifth largest and

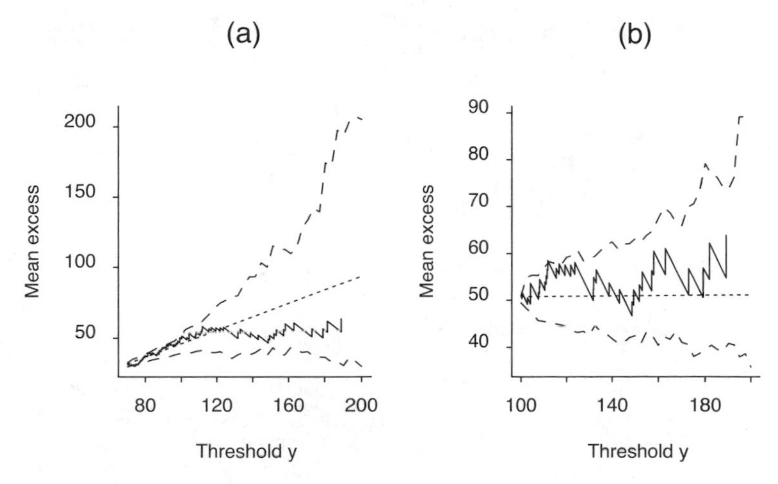

Figure 1.13 *Mean excess over threshold plots for Nidd data, with Monte Carlo confidence bands, relative to threshold 70 (a) and 100 (b).*

fifth smallest values of

$$\hat{\mu}^{(j)}(y) - \frac{\hat{\sigma}^{(j)} + \hat{\xi}^{(j)}(y-u)}{1-\hat{\xi}^{(j)}} + \frac{\hat{\sigma} + \hat{\xi}(y-u)}{1-\hat{\xi}} \tag{1.28}$$

as index j ranges over the random samples $1, \ldots, 99$. Then these values form, for each y, approximate 5% upper and lower confidence bounds on $\hat{\mu}(y)$, if the GPD is correct.

It should be pointed out that this is only a pointwise test, i.e., the claimed 90% confidence level is true for any given y, but not simultaneously over all y. Therefore, the test needs some caution in its interpretation — if the plot remains within the confidence bands over most of its range but strays outside for a small part of its range, that does not necessarily indicate lack of fit of the GPD. Nevertheless, this Monte Carlo procedure can be very useful in gauging how much variability to expect in the mean excess plot.

Figure 1.13 shows the mean excess plot, with confidence bands, for the Nidd data, based on all exceedances over thresholds $u = 70$ (plot (a)) and $u = 100$ (plot (b)). The dotted straight line is the estimated theoretical mean excess assuming the GPD at threshold u, and the jagged dashed lines are the estimated confidence bands based on (1.28). Both plots lie nearly everywhere inside the confidence bands, but plot (a) appears to show more systematic departure from a straight line than (b) (note, in particular, the change of slope near $y = 120$), adding to the evidence that threshold 100 is a better bet than 70.

As another example, we consider three windspeed data sets for cities in North Carolina, based on 22 years of daily windspeed data at Raleigh, Greensboro, and Charlotte. For the moment we shall assess these series solely from the point of view of their suitability for a threshold analysis; later (Section 1.5.2) we shall consider the influence of seasonality and the possibility of long-term trends.

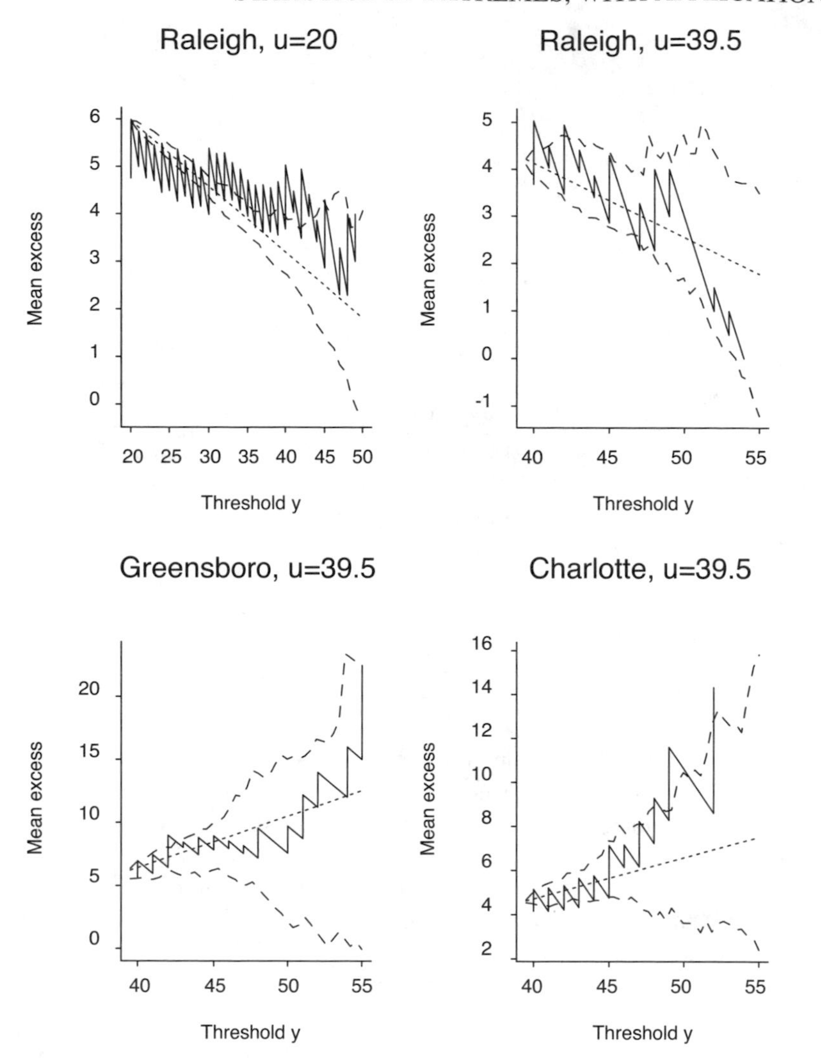

Figure 1.14 *Mean excess plots (with Monte Carlo confidence bands) for windspeed data sets.*

Figure 1.14 shows mean excess plots constructed with confidence bands for the three series. The jagged appearance of the plots occurs with all plots of this nature, due to the discretization of the data. The first plot shows the plot for the Raleigh data computed relative to the threshold $u = 20$ knots. It is clear that there are problems with this plot, since it crosses the upper boundary of the Monte Carlo confidence limits in numerous places. The remaining three plots are all calculated relative to threshold 39.5 (just below 40, to ensure that windspeeds of exactly 40 knots would be included in the analysis) and show no major problems with the fit, though for all three series there are isolated values of the mean excess plot that cross the Monte Carlo confidence limits.

The conclusion in this case is that (for Raleigh at least) 20 knots is clearly too low a threshold for the GPD to be a credible model, but for a threshold of 39.5 knots, it seems to be reasonable.

Other examples of mean excess plots are given in Section 1.5.3.

1.4.4 Z- and W-statistic plots

Consider the nonstationary version of the point process model (Section 1.2.5) with μ_t, ψ_t, ξ_t dependent on t. This is the most general form of threshold model that we have seen so far. We denote the exceedance times $T_1, T_2, \ldots$, $(T_0 = 0)$ and the corresponding excess values $Y_1, Y_2, \ldots$, where Y_k is the excess over the threshold at time T_k.

The Z statistic is based on intervals between exceedances T_k:

$$Z_k = \int_{T_{k-1}}^{T_k} \lambda_u(t)dt, \quad \lambda_u(t) = \{1 + \xi_t(u - \mu_t)/\psi_t\}^{-1/\xi_t}. \tag{1.29}$$

The idea is that if $\{T_1, T_2, \ldots,\}$ are viewed as a one-dimensional point process in time, they form a nonhomogeneous Poisson process with intensity function $\lambda_u(\cdot)$; the transformation then ensures that $Z_1, Z_2, \ldots$, are i.i.d. exponential random variables with mean 1. In practice this will only be an approximate result, not exact, because we do not know the true values of μ_t, ψ_t, ξ_t and have to use estimates.

The W statistic is based on the excess values:

$$W_k = \frac{1}{\xi_{T_k}} \log\left\{1 + \frac{\xi_{T_k} Y_k}{\psi_{T_k} + \xi_{T_k}(u - \mu_{T_k})}\right\}. \tag{1.30}$$

This is equivalent to a probability integral transformation on the excesses over a threshold: if the model is exact, $W_1, W_2, \ldots$ are also i.i.d. exponential variables with mean 1. Again, in practice this is only an approximation because the parameters are estimated.

The Z and W statistics can be tested in various ways to examine how well they agree with the i.i.d. exponential assumption. As an example, Figure 1.15 shows three types of plots computed for the Charlotte windspeed data (the precise analysis from which these plots were constructed includes seasonal factors and is described in Section 1.5.2).

Plots (a) and (d) show the Z and W statistics respectively plotted against time of exceedance, i.e., either Z_k (in plot (a)) or W_k (in plot (d)) is plotted against T_k. The idea here is to observe a possible time trend in the observations. To aid in judging this, a simple fitted curve (using the "lowess" function in S-Plus) is superimposed on the plot. In neither case is there evidence of a systematic trend.

Plots (b) and (e) are QQ plots of the Z and W statistics. Since the hypothesized distribution G is unit exponential, for which $G^{-1}(p) = -\log(1-p)$, this means plotting either $Z_{i:N}$ or $W_{i:N}$ (the ith smallest of N ordered values) against $-\log(1 - p_{i:N})$, where $p_{i:N} = (i - \frac{1}{2})/N$. The results in this case show no reason to question the assumption of an exponential distribution with mean 1.

Plots (c) and (f) are plots of the first ten sample autocorrelations for the Z and W values respectively, with approximate confidence bands at $\pm 2/\sqrt{N}$ where N is the sample size. This is a standard plot used in time series analysis and is used here as

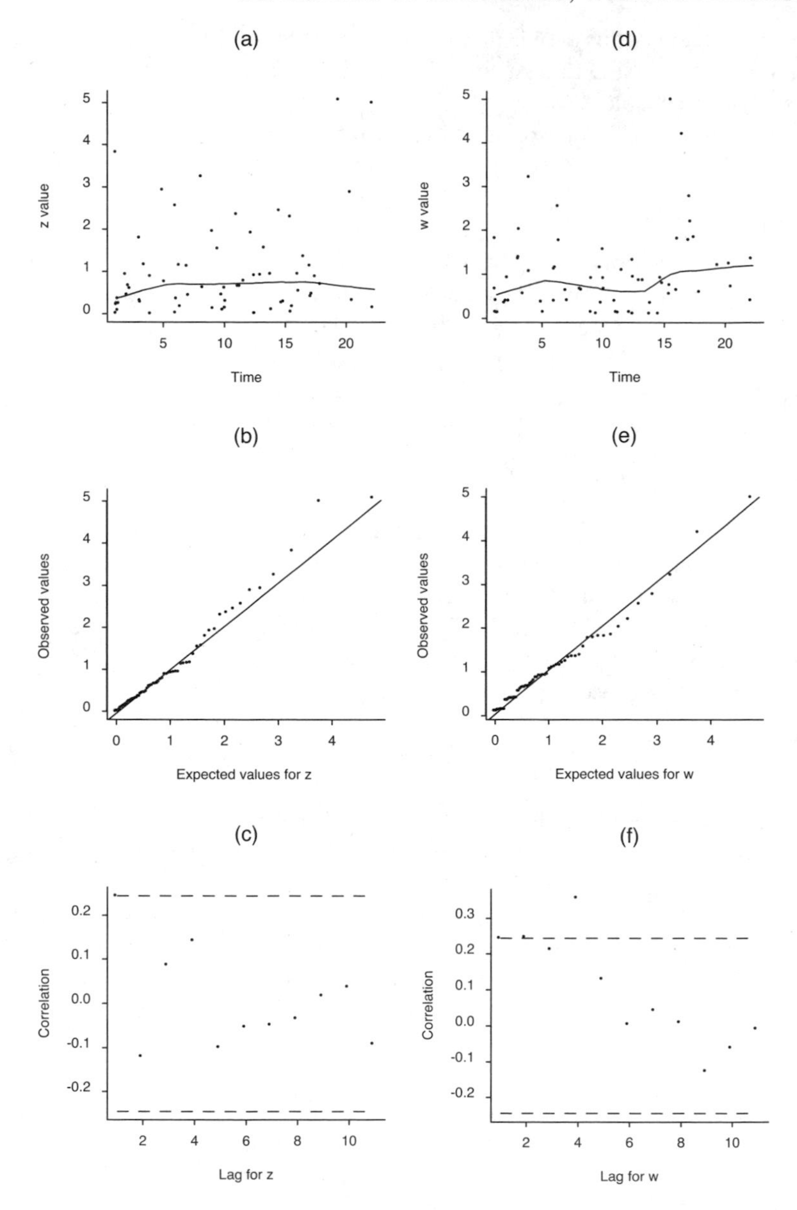

Figure 1.15 *Diagnostic plots based on Z and W statistics for Charlotte seasonal Model 1 (Section 1.5.2), (a,d): plots against time with fitted LOESS curve, (b,e): QQ plots, and (c,f): autocorrelations. The Z plots are in (a,b,c), W plots in (d,e,f).*

an indicator of possible serial correlation in either the Z or W series. The only value outside the confidence bands is the fourth autocorrelation coefficient for the W series (plot (f)); in conjunction with moderately high values for the first three autocorrelations, this does perhaps give some suggestion that there is some autocorrelation among the daily excess values of the windspeed series. Given that meteorological phenomena often persist over several days, this is not a surprising conclusion. The threshold methods described so far do not take account of short-term autocorrelations, a theme we return to in Section 1.8.

The calculation in (1.29) and (1.30) assumes u is a constant. This is another assumption that could be generalized, allowing u to be time-dependent. Thus, we could replace u by u_t in (1.29) or u_{T_k} in (1.30) to indicate the dependence on time.

1.5 Environmental extremes

This section describes three examples of extreme value theory applied to the environment. Section 1.5.1 is an analysis of ozone extremes, motivated by the problem of determining whether exceedances of an ozone standard are becoming more or less frequent. Section 1.5.2 treats a similar problem related to windspeed extremes — in this case, the question of exceeding a standard does not arise, but it is a natural question in the context of climate change, to determine whether extreme events are becoming more frequent. Section 1.5.3 and Section 1.5.4 extend this discussion further, first by asking the same question with respect to rainfall extremes, and then considering how to combine the information from a large number of rainfall stations. Thus, this section uses techniques of spatial statistics as well as extreme value theory.

1.5.1 Ozone extremes

As a first example of the use of regression models for extreme value data, we describe an analysis of ozone extremes given originally by Smith and Shively (1995).

The U.S. Environmental Protection Agency (EPA) establishes standards for a number of atmospheric pollutants including ozone. This standard applies to ground-level ozone, which is harmful to human health, and has nothing to do with the stratospheric ozone layer, which has the beneficial effect of blocking out certain kinds of radiation.

At the time of this study, the EPA standard for ground-level ozone was based on a daily maximum level of 120 ppb (parts per billion). The question of interest here is to try to determine whether the EPA standards have had any beneficial effect — which, for the present study, is interpreted to mean whether there is any downward trend in either the frequency or the size of exceedances over the threshold. However, a major complication in assessing trends is the influence of meteorology.

Ozone is produced by various photochemical processes in the atmosphere, and these photochemical processes are highly influenced by the weather (in particular, ozone is likely to be high on windless sunny days in the summer). There is therefore some interest in separating out the trend effects that might be due to genuine trends in the emissions of the so-called ozone precursors — gases such as nitrogen oxides, carbon monoxides, and other pollutants typically produced by industrial processes and by automobiles — and those effects that might be attributed to meteorological variation.

The main objective of the present analysis is to build a regression model in which relevant meteorological variables and time are all considered covariates in the regression, so that the coefficient of time may be interpreted as an adjusted time trend after taking the meteorological effects into account.

The data source in this case consists of ozone exceedances from a single station in Houston from 1983 to 1992. The assumed model has the following components:

1. The probability of an exceedance of a given level u (in practice taken as 120 ppb, the then-current standard) on day t is $e^{\alpha(t)}$, where

$$\alpha(t) = \alpha_0 + \alpha_1 s(t) + \sum_{j=2}^{p} \alpha_j w_j(t), \tag{1.31}$$

$s(t)$ being the calendar year in which day t falls and $\{w_j(t),\ j = 2, \ldots, p\}$ the values of $p - 1$ weather variables on day t.

2. Given an exceedance of level u on day t, the probability that it exceeds $u + x$, where $x > 0$, is

$$\{1 + \xi\beta(t)x\}^{-1/\xi}, \tag{1.32}$$

where

$$\beta(t) = \beta_0 + \beta_1 s(t) + \sum_{j=2}^{p} \beta_j w_j(t). \tag{1.33}$$

Note that for (1.31) and (1.33) to make sense, we require $\alpha(t) < 0$, $\beta(t) > 0$, for all t. There is a theoretical possibility that these conditions could be violated, but this was not a practical difficulty in the example under discussion. Also, a further generalization would also allow ξ to be dependent on covariates — in other words, replace ξ by $\xi(t)$ where $\xi(t)$ is given by another formula of the form (1.31) or (1.33). This extension was not used in the paper (in fact the authors took $\xi = 0$, as discussed further below).

The meteorological variables were selected after consultation with Texas air control experts who had extensive experience of the kinds of effects that are relevant in the Houston area. A specific list of variables was as follows:

TMAX. Maximum hourly temperature between 6 a.m. and 6 p.m.

TRANGE. Difference between maximum and minimum temperature between 6 a.m. and 6 p.m. This is considered to be a proxy for the amount of sunlight.

WSAVG. Average wind speed from 6 a.m. to 6 p.m. Higher windspeeds lead to lower ozone levels because of more rapid dispersion of ozone precursors.

WSRANGE. Difference between maximum and minimum hourly windspeeds between 6 a.m. and 6 p.m.

NW/NE. Percentage of time between 6 a.m. and 6 p.m. that the wind direction was between NW and NE.

NE/ESE. Percentage of time between 6 a.m. and 6 p.m. that the wind direction was between NE and ESE.

ESE/SSW. Percentage of time between 6 a.m. and 6 p.m. that the wind direction was between ESE and SSW.

SSW/NW. Percentage of time between 6 a.m. and 6 p.m. that the wind direction was between SSW and NW.

Table 1.3 *Coefficient and standard errors for $\alpha(t)$.*

Variable	Coefficient	Standard error
$s(t)$	–0.149	0.034
TRANGE	0.072	0.016
WSAVG	–0.926	0.080
WSRANGE	0.223	0.051
NW/NE	–0.850	0.408
NE/ESE	1.432	0.398

Table 1.4 *Coefficient and standard errors for $\beta(t)$, assuming $\xi = 0$.*

Variable	Coefficient	Standard error
$s(t)$	0.035	0.011
TRANGE	–0.016	0.005
WSAVG	0.102	0.019
NW/NE	0.400	0.018

The wind directions are important for Houston because they determine the level of industrial pollution — for instance, there is a lot of industry to the south of Houston, and ozone levels tend to be higher when the wind direction is in the ESE/SSW sector.

In most analyses, the variable SSW/NW was omitted because of the obvious collinearity with the other wind directions.

The models (1.31) through (1.33) were fitted by numerical maximum likelihood, and standard variable selection procedures were adopted, with variables being dropped from the analysis if their coefficients were not statistically significant. The results are shown in Table 1.3 and Table 1.4. In (1.32), it was found that the parameter ξ was not significantly different from 0, and of course (1.32) reduces to $e^{-\beta(t)x}$ if $\xi = 0$, so that was the form adopted for the reported analysis.

The results show that in both cases the coefficient of $s(t)$ is statistically significant, the sign (negative in Table 1.3, positive in Table 1.4) being consistent with the interpretation of an overall decrease in the extreme ozone levels, which was the hoped-for conclusion.

As a comparison, Smith and Shively also fitted the same model for trend alone, ignoring meteorological covariates. In this case, the estimated coefficients of $s(t)$ were -0.069 (standard error 0.030) in $\alpha(t)$ and 0.018 (standard error 0.011) in $\beta(t)$. The coefficients are thus much smaller in magnitude if the model is fitted without any meteorology. This confirms the significance of the meteorological component and shows how the failure to take it into account might obscure the real trend.

Fixing $\xi = 0$ in this analysis would probably not be desirable if the objective was to estimate probabilities of extreme events (or the quantiles associated with specified small return probabilities). This is because the estimates tend to be biased if ξ is assumed to be 0 when it is not, and for this kind of analysis, there is no obvious reason to treat $\xi = 0$ as a null hypothesis value. On the other hand, the main emphasis in this example was on the regression coefficients associated with the extreme events, and especially on the coefficients of $s(t)$ in the expressions for both $\alpha(t)$ and $\beta(t)$, and for that purpose, assuming $\xi = 0$ seems less likely to bias the results. Another possible extension of the analysis would be to allow the threshold u to be time-dependent, though in this example the natural value is 120 ppb because at the time this analysis was originally conducted, that was the U.S. ozone standard.

1.5.2 Windspeed extremes

This section illustrates an alternative approach to searching for trends in extreme value data, using the point process approach. The method is applied to the North Carolina windspeed data introduced in Section 1.4. In this kind of example, it is of interest, especially to insurance companies, to determine whether there is any evidence of a long-term increase in the frequency of extreme weather events. There has been much recent discussion among climatologists (see Section 1.5.3 for further discussion) that the world's weather is becoming more extreme as a possible side-effect of global warming, so when considering a series of this form, it is natural to look for any possible evidence of a trend. However, another and much more obvious effect is seasonality, and the analysis must also reflect that.

The analysis uses the nonstationary form of the point process model (Section 1.2.5) in which the GEV parameters are represented by (μ_t, ψ_t, ξ_t) to emphasise the dependence on time t. A typical model is of the form

$$\mu_t = \sum_{j=0}^{q_\mu} \beta_j x_{jt}, \quad \log \psi_t = \sum_{j=0}^{q_\psi} \gamma_j x_{jt}, \quad \xi_t = \sum_{j=0}^{q_\xi} \delta_j x_{jt}, \tag{1.34}$$

in terms of covariates $\{x_{jt},\ j = 0, 1, 2, \ldots\}$ where we usually assume $x_{0t} \equiv 1$. In the present analysis, we consider models in which only μ_t depends on covariates, so $q_\psi = q_\xi = 0$.

The log likelihood for this model may be derived from the joint density (1.18) and is maximized numerically to obtain estimates of the unknown parameters and their standard errors.

We illustrate the methodology by applying it to the North Carolina windspeed data. The model (1.34) was fitted to the Raleigh data based on exceedances over 39.5 knots. It was restricted to the case $q_\psi = q_\xi = 0$, largely because it simplifies the model not to have too many regression components, and it is natural to treat the location parameter μ_t, rather than ψ_t or ξ_t, as the one dependent on covariates.

The covariates x_{jt} were taken to be polynomial functions of time (t, t^2, etc.) and sinusoidal terms of the form $\sin \frac{2\pi kt}{T_0}$, $\cos \frac{2\pi kt}{T_0}$, where $k = 1, 2, \ldots$, and T_0 represents one year. In practice, no model fitted required sinusoidal terms beyond $k = 2$, and none of the polynomial trend terms were significant. As an example, Table 1.5 and Table 1.6 give the models fitted to the Raleigh data that involved no covariates (Model 0), and the covariates $\sin \frac{2\pi t}{T_0}$, $\cos \frac{2\pi t}{T_0}$, corresponding to a single sinusoidal curve (Model 1). The difference between negative log likelihoods ($X^2 = 2 \times (114.8 - 103.1) = 23.4$)

Table 1.5 *Raleigh data, Model 0: no seasonality. NLLH = 114.8*

Variable	Coefficient	Standard error
β_0	42.4	0.9
γ_0	1.49	0.16
δ_0	–0.19	0.19

Table 1.6 *Raleigh data, Model 1: Single sinusoidal component. NLLH = 103.1*

Variable	Coefficient	Standard error
β_0	40.9	0.9
β_1	0.90	1.07
β_2	5.29	1.39
γ_0	1.43	0.13
δ_0	–0.12	0.10

is clearly significant considering that X^2 has an approximate χ_2^2 distribution when Model 0 is correct. However, when other covariates are added to the model and the likelihood ratio statistics computed, the results are not significant.

Figure 1.16 shows QQ plots for the W statistics for the Raleigh data just considered, and corresponding results for Greensboro and Charlotte. These are based on Model 1 except for plot (d), which is based on Model 0 for Charlotte. Each of plots (a), (b), and (c) show an excellent fit to the model. Plot (d) is more questionable because the largest two observations appear to be outliers, which is another argument in favor of the seasonal model.

One conclusion from this analysis is that there do not appear to be overall trends in the series. This is not conclusive in itself because, as we shall see in the context of rainfall extremes in Section 1.5.4, trends estimated at individual stations tend to vary widely, and it is only when a large number of stations are examined together that an overall positive trend emerges. The other point of debate is whether hurricanes should be treated as separate events from the rest of the data. Meteorologically this makes sense, because a hurricane arises from quite different physical processes than ordinary storm events. Nevertheless, it is not clear how this should affect the data analysis. In the present three series, the two largest windspeeds for each of Greensboro and Charlotte are hurricane events (the largest for Charlotte is Hurricane Hugo in 1989) but there are no hurricanes in this section of the Raleigh data series (Hurricane Fran produced windspeeds up to 79 mph in Raleigh in 1996, but that was after the period covered by the current data set). The results for Greensboro and Charlotte suggest that when a threshold model is fitted to the whole of the data, the hurricanes do not appear as exceptional outliers. On the other hand, the fitted models for Greensboro and Charlotte have long-tailed distributions (though not given in detail here, both had $\hat{\xi}$ values around 0.2) whereas Raleigh had $\hat{\xi} = -0.19$, indicating a short-tailed distribution.

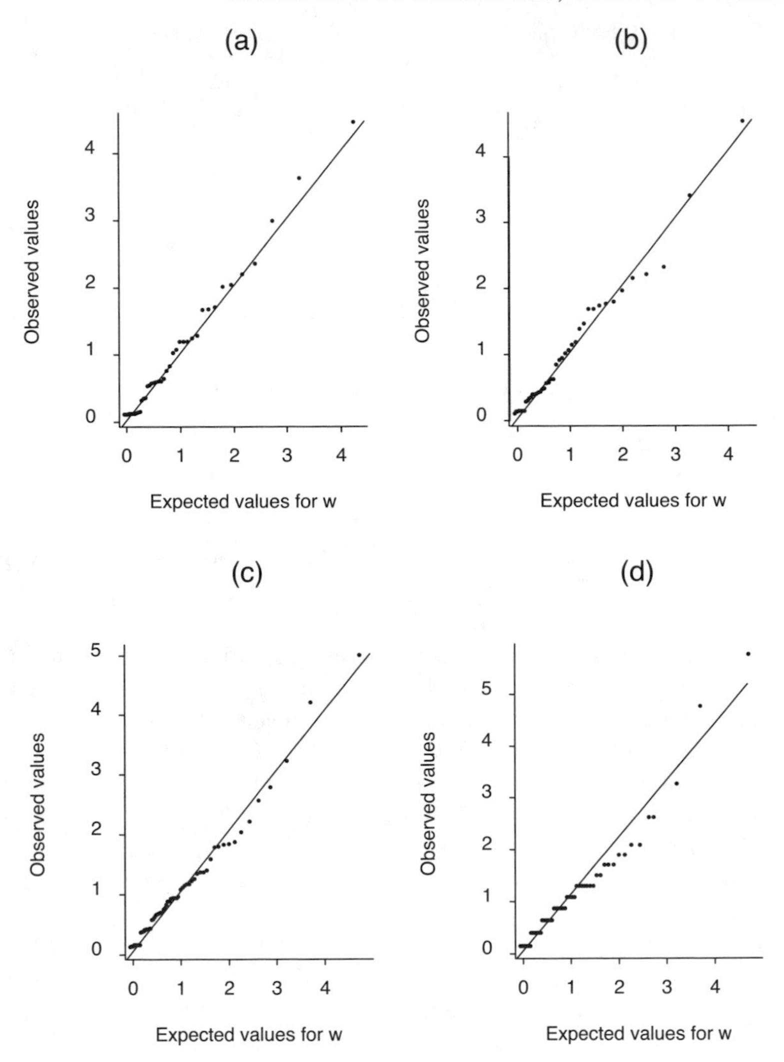

Figure 1.16 *QQ plots of residuals for Raleigh (a), Greensboro (b), and Charlotte (c and d), based on Model 1 except for (d) which uses Model 0.*

The standard error for $\hat{\xi}$ in Raleigh is also 0.19, indicating that the estimate is not significantly different from 0. Because of their geographical proximity, one would expect the parameters of the three series to be similar, and in Section 1.5.4 we shall propose a specific model to take that feature into account. For the current data sets, it is as far as we can go to say that the point process model appears to fit adequately to the extremes of all three cities, and that the absence of hurricane events in Raleigh, for the time period of the analysis, does not necessarily mean that the distribution of extreme windspeeds is significantly different from those of Greensboro and Charlotte.

1.5.3 Rainfall extremes

This section is based on the preprint of Smith (1999) which examined, from a rather broader perspective than the examples so far, the question of whether there is an overall increasing tendency in extreme rainfall events in the U.S., a popular hypothesis among climatologists.

The data base consisted of 187 stations of daily rainfall data from the Historical Climatological Network (HCN), which is part of the data base maintained by the National Climatic Data Center. Most stations start from 1910, but this analysis is restricted to 1951 through 1997 during which coverage percentage is fairly constant (Figure 1.17).

The strategy adopted in this analysis was to look at four stations in detail, which are widely separated geographically, and then to attempt a spatial analysis over the entire network. The locations of the four stations are shown in Figure 1.18.

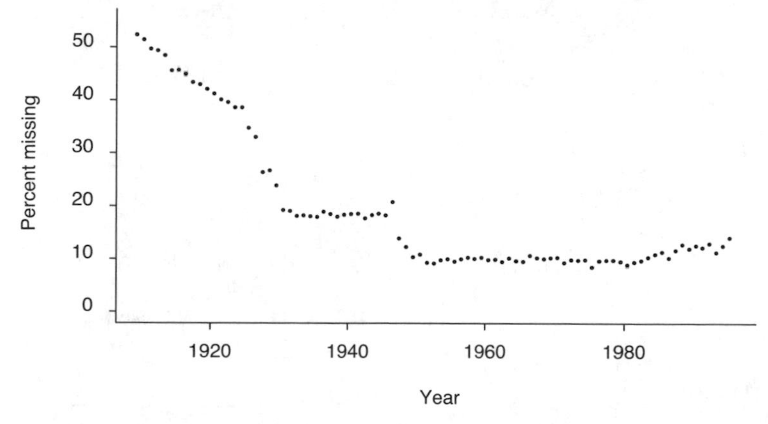

Figure 1.17 *Proportion of missing data over the whole network, for each year from 1910 to 1996.*

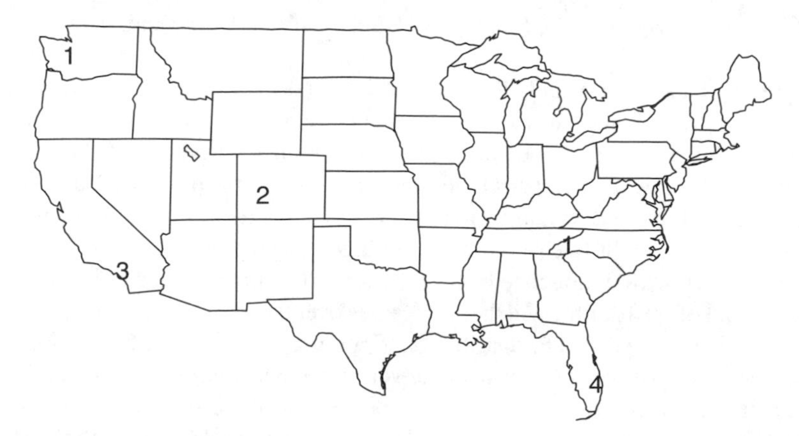

Figure 1.18 *Location of four test stations: (1) Station 319147, Waynesville, NC, (2) Station 53662, Gunnison, CO, (3) Station 49087, Tustin Irvine, CA, and (4) Station 80611, Belle Glade, FL.*

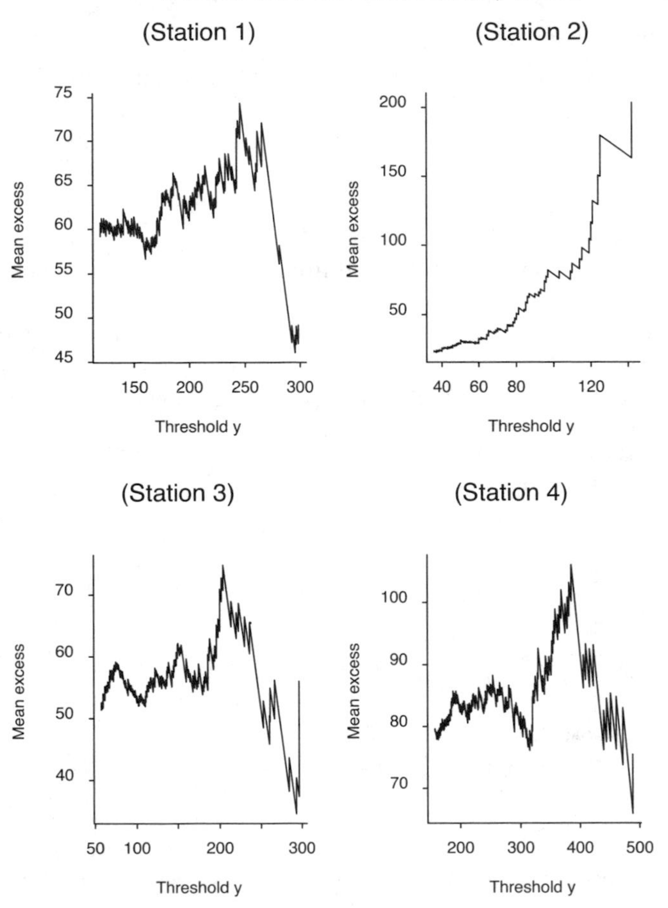

Figure 1.19 *Mean excess plots for four rainfall stations.*

Figure 1.31 and Figure 1.32 illustrate some rather deceptive features of mean excess plots. In these plots, the units of measurement are $\frac{1}{100}$ inch, and the base threshold relative to which the rest of the plot is calculated has been arbitrarily fixed at the 98th percentile of the raw data at each of the sites. From the simple plots without confidence bands (Figure 1.19) one would be tempted to conclude that only Station 2 is reasonably close to a straight line, the others looking very jagged. In fact, when the same plots are drawn with confidence bands (Figure 1.20), it becomes clear that Station 2 is the one out of the four for which the excesses over the assumed threshold clearly do *not* fit the generalized Pareto distribution. Station 2 is Gunnison, Colorado, and the dominant feature of the series is a single very large observation, of around 8 inches, more than three times the second largest observation in the series. When this observation is removed, the mean excess plot looks quite similar to that for the other three stations.

We now consider some possible models to fit to the data at individual stations. All the models are of the nonstationary point process structure of Section 1.2.5, and we

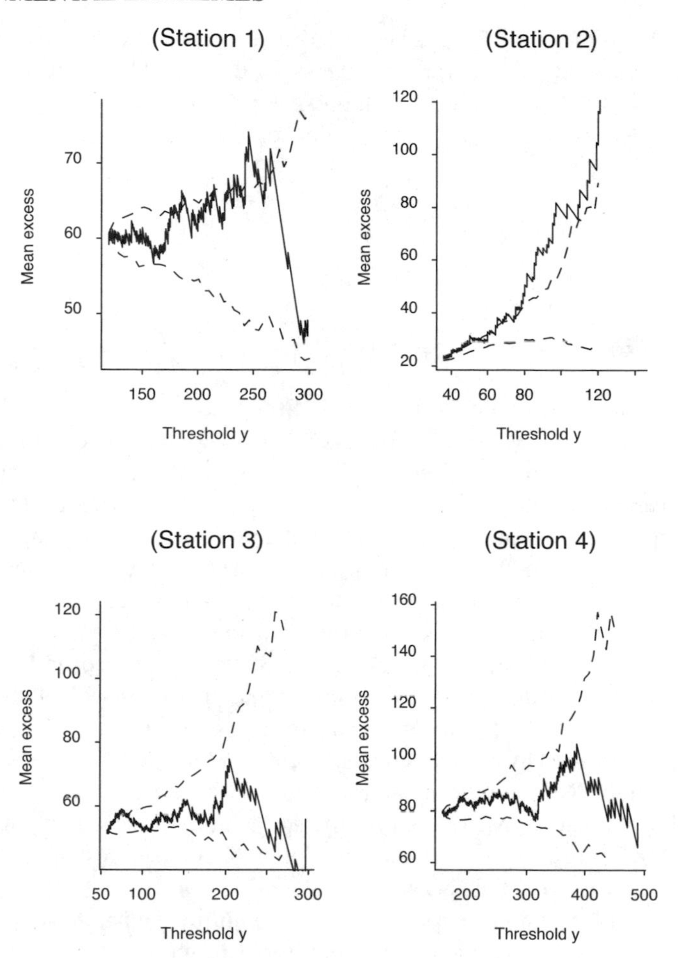

Figure 1.20 *Mean excess plots with confidence bands.*

consider two specific model classes:

$$\mu_t = \mu_0 + v_t, \quad \psi_t = \psi_0, \quad \xi_t = \xi_0, \quad \text{(Model 1)}; \tag{1.35}$$

$$\mu_t = \mu_0 e^{v_t}, \quad \psi_t = \psi_0 e^{v_t}, \quad \xi_t = \xi_0, \quad \text{(Model 2)}. \tag{1.36}$$

In either (1.35) or (1.36), v_t is a regression term of the form

$$v_t = \sum x_{tj}\beta_j,$$

where $\{x_{tj}\}$ are known covariates, and the $\{\beta_j\}$ are coefficients to be estimated.

For the regression terms x_{tj}, we consider a combination of linear time trends ($x_{tj} = t$), sinusoidal terms of the form $\cos \omega t$, $\sin \omega t$ to represent seasonal effects, and some external signals — as in the analysis of Section 1.6.4, both the Southern Oscillation Index (SOI) and the North Atlantic Oscillation (NAO) were considered as possible external influences.

Model 2 has some attractiveness because of the following interpretation. Suppose for a moment we ignore the covariates and just use the model to find the n-year return level y_n (i.e., the level that in any one year is exceeded with probability $\frac{1}{n}$) as a function of (μ, ψ, ξ). This is given (approximately) by solving the equation

$$\left(1+\xi\frac{y_n-\mu}{\psi}\right)^{-1/\xi}=\frac{1}{n},$$

leading to the solution

$$y_n=\mu+\psi\left(\frac{n^{\xi}-1}{\xi}\right). \tag{1.37}$$

Now consider the case in which μ and ψ are both dependent on time through a function of the form $\mu_t = \mu_0 e^{\beta_1 t}$, $\psi_t = \psi_0 e^{\beta_1 t}$, including the linear covariate on time but ignoring other covariates. Substituting into (1.37), this implies that the n-year return level is itself increasing with time proportional to $e^{\beta_1 t}$. If β_1 is small, this has the interpretation "the extreme rainfall amounts are increasing at a rate $100\beta_1\%$ per year." In contrast, if we take $v_t = \beta_1 t$ in model (1.35), this does not have such a direct interpretation.

For this reason, the analysis uses model 2 as the main model of interest, though it is not clear that this is actually the best-fitting model overall.

Table 1.7 shows a variety of model fits for Station 1. The covariates tried here included seasonal effects represented by either one or two sinusoidal curves, linear trend LIN, as well as SOI and NAO. The results show that seasonal variation is adequately represented by a single sinusoidal curve, SOI is marginally significant but NAO is not, and LIN is significant. Although our main analysis in this section is for Model 2, the table shows that Model 1 fits the data a little better for this station, raising a possible conflict between choosing the model that has the easiest interpretation and the one that appears to fit the data best.

In subsequent discussion we use Model 2 and multiply the parameter β_1 by 100 so that it has the rough interpretation of "percent rise in the most extreme levels per year."

Table 1.7 *NLLH and AIC values for nine models fitted to Station 1 (Waynesville, North Carolina).*

Model type	Covariates	NLLH	DF	AIC
2	None	1354.0	3	2714.0
2	Seasonal (1 component)	1350.3	5	2710.6
2	Seasonal (2 components)	1348.8	7	2711.6
2	Seasonal (1 component) + NAO	1349.6	6	2711.2
2	Seasonal (1 component) + SOI	1348.3	6	2708.6
2	Seasonal (1 component) + LIN	1346.9	6	2705.8
2	Seasonal (1 component) + LIN + SOI	1343.5	7	2701.0
1	Seasonal (1 component) + LIN	1344.8	6	2701.6
1	Seasonal (1 component) + LIN + SOI	1341.3	7	2696.6

For Waynesville, the estimate of this parameter is .074 if SOI is not included, .077 if SOI is included, both with standard error .035, indicating a significant positive trend.

Summary results for the other three stations (each based on Model 2):

In Station 2 (Gunnison, CO), there is a large outlier present, but it is not "influential" in the sense of affecting the parameter estimates. NAO and SOI are not significant. There is a strong negative linear trend (estimate –.72, standard error .21).

In Station 3 (Tustin Irvine, CA) we fit one seasonal component, the SOI is signal stronger than NAO, and the linear trend has estimate .67 without the SOI adjustment, .50 with, each with standard error .26.

In Station 4 (Belle Glade, FL), there is one seasonal component, and none of the trend terms are significant.

QQ plots based on the Z statistics for the four stations are shown in Figure 1.21, and those based on the W statistics are in Figure 1.22. For these plots, the outlier

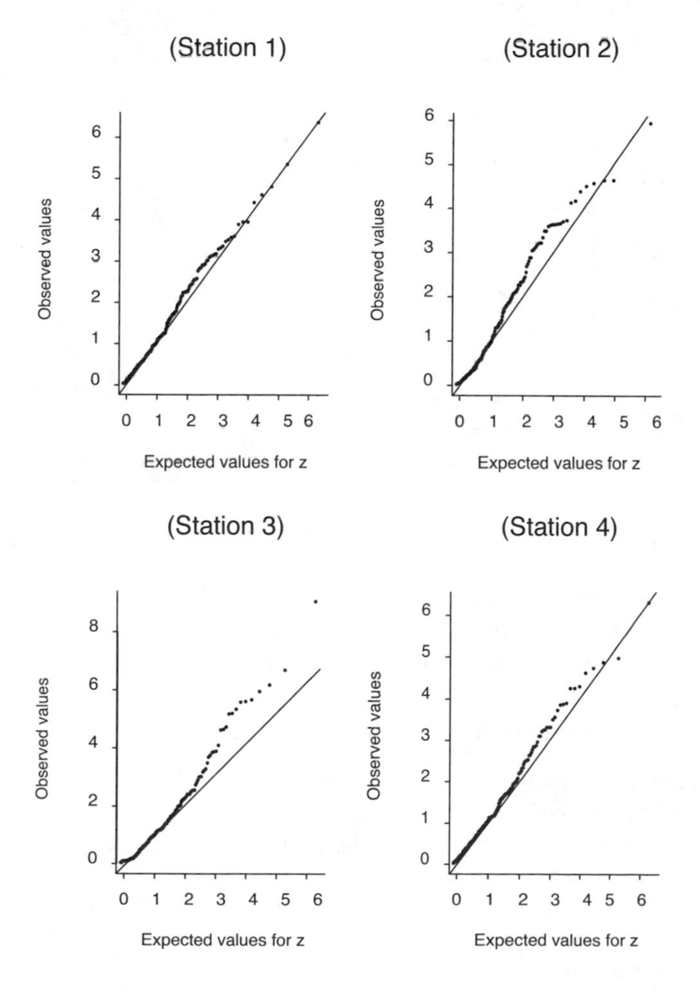

Figure 1.21 *Z plots for four rainfall stations.*

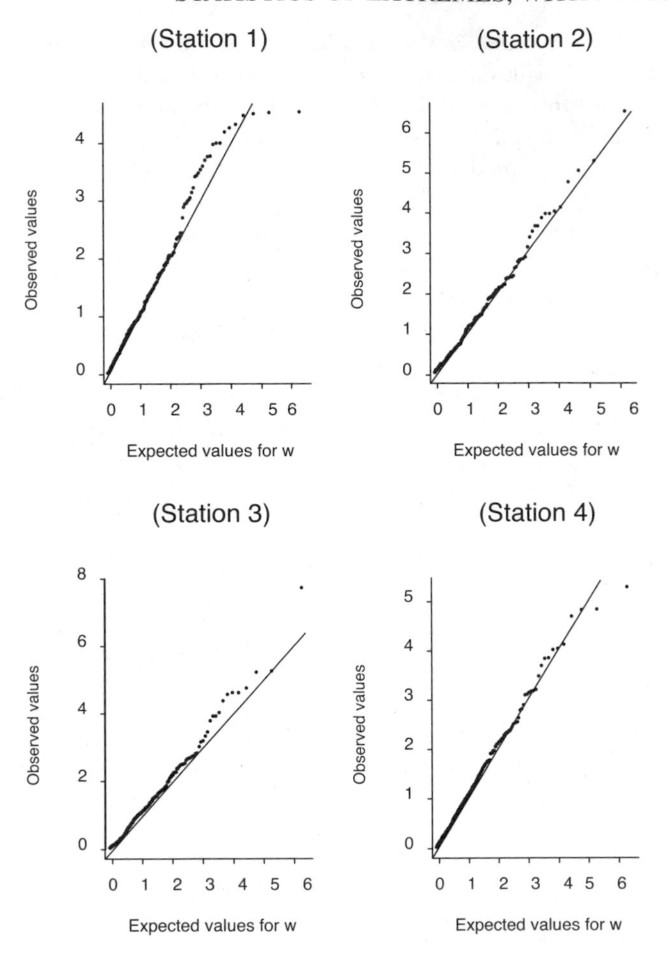

Figure 1.22 *W plots for four rainfall stations.*

in Station 2 has been removed. In general the plots indicate a good fit, but there are some features that might justify further investigation, e.g., both the Z and W plots for Station 3 seem somewhat discrepant at the upper end of the plot.

1.5.4 Combining results over all stations

After the preliminary analyses described in Section 1.5.3, Model 2 was fitted to each of the 187 stations (with successful fits in 184 cases), including a linear trend and two-component seasonal terms, but not SOI or any other external covariate. The threshold for each station was defined as the 98% empirical quantile for the data at that station. The main focus was on the parameter β_1 representing the overall trend, but the analysis also kept track of other parameters including the shape parameter ξ. Apart from the parameter estimates themselves, also computed were the t statistics, calculated by dividing the parameter estimate by its standard error.

Table 1.8 *Summary table of t statistics (parameter estimate divided by standard error) for extreme value model applied to 187 stations and 98% threshold. Tabulated values are percentages out of 184 stations analyzed.*

	$\hat{\beta}_1$	$\hat{\xi}$
$t > 2$	14%	40%
$t > 1$	40%	73%
$t > 0$	68%	88%
$t < 0$	32%	12%
$t < -1$	11%	2.7%
$t < -2$	5%	0.5%

As an example of the difficulty in determining some overall "significance" of a trend, Table 1.8 summarizes the t statistics for both the β_1 and ξ parameters. For example, if we take $|t| > 2$ as the criterion for a significant result, based on β_1, only 25 of the 184 stations have a significant positive trend, and 10 have a significant negative trend. On the other hand, for 125 stations we have $\hat{\beta}_1 > 0$, compared with only 59 for which $\hat{\beta}_1 < 0$. Thus, it is clear that there is an overall preponderance of stations with positive trends, and more detailed examination of the spatial distribution of $\hat{\beta}_1$ cofficients suggest a fairly random pattern, rather than, for example, the positive coefficients being concentrated in one part of the country and the negative coefficients in another.

The corresponding results for $\hat{\xi}$ in Table 1.8 show a preponderance of stations for which $\hat{\xi} > 0$. This is also somewhat contradicting conventional wisdom about rainfall distributions, since the gamma distribution is very often used in this context, and the gamma distribution is in the domain of attraction of the Gumbel distribution (Section 1.2), which would imply $\xi = 0$. The present analysis confirms that most empirical rainfall distributions are somewhat more long-tailed than the gamma.

To integrate the results across the 184 stations, a spatial smoothing technique is adopted, similar to Holland et al. (2000).

Suppose there is an underlying "true" field for β_1, denoted $\beta_1(s)$ to indicate the dependence on spatial location s. In other words, we assume (hypothetically) that with infinite data and the ability to determine $\beta_1(s)$ precisely for any location s, we could measure a smooth spatial field. However, in practice we are performing regression analysis and only estimating $\beta_1(s)$, as $\hat{\beta}_1(s)$, at a finite set of locations s. Assuming approximate normality of both the underlying field and the estimates, we can represent this as a two-stage process, with β_1 and $\hat{\beta}_1$ denoting the vector of values at measured locations s,

$$\beta_1 \sim N[X\gamma, \Sigma], \quad \hat{\beta}_1 \,|\beta_1 \sim N[\beta_1, W]. \tag{1.38}$$

Combining the two parts of (1.38) into one equation,

$$\hat{\beta}_1 \sim N[X\gamma, \Sigma + W]. \tag{1.39}$$

In these equations, Σ is interpretated as the spatial covariance matrix of the unobserved field β_1 — for the present application, this is taken to be the Matérn covariance function which is popular in spatial analysis (Cressie, 1993). Also, $X\gamma$ represents systematic variation in space of the β_1 field: in the present application, after examining various alternatives, the components of the X matrix were taken to be linear and quadratic terms in the latitude and longitude of a station, leading to a quadratic surface as the "mean field." Finally the matrix W is interpreted as the covariance matrix of errors arising from the extreme value regression procedures: this is assumed to be known, and a diagonal matrix, with the variance of $\hat{\beta}_1$ at each station taken as the square of the standard error obtained as part of the extreme value fitting procedure. Equation (1.39) is then fitted to the vector of $\hat{\beta}_1$ coefficients, the regression parameters γ and the unknown parameters of Σ estimated by numerical maximum likelihood. The final step of the fitting procedure uses the fact that once the model parameters are estimated, the relationships (1.38) can be inverted to obtain the conditional distribution of β_1 given $\hat{\beta}_1$: this is applied to obtain a smoothed predictive estimate of the β_1 field, together with estimated prediction covariances, in a manner analogous though not identical to the spatial interpolation procedure known as kriging.

One set of results of this procedure is shown in Figure 1.23, which depicts contour and perspective plots of the smoothed surface. Even after smoothing, the surface is somewhat irregular, but it is much smoother than the set of raw $\hat{\beta}_1$ values before any smoothing.

Finally, we consider the computation of regional averages. Figure 1.24 (from Grady, 2000), who used the same regions in a slightly different context) shows the 187 stations divided into five regions. In Table 1.9, we give estimated "regional trends," with standard errors. The regional trends are derived by averaging the smoothed β_1 estimates over a very dense grid of points within the region of interest, and the standard errors are the square roots of the estimated variances of those averages, given the $\hat{\beta}_1$ values. These quantities are computed for each of the five regions and also overall, i.e., combining the five regions into one.

When interpreted in this way, the results show that each of the five regions has a positive (and statistically significant) trend in the rainfall extremes, but there are also significant differences among the five regions, region 3 in the center of the country having the largest trend and the two coastal regions, region 1 in the west and region 5 in the southeast, having the smallest trends.

1.6 Insurance extremes

This section is concerned with two examples of extremes arising in the insurance industry. Section 1.6.1 through Section 1.6.3 extend the discussion of the oil company data set discussed in Section 1.1. As noted there, after some initial preprocessing of the data, there are 393 claims over a nominal threshold 0.5.

In Section 1.6.1, we consider the results of the GPD and homogeneous point process model fitted to data above different thresholds. Section 1.6.2 then extends the discussion to include Bayesian predictive distributions for future losses, drawing on our discussion of Bayesian predictive distributions in Section 1.3. Section 1.6.3 is about hierarchical models, in particular arguing the case in favor of treating the different types of claims in Table 1.2 as coming from separate distributions rather than all arising from a single distribution of claims.

Table 1.9 *Regionally averaged trends and standard errors for five regions of U.S. and overall.*

Region	Extreme rainfall trend	S.E.
1	.055	.024
2	.092	.017
3	.115	.014
4	.097	.016
5	.075	.013
All	.094	.007

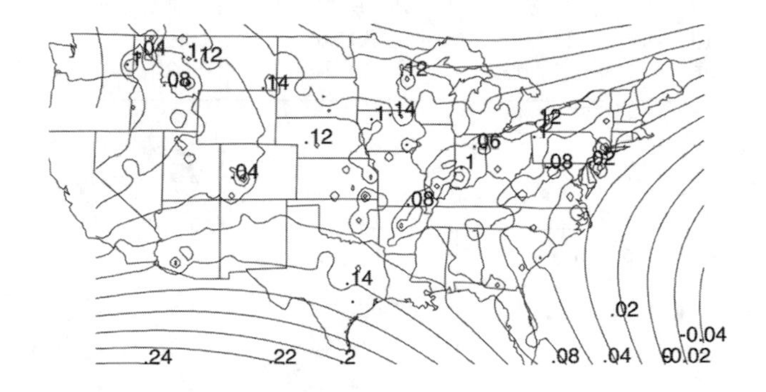

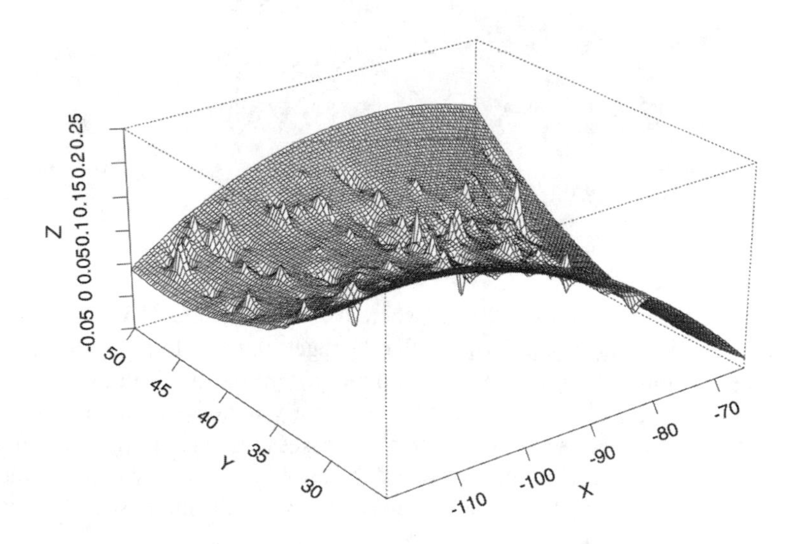

Figure 1.23 *Contour and perspective plots for spatial distribution of rainfall extreme trend coefficients.*

Table 1.10 *Fit of the generalized Pareto distribution to the excesses over various thresholds. The threshold is denoted u, and N_u the number of exceedances over u; we also tabulate the mean of all excesses over u and the maximum likelihood estimates of the GPD parameters σ and ξ. Standard errors are in parentheses.*

u	N_u	Mean excess	$\hat{\sigma}$	(S.E.)	$\hat{\xi}$	(S.E.)
0.5	393	7.11	1.02	(0.10)	1.01	(0.10)
2.5	132	17.89	3.47	(0.59)	0.91	(0.17)
5	73	28.90	6.26	(1.44)	0.89	(0.22)
10	42	44.05	10.51	(2.76)	0.84	(0.25)
15	31	53.60	5.68	(2.32)	1.44	(0.45)
20	17	91.21	19.92	(10.42)	1.10	(0.53)
25	13	113.70	33.76	(18.93)	0.93	(0.55)
50	6	37.97	150.80	(106.30)	0.29	(0.57)

Figure 1.24 *Five regions for regional analysis.*

Section 1.6.4 is on a different theme, where a 75-year series of extreme storms in the U.K. were assessed on the basis of the damage they would have caused under present-day insurance policies. By standardizing to present-day values, we take account not merely of price inflation, but of the greatly increased quantity and value of property insured. Such a series is valuable in assessing whether there might have been long-term trends in insurance claims that could be caused by external factors. We again use Bayesian methods to assess risks under various models for a long-term trend.

1.6.1 Threshold analyses with different thresholds

The first analysis was to fit the GPD to the excesses over various thresholds. The results are shown in Table 1.10.

Table 1.11 *Fit of the homogenous point process model to exceedances over various thresholds u. Standard errors are in parentheses.*

u	N_u	$\hat{\mu}$	(S.E.)	$\log \hat{\psi}$	(S.E.)	$\hat{\xi}$	(S.E.)
0.5	393	26.5	(4.4)	3.30	(0.24)	1.00	(0.09)
2.5	132	26.3	(5.2)	3.22	(0.31)	0.91	(0.16)
5	73	26.8	(5.5)	3.25	(0.31)	0.89	(0.21)
10	42	27.2	(5.7)	3.22	(0.32)	0.84	(0.25)
15	31	22.3	(3.9)	2.79	(0.46)	1.44	(0.45)
20	17	22.7	(5.7)	3.13	(0.56)	1.10	(0.53)
25	13	20.5	(8.6)	3.39	(0.66)	0.93	(0.56)

A second analysis was to fit the homogenous point process model of Section 1.2.5, again varying the threshold. In the case of the scale parameter ψ, this is reparametrized as $\log \psi$ for greater numerical stability. The results are in Table 1.11.

For any given threshold, the models of Table 1.10 and Table 1.11 are mathematically equivalent, but in comparing results across thresholds, Table 1.11 is easier to interpret because if the model fits, the (μ, ψ, ξ) parameters will not change with the threshold. In this case, taking into account the standard errors, there seems good agreement among the different parameter estimates across a range of thresholds, confirming the good fit of the model. This is further confirmed by the diagnostic plots of Figure 1.25, which show the Z and W statistics (analogous to Figure 1.15) for threshold 5. Based on this, one could recommend threshold 5 as the one to be used for subsequent calculations of extreme event probabilities, though given the near-constancy of parameter estimates across different thresholds, it appears that the results are not very sensitive to the choice of threshold. Note, however, that all the estimated distributions are very long-tailed, with estimates of ξ typically close to 1.

1.6.2 Predictive distributions of future losses

One question of major interest to the company is how much revenue it needs to put aside to allow for possible losses in some future period. This is a question of predictive inference about future losses, and as noted already in Section 1.3, there may be good reasons for taking a Bayesian approach to such questions. We approach this here from the point of view of losses over a fixed time period, taken for definiteness to be one year.

Let Y be future total loss in a given year. We write the distribution function of Y as $G(y; \mu, \psi, \xi)$ to indicate the dependence on unknown parameters.

As already noted in Section 1.3, the traditional frequentist approach to predictive inference is based on the "plug-in" approach in which maximum likelihood estimates are substituted for the unknown parameters,

$$\hat{G}(y) = G(y; \hat{\mu}, \hat{\psi}, \hat{\xi}).$$

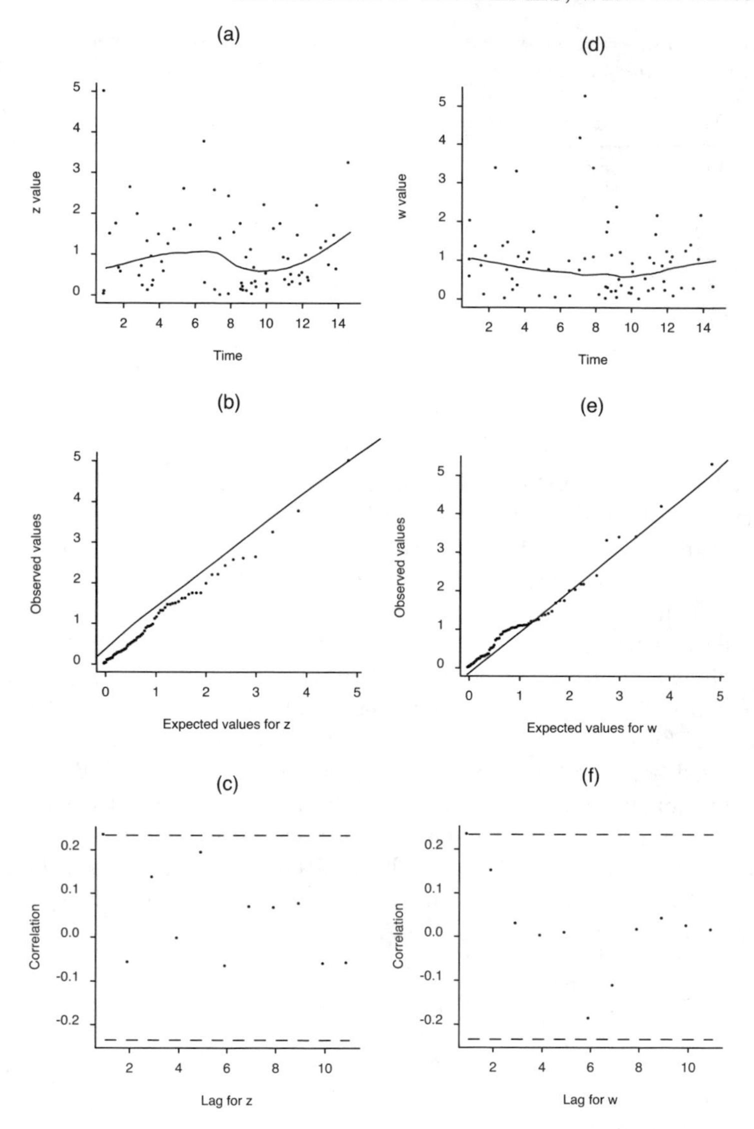

Figure 1.25 *Diagnostic plots based on $u = 5$.*

The alternative Bayesian approach leads to

$$\tilde{G}(y) = \int G(y; \mu, \psi, \xi)\pi(\mu, \psi, \xi \mid \mathbf{X}) d\mu d\psi d\xi \tag{1.40}$$

where $\pi(\cdot \mid \mathbf{X})$ denotes the posterior density given data $\mathbf{X}$. For the prior density, we take a diffuse but proper prior.

Equation (1.40) is evaluated numerically using a Markov chain Monte Carlo approach. Within this Monte Carlo calculation, the distribution function $G(y; \mu, \psi, \xi)$

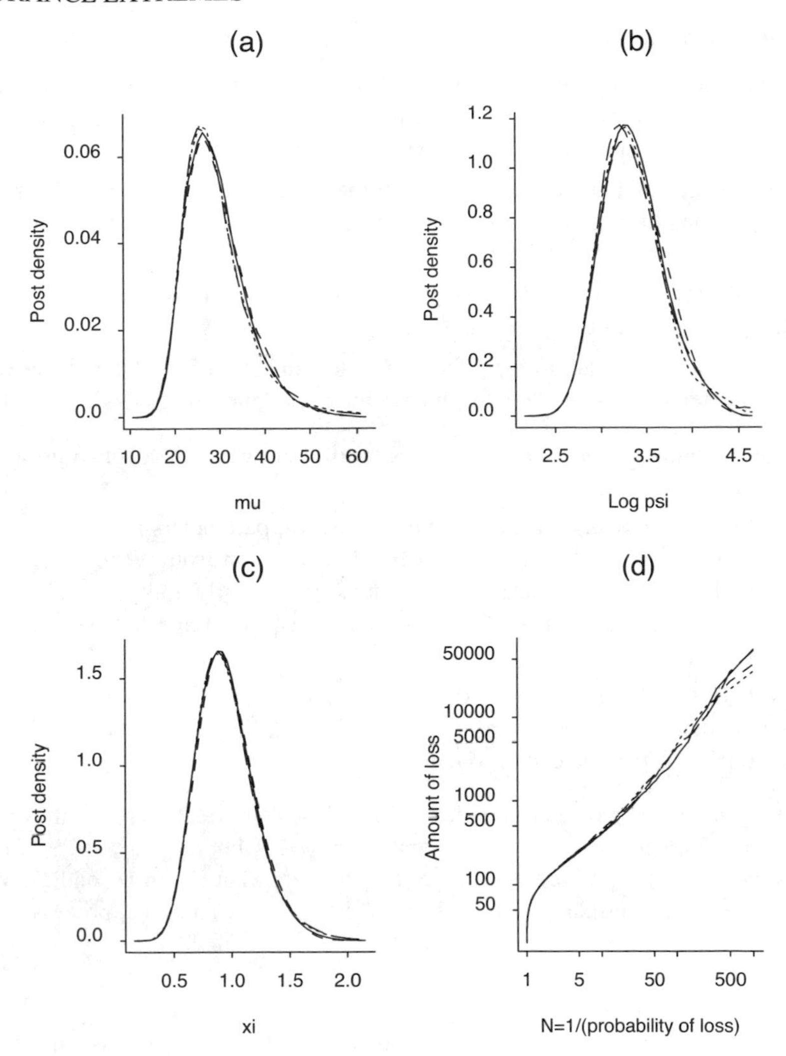

Figure 1.26 *Estimated posterior densities for the three parameters and for the predictive distribution function (1.40). Four independent Monte Carlo runs are shown for each plot.*

is also evaluated by simulation, since it represents the accumulation of many different claims. However, we can assume that the major contribution to the total loss comes from claims over the threshold, so the distribution may be represented as a function of the extreme value parameters μ, ψ, and ξ.

In Figure 1.26, the posterior densities of μ, ψ, and ξ are shown, together with the estimated predictive distribution function for the future losses, calculated from (1.40). To examine the influence of the Monte Carlo error, four independent simulations are performed for each plot. The four estimates within each plot are very close to one another, implying that the Monte Carlo-based error is small.

1.6.3 Hierarchical models for claim type and year effects

In this section, we extend the preceding analysis to allow for the different claim types, and also examine the possibility of a year effect, that might represent some overall trend. Preliminary analyses of these aspects indicate:

1. When separate GPDs are fitted to each of the six main types, there are clear differences among the parameters. (We omit type seven from this discussion because it contains only two small claims.)
2. The rate of high-threshold crossings does not appear uniform over the different years, but peaks around years 10 to 12.

To try to take the claim-type effects into account, Smith and Goodman (2000) proposed extending the model into a hierarchical model of the following structure:

Level I. Parameters $m_\mu, m_\psi, m_\xi, s_\mu^2, s_\psi^2$, and s_ξ^2 are generated from a prior distribution.

Level II. Conditional on the parameters in level I, parameters $\mu_1, \ldots, \mu_J$ (where J is the number of types) are independently drawn from $N(m_\mu, s_\mu^2)$, the normal distribution with mean m_μ, variance s_μ^2. Similarly, $\log\psi_1, \ldots, \log\psi_J$ are drawn independently from $N(m_\psi, s_\psi^2)$, $\xi_1, \ldots, \xi_J$ are drawn independently from $N(m_\xi, s_\xi^2)$.

Level III. Conditional on level II, for each $j \in \{1, \ldots, J\}$, the point process of exceedances of type j is a realization from the homogeneous point process model with parameters μ_j, ψ_j, and ξ_j.

This model may be further extended to include a year effect. Suppose the extreme value parameters for type j in year k are not μ_j, ψ_j, ξ_j but $\mu_j + \delta_k, \psi_j, \xi_j$. In other words, we allow for a time trend in the μ_j parameter, but not in ψ_j and ξ_j. We fix $\delta_1 = 0$ to ensure identifiability, and let $\{\delta_k,\ k > 1\}$ follow an AR(1) process:

$$\delta_k = \rho\delta_{k-1} + \eta_k, \quad \eta_k \sim N(0, s_\eta^2) \tag{1.41}$$

with a vague prior on (ρ, s_η^2).

These models are estimated by Markov chain Monte Carlo methods, an extension of the Monte Carlo methods used for computing posterior and predictive densities for single distributions. Figure 1.27 shows boxplots based on the Monte Carlo output, for each of $\mu_j, \log\psi_j, \xi_j, j = 1, \ldots, 6$ and for $\delta_k,\ k = 2, \ldots, 15$. Taking the posterior interquartile ranges (represented by vertical bars) into account, it appears that there are substantial differences among the six claim types for the μ parameter, and to a lesser extent for the ψ parameter. In contrast, the six types seem homogeneous in the ξ parameter, though it is notable that the posterior means of $\xi_1, \ldots, \xi_6$ are all in the range 0.7 to 0.75, compared with a value close to 1 when the six types are combined into one distribution, as in Figure 1.26. This shows one effect of the disaggregation into different types — when the types of claims are separated, the apparent distributions are less long-tailed than if the types are aggregated. The effect this has on the predictive distributions will be seen later (Figure 1.29). In contrast, the estimates of the δ_k parameters seem fairly homogeneous when the posterior interquartile ranges are taken

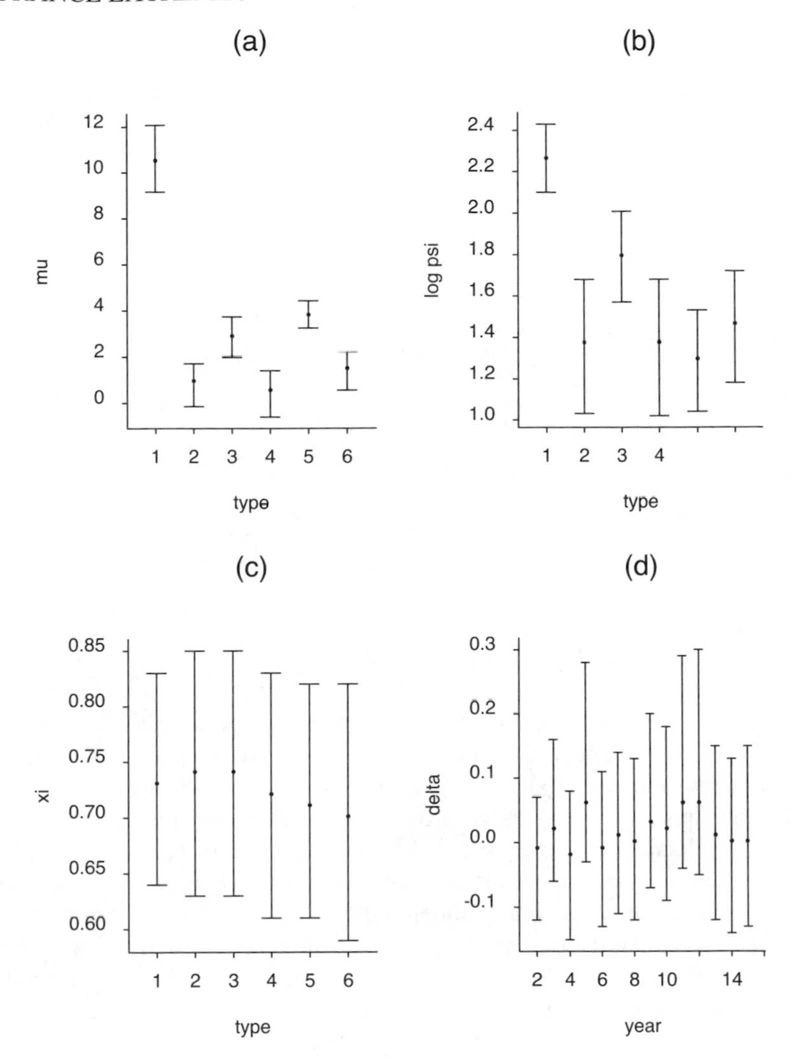

Figure 1.27 *Boxplots for the posterior distributions of* μ_j, $\log \psi_j$, ξ_j $(j = 1, \ldots, 6)$ *and for* δ_k $(k = 2, \ldots, 15)$, *based on the hierarchical model. The central dots represent posterior means and the horizontal bars the first and third quartiles of the posterior distribution, taken from the Monte Carlo output.*

into account, suggesting that trends over the years are not a significant feature of this analysis.

Figure 1.28 shows the estimated posterior density of ρ from (1.41), which is fairly flat over the region of stationarity $(-1, 1)$. Finally, Figure 1.29 shows the estimated loss curves for future total loss over a one-year period, similar to (1.40) but integrating over all the unknown parameters in the hierarchical model. Four loss curves are computed: curve A based on the homogenous model with all data combined; curve B taking into

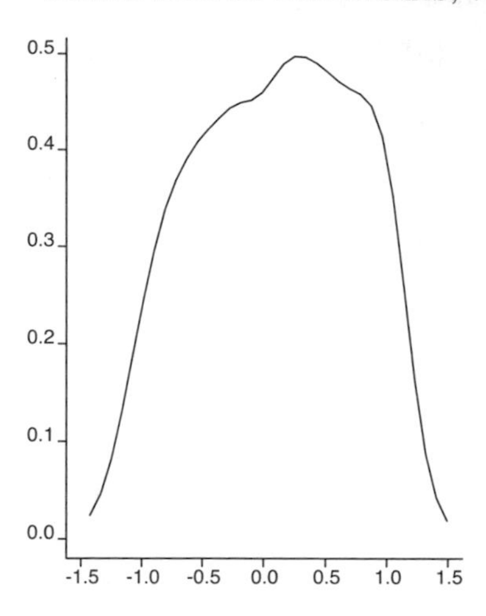

Figure 1.28 *Posterior density of* ρ.

account the claim-type effects but not year effects; curve C taking into account both claim-type and year effects; and curve D which is computed the same way as curve C, but omitting the two outliers which were mentioned in the earlier discussion. The sharp contrast between curves A and B highlights the advantages of including claim-type effects: we obtain much less extreme predicted losses, a consequence of the fact that the data appear less long-tailed if the claim types are separated than if they are combined. Including the year effects (curve C) makes little difference to the predicted curve for a typical future year. Curve D was included largely because there was some interest within the company in separating out the "total loss" events (recall from the earlier discussion that the two largest claims in this series were the only claims that represent the total loss of a facility). There is a substantial difference between curves C and D, highlighting that these very extreme events do have a significant influence on future predictions — from the company's point of view, a predicted curve that includes total losses can be expected to differ from one that excludes total losses, and this is reflected in the figure here.

1.6.4 Analysis of a long-term series of U.K. storm losses

This example is based on a preliminary analysis of a data set constructed by the U.K. insurance company Benfield-Greig. The objective of this analysis is to study long-term trends in insurance claims. However, it is recognized that there are many reasons, besides possible climate-change effects, why insurance claims are rising over time — inflation is an obvious reason, but also, with more property being insured, and insured to higher values in real terms than in earlier times, it is inevitable that insurance claims will increase with time as well. The interest here, however, lies in separating out the effects that might possibly be due to long-term climate change.

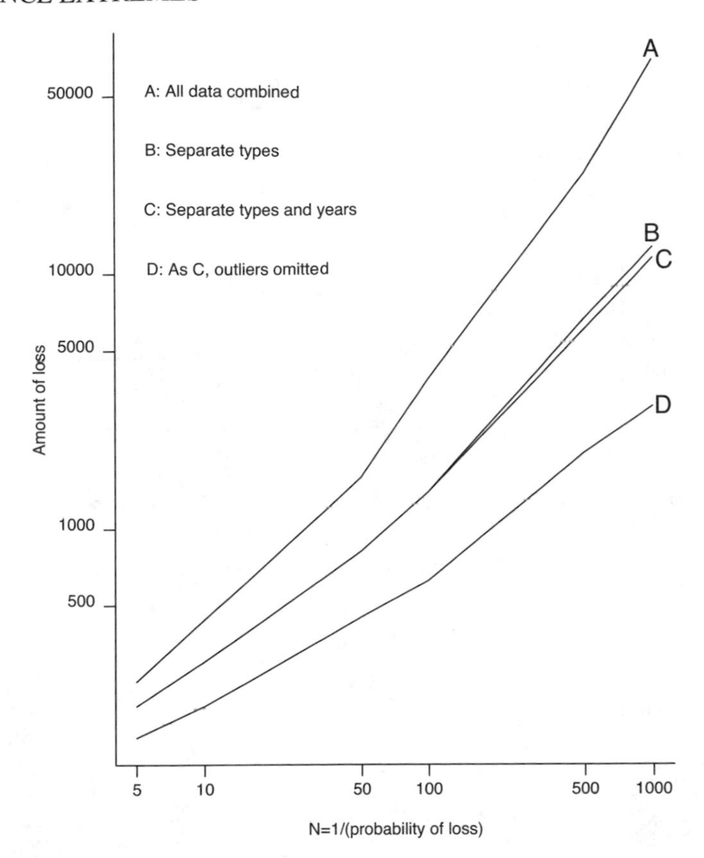

Figure 1.29 *Computations of posterior predictive distribution functions (plotted on a log-log scale) corresponding to the homogenous model (curve A) and three different versions of the hierarchical model.*

With this in mind, Benfield-Greig compiled a list of 57 historical storm events in the period 1920 to 1995, and assessed their insurance impact based on the likely total loss due to each storm that would have occurred if the same storm event had occurred in 1998. If *this* series contains a trend, it seems reasonable to conclude that it is due to climatic influences rather than either inflation or changes in insurance coverage.

The analysis adopted here is again of the point process form (1.34) with $q_\psi = q_\xi = 0$, and with μ_t modelled as a function of time t through various covariates:

- Seasonal effects (dominant annual cycle);
- Polynomial terms in t;
- Nonparametric trends; and
- Dependence on climatic indicators such as SOI and NAO (discussed later).

Other models in which ψ_t and ξ_t depend on t were also tried, but did not produce significantly different results.

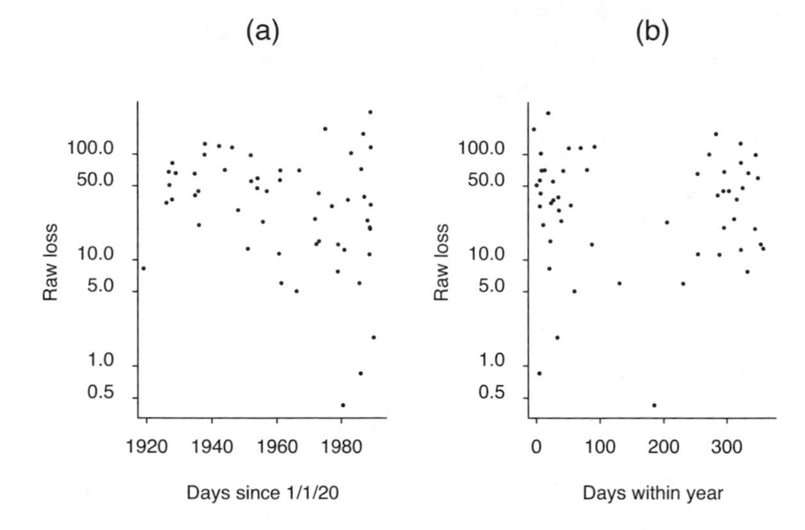

Figure 1.30 *Plots of estimated storm losses against (a) time measured in years, and (b) day within the year.*

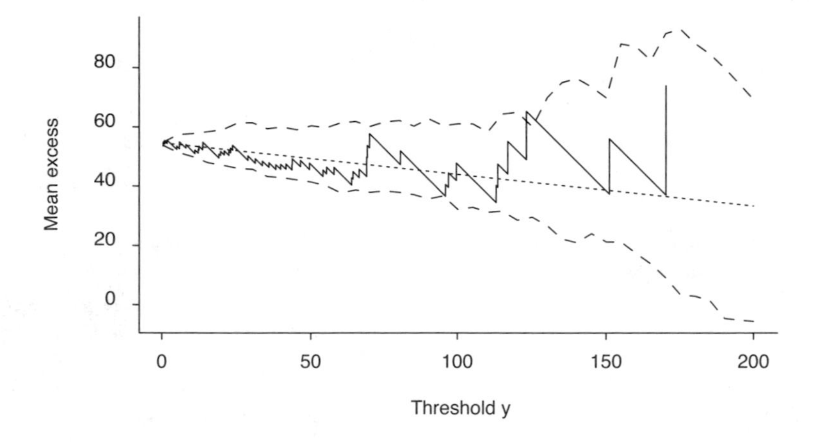

Figure 1.31 *Mean excess plot with Monte Carlo confidence bands.*

Figure 1.30 shows scatterplots of the losses against (a) year, and (b) day within year — the latter plot shows an obvious and expected seasonal effect, but the former is somewhat ambiguous as to whether there is a trend. Figure 1.31 shows a mean excess plot, with Monte Carlo confidence bands, to the raw data assuming no trend — since the slope of the fitted straight line is very near 0, this appears fully consistent with an exponential distribution for exceedances over a high threshold (as noted in Section 1.4.3, the slope of the mean excess plot may be used as a diagnostic for the sign of ξ).

To fit the data, a number of different trend effects were included. Seasonal effects were modelled by a simple sinusoid of period one year (higher-order harmonics were

Table 1.12 *NLLH and AIC values (AIC = 2 NLLH + 2 p) for several models.*

Model	p = number of parameters	NLLH	AIC
Simple GPD	3	312.5	631.0
Seasonal	5	294.8	599.6
Seasonal + cubic	8	289.3	594.6
Seasonal + spline	10	289.6	599.2
Seasonal + SOI	6	294.5	601.0
Seasonal + NAO	6	289.0	590.0
Seasonal + NAO + cubic	9	284.4	586.8
Seasonal + NAO + spline	11	284.6	591.2

tested but found not to be significant). Polynomial trends of various orders were tried, the best fit being a cubic curve, and also various nonparametric representations of the long-term trend, using spline functions with various degrees of freedom. Also tried were some meteorologically-based covariates that are believed to influence British storm events: the Southern Oscillation Index (pressure difference between Darwin and Tahiti, henceforth abbreviated to SOI) is very widely used as a measure of the strength of the El Niño effect, but more relevant to British meteorology is the North Atlantic Oscillation or NAO (pressure difference between Iceland and the Azores). A series of different models, with the corresponding values of the negative log likelihood (NLLH) and Akaike Information Criterion (AIC) are given in Table 1.12. This confirms that the seasonal effect and the NAO are both statistically significant, though not the SOI. The spline trend used here has five degrees of freedom and fits the data slightly worse than the cubic trend which has only three degrees of freedom; however, given the commonsense conclusion that a nonparametric representation of the long-term trend is likely to prove more satisfactory than the probably accidental fit of a polynomial function, we retain both trend terms for future study.

For the model including the seasonal effect and NAO only, Figure 1.32 shows the diagnostic plots based on the Z and W statistics, analogous to our earlier plots in Figure 1.15 and Figure 1.25. There is no evidence in these plots against the fit of the proposed model.

To assess the impact of the trend terms on return levels, Figure 1.33(a) shows the 10-year, 100-year, and 1000-year return levels for January, under a model with both seasonal and NAO effects. To avoid trying to represent both the seasonal and NAO effects on a single graph, the plot here just represents the loss curve for January. The 10-year return level is here defined as the value that is exceeded on any one day with probability 1/3652 and similar definitions for 100-year and 1000-year return levels. The plot brings out the differences among the 10-year, 100-year, and 1000-year return levels, but taking just the middle one of these and plotting it along with its confidence limits in Figure 1.33(b) makes clear that the uncertainty of these estimates is higher than the differences among the three curves. For models in which either a cubic or a spline-based trend was included, the corresponding curves are shown as Figure 1.34

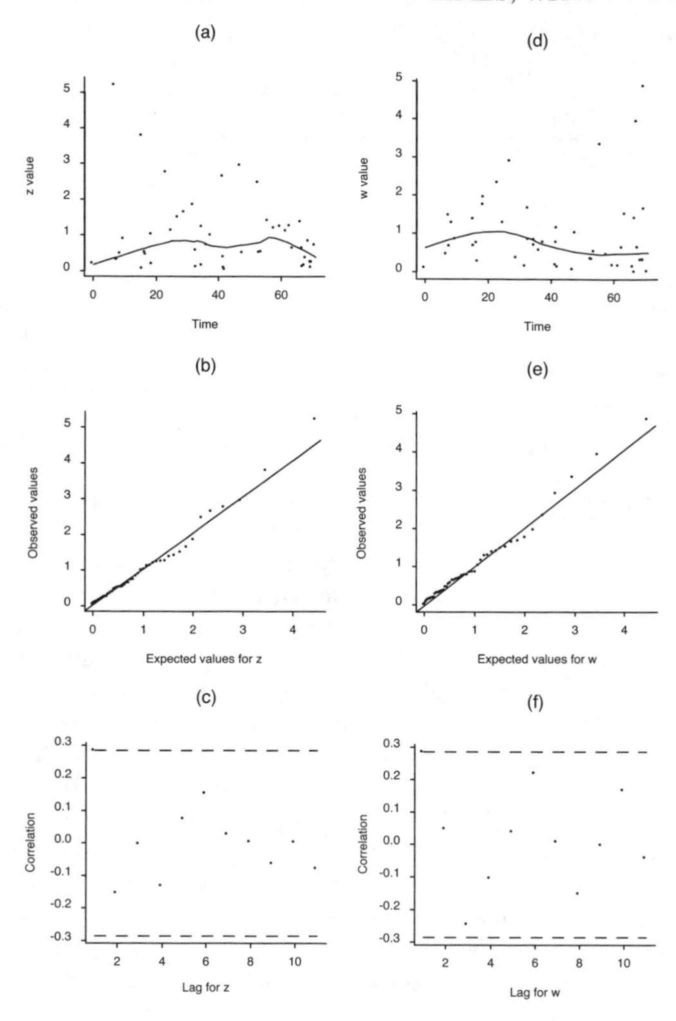

Figure 1.32 *Diagnostic plots based on the Z (plots a,b,c) and W (plots d,e,f) statistics.*

and Figure 1.35. Both curves show a dip in the loss curve in the middle of the century, followed by a rise in the period following 1975. Figure 1.36 is similar to Figure 1.33 but computed for July — of course, the return levels are much lower in this case. Taken together, these plots illustrate the influence of different types of trend, demonstrating that NAO has a strong influence on the return levels but also that there seem to be persistent long-term trends.

Finally, we consider the question that is of most interest to the insurance industry — the likelihood of possibly greater losses in the future. The largest value in the present data set was the January 1990 windstorm that caused damage across a very large area of the U.K. (worse than the storm of October 1987, which involved stronger winds, but was confined to a much smaller geographic area). We consider this question by

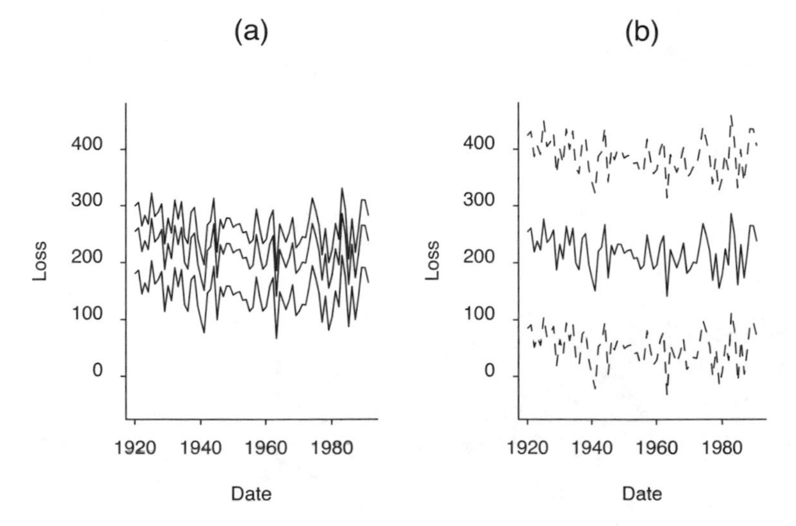

Figure 1.33 *(a) Estimates of 10-year (bottom), 100-year (middle), and 1000-year (top) return levels based on the fitted model for January, assuming long-term trend based on NAO, and (b) 100-year return level with confidence limits.*

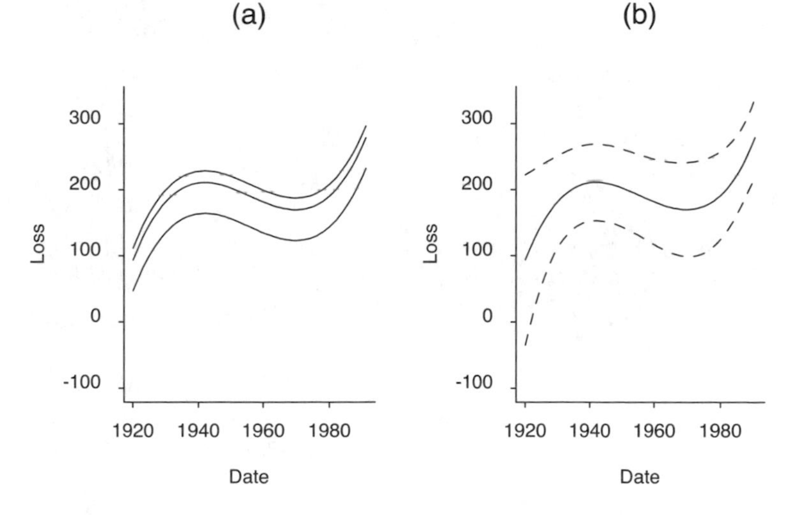

Figure 1.34 *(a) Estimates of 10-year (bottom), 100-year (middle), and 1000-year (top) return levels based on the fitted model for January, assuming long-term trend based on a cubic polynomial, and (b) 100-year return level with confidence limits.*

computing the return period associated with the 1990 event, calculated under various models (Table 1.13). The first row of this table assumes the simple Pareto distribution — this distribution does not fit the data at all, since as already seen from Figure 1.31, the true distribution of the tail is much closer to exponential than Pareto. The return period under the exponential distribution, fitted without any seasonality or trend, is

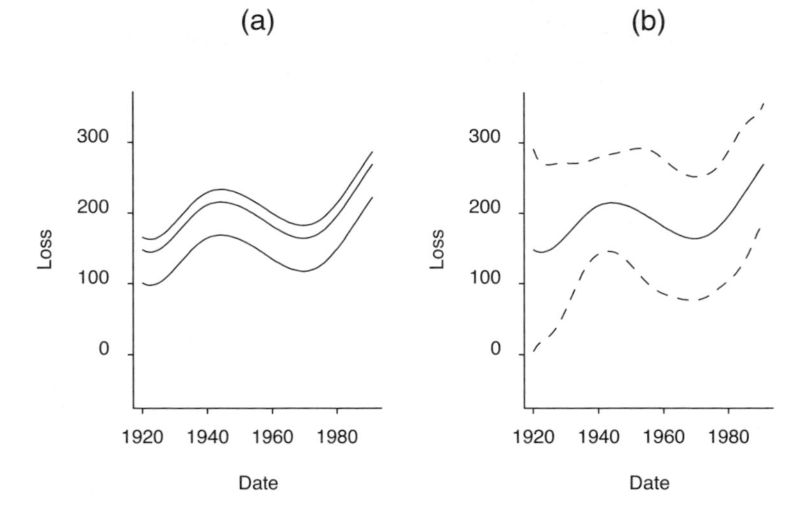

Figure 1.35 *(a) Estimates of 10-year (bottom), 100-year (middle), and 1000-year (top) return levels based on the fitted model for January, assuming long-term trend based on a cubic spline with 5 knots, and (b) 100-year return level with confidence limits.*

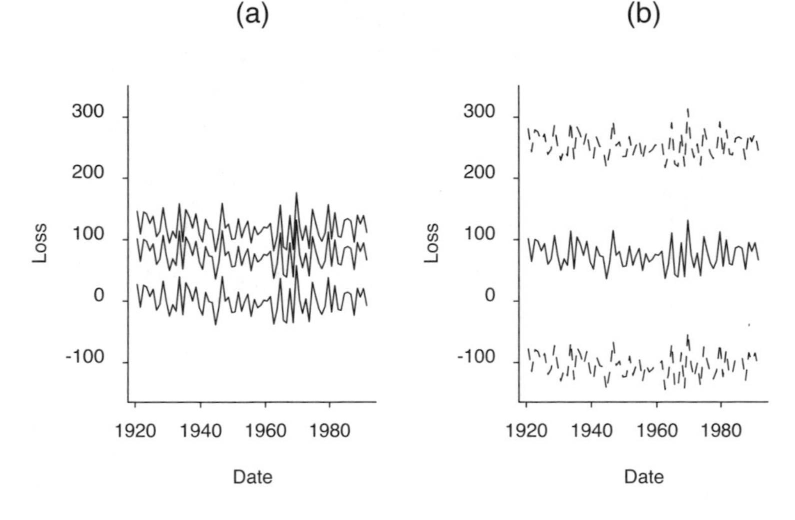

Figure 1.36 *(a) Estimates of 10-year (bottom), 100-year (middle), and 1000-year (top) return levels based on the fitted model for July, assuming long-term trend based on NAO, and (b) 100-year return level with confidence limits.*

in the second row of the table. Fitting the GPD instead of the Pareto, and then adding a seasonal effect (but for the return level calculations, combining the entire year into a single calculation) has the effect of reducing the estimated return period, compared with the exponential model, but not to a very great extent. The next three rows take the NAO effect into account, first for a "high NAO" year, then for a "low NAO" year, and

Table 1.13 *Estimated return periods (computed from maximum likelihood estimates) associated with the 1990 storm event for several different models.*

Model	Return period (years)
Pareto	18.7
Exponential	487
Generalized Pareto	333
GPD with seasons	348
GPD, high NAO year	187
GPD, low NAO year	1106
GPD with random NAO	432
Current industry estimate?	<50?

finally for "random NAO" — in other words, NAO is accounted for in the model but for the return period calculations, it is averaged over the distribution of NAO. In all cases, the quantity actually being estimated is the probability of a loss over the 1990 value occurring in one year; the cited "return period" is the reciprocal of this. The final row of this table represents the collective wisdom of a group of British insurance industry executives at a workshop in Cambridge, who were shown this table and then asked to give their own estimate of the likelihood of a recurrence of the 1990 event. Whether they were being more pessimistic or simply more realistic than the statistical analysis is a matter on which the reader is left to decide.

1.7 Multivariate extremes and max-stable processes

Multivariate extreme value theory is concerned with the joint distribution of extremes in two or more random variables. Max-stable processes arise from the extension of multivariate extreme value theory to infinite dimensions. Although there is by now a large amount of literature on multivariate extreme value theory (see, e.g., Chapter 7 by Fougères (2004) in the present volume), there has been much less work on statistical approaches to max-stable processes. In the present section we review these ideas and suggest a particular approach based on so-called M4 processes. In Section 1.8, we consider a possible application of this approach to financial time series.

1.7.1 Multivariate extremes

Suppose $\{X_i = (X_{i1}, \ldots, X_{iD}),\ i = 1, 2, \ldots,\}$ are i.i.d. multivariate random vectors, and define vector maxima $M_n = (M_{n1}, \ldots, M_{nD})$ where $M_{nd} = \max\{X_{id} : 1 \le i \le n\}$ for $1 \le d \le D$. Suppose there exist normalizing constants $a_{nd} > 0,\ b_{nd}$ such that

$$\frac{M_{nd} - b_{nd}}{a_{nd}}$$

converges in distribution to a nondegenerate limit for each d. In its original form, multivariate extreme value theory was concerned with limiting forms of the joint distribution, i.e., with results of the form

$$\Pr\left\{\frac{M_{nd}-b_{nd}}{a_{nd}} \le y_d,\ 1 \le d \le D\right\} = H(y_1,\ldots,y_D), \tag{1.42}$$

where H is some nondegenerate D-dimensional distribution function. Of course, it is a consequence of the theory of Section 1.2 that the marginals of H must be one of the three types of univariate extreme value distributions, but our interest here is in the kind of joint distributions that can arise.

Given that we have already developed an extensive theory for fitting univariate distributions above a high threshold, we can use this to transform the univariate marginal distributions any way we want. The results we are going to develop are most conveniently expressed if we assume the marginal distributions are unit Fréchet, i.e.,

$$\Pr\{X_{id} \le x\} = e^{-1/x} \text{ for each } d, \tag{1.43}$$

so henceforth we assume (1.43). Note also that in this case, we may take the normalizing constants to be $a_{nd} = n, b_{nd} = 0$.

In practice, we may not be able to achieve (1.43) exactly, but we can perform a probability integral transformation based on the distribution fitted above a high threshold. Suppose we select a threshold u_d for the dth component, and fit the GPD to the excesses of X_{id} above u_d. Then we can estimate parameters $\lambda_d, \sigma_d, \xi_d$ such that

$$\begin{aligned}\Pr\{X_{id} \ge u_d\} = \lambda_d, \quad & \Pr\{X_{id} \ge u_d + y \mid X_{id} \ge u_d\} \\ &= \left(1+\xi_d\frac{y}{\sigma_d}\right)^{-1/\xi_d},\end{aligned}$$

valid over the range where $1+\xi_d y/\sigma_d > 0$. In that case, the transformation

$$Z_{id} = \left[-\log\left\{1-\lambda_d\left(1+\xi_d\frac{X_{id}-u_d}{\sigma_d}\right)^{-1/\xi_d}\right\}\right]^{-1} \tag{1.44}$$

achieves the result

$$\Pr\{Z_{id} \le z\} = e^{-1/z}, \quad z \ge \{-\log(1-\lambda_d)\}^{-1}. \tag{1.45}$$

With the transformation to unit Fréchet margins, the key property that characterises multivariate extreme value distributions is max-stability. Suppose, for each i, $(Y_{i1},\ldots,Y_{iD})$ is an independent random vector with distribution function H. For each d, we know that

$$Y^*_{nd} = \frac{1}{n}\max\{Y_{1d},\ldots,Y_{nd}\}$$

has the same univariate distribution as Y_{1d}. The property that characterizes max-stability is that the same thing should be true for the multivariate distributions:

$$\Pr\{Y^*_{n1} \le y_1,\ldots,Y^*_{nD} \le y_D\} = H(y_1,\ldots,y_D). \tag{1.46}$$

Equation (1.46) is equivalent to

$$H(ny_1, \ldots, ny_D)^n = H(y_1, \ldots, y_D) \tag{1.47}$$

so (1.47) may be taken as the definition of a multivariate extreme value distribution with unit Fréchet margins. There is extensive literature on the characterizations and properties of multivariate extreme value distributions, which has been very well reviewed by Fougères (2004).

1.7.2 Max-stable processes

Consider now the case of a discrete-time stochastic process, i.e., a dependent sequence $\{Y_i,\ i = 1, 2, \ldots\}$. If we are interested in the extremal properties of this process, then for the same reasons as with multivariate extremes, it suffices to consider the case where all the marginal distributions have been transformed to unit Fréchet. Such a process is called max-stable if all the finite-dimensional distributions are max-stable, i.e., for any $n \geq 1, r \geq 1$

$$\Pr\{Y_1 \leq ny_1, \ldots, Y_r \leq ny_r\}^n = \Pr\{Y_1 \leq y_1, \ldots, Y_r \leq y_r\}.$$

In fact we are interested in D-dimensional processes so it makes sense to consider this case as well. A process $\{Y_{id} : i = 1, 2, \ldots; 1 \leq d \leq D\}$ with unit Fréchet margins is max-stable if for any $n \geq 1, r \geq 1$,

$$\begin{aligned} &\Pr\{Y_{id} \leq ny_{id} :\ 1 \leq i \leq r;\ 1 \leq d \leq D\}^n \\ &\quad = \Pr\{Y_{id} \leq y_{id} :\ 1 \leq i \leq r;\ 1 \leq d \leq D\}. \end{aligned} \tag{1.48}$$

Corresponding to this, a process $\{X_{id} :\ i = 1, 2, \ldots;\ 1 \leq d \leq D\}$ is said to be in the domain of attraction of a max-stable process $\{Y_{id} :\ i = 1, 2, \ldots;\ 1 \leq d \leq D\}$ if there exist normalizing constants $a_{nid} > 0,\ b_{nid}$ such that for any finite r

$$\begin{aligned} &\lim_{n\to\infty} \Pr\left\{ \frac{X_{id} - b_{nid}}{a_{nid}} \leq ny_{id} : 1 \leq i \leq r; 1 \leq d \leq D \right\}^n \\ &\quad = \Pr\{Y_{id} \leq y_{id} : 1 \leq i \leq r; 1 \leq d \leq D\}. \end{aligned} \tag{1.49}$$

If we assume *a priori* that the X process also has unit Fréchet margins, then we may again take $a_{nid} = n,\ b_{nid} = 0$.

Smith and Weissman (1996) made a connection between max-stable processes and the limiting distributions of extreme values in dependent stochastic processes. In the one-dimensional case, again assuming for simplicity that the marginal distributions are unit Fréchet, there is a large amount of literature concerned with results of the form

$$\Pr\left\{ \max_{1\leq i\leq n} X_i \leq nx \right\} \to e^{-\theta/x} \tag{1.50}$$

where $\{X_i,\ i = 1, 2, \ldots,\}$ is a stationary stochastic process, and θ is known as the extremal index. See for example Leadbetter et al. (1983), Leadbetter (1983), and O'Brien (1987). If θ exists, it must be in the range [0, 1] and is a constant for the process, i.e., it does not depend on x in (1.50).

For D-dimensional processes, the corresponding quantity is the *multivariate extremal index* defined by Nandagopalan (1990, 1994). For a stationary process $\{X_{id}, d = 1, \ldots, D, i = 1, 2, \ldots,\}$ with unit Fréchet margins in each of the D components, we assume

$$\Pr\{X_{id} \le ny_d, \; 1 \le d \le D\}^n \to H(y_1, \ldots, y_D),$$

analogous to (1.42). The result corresponding to (1.50), for the joint distributions of sample maxima in the dependent process, is

$$\Pr\left\{\max_{1 \le i \le n} X_{id} \le ny_d, \; 1 \le d \le D\right\} \to H(y_1, \ldots, y_D)^{\theta(y_1, \ldots, y_D)}. \tag{1.51}$$

Equation (1.51), when it is true, defines the multivariate extremal index $\theta(y_1, \ldots, y_D)$.

As with univariate stochastic processes, the multivariate extremal index satisfies $0 \le \theta(y_1, \ldots, y_D) \le 1$ for all $(y_1, \ldots, y_D)$. Unlike the univariate case, it is not a constant for the whole process, but it is true that $\theta(cy_1, \ldots, cy_D) = \theta(y_1, \ldots, y_D)$ for any $c > 0$. Although it is by no means the only quantity of interest in studying extreme values of discrete-time stationary processes, because of its role in limit theorems such as (1.50) and (1.51), the (univariate or multivariate) extremal index is widely regarded as a key parameter.

The connection between max-stable processes and the multivariate extremal index is as follows. Suppose $\{X_{id}\}$ is a stationary D-dimensional process whose finite-dimensional distributions are in the domain of attraction of a max-stable process — in other words, there is a max-stable process $\{Y_{id}\}$ for which (1.49) is true for every $r \ge 1$. We also assume two technical conditions as follows:

1. For a given sequence of thresholds $u_n = (u_{nd}, 1 \le d \le D)$ and for $1 \le j \le k \le n$, let $\mathcal{B}_j^k(u_n)$ denote the σ-field generated by the events $\{X_{id} \le u_{nd}, \; j \le i \le k\}$, and for each integer t let

$$\alpha_{n,t} = \sup\left\{|P(A \cap B) - P(A)P(B)| : A \in \mathcal{B}_1^k(u_n), \quad B \in \mathcal{B}_{k+t}^n(u_n)\right\}$$

where the supremum is taken not only over all events A and B in their respective σ-fields but also over k such that $1 \le k \le n - t$. Following Nandagopalan (1994), $\Delta(u_n)$ is said to hold if there exists a sequence $\{t_n, n \ge 1\}$ such that

$$t_n \to \infty, t_n/n \to 0, \alpha_{n,t_n} \to 0 \text{ as } n \to \infty.$$

2. Assume $\Delta(u_n)$ holds with respect to some sequence t_n, and define a sequence k_n so that as $n \to \infty$,

$$k_n \to \infty, \quad k_n t_n/n \to 0, \quad k_n \alpha_{n,t_n} \to 0.$$

We assume that, with $r_n = \lfloor n/k_n \rfloor$,

$$0 = \lim_{r \to \infty} \lim_{n \to \infty} \sum_{i=r}^{r_n} \sum_{d=1}^{D} \Pr\left\{X_{id} > u_{nd} \middle| \max_d \left(\frac{X_{1d}}{u_{nd}}\right) > 1\right\}. \tag{1.52}$$

Then we have:

Theorem 1.7.1 *Suppose the processes* $\mathbf{X} = \{X_{id}\}$ *and* $\mathbf{Y} = \{Y_{id}\}$ *are each stationary with unit Fréchet margins, that* $\mathbf{Y}$ *is max-stable, and (1.49) holds. For a given* $(y_1, \ldots, y_D)$ *and* $u_{nd} = ny_d$, $d = 1, \ldots, D$, *suppose* $\Delta(u_n)$ *and (1.52) hold for both* $\mathbf{X}$ *and* $\mathbf{Y}$. *Then the two processes* $\mathbf{X}$ *and* $\mathbf{Y}$ *have the same multivariate extremal index* $\theta(y_1, \ldots, y_D)$.

The practical implication of this is that in calculating the extremal index of $\mathbf{X}$, it suffices to look at the limiting max-stable process, assuming this exists.

1.7.3 Representations of max-stable processes

Suppose $\{\alpha_k, -\infty < k < \infty\}$ is a doubly-infinite sequence with $\alpha_k \geq 0$, $\sum_k \alpha_k = 1$, and let $\{Z_i\}$ denote i.i.d. unit Fréchet random variables. Consider the moving maximum process

$$Y_i = \max_{-\infty<k<\infty} \alpha_k Z_{i-k}.$$

It is readily checked that Y_i is a stationary stochastic sequence with unit Fréchet margins, and moreover, is a max-stable process. The latter is most easily seen from the identity

$$\Pr\{Y_1 \leq y_1, \ldots, Y_r \leq y_r\} = \exp\left\{-\sum_{s=-\infty}^{\infty} \max_{1-s\leq k\leq r-s} \frac{\alpha_k}{y_{s+k}}\right\},$$

which satisfies (1.48) with $D = 1$.

Deheuvels (1983) defined a class of processes which, translated to the present context, we may call maxima of moving maxima, or M3, processes, as

$$Y_i = \max_{\ell\geq 1} \max_{-\infty<k<\infty} \alpha_{\ell,k} Z_{\ell,i-k},$$

where now $\{\alpha_{\ell,k}\}$ is a double sequence of constants satisfying $\alpha_{\ell,k} \geq 0$ and $\sum_\ell \sum_k \alpha_{\ell,k} = 1$, and $Z_{\ell,i}$ is a double sequence of independent unit Fréchet random variables. It is readily checked that this process is also max-stable.

Smith and Weissman (1996) generalized the M3 process to a multivariate context, the class of multivariate maxima of moving maxima processes, or M4 for short. In this definition, we assume a double sequence of independent unit Fréchet random variables $Z_{\ell,i}$ and a triple sequence of constants $\{\alpha_{\ell,k,d} : \ell \geq 1; -\infty < k < \infty, 1 \leq d \leq D\}$, such that $\alpha_{\ell,k,d} \geq 0$,

$$\sum_\ell \sum_k \alpha_{\ell,k,d} = 1 \text{ for each } d.$$

Then define the M4 process $\mathbf{Y} = \{Y_{id}\}$ by

$$Y_{id} = \max_\ell \max_k \alpha_{\ell,k,d} Z_{\ell,i-k}.$$

For this process, we have

$$\Pr\{Y_{id} \le y_{id},\ 1 \le i \le r,\ 1 \le d \le D\}$$
$$= \exp\left\{-\sum_{\ell=1}^{\infty}\sum_{s=-\infty}^{\infty} \max_{1-s\le k\le r-s}\ \max_{1\le d\le D} \frac{\alpha_{\ell,k,d}}{y_{s+k,d}}\right\}. \tag{1.53}$$

From (1.53), it is readily checked that **Y** is max-stable, and moreover, we can also show that the multivariate extremal index is given by the formula

$$\theta(y_1, \ldots, y_D) = \frac{\sum_\ell \max_k \max_d \alpha_{\ell,k,d} y_d}{\sum_\ell \sum_k \max_d \alpha_{\ell,k,d} y_d}. \tag{1.54}$$

The key characterization result is the following: under some additional technical conditions which we shall not specify here, any D-dimensional stationary max-stable process with unit Fréchet margins may be approximated arbitrarily closely as the sum of an M4 and a deterministic process. This result was given by Smith and Weissman (1996) and is a direct generalization of the result of Deheuvels (1983) for one-dimensional processes.

The deterministic process mentioned here is one that cycles infinitely through some finite set of values; since this seems unreasonable behaviour for most real observed processes, we usually assume it is absent. In that case, the practical interpretation of the result is that we can approximate any max-stable process with unit Fréchet margins by an M4 process.

Combined with the results of Section 1.7.2, this suggests the following strategy for analyzing extremal properties of multivariate stationary processes. First, fit the GPD or the point process model of Section 1.2.5 to the exceedances of the process above a threshold (possibly a different threshold for each component). Second, apply the transformation (1.44) so that the margins of the transformed process may be assumed to be of unit Fréchet form above a threshold (1.45). Third, assume as an approximation that the joint distributions of the transformed process, above the relevant thresholds, are the same as those of an M4 process. Fourth, estimate the parameters of the M4 process $\{\alpha_{\ell,k,d}\}$. The multivariate extremal index of the process is then given by (1.54) and other extremal properties of the process may be calculated directly from the formula (1.53).

1.7.4 Estimation of max-stable processes

So far, we have argued that max-stable processes form a suitable class of processes by which to study the extremal properties of D-dimensional stationary time series, and that after transforming the tail distributions to unit Fréchet margins, max-stable processes may be approximated by those in the M4 class. There are still some major difficulties, however, in estimating these processes in practice.

Since the M4 process is defined by an infinite set of parameters $\alpha_{\ell,k,d}$, to make progress we must in practice reduce these to a finite set. We therefore assume $\alpha_{\ell,k,d} = 0$ outside the range $1 \le \ell \le L$, $-k_1 \le k \le k_2$ where L, k_1, and k_2 are known. In the financial time series example to be discussed in Section 1.8, we have $D = 3$ and somewhat arbitrarily assume $L = 25$, $k_1 = k_2 = 2$.

Next, consider a situation where some value of $Z_{\ell,i}$, say Z_{ℓ^*,i^*}, is much larger than its neighbors $Z_{\ell,i}$ when $1 \le \ell \le L$, $i^* - k_1 \le i \le i^* + k_2$. Within this range, we will have

$$Y_{i,d} = \alpha_{\ell^*,i-i^*,d} Z_{\ell^*,i^*}, \quad i^* - k_1 \le i \le i^* + k_2, \quad 1 \le d \le D.$$

Define

$$S_{i,d} = \frac{Y_{i,d}}{\max_{i^*-k_1 \le i \le i^*+k_2,\ 1 \le d \le D} Y_{i,d}}, \quad i^* - k_1 \le i \le i^* + k_2,\ 1 \le d \le D. \tag{1.55}$$

Then

$$S_{i,d} = \frac{\alpha_{\ell^*,i-i^*,d}}{\max_{i^*-k_1 \le i \le i^*+k_2,\ 1 \le d \le D} \alpha_{\ell^*,i-i^*,d}}, \quad i^* - k_1 \le i \le i^* + k_2,\ 1 \le d \le D. \tag{1.56}$$

The variables $\{S_{i,d}\}$ define a signature pattern, which specifies the shape of the process near its local maximum.

For any (ℓ^*, i^*), there is positive probability that (1.56) holds exactly. There are L such deterministic signature patterns (one defined by each ℓ^*), and when looking along the whole process for $-\infty < i < \infty$, it follows from elementary probability arguments that each of the L signature patterns will be observed infinitely often. Zhang (2002) has given some precise formulations and alternative versions of these results.

From an estimation point of view, this is both good and bad. On the one hand, if we were indeed to observe a process of exactly M4 form for a sufficiently long time period, we would be able to identify all L signature patterns exactly, and hence exactly estimate all the $\{\alpha_{\ell,k,d}\}$ parameters. However, in practice we would not expect to observe an exact M4 process, and even the theoretical results only go as far as justifying the M4 process as an approximation to the general max-stable process. Moreover, the presence of deterministic signature patterns means that the joint densities of the M4 processes contain singularities, which render the method of maximum likelihood unsuitable for such processes. Therefore, we must seek alternative methods of estimation.

Davis and Resnick (1989) developed many properties of max-ARMA processes (a special class of moving maxima processes), but did not devise a good statistical approach.

Hall, Peng, and Yao (2002) proposed an alternative estimation scheme based on multivariate empirical distributions, for moving maxima processes. Zhang (2002) has generalized this to M4 processes.

It may also be possible to approach this question by assuming an unobserved process that is exactly of M4 form, and an observed process derived by adding noise, filtering the former from the latter by a Monte Carlo state-space approach. This has not so far been tried, however.

Here we propose a simple intuitive method for which we claim no optimality properties but which appears to be a useful practical approach. The steps of this new approach are as follows:

1. For each univariate series, fit the standard extreme value model to exceedances above a threshold and transform the margins to unit Fréchet via (1.44).
2. Fix the values of L, k_1, and k_2 such that $\alpha_{\ell,k,d} = 0$ except when $1 \le \ell \le L$, $-k_1 \le k \le k_2$. In our example we take $L = 25$ and $k_1 = k_2 = 2$.
3. For each local maximum above the threshold, define the signature pattern (1.55). In practice, some of the $Y_{i,d}$ values in (1.55) will be below their corresponding thresholds; in that case, set $S_{i,d} = 0$ for those values. In our example, $D = 3$ and there are 607 local maxima, so we end up with 607 candidate signature patterns in 15 dimensions.

Note that the theory specifies that L signature patterns occur infinitely often and these are the ones that identify the coefficients $\{\alpha_{\ell,k,d}\}$. In practice we will observe many more than L signature patterns, all different. The idea pursued here is to use a clustering algorithm to approximate the process by one in which there are only L signature patterns. Therefore:

4. Use the "K-means clustering" procedure to group the signature patterns into L clusters. K-means clustering is implemented in S-Plus and took less than a minute for the example given here.
5. Assume each cluster of signature patterns is represented by one signature pattern corresponding to the cluster mean, and estimate the $\alpha_{\ell,k,d}$ coefficients from the L cluster mean signature patterns.

In Section 1.8, we shall see the results of this procedure applied to financial data.

1.8 Extremes in financial time series

The 1990s were mostly a boom time for the stock market, but some well-publicized catastrophes (e.g., Barings Bank, Orange County, Long-Term Capital Management) made investors aware of the dangers of sudden very sharp losses.

In response to this, there grew a new science of risk management. The best known tool is Value at Risk (VaR), defined as the value x which satisfies

$$\Pr\{X_T > x\} = \alpha,$$

where X_T is the cumulative loss over given time horizon T (e.g., 10 trading days), and α is a given probability (typically .05 or .01). However, X_T is typically calculated from a portfolio involving a large number of stocks, so the analysis of high-dimensional time series is involved.

Various tools have been devised to estimate VaR. The best known, but also the crudest, is RiskMetrics. Although this contains many complicated details, the main principles are:

Table 1.14 *Parameters of the point process model of Section 1.2.5, fitted to each of the three financial time series based on threshold $u = .02$.*

Series	Number of exceedances	μ (SE)	$\log \psi$ (SE)	ξ (SE)
Pfizer	518	.0623	−4.082	.174
		(.0029)	(.132)	(.051)
GE	336	.0549	−4.139	.196
		(.0029)	(.143)	(.062)
Citibank	587	.0743	−3.876	.164
		(.0036)	(.119)	(.012)

1. The covariance matrix of daily returns on the stocks of interest is estimated from a fixed period (e.g., one month) of recent prices.
2. The distribution is assumed to be multivariate normal.

The purpose of the present discussion is to explore whether we could do better using extreme value theory, based on a specific example.

We use negative daily returns from closing prices of 1982 to 2001 stock prices in three companies, Pfizer, General Electric, and Citibank. The data were plotted in Figure 1.2.

Univariate extreme value models were fitted to threshold $u = .02$ with results given in Table 1.14.

Diagnostic plots based on the Z and W statistics are shown for the Pfizer data in Figure 1.37. The clear problem here is seen in the QQ plot for the Z statistics — this does not stay close to the assumed straight line, indicating that the point process of exceedance times is not well described by a uniform Poisson process. The reason for this is obvious to anyone familiar with the literature on financial time series: like all series of this nature, the Pfizer series goes through periods of high and low volatility, and extreme values are much more likely to occur in times of high volatility than low volatility.

For the present study, volatility was estimated by fitting a GARCH(1,1) model to each series. This is a familiar model in econometrics, see e.g., Shephard (1996). If y_t denotes the observed series (in this case, the observed daily return) on day t, assumed standardized to mean 0, then the model represents y_t in the form

$$y_t = \sigma_t \epsilon_t,$$

where $\{\epsilon_t\}$ are i.i.d. $N[0, 1]$ random variables, and the volatility σ_t is assumed to satisfy an equation of form

$$\sigma_t^2 = \alpha_0 + \alpha_1 y_{t-1}^2 + \beta_1 \sigma_{t-1}^2.$$

For the purpose of the present analysis, a GARCH(1,1) model was fitted to each of the three series, and a standardized series y_t/σ_t computed. The threshold analysis was then repeated for the standardized series, using threshold $u = 1.2$. Results from this analysis are shown in Table 1.15. The diagnostic plots are now satisfactory (plots for the Pfizer series are shown in Figure 1.38; the others are similar).

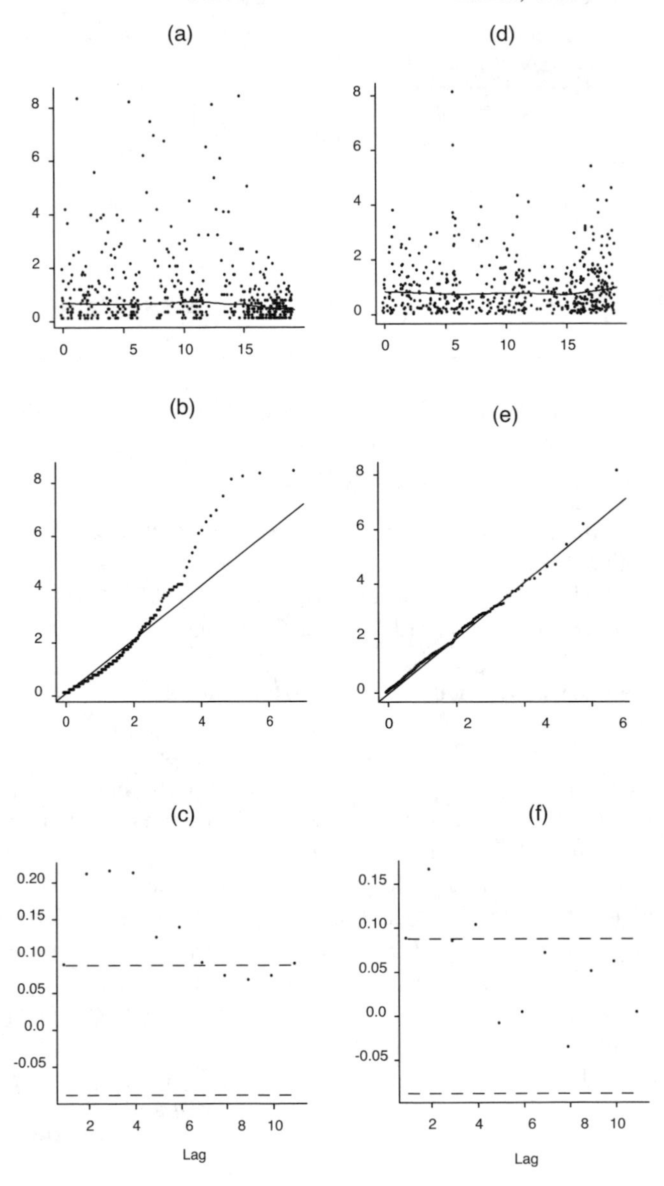

Figure 1.37 *Diagnostic plots for the extreme value model fitted to the Pfizer daily returns.*

After fitting a univariate extreme value model to each series, the exceedances over the threshold for each series are transformed to have marginal Fréchet distributions. On the transformed scale, the data in each series consist of all values in excess of threshold 1. The resulting series are shown in Figure 1.39.

Up to this point in the analysis, although we have taken into account long-range dependencies in the data through the volatility function, we have taken no account of

Table 1.15 *Parameters of the point process model of Section 1.2.5, fitted to each of the three financial time series based on threshold 1.2, after standardizing for volatility.*

Series	Number of exceedances	μ (SE)	$\log \psi$ (SE)	ξ (SE)
Pfizer	411	3.118	−.177	.200
		(.155)	(.148)	(.061)
GE	415	3.079	−.330	.108
		(.130)	(.128)	(.053)
Citibank	361	3.188	−0.118	.194
		(.157)	(.126)	(.050)

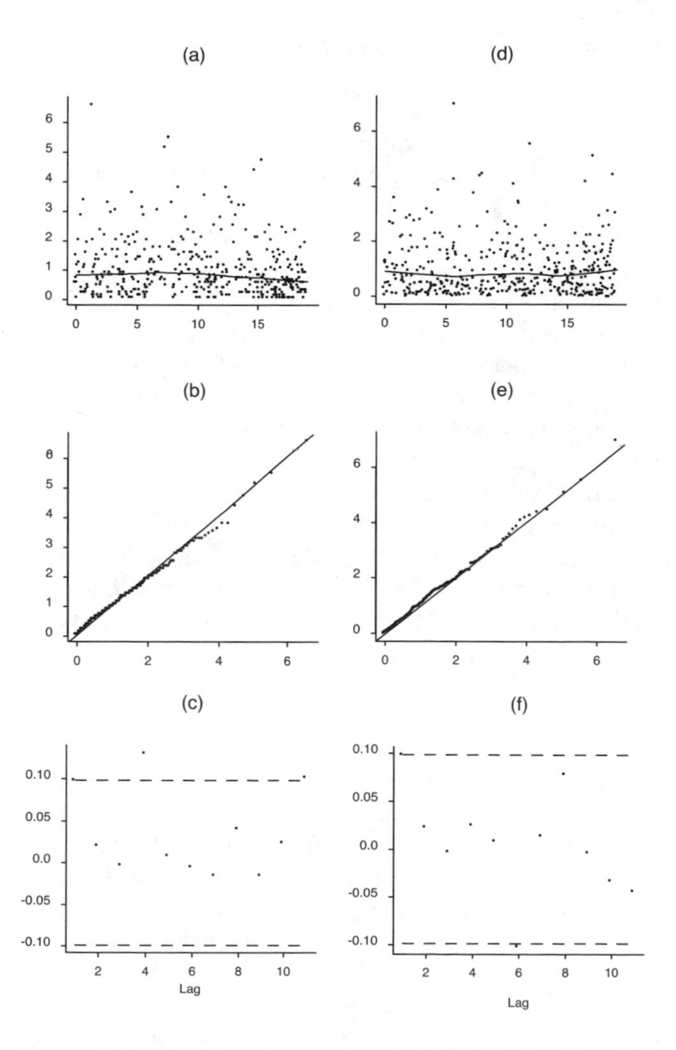

Figure 1.38 *Diagnostic plots for the extreme value model fitted to the Pfizer daily returns, after standardizing for volatility.*

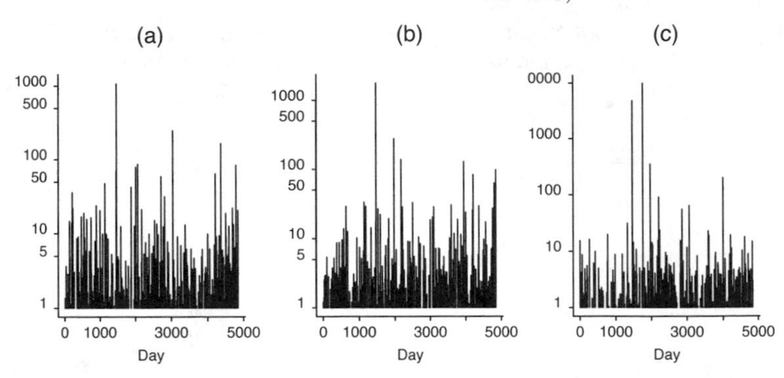

Figure 1.39 *Plot of the threshold exceedances of standardized daily returns after transforming to unit Fréchet marginal distributions.*

possible short-term dependencies (e.g., serial correlation between values on one day and neighboring days), nor have we considered any form of dependence among the three series.

In Figure 1.40, pairwise scatterplots are shown of the three transformed series against each other on the same day (top three plots), and against series on neighboring days. The two numbers on each plot show the expected number of joint exceedances based on an independence assumption and the observed number of joint exceedances.

Figure 1.41 shows a plot of Fréchet exceedances for the three series on the same day, normalized to have total one, plotted in barycentric coordinates. The three circles near the corner points P, G, and C correspond to days for which that series alone had an exceedance.

These plots provide empirical evidence that there is dependence among the values of the three transformed series on the same day, and less clear evidence that there is also some serial correlation among the values on neighboring days. These are the kinds of dependencies that the M4 series is intended to model, and the next step in the analysis is therefore to fit an M4 model to the threshold exceedances of the three transformed series. The method of fitting the M4 model follows the recipe given at the end of Section 1.7.

One test of whether the fitted M4 process is a realistic representation of the Fréchet-transformed time series is whether the sample paths simulated from the fitted process look similar to those from the original series. As a test of this, a Monte Carlo sample from the fitted model was generated, and Figure 1.42 through Figure 1.44 were drawn in the same way as Figure 1.39 through Figure 1.41 from the original series. One point to note here is that although the data were generated so that the marginal distributions were exactly unit Fréchet, in order to provide a fair comparison with the original estimation procedure, the marginal distributions were reestimated, and transformed according to the estimated parameters, before drawing the scatterplots in Figure 1.43 and Figure 1.44. As a result, the signature patterns in the transformed simulated series are no longer the exact signature patterns of the M4 model, and the scatterplots are less clumpy than they would have been without the transformation. They are still more clumpy than the original Figure 1.40 and Figure 1.41. Despite this,

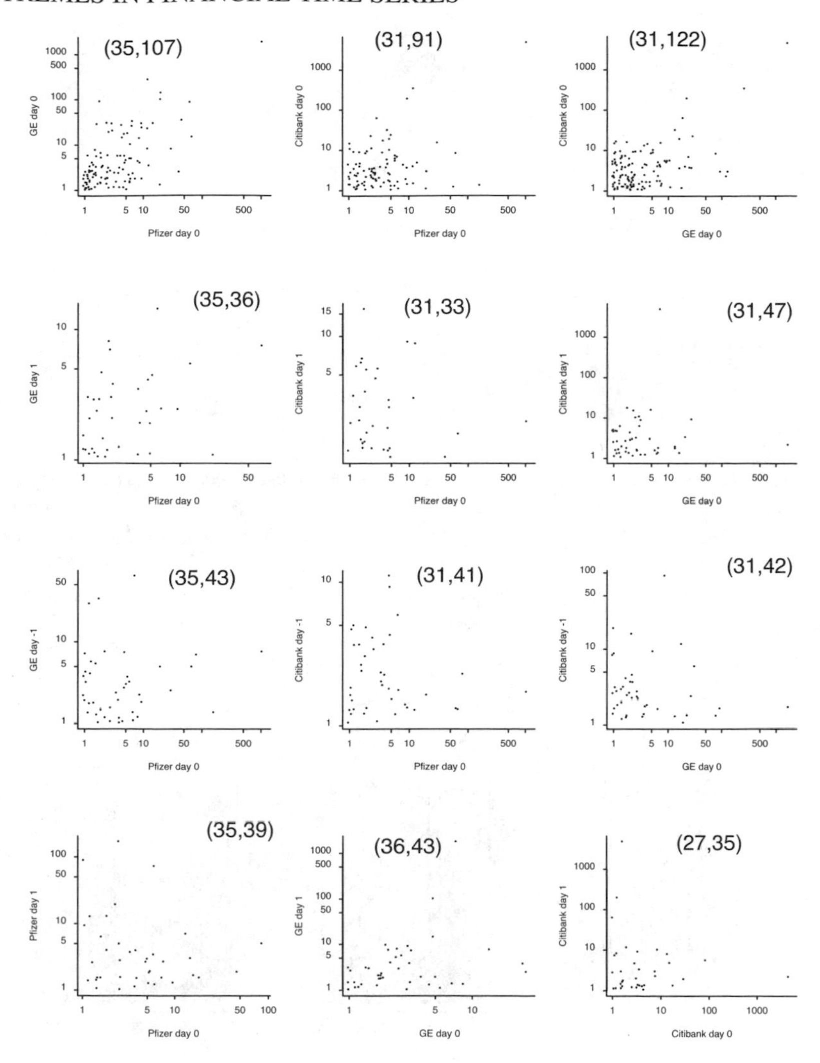

Figure 1.40 *Scatterplots of the values over the threshold in the Fréchet transformed series. For each series, a scatterplot is shown against either the values of the other two series on the same day, or the values of all three series on either the day before or the day after. Of the two numbers shown on each plot, the second is the observed number of joint exceedances in the two series, i.e., the actual number of points in the scatterplot, while the first represents a calculation of the expected number of joint exceedances if the two series were independent (or for the case of one series plotted against itself, assuming the daily values are independent).*

we would argue that Figure 1.42 through Figure 1.44 provide a reasonable indication that the simulated series are similar to the original Fréchet-transformed time series.

Finally, we attempt to validate the model by calibrating observed versus expected probabilities of extreme events under the model. The "extreme event" considered is that there is at least one exceedance of a specific threshold u, on the scale of the

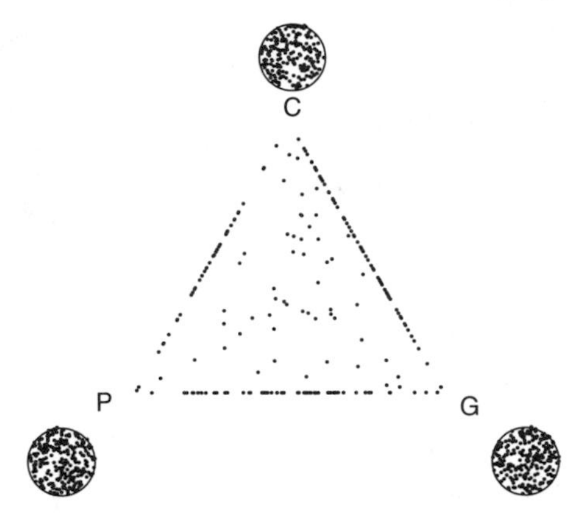

Figure 1.41 *Representation of the trivariate distribution of threshold exceedances on the same day for the three Fréchet-transformed series. The letters P, G, and C refer to the Pfizer, GE, and Citibank series. The circle at each vertex of the triangle represent all the days that there was an exceedance in that series but neither of the other two series. For days on which at least two of the series had exceedances, the values were normalised to sum to one, and plotted in barycentric coordinates.*

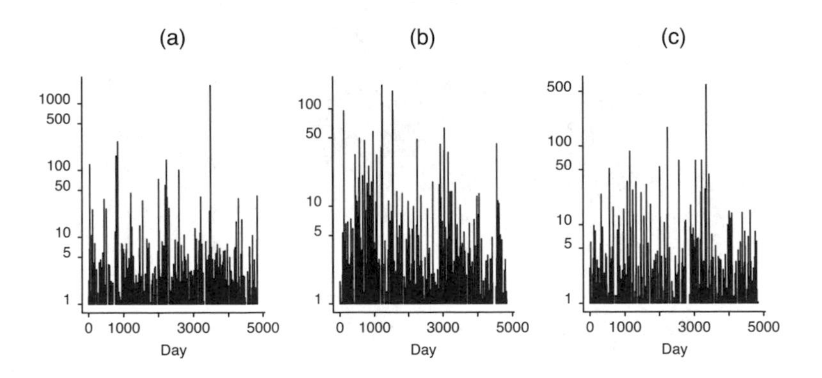

Figure 1.42 *Same as Figure 1.39, but with data simulated from the fitted M4 model.*

original daily return series, by one of the three series in one of the next 10 days after a given day. It is fairly straightforward to write down a theoretical expression for this, given the M4 model. To make the comparison honest, the period of study is divided into four periods each of length just under 5 years. The univariate and multivariate extreme value model is fitted to each of the first three 5-year periods, and used to predict extreme events in the following period.

Figure 1.45 shows observed (dashed lines) and expected (solid lines) counts for a sequence of thresholds u. There is excellent agreement between the two curves except for the very highest thresholds, when the calculation may be expected to break down.

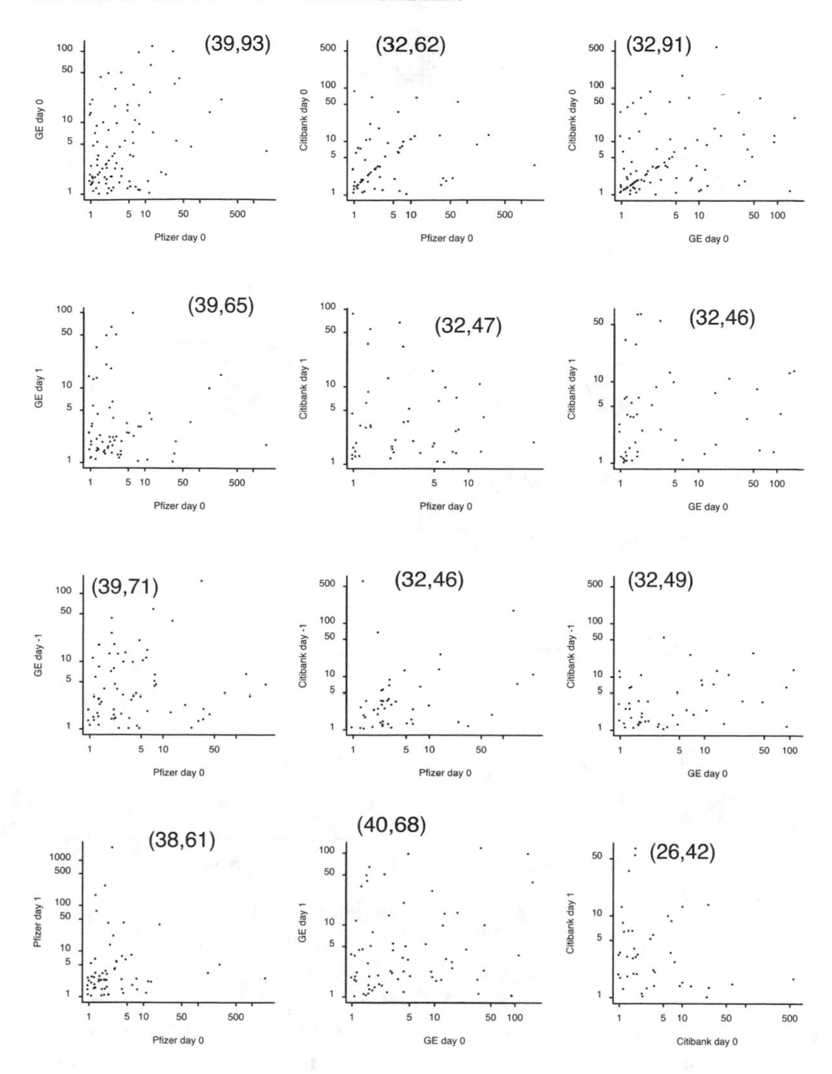

Figure 1.43 *Same as Figure 1.40, but calculated from the simulated data.*

We conclude with some summary remarks.

The representation in terms of M4 processes contains the possibility of estimating both within-series and between-series dependence as part of the same model.

The key step in this method is the use of K-means clustering to identify a measure in a high-dimensional simplex of normalized exceedances. In contrast, existing methods of estimating multivariate extreme value distributions usually only work in low dimensions (up to five or so). However, this method of estimation is itself very ad hoc, and the field is still open for more systematic methods of estimation with more clearly defined mathematical properties. The recent thesis of Zhang (2002) may point the way forward to some more general and powerful methods.

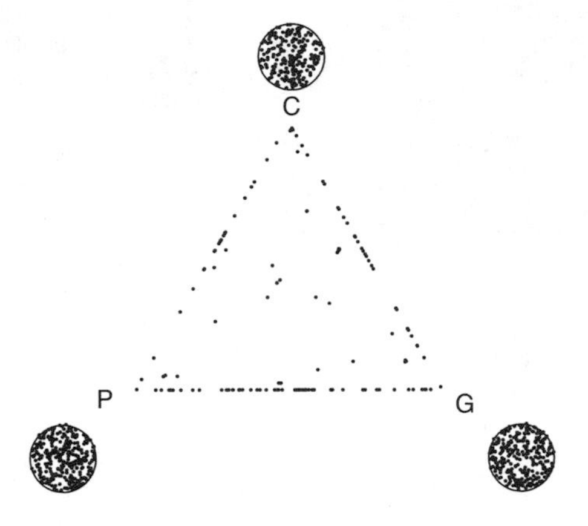

Figure 1.44 *Same as Figure 1.41, but calculated from the simulated data.*

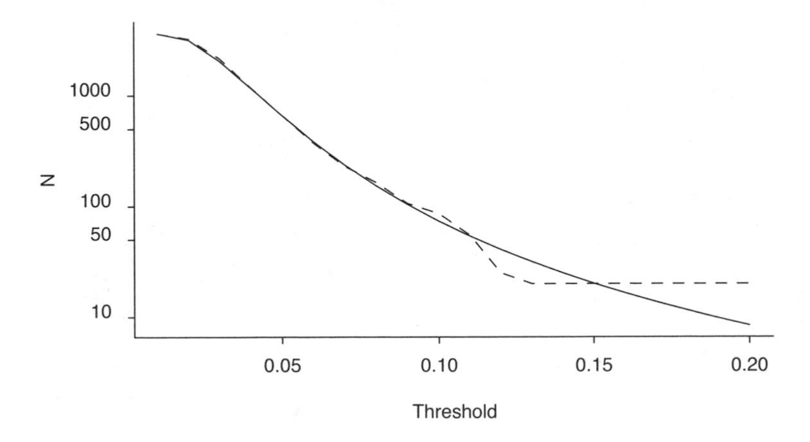

Figure 1.45 *Representation of the probability* $1/N$ *that there is at least one threshold exceedance by one of the three series in a 10-day period, as a function of the threshold, on the scale of the original data. The solid curve represents the theoretical calculation based on the models fitted to the data. The dashed curve represents an empirical calculation of the same event, by counting observed exceedances.*

Ultimately the test of such methods will be whether they can be used for more reliable risk calculations than established methods such as RiskMetrics. The numerical example at the end shows that good progress has been made, but there are also many variations on the basic method that deserve to be explored.

Acknowledgments

The material in this paper formed the content of three lectures given as part of the SEMSTAT conference on Extreme Value Theory in Gothenburg, Sweden, December 10 through 15, 2001.

I am grateful to the organizers of the conference, particularly Claudia Klüppelberg, Bärbel Finkenstädt, and Holger Rootzén, for inviting me to speak and for their efficient organization of the meeting, and to the rest of the participants for their feedback during the meeting.

Part of the research reported here was carried out during the program entitled *Managing Risk*, held at the Isaac Newton Institute in Cambridge, in July and August, 2001. I am grateful to the organizers of the program, in particular Dr. Dougal Goodman of the (U.K.) Foundation for Science and Technology, and to the Director and staff of the Newton Institute for their support of this program. The Newton Institute is supported by a grant from the (U.K.) Engineering and Physical Sciences Research Council. I would also like to acknowledge the support of my personal research through the National Science Foundation, in particular grants DMS-9971980 and DMS-0084375.

References

Cohen, J. P. (1982a), The penultimate form of approximation to normal extremes. *Adv. Appl. Prob.* 14, 324–339.

Cohen, J. P. (1982b), Convergence rates for the ultimate and penultimate approximations in extreme value theory. *Adv. Appl. Prob.* 14, 833–854.

Coles, S. G. (2001), *An Introduction to Statistical Modeling of Extreme Values.* Springer Verlag, New York.

Cressie, N. (1993), *Statistics for Spatial Data.* Second edition, John Wiley, New York.

Davis, R. A. and Resnick, S. I. (1989), Basic properties and prediction of max-ARMA processes. *Ann. Appl. Probab.* 21, 781–803.

Davison, A. C. and Smith, R. L. (1990), Models for exceedances over high thresholds (with discussion). *J.R. Statist. Soc. B*, 52, 393–442.

Deheuvels, P. (1983), Point processes and multivariate extreme values. *J. Multivar. Anal.* 13, 257–272.

Embrechts, P., Klüppelberg, C., and Mikosch, T. (1997), *Modelling Extremal Events for Insurance and Finance.* Springer, New York.

Feller, W. (1968), *An Introduction to Probability Theory and Its Applications, Vol. I.* (3rd ed.) Wiley, New York.

Fisher, R. A. and Tippett, L. H. C. (1928), Limiting forms of the frequency distributions of the largest or smallest member of a sample. *Proc. Camb. Phil. Soc.* 24, 180–190.

Fougères, A.-L. (2004), Multivariate extremes. In *Extreme Values in Finance, Telecommunications, and the Environment*, edited by Finkenstädt, B. and Rootzén, H. Chapman & Hall/CRC Press, Boca Raton, Ch. 7.

Gamerman, D. (1997), *Markov Chain Monte Carlo: Stochastic Simulation for Bayesian Inference.* Texts in Statistical Science, Chapman & Hall/CRC Press, Boca Raton.

Gnedenko, B. V. (1943), Sur la distribution limite du terme maximum d'une série aléatoire. *Ann. Math.* 44, 423–453.

Grady, A. (2000), *A Higher Order Expansion for the Joint Density of the Sum and the Maximum with Applications to the Estimation of Climatological Trends.* Ph.D. dissertation, Department of Statistics, University of North Carolina, Chapel Hill.

Gumbel, E. J. (1958), *Statistics of Extremes.* Columbia University Press.

Hall, P., Peng, L., and Yao, Q. (2002), Moving-maximum models for extrema of time series. *Journal of Statistical Planning and Inference* 103, 51–63.

Holland, D. M., De Oliveira, V., Cox, L. H., and Smith, R. L. (2000), Estimation of regional trends in sulfur dioxide over the eastern United States. *Environmetrics* 11, 373–393.

Leadbetter, M. R. (1983), Extremes and local dependence in stationary sequences. *Z. Wahrsch. v. Geb.* 65, 291–306.

Leadbetter, M. R., Lindgren, G., and Rootzén, H. (1983), *Extremes and Related Properties of Random Sequences and Series.* Springer Verlag, New York.

Nandagopalan, S. (1990), *Multivariate extremes and the estimation of the extremal index.* Ph.D. dissertation, Department of Statistics, University of North Carolina, Chapel Hill.

Nandagopalan, S. (1994), On the multivariate extremal index. *J. of Research, National Inst. of Standards and Technology* 99, 543–550.

O'Brien, G. L. (1987), Extreme values for stationary and Markov sequences. *Ann. Probab.* 15, 281–291.

Pickands, J. (1975), Statistical inference using extreme order statistics. *Ann. Statist.* 3, 119–131.

Resnick, S. (1987), *Extreme Values, Point Processes, and Regular Variation.* Springer Verlag, New York.

Robert, C. P. and Casella, G. (2000), *Monte Carlo Statistical Methods.* Springer Texts in Statistics, Springer Verlag, New York.

Robinson, M. E. and Tawn, J. A. (1995), Statistics for exceptional athletics records. *Applied Statistics* 44, 499–511.

Shephard, N. (1996), Statistical aspects of ARCH and stochastic volatility. In *Time Series Models: In econometrics, finance, and other fields*, edited by D. R. Cox, D. V. Hinkley, and O. E. Barndorff-Nielsen. Chapman & Hall, London, 1–67.

Smith, R. L. (1985), Maximum likelihood estimation in a class of nonregular cases. *Biometrika* 72, 67–90.

Smith, R. L. (1986), Extreme value theory based on the *r* largest annual events. *J. Hydrology* 86, 27–43.

Smith, R. L. (1989), Extreme value analysis of environmental time series: An application to trend detection in ground-level ozone (with discussion). *Statistical Science* 4, 367–393.

Smith, R. L. (1990), Extreme value theory. In *Handbook of Applicable Mathematics* 7, edited by W. Ledermann. John Wiley, Chichester, 437–471.

Smith, R. L. (1997), Statistics for exceptional athletics records: Letter to the editor. *Applied Statistics* 46, 123–127.

Smith, R. L. (1999), Trends in rainfall extremes. Preprint, University of North Carolina, Chapel Hill.

Smith, R. L. and Goodman, D. (2000), Bayesian risk analysis. Chapter 17 of *Extremes and Integrated Risk Management*, edited by P. Embrechts. Risk Books, London, 235–251.

Smith, R. L. and Shively, T. S. (1995), A point process approach to modeling trends in tropospheric ozone. *Atmospheric Environment* 29, 3489–3499.

Smith, R. L. and Weissman, I. (1996), Characterization and estimation of the multivariate extremal index. Preprint, under revision.

Tawn, J. A. (1988), An extreme value theory model for dependent observations. *J. Hydrology* 101, 227–250.

Zhang, Z. (2002), *Multivariate Extremes, Max-Stable Process Estimation, and Dynamic Financial Modeling.* Ph.D. dissertation, Department of Statistics, University of North Carolina, Chapel Hill.

CHAPTER 2

The Use and Misuse of Extreme Value Models in Practice

Stuart Coles
University of Bristol

Contents

Extreme value models and techniques are widely applied in environmental settings for designing protection systems against the effects of extreme levels of environmental processes. Asymptotic arguments lead to models that give an approximate description of the stochastic behavior of extremes under idealized conditions. This approach has widespread support, at least for its pragmatism. There is less common ground on how to use the models — decisions of model choice, mode of inference, method of interpretation, and so on. We consider these issues in the context of a specific data series of daily rainfall recorded at a single location in Venezuela. For these data such choices turn out to have a profound effect on the conclusions drawn, and the effect of miscalculation has already been badly felt: a daily rainfall event in 1999, which was thought to be virtually impossible on the basis of an extreme value analysis of historical data, caused widespread destruction and loss-of-life. In this chapter we review the various options available for extreme value analyses, and argue that four components are necessary to attain a satisfactory analysis. These comprise an appropriate choice of asymptotic model, an exploitation of all appropriate information, a proper allowance for estimation uncertainty, and the modeling of nonstationary effects. We show that the 1999 rainfall event in Venezuela, while still exceptional, is considerably less surprising once such issues are properly handled in an extreme value analysis.

1-58488-411-8/04/$0.00+$.50

2.1 Introduction: Classical analyses of annual maximum data

Throughout this chapter we examine a series of daily rainfalls recorded at Maiquetia International Airport in Venezuela. For reasons that will become clear shortly, these data are of particular interest from an extreme value point of view and represent a challenge to practitioners of the theory. Annual maxima of the series over the period 1951 to 1998 are shown in Figure 2.1. Suppose that having available these data in 1998 your task, as a hydrologist, is to assess what levels to allow for in order to afford protection against the maximum daily rainfall in a future year, say 1999. Though oversimplified for purposes of presentation, this is precisely the sort of problem that typifies an extreme value analysis: historical data on extremes are available, but requirements are made to estimate the likelihood of events that are even more extreme than those already observed.

Extreme value theory and the models it leads to are often regarded as a panacea for resolving such difficulties. The basic arguments are well-rehearsed: in the simplest case we have a process of independent observations $X_1, \ldots, X_n$ with a common distribution function F, and wish to estimate the behavior of $F(x)$ for large values of x. Naturally, empirical estimates are poor, especially for values of x that exceed the largest of the sample values of the X_i. Equally, although the exact distribution of large order statistics from the X_i sequence is determined precisely by F — for example, the distribution function of $M_n = \max\{X_1, \ldots, X_n\}$ is $F^n(x)$ — this is of little help in practice, since F itself is unknown, and misspecification of F can lead to gross error in the distribution of order statistics.

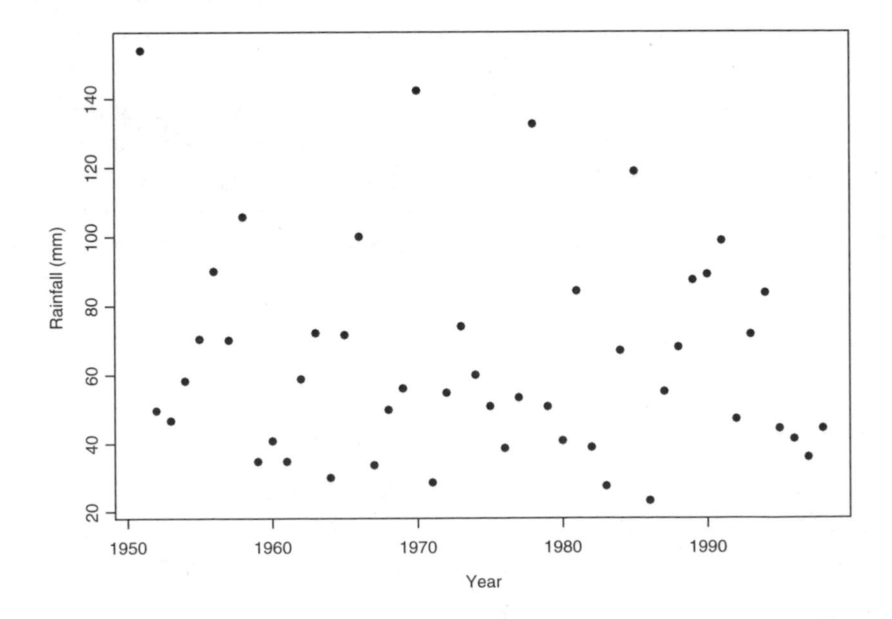

Figure 2.1 *Annual maximum daily rainfall values recorded in Venezuela.*

In the absence of more direct alternatives, it is usual to turn to asymptotic arguments as a means of generating models for extremes. For example, in the case of M_n, it is not difficult to establish that the complete class of limiting distributions of the variable

$$M_n^* = \frac{M_n - b_n}{a_n},$$

which includes a linear renormalization to avoid degeneracy, is the generalized extreme value (GEV) family

$$G(z) = \exp\left\{ -\left[1 + \xi\left(\frac{z-\mu}{\sigma}\right)\right]_+^{-1/\xi} \right\} \tag{2.1}$$

for parameters $\mu, \sigma(> 0)$, and ξ. In fact, this is the limit family for M_n^* under much weaker dependence assumptions than independence: the limit must be of the form stated if the series satisfies a requirement for weak long-range dependence at extreme levels; see Leadbetter et al. (1983) for precise definitions and results or Coles (2001) for a less formal overview. This model is usually applied in practice by adopting the limit family (2.1) as a set of candidate models for the distribution of M_n based on large enough sample sizes. This is justified since the unknown coefficients a_n and b_n can be absorbed into the GEV location and scale parameters, so that if M_n^* is approximately GEV, so is M_n, albeit with different values of μ and σ. With daily observations, for example, and with sample size $n = 365$, the variable M_n corresponds to the annual maximum, and inference amounts to fitting the GEV distribution to a sequence of observed annual maxima.

The first derivation of the limit laws dates back to Fisher and Tippett (1928), though the characterization was completed by Gnedenko (1943). However, the limit results were not originally developed or expressed in the form (2.1). Instead, three families of limit distributions were given: the so-called Weibull, Gumbel, and Fréchet types. Although these families of distributions correspond simply to the GEV subclasses with $\xi < 0, \xi = 0$, and $\xi > 0$ respectively, it is not immediately obvious from their conventional parameterizations that they can be unified into a smooth parametric family. The recognition that this could be achieved with the GEV family — first observed independently by von Mises (1954) and Jenkinson (1955) — represented an important advance in the utility of extreme value models for practical application. Prior to this, when modeling extreme value data, it was commonplace to try out all three subfamilies and to make a subjective choice about which of the three provided the best-fitting model. By unifying the limit distributions in a single family, the GEV model has several advantages: just one inference needs to be made, arbitrary discrimination between models is avoided, and uncertainty about model type is implicit in any inference. In particular, through inference within the GEV model, lack of knowledge about the shape parameter ξ is properly accounted for and quantified.

A legacy remains, however, from the partitioning of the limit family into three separate subclasses. This concerns the Gumbel family, derived by taking the limit

$\xi \to 0$ in the GEV family, which leads to the family of distribution functions

$$G(z) = \exp\left[-\exp\left\{-\left(\frac{z-\mu}{\sigma}\right)\right\}\right], \quad -\infty < z < \infty.$$

It is still common practice for this family either to be adopted outright as a model for block maxima in preference to the GEV family, or at least to be considered as a candidate for model reduction from the full GEV family. Additional arguments are sometimes adopted to justify this procedure, hinging on the fact that if the X_i belong to any of a wide class of standard families including the normal, exponential, gamma, and Weibull distributions, then the limit distribution of M_n^* is Gumbel.

The issue of reduction to the Gumbel model is one of four fundamental aspects of extreme value modeling that we address in this chapter by studying in detail the Venezuelan rainfall data. Our contention is that reduction to the Gumbel model is a risky strategy, even if it has empirical support from historical data. Two of the other aspects we address also concern model choice; first the option of modeling a greater extent of the available extremal data by means of threshold models, and second the importance of modeling nonstationary effects. The remaining aspect, and possibly the most important, is the management of uncertainty in an extreme value analysis. We will argue that quantifying uncertainty is an essential (perhaps *the* essential) component of an extreme value analysis, and that this is most coherently managed by adopting a Bayesian analysis. We begin, however, with a classical maximum likelihood comparison of the GEV and Gumbel models, which will in itself highlight many of the pertinent issues.

2.2 Inference for extreme value models

A variety of techniques have been proposed for estimating extreme value models, including graphical procedures, moment-based estimators, estimators based on order statistics and likelihood-based estimators. Doubts about the regularity of the maximum likelihood estimator due to the dependence of the distributional support on the parameters of the distribution were resolved by Smith (1985), who showed that despite the violation of the "standard regularity conditions," the maximum likelihood estimator has the usual property of asymptotic normality provided the shape parameter ξ is larger than -0.5. This condition, which excludes very light-tailed distributions with finite right endpoint, is almost always satisfied in practice by physical processes, at least in environmental applications. With this potential difficulty resolved, there are several reasons for preferring maximum likelihood to infer extreme value models. The log-likelihood function is easy to evaluate and to maximize numerically, asymptotic theory provides simple approximations for standard errors and confidence intervals, and the likelihood can be generalized to more complex model structures. The third of these points is perhaps the most important in practice, because it widens the scope of extreme value models to enable inferences on, for example, nonstationary processes. In the context of model selection, one special case of the second point is that the test statistic

$$D = 2\{\ell(\widehat{\theta}) - \ell(\widehat{\theta}^*)\},$$

Table 2.1 *Maximum likelihood estimates, standard errors (in parentheses) and negative log-likelihood (nllh) of Gumbel and GEV models fitted to Venezuelan annual maximum rainfall series.*

	Gumbel	GEV
μ	50.9 (3.3)	49.0 (3.4)
σ	21.5 (2.5)	19.9 (2.7)
ξ	—	0.17 (0.14)
nllh	224.9	224.1

where $\ell(\widehat{\theta})$ and $\ell(\widehat{\theta}^*)$ denote the maximized log-likelihoods within the GEV and Gumbel models respectively, has a null χ_1^2 distribution under the Gumbel hypothesis ($\xi = 0$); see Cox and Hinkley (1974) for example. Hence, a formal test can be derived based on likelihood principles for model reduction to the Gumbel distribution from the full GEV family.

Assessment of extreme value models, especially in terms of extrapolation properties, is usually made via quantiles:

$$z_p = \begin{cases} \mu - \frac{\sigma}{\xi}\left[1 - \{-\log(1-p)\}^{-\xi}\right], & \text{for } \xi \neq 0, \\ \mu - \sigma \log\{-\log(1-p)\}, & \text{for } \xi = 0, \end{cases}$$

where $G(z_p) = 1 - p$. With parameter values replaced by estimates, results are often presented as a plot of z_p against $-\log(1-p)$ on a logarithmic scale, a so-called return level plot, which has the virtue of compressing the tail of the distribution for ease of examination, and generates linear plots in the Gumbel case $\xi = 0$.

Details of maximum likelihood inferences for both the Gumbel and GEV models are given in Table 2.1. In particular, the test statistic for model reduction to the Gumbel distribution is $D = 2(224.9 - 224.1) = 1.6$, which is not exceptional on the scale of a χ_1^2 distribution. Hence, by significance testing at any conventional level, the Gumbel model would be accepted as a suitable reduction from the GEV family for these data. The consequences can be seen in the return level plots of Figure 2.2. Added to the plot are approximate 95% confidence intervals obtained by standard manipulation of the asymptotic variance-covariance matrix of the maximum likelihood estimator through the delta method. It is apparent that for short- to medium-term extrapolations, the return level plots of the GEV and Gumbel models are similar. Moreover, the plot for each model is reasonably consistent with the corresponding empirical estimate, lending support to the viability of either model.

But let us take stock of the situation. The point of the analysis is likely to be to set a design level for protection against future extreme levels of the process. Suppose, for argument's sake, that we seek a design level v_p such that the probability of experiencing a daily rainfall that exceeds v_p in a given year is p or less, for some small value of p. In other words,

$$G(v_p) = 1 - p \approx \log(1/p). \qquad (2.2)$$

Table 2.2 *Design levels v_p in millimetres. The MLE estimates are given by $\widehat{v}_p = \widehat{G}^{-1}(-\log p)$, where $\widehat{G}$ is the estimated distribution function. The '95% CI' estimates are given by $\widehat{v}_p = \tilde{G}^{-1}(-\log p)$, where $\tilde{G}(p)$ is the upper 95% confidence limit for v_p as a function of p.*

	MLE		95% CI	
p	Gumbel	GEV	Gumbel	GEV
.01	150	186	176	277
.002	184	265	217	473

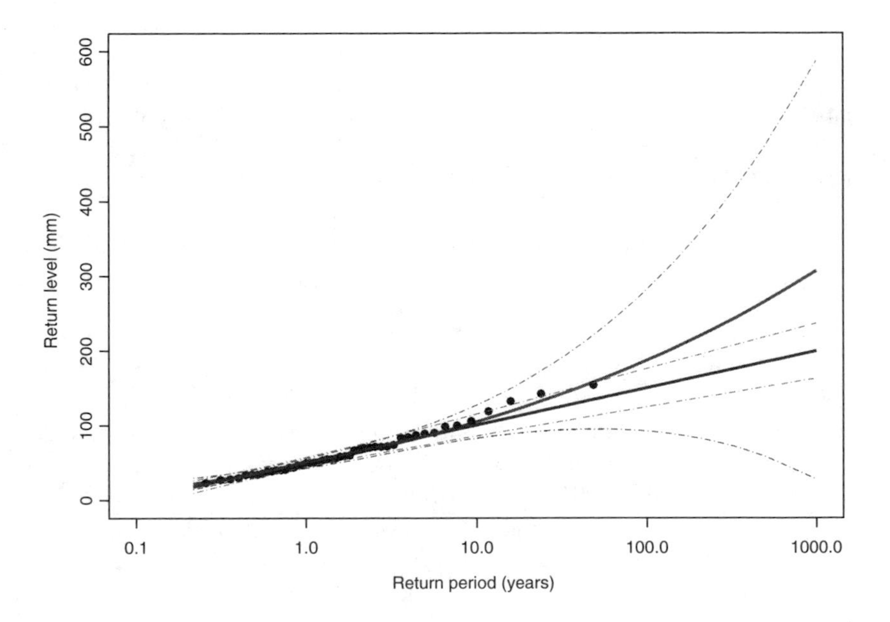

Figure 2.2 *Return level plots for Gumbel and GEV models of Maiquetia annual maximum daily rainfall. Dashed curves correspond to 95% confidence intervals; points correspond to empirical estimates including the 1999 datum.*

With $p = 1/100$ and $p = 1/500$ the maximum likelihood estimates of v_p, read-off directly from Figure 2.2, are given in Table 2.2. Due to the extrapolation involved, the choice between the Gumbel and GEV models starts to make a difference: the differences in the estimates are around 3.5 and 8 cm for $p = .01$ and $p = .002$, respectively. So despite the similarity in the GEV and Gumbel models for describing the statistical behavior of the historical series, the models may provide materially different estimates of design parameters.

This point becomes more critical once parameter uncertainty is taken into account. The handling of uncertainty due to model estimation is not an issue that has been dealt with universally well in the extreme value literature. Models often extrapolate to such uncomfortably high levels that practitioners have preferred to turn a blind eye to the additional increase in design level that would be implied by the incorporation of sampling error. But the uncertainties are there whether we choose to account for them or not, and not doing so may lead to catastrophically poor design parameters.

There seems no entirely natural way to handle uncertainty in the context of a maximum likelihood analysis. We argue below that a Bayesian formulation of the problem leads to a more satisfactory resolution, but suppose temporarily that we adopt an ad hoc criterion of designing to the upper end-point of a $(1-\alpha)\%$ confidence interval for v_p, with p and α prespecified at some small values. The rationale for this is that the criterion determined by (2.2) leads to a design parameter that is the maximum likelihood estimate for v_p. In the understood 'repeated sampling' sense, it follows that the inferred confidence interval for v_p contains the true value with probability $(1-\alpha)$. Hence, designing to the upper end-point of such an interval builds conservatism into the estimate. Notwithstanding the absence of a formalism for such a procedure, and sidestepping questions about the choice of p and α, it is at least clear that this procedure goes some way to providing an inference that takes parameter uncertainty into account.

The estimates obtained in this case, with $\alpha = 0.05$ and for both $p = 1/100$ and $p = 1/500$, are included in Table 2.2. The difference between the estimates based on the GEV and Gumbel models, respectively, are now around 10 and 25 cm, respectively. Hence, the GEV and Gumbel models become markedly different once uncertainty is taken into account. This was also clear from Figure 2.2 directly: the relative size of confidence intervals on v_p for the GEV and Gumbel models grows without limit. This should not be surprising, as the major component of uncertainty in extreme value modeling is in the rate of tail decay, i.e., the shape parameter ξ. In the Gumbel family this value is assumed to be known ($\xi = 0$), thus eliminating the greatest part of uncertainty from the inference. It follows that even if the maximum likelihood estimate on ξ is close to zero, so that the maximum likelihood model within the GEV is "near-Gumbel," the inferences will differ to a far greater extent in terms of their precision.

So, is it appropriate to use the Gumbel model, which would offer considerable savings in terms of design specifications? A personal view is that, even when formal tests support reduction to the Gumbel family, as with the Venezuelan rainfall data, it is better to exploit the conservatism induced by acknowledgement of the uncertainty of ξ in the GEV model.

To complete the story on the Venezuelan data: in December, 1999, large areas of the country were devastated by a torrential daily rainfall event. Widespread damage was caused and there was a considerable loss of life — some estimates put the number of dead at around 50,000. At Maiquetia a daily rainfall of more than 410 mm was recorded, an event that was almost three times greater than any other on record. Further details of the extent of the damage caused are given by Larsen et al. (2001). There is also a large amount of information available on the web: http://pr.water.usgs.gov/public/venezuela is a good starting point, giving both geological background information as well as photographs of damage caused in the Maiquetia vicinity as a consequence of the 1999 event.

While it is easy to be wise after the event, it is constructive to ask whether an analysis of the pre1999 data could, or even should, have pointed to the necessity to design against such an extreme event. Looking back at our analyses, it is clear that the Gumbel model would have failed completely in anticipating the 1999 event, even if conservatism were built in by designing to the upper 95% confidence limit of any reasonable return level v_p. Designing to the maximum likelihood estimate of the GEV model would have afforded little extra protection against the 1999 event, but designs made to the upper 95% confidence limit of GEV return levels start to approach this value. Indeed, had designs been made to the upper limit of the 95% confidence interval for the 500-year return level, then the 1999 event would have been protected against.[1]

2.3 Bayesian analysis of annual maxima data

In subsequent sections we assess whether greater accuracy in estimating the extremal behavior of the Venezuelan rainfall process can be achieved by exploiting more of the available extreme data than just the annual maxima. First we consider in slightly more detail the issue of estimation uncertainty. Allowance for parameter uncertainty by designing to confidence interval limits is not entirely satisfactory — the design parameter obtained in this way has no simple probabilistic interpretation, and there is the necessity to set two constants, α and p, each of which is influential in the design parameter estimation. A better approach is provided by a Bayesian inference of the model.

Bayes' theorem converts a prior distribution for a parameter (vector) θ, $f(\theta)$, on the availability of historical data $\mathcal{H}$, into a posterior distribution, $f(\theta \mid \mathcal{H})$. This takes the form

$$f(\theta \mid \mathcal{H}) \propto f(\mathcal{H} \mid \theta) \times f(\theta), \tag{2.3}$$

where $f(\mathcal{H} \mid \theta)$ is the likelihood for the historical data. The difficulty in practice is that the proportionality constant is unknown and potentially difficult to evaluate. Analytical calculation could be avoided if it were possible to simulate from the posterior distribution, but this is also unfeasible for most models. Markov chain Monte Carlo (MCMC) has emerged as a flexible alternative simulation method for Bayesian models. Details of the algorithm are well-known (see Gilks et al. 1996, for example), but for completeness we give a brief summary.

The basic idea of MCMC is to iterate a sequence of values — technically a Markov chain, since the iteration to generate the next term in the sequence depends only on the current value — whose stationary distribution is the target posterior distribution. In slightly more detail, we simulate a sequence $\{\theta_i\}$ by setting an initial value θ_1 and specifying an essentially arbitrary probability rule $q(\theta^* \mid \theta_i)$ to generate a proposal for the next term in the sequence. Setting

$$\alpha_i = \min\left\{1, \frac{f(\mathbf{x} \mid \theta^*) f(\theta^*) q(\theta_i \mid \theta^*)}{f(\mathbf{x} \mid \theta_i) f(\theta_i) q(\theta^* \mid \theta_i)}\right\},$$

[1] As an aside, we should point out that the impact of the 1999 event on such a geologically fragile region means it is unlikely that any sort of preemptive action could have afforded genuine protection to its effects. Nonetheless, an accurate assessment of the risk of a region to extreme rainfall, coupled with an understanding of the geophysical effects such events may have, would at least have signaled the vulnerability of the region to the type of catastrophic event that subsequently occurred.

we complete the transition by letting

$$\theta_{i+1} = \begin{cases} \theta^* & \text{with probability } \alpha_i, \\ \theta_i & \text{with probability } 1 - \alpha_i. \end{cases}$$

By construction, the generated sequence is a Markov chain which, under simple regularity assumptions, can be shown to have a marginal stationary distribution that is the target posterior distribution. This implies that, for a large enough value of k, the sequence $\theta_{k+1}, \theta_{k+2}, \ldots$ is approximately stationary, with marginal distribution given by (2.3). This sequence can therefore be used to estimate, for example, posterior means, variances and densities, in a way similar to a sequence of independent values.

Application of the MCMC procedure requires specification of the prior distribution $f(\theta)$ and the proposal transition density $q(\cdot)$. Prior choice is intrinsic to the model: different choices will lead to different inferences. In principle, choices can be made that add genuine information to the analysis (see Coles and Tawn 1996 for a discussion of this issue in the context of extreme value analyses). When such knowledge is unavailable, it is usual to adopt arbitrary prior distributions that have a large variance, so as to reflect prior ignorance about parameter values, though sensitivities to arbitrary choices should always be checked. In contrast (subject to regularity) inferences are invariant to the choice of proposal density q for the MCMC algorithm. Common choices include random walk densities, or "independence" models, in which $q(\theta^* \mid \theta_i)$ has fixed form independent of θ_i. However, despite this inferential invariance, selection of q can have a profound effect on the efficiency of the MCMC algorithm, which is determined by how adequately the posterior space is explored by a finite series $\theta_{k+1}, \ldots, \theta_m$. If the variance of q is too small, proposals θ^* are likely to be close to the current value θ_i; if the variance of q is too large, proposals θ^* are likely to move from regions of high posterior density to regions of low posterior density, and the rejection step of the algorithm rejects the proposal with high probability. In either case, the correlation of successive values in the θ_i process will be large, and a long sequence of the process will be required before the entire space is reasonably covered. Usually θ is multidimensional, which complicates further the selection of proposal transition densities that will have adequate efficiency. Some theoretical guidance based on asymptotic approximations for special models is available in the literature, but it is more common in practice to adapt by trial-and-error to obtain a chain whose mixing properties seem reasonable based on the empirical evidence, usually supported by a range of diagnostic procedures.

For the analysis of the Venezuelan data we took normal and (in the case of σ) lognormal prior distributions, assumed to be independent and with large variance, for μ, σ, and ξ.[2] We also generalized the MCMC algorithm in a standard way by using successive proposal transitions of the type $q_1(\mu^* \mid \mu_i, \sigma_i, \xi_i)$, $q_2(\sigma^* \mid \mu_{i+1}, \sigma_i, \xi_i)$, $q_3(\xi^* \mid \mu_{i+1}, \sigma_{i+1}, \xi_i)$. Standard random walks with Gaussian steps were used for each of these terms. A small amount of trial-and-error was needed to adapt the variances of the random walk terms to obtain a sequence that seemed to have reasonable mixing properties.

[2] Repeat analyses with modified prior parameter values led to near-identical results, suggesting a lack of sensitivity to prior choice provided the variances are reasonably large.

Denoting the GEV (or Gumbel) distribution function by G and the posterior density for $(\mu, \sigma, \xi) \mid \mathcal{H}$ by h,

$$\Pr(Z \leq z \mid \mathcal{H}) = \int G(z \mid \mu, \sigma, \xi) h(\mu, \sigma, \xi \mid \mathcal{H}) d(\mu, \sigma, \xi)$$

is defined to be the predictive distribution of Z, a future annual maximum random variable. The fact that inference is carried out by MCMC enables simple approximate evaluation of the predictive distribution from the generated sample:

$$\Pr(Z \leq z \mid \mathcal{H}) \approx \frac{1}{m-k} \sum_{j=k+1}^{m} G(z \mid (\mu, \sigma, \xi)_i)$$

where $(\mu, \sigma, \xi)_i$ denotes the parameter set simulated at iteration i of the algorithm.

It follows that a plot of z_p against $-\log(1-p)$ on a logarithmic scale, where $\Pr(Z \leq z_p \mid \mathcal{H}) = 1 - p$, may be interpreted as a Bayesian equivalent of the conventional return level plot. In particular, for any choice of p, the corresponding value of z_p is the level that is expected to be exceeded by Z with probability p, after allowance for parameter uncertainty. It therefore seems reasonable, for example, to design to a level such as $z_{0.01}$, because this parameter has a similar interpretation to the 100-year return level, but includes an implicit allowance for parameter uncertainty.

For the Venezuelan rainfall data the posterior means and standard deviations for the GEV and Gumbel models are given in Table 2.3. The corresponding predictive return level plots are shown in Figure 2.3. Included on these plots are empirical estimates based on all the annual maxima, including the 1999 datum. It is apparent that although both models attach low predictive probability to the actual 1999 event, the extra allowance for parameter uncertainty contained within the GEV model means that the event has much higher predictive probability under this model relative to the GEV model: the exceedance probability of 410.4 mm is estimated as 0.00243 under the GEV model and 5.24×10^{-7} under the Gumbel model. It is also clear that designing to predictive return levels rather than maximum likelihood estimates would have afforded greater protection against the 1999 event. In other words, estimation uncertainty is again accommodated, but now with a clearer probabilistic interpretation of the design level obtained.

Table 2.3 *Posterior means and standard deviations for GEV and Gumbel analysis of daily rainfall annual maxima.*

	Gumbel		GEV	
	Mean	SD	Mean	SD
μ	50.8	3.4	49.0	3.6
σ	22.5	2.8	21.1	3.0
ξ	—	—	0.182	0.15

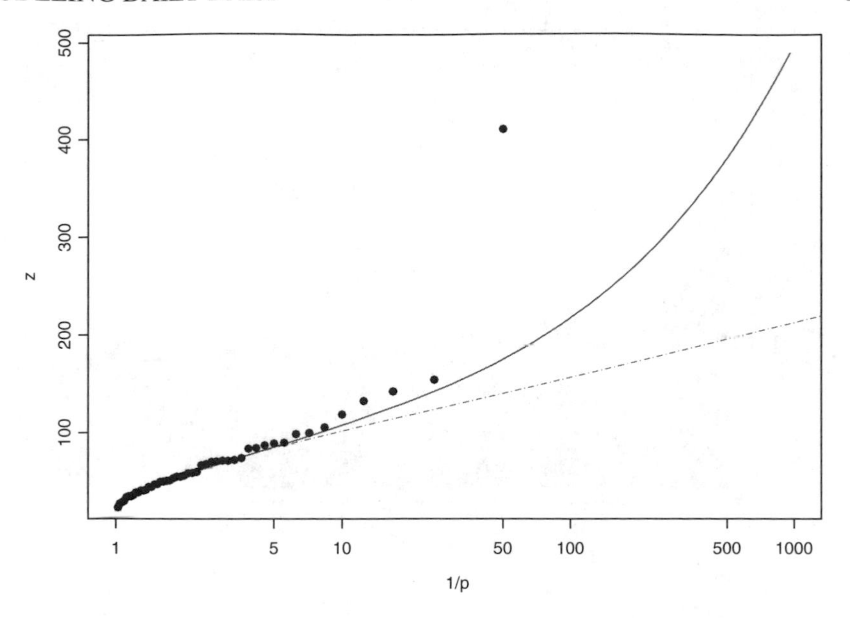

Figure 2.3 *Predictive return level plots for Gumbel and GEV models of Venezuelan annual maximum daily rainfall. Solid curve is GEV model; dashed curve is Gumbel model. Points correspond to empirical estimates including the 1999 datum.*

2.4 Modeling daily data

With the exception of a two-month period for which data are missing, complete records of daily rainfall observations are available for the Maiquetia series over the period 1961 to 1999. The time series of these data for the pre1999 period is shown in Figure 2.4. Modeling the extremes of this series enables a more efficient usage of extreme value information than that given by an analysis of annual maxima data, which excludes from inference many extreme events that did not happen to be the largest annual event; see Weissman (1984) for a theoretical treatment of this issue.

The usual method for handling such data is the peaks over threshold technique. As a statistical modeling technique this procedure was popularized by Davison and Smith (1990), though the original asymptotic characterization is due to Pickands (1975). Assuming the daily data to be independent with common distribution function F, the conditional distribution of excesses of a threshold u is determined by

$$\Pr(X \leq u + y \mid X > u) = 1 - \frac{1 - F(u + y)}{1 - F(u)}, \quad y > 0. \tag{2.4}$$

Renormalizing and letting $u \to \infty$ leads to an approximate family of distributions for (2.4) given by

$$G(y) = 1 - (1 + \xi(y - u)/\sigma^*)_+^{-1/\xi}, \tag{2.5}$$

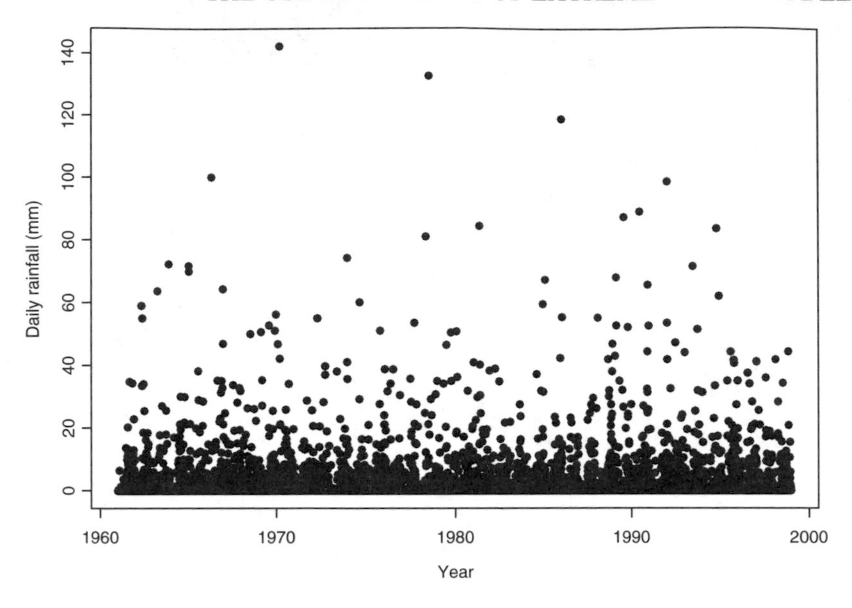

Figure 2.4 *Daily rainfall values recorded at Maiquetia, Venezuela.*

the generalized Pareto family. This family describes all possible nondegenerate limiting distributions of the scaled excess-of-threshold distributions, in the same way as the GEV family describes all possible nondegenerate limiting distributions of the centered and normalized sample maximum. A direction connection between the generalized Pareto and GEV models is that if the number of exceedances of u per year has a Poisson distribution with mean τ, and if the exceedance distribution falls within the generalized Pareto family with shape parameter ξ, then the annual maximum distribution falls within the GEV family with the same shape parameter. There is also a one-one correspondence between (τ, σ^*) and (μ, σ), the location and scale parameters of the annual maximum GEV distribution.

Inference proceeds by deriving a likelihood for σ^* and ξ based on model (2.5), which can be applied to the observed excesses of a specified high threshold u. Inference on τ is also straightforward from the Poisson assumption, so by transformation inference can also be made on the associated GEV annual maximum distribution parameters. The essential point is that inference on the annual maximum distribution is now derived from a greater amount of information, leading to increased precision. Naturally, the issue of threshold choice involves a compromise between bias and variance. Low thresholds generate many exceedances, so variance is low, but the asymptotic support for the model is weak, leading to large bias. Modeling with high thresholds has stronger asymptotic support, so bias is low, but there are fewer exceedances and the variance is potentially high. In practice, a range of diagnostics are available to assist with the selection (as far as possible) of low thresholds that are consistent with model structure (Davison and Smith 1990).

There are advantages in reformulating the threshold exceedance model as a point process model. This approach exploits representations of extremal behavior set out in Leadbetter et al. (1983), but was again popularized for statistical application by Smith (1989). Assume the variables $X_1, X_2, \ldots, X_n$ are independent and identically distributed. For sufficiently high thresholds u the process of points $\{(i/n, X_i) : i = 1, \ldots, n\}$ approximates a nonhomogeneous Poisson process when restricted to the region $(0, 1) \times [u, \infty)$. Furthermore, analysis of the extremal behavior based on standard asymptotic arguments implies that the intensity function of the approximating process falls within the parametric family

$$\Lambda((t_1, t_2) \times [z, \infty)) = (t_2 - t_1) n_y \left[1 + \xi\left(\frac{z - \mu}{\sigma}\right)\right]^{-1/\xi}. \tag{2.6}$$

In this formulation, the parameters (μ, σ, ξ) correspond to the approximating GEV distribution of block maxima of the process. Furthermore, choosing the scaling coefficient n_y to be the number of years of observation ensures that the unit measurement is one year. In other words, μ, σ, ξ are simply the GEV parameters of the annual maximum distribution.

As with the generalized Pareto peaks-over-threshold model, the point process model offers the advantage that all exceedances of u contribute to the inference on extremes. However, the point process model has the additional advantage that the parameterization coincides exactly with that of the GEV distribution for annual maxima. This is advantageous for reasons beyond the extra simplicity of interpretation. First, it may be desirable to build nonstationarity into the model by modeling parameters as functions of time; this approach is more restrictive with the generalized Pareto model due to the way τ is modeled separately from the remainder of the model. Second, if there is nonstationarity, it may be appropriate to specify time-varying thresholds. For example, a seasonally varying process may generate exceedances of a uniform threshold only in one season. Finally, because of the commonality of parameterization, the point process likelihood can be supplemented with historical information that might consist only of annual maximum data. This is helpful in the case of the Venezuelan rainfall series, because the recording of annual maxima began 10 years before that of daily values. Denote by $z_1, \ldots, z_m$ the annual maxima in those years for which they are the only data available, and by $x_1, \ldots, x_k$ the exceedances of the threshold u when the daily data are available. The log-likelihood is

$$\begin{aligned} \ell(\mu, \sigma, \xi) &= -(m + k)\log\sigma - (1 + 1/\xi)\left\{\sum_{i=1}^{m} \log\left[1 + \xi\left(\frac{z_i - \mu}{\sigma}\right)\right]\right. \\ &\quad \left. + \sum_{i=1}^{k} \log\left[1 + \xi\left(\frac{x_i - \mu}{\sigma}\right)\right]\right\} - \sum_{i=1}^{m}\left[1 + \xi\left(\frac{z_i - \mu}{\sigma}\right)\right]^{-1/\xi} \\ &\quad - n_y\left[1 + \xi\left(\frac{u - \mu}{\sigma}\right)\right]^{-1/\xi}, \end{aligned} \tag{2.7}$$

where n_y is the number of years of observations for which the complete daily data are available. This is simply the GEV log-likelihood for the years in which only annual maxima are available added to the log-likelihood derived from a Poisson process with intensity in the family (2.6).

For the Venezuelan daily rainfall data, based on a threshold of $u = 10$ mm, applying this model to both the first 10 years of annual maximum data and the daily data from 1961 to 1998 leads to maximum likelihood parameter estimates (with standard errors in parentheses)

$$\widehat{\mu} = 49.2\ (2.5),\ \widehat{\sigma} = 20.7,\ (2.3),\ \widehat{\xi} = 0.26\ (0.06).$$

In this case the hypothesis $H : \xi = 0$ within the generalized Pareto family, corresponding to an exponential distribution of threshold excesses, is comfortably rejected by a test at any conventional level of significance. In other words, the evidence for an exponential tail of the daily rainfall distribution, which could not be rejected by a test based on the annual maximum data, is not strong once the additional daily information is included. Furthermore, the increased value of $\widehat{\xi}$ in this model, compared with the corresponding value in the GEV model, suggests we may do better in terms of anticipating the 1999 event, although the increased precision could offset some of the gain if confidence intervals are taken into account. A return level plot based on this analysis is shown in Figure 2.5. From the return levels, comparable design parameters to those obtained for the GEV model in Table 2.2 can be calculated; these are given in Table 2.4. Comparison of the two tables confirms that designs based on this model, with or without an allowance for parameter uncertainty, would have afforded greater protection against the 1999 event.

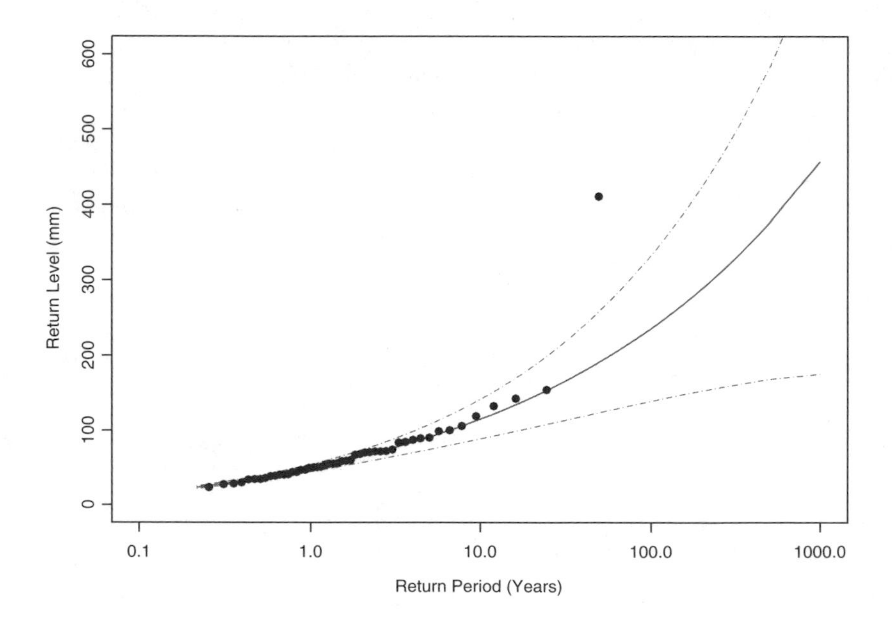

Figure 2.5 *Return level plot for point process/GEV model fitted to Venezuelan rainfall data. Solid line is maximum likelihood estimate; dashed curves correspond to 95% confidence intervals.*

Table 2.4 *Design levels v_p in millimetres based on criterion (2.2) using the point process model. The 95% CI corresponds to estimates based on designing to the upper 95% confidence limit of v_p.*

p	MLE	95% CI
.01	236	371
.002	331	589

The point process model can also be inferred as a Bayesian model using a similar prior formulation and MCMC procedure to that described for the GEV annual maximum model. There is a complication, however, due to the model parameterization. The parameter $\lambda = \tau/366$, corresponding to the crossing rate of the threshold u, can be estimated with high precision, and so has low variance. However, λ is a function of μ, σ, and ξ, so there is a near-degeneracy in the (μ, σ, ξ) space along which the density of the posterior is concentrated. The previously described MCMC algorithm makes proposal transitions orthogonal to the μ-, σ-, and ξ-axes, so that proposal transitions are unable to remain in areas of high posterior density, and a very large number of iterations are required to cover the space adequately. This is not just a point of aesthetics: we obtained sequences of wildly divergent parameter values in the MCMC output, presumably because the algorithm was searching over spaces of virtually zero posterior density.

The difficulty is easily avoided by reparameterizing the model in terms of the conventional generalized Pareto peaks-over-threshold parameters (λ, σ^*, ξ). Formally, we let

$$\lambda = 1 - \exp\left[-\frac{1}{366}\left\{1 + \xi\left(\frac{u-\mu}{\sigma}\right)\right\}^{-1/\xi}\right],$$
$$\sigma^* = \log\{\sigma + \xi(u - \mu)\}.$$

To first order, λ and $\exp(\sigma^*)$ are, respectively, the threshold exceedance probability and scale parameter in model (2.5). This works because the curve of high density in the (μ, σ, ξ) space has been transformed to lie orthogonal to the λ axis in the new space. Hence, random walk proposals in the respective λ, σ^*, and ξ directions enable transitions across the region of high posterior density, enabling high coverage with relatively few simulations. Indeed, with this adjustment, the simulated series displayed fast convergence and good quality mixing properties.

Based on this scheme, and using large-variance priors, we obtained the following posterior means and standard deviations (in parentheses):

$$\widehat{\mu} = 51.7\ (3.0),\ \widehat{\sigma} = 23.1,\ (2.9),\ \widehat{\xi} = 0.26\ (0.06).$$

These values are slightly different from the corresponding maximum likelihood estimates and standard errors, partly because of the substantive differences between a Bayesian and classical analysis, and partly because of the convenience in summarizing a Bayesian inference obtained through MCMC by the marginal posterior means rather than the posterior mode. Under the Bayesian point process model, $\Pr(Z > 410.4 \mid \mathcal{H}) = 0.0038$, so although the probability of an event as extreme as the 1999 value remains low, its value is around 60% greater than the corresponding estimate of 0.00243 under the GEV model. Nonetheless, the scale of the 1999 event is still surprising, and it is unlikely that an event with such low probability would have been anticipated, even if this analysis suggests it to have been a much less surprising result than earlier analyses implied.

2.5 Seasonal models

Another advantage of working with daily data rather than annual maxima is that it creates the opportunity to model within-year nonstationary effects. This is usually achieved by assuming the appropriate extreme value model structure to be valid, but with parameters substituted by functions of time. For seasonal effects there are two simple possibilities: one is a model with continuously varying parameters $\mu(t)$, $\sigma(t)$, and $\xi(t)$, probably chosen to be periodic with period one-year. The other possibility is a "separate seasons" model, in which distinct parameters are assumed in distinct seasons, say monthly or three-monthly. Though the first of these model structures seems more natural — meteorological processes might be expected to change smoothly through time — it is often difficult to parameterize such time variation in a suitable way. Moreover, meteorological processes are often observed to have sudden changes, perhaps due to the passage from one weather system to another. For each of these reasons a separate seasons model is often preferred.

An exploratory assessment of the Venezuelan data suggests there is a seasonal variation, but that it may be difficult to capture using smoothly varying parametric models. For example, Figure 2.6 shows monthly stratified boxplots of log-daily rainfall for events exceeding 10 mm. Though a typical event in the winter months is more extreme than an event in the summer months, some of the summer events have been among the most extreme. This suggests that a parsimonious parameterization that can adequately capture seasonal variation is likely to be difficult to obtain. Hence, we focus on separate-season models. A difficulty in implementing such models however, at least through maximum likelihood, is the necessity to define both the number and timing of seasons prior to analysis. Documentation of the meteorological activity in Venezuela suggests a two-season pattern to the rainfall process, though there is little guidance in the literature as to the precise timing of the seasonal changes. Our initial investigations consisted of a likelihood analysis based on prespecified seasonal timings. In particular we compared models based on a three-season structure (November through February, March through June, and July through October) with reduced models in which pairs of these seasons were combined. We found, by standard likelihood ratio tests, the three-season model to be superior, even though this is at odds with the perceived hydrological wisdom.

To examine this issue further, we considered as an alternative a change-point model in which the year is partitioned into two seasons: $\mathcal{I}_1 = [1, k_1] \cup [k_2, 366]$ and $\mathcal{I}_2 = [k_1, k_2]$. Loosely, these can be thought of as winter and summer seasons, respectively. The point process model is assumed within each of the $\mathcal{I}_j$, but with distinct

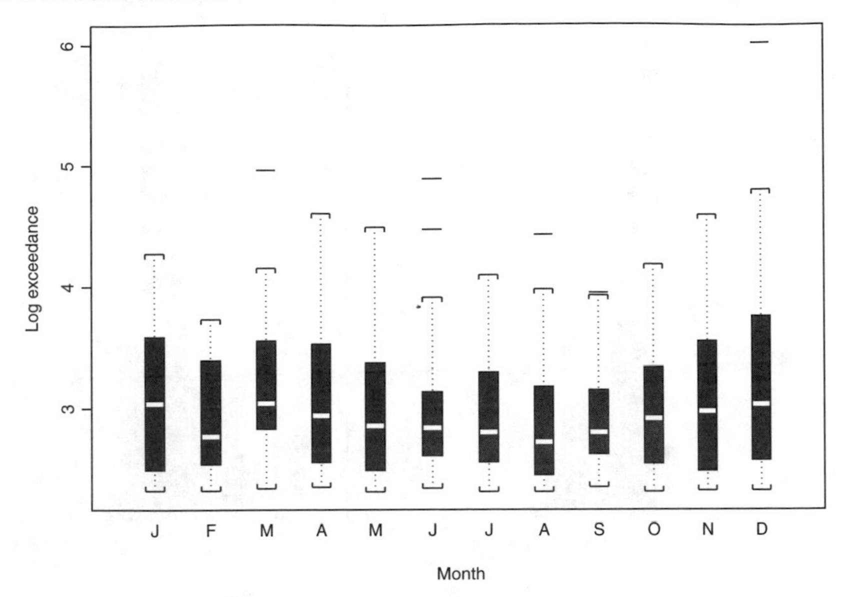

Figure 2.6 *Boxplots of log-daily rainfall for events exceeding 10 mm stratified by month.*

parameters in periods $\mathcal{I}_1$ and $\mathcal{I}_2$. Moreover, the changepoints k_1 and k_2 are regarded as model parameters to be inferred.

In application we excluded the first 10 years of data in which only the annual maxima data are available, since the timings of the events are not recorded. If this information were available, these data could also be incorporated in the analysis. Denoting the parameters in period $\mathcal{I}_j$ by (μ_j, σ_j, ξ_j) respectively, the log-likelihood for the model is

$$\ell = -\sum_{j=1}^{2}\left\{\sum_{i\in\mathcal{I}_j}\left(\log\sigma_j + (1+1/\xi_j)\log\left[1+\xi_j\left(\frac{x_i-\mu_j}{\sigma_j}\right)\right]\right)\right.$$
$$\left. + \frac{n_y|\mathcal{I}_j|}{366}\left[1+\xi_j\left(\frac{u-\mu_j}{\sigma_j}\right)\right]^{-1/\xi_j}\right\}, \tag{2.8}$$

where $|\mathcal{I}_j|$ is the number of days in period $\mathcal{I}_j$. Because of the variable season structure, maximum likelihood is not a viable option for fitting this model. However, the model is straightforward to implement within a Bayesian setting using a variation on the earlier MCMC algorithm.

A reparameterization to conventional generalized Pareto parameters is again necessary to obtain reasonable mixing. It is also necessary to impose some limited structure on the changepoint parameters, minimally to ensure that $k_1 < k_2$. In fact we imposed a slightly stronger structure to incorporate at an approximate level the perceived knowledge about timing of seasonal changes. In particular, we imposed discrete uniform priors on both k_1 and k_2, with $k_1 \sim$ U[50, 250] and $k_2 \sim$ U[200, 366],

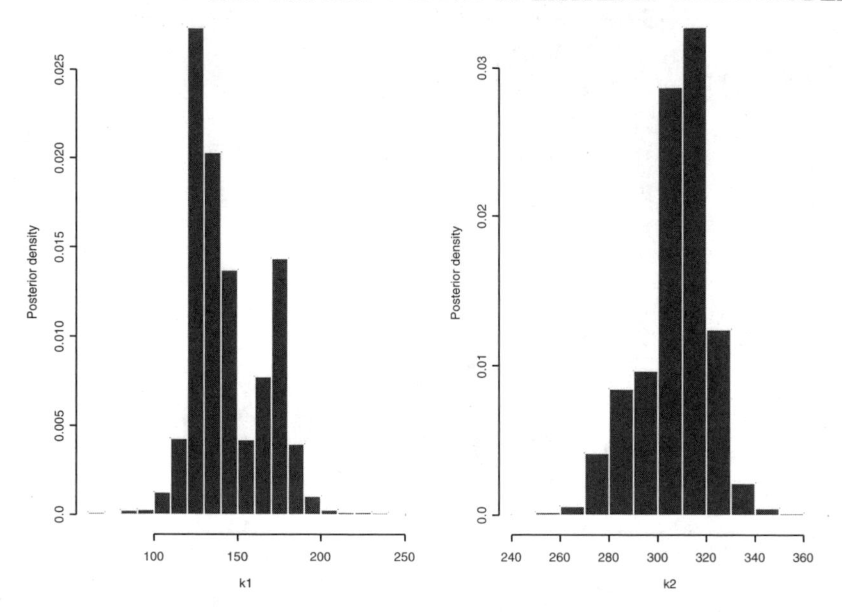

Figure 2.7 *Posterior distributions of seasonal changepoint parameters k_1 (left panel) and k_2 (right panel).*

together with a constraint that $k_1 < k_2$. Inclusion of each of these parameters into the Metropolis-Hastings algorithm was achieved with simple discrete random walks.

The posterior distributions on k_1 and k_2 are shown in Figure 2.7. Multiple runs of the MCMC algorithm confirm that the bimodality in the distribution of k_1 is a genuine feature and not attributable to sampling error in the Monte Carlo estimation. Nonetheless, the lack of uniformity in each of the distributions is strongly supportive of a well-defined seasonal effect, although the precise timing of the optimal change from winter to summer is less clear-cut than that of summer to winter. In brief summary, the evidence points to a seasonal breakdown that constitutes mid-November to April as the winter period, and the remaining months as the summer period.

A summary of the seasonal model parameters is given in Table 2.5. There is clearly a difference in the model parameters across the seasons, especially once the 1999 data are included in the inference. Notice also that in both seasons the posterior mean for ξ is some distance (in probability terms) from zero, suggesting that $\xi = 0$ remains an inappropriate model reduction for either season. This holds true even without inclusion of the 1999 data. The net effect of these parameter estimates on return level estimation is again best assessed via the predictive distribution, which is shown in Figure 2.8. Relative to the previous models, the 1999 annual maximum appears less surprising. The inclusion of the 1999 data produces a model that is entirely consistent with that datum, but even without the inclusion of these data the 1999 event no longer seems

Table 2.5 *Posterior means and standard deviations (in parentheses) of point process model parameters in two-seasonal model fitted to Venezuelan daily rainfall series, with and without 1999 value respectively.*

	1999 excluded		1999 included	
	$j=1$	$j=2$	$j=1$	$j=2$
μ	62.8 (8.3)	41.3 (2.8)	73.7 (12.0)	41.6 (2.8)
σ	31.9 (8.9)	14.3 (2.4)	45.1 (14.1)	14.4 (2.4)
ξ	0.37 (0.13)	0.14 (0.08)	0.52 (0.13)	0.14 (0.08)

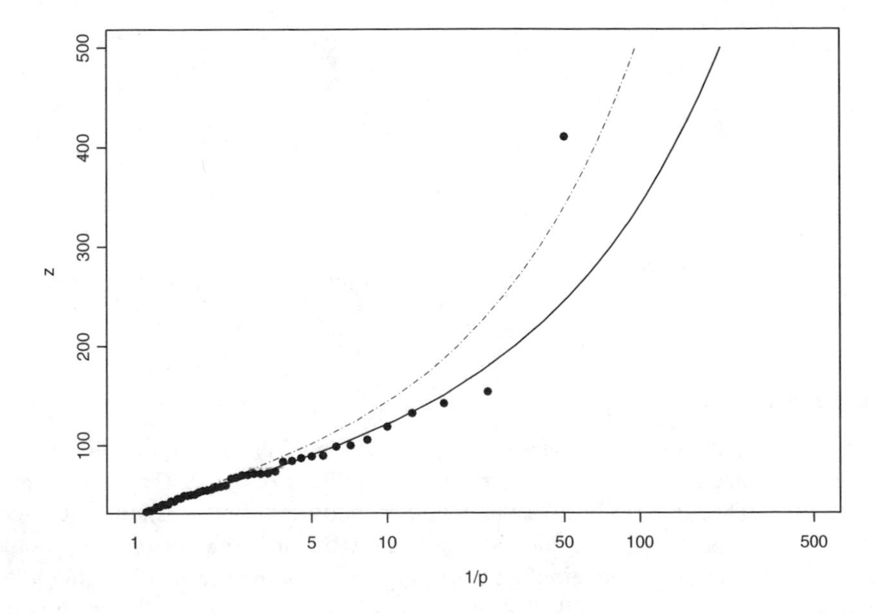

Figure 2.8 *Predictive distributions for seasonal models fitted to Venezuelan rainfall data. Solid line excludes 1999 data; dot-dash line includes 1999 data. Points correspond to empirical estimates based on the 49 available annual maxima.*

so exceptional: $\Pr(Z > 410.4 \mid \mathcal{H}) = 0.0067$, corresponding to another substantial increase on the previous model.

We note also that although the probability that an individual year having a daily rainfall that exceeds the 1999 level of 410.4 mm is still low, the corresponding probability that any one of 49 annual maxima exceeds this level is $1-(1-0.0067)^{49} = 0.281$. Judged this way the 1999 event seems even less remarkable, and in any case not inconsistent with the overall model structure.

In conclusion we present the predictive distributions of the annual maximum distribution for the various models applied to the pre-1999 data in Figure 2.9.

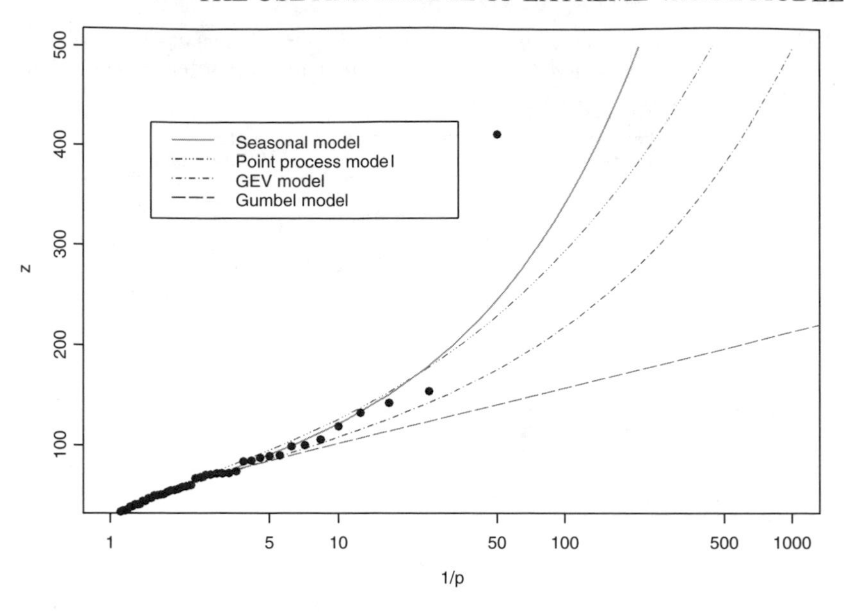

Figure 2.9 *Predictive distributions for various models fitted to the Venezuelan rainfall data. Points correspond to empirical estimates.*

2.6 Discussion

Our general conclusion is that analysis of the historical Venezuelan rainfall data requires four components before an event of the magnitude of the 1999 event becomes foreseeable. The full family of asymptotically motivated models should be used; all available data should be exploited; uncertainty due to model estimation should be accounted for; and population heterogeneity (in this case, seasonality) should be modeled. The effect of several of these aspects is seen in Figure 2.9, which compares the predictive distributions of the various fitted models. Though our analysis in this chapter has been restricted to the single dataset, similar analyses on other datasets suggest that these themes are also likely to be important in any extreme value analysis.

An extended discussion of the Venezuelan case study is given by Coles and Pericchi (2001). In particular, they look in greater detail at the extent to which models adapted to the exceptional 1999 event once it occurred. It emerges that the Gumbel model is very poor at adapting to the new information, confirming the model's inappropriateness for this process. The GEV, point process and seasonal models all adapt well to the new information, providing analyses that give adequate fits to all data including the 1999 data. (We have shown such a plot here only for the seasonal model in Figure 2.8.) Other aspects of the Venezuelan rainfall process are currently under investigation. These include the issue of temporal dependence, which we have argued here to be sufficiently weak to have had little or no impact on extreme value inferences, and the seasonal aspect for which we are assessing the impact of the two-seasonal changepoint model adopted here. We hope to report on each of these aspects elsewhere in due course.

Finally, we note that the discussion throughout this article has assumed that an asymptotic model for the Venezuelan rainfall process is appropriate. This may not, however, be the case. First, there is the possibility of some sort of climate change, implying that the historical data are not representative of the rainfall process as observed in 1999. If this were the case, then any analysis based solely on the historical data would inevitably fail to anticipate the exceptional 1999 value. However, there is no evidence of any change in the years preceding 1999, nor any strong meteorological support for a sudden change, so it seems not unreasonable to assume that the 1999 rainfall process did have identical stochastic structure to that of earlier years.

Another possibility is simply that extreme value models are hopelessly inadequate for modeling this process, at least at the levels to which they are being applied. Suppose, for example, that with probability $\lambda \approx 1$ the daily rainfall is generated from a distribution function G_1, but with probability $1 - \lambda > 0$ the daily rainfall is derived from a distribution function with a much heavier tail, G_2 say. This corresponds to a mixture model in which the overall distribution of daily rainfall is given by

$$G(x) = \lambda G_1(x) + (1 - \lambda)G_2(x). \tag{2.9}$$

Now, because G_2 has a much heavier tail than G_1, asymptotic models for G will be determined by G_2. Intuitively, in the long run we will expect heavier rainfall from the G_2 population, even if the occurrence of such events is rare. But because of the rarity of such events, it may be that an observed history of the process consists of no events generated from the G_2 population. If this were the case, no analysis of the historical data would give clues about the G_2 distribution, and extrapolations based on modeling the data from G_1 would be bound to underestimate extreme return levels. In the case of the Venezuelan rainfall data, the arguments for a model structure of the type (2.9) seem compelling after the 1999 event, though there seemed no need for such complexity prior to that.

What we aim to have shown in this chapter is that careful implementation of a single-population extreme value model would also have signaled concern about the plausible magnitude of future extreme events. To partially quantify this, based on the pre1999 data, the 1999 event had a 17.6 million year return period according to the fitted Gumbel model. Based on the same data, our final model attaches a probability of around 30% to the event comprising an occurrence of a daily rainfall of the 1999 annual maximum magnitude at some time within a 49 year period, suggesting that the risk of catastrophic flood events was orders of magnitude greater than the adopted Gumbel model had implied.

Acknowledgments

The Venezuelan case study on which much of this chapter is based was analyzed jointly with Luis Pericchi of the University of Puerto Rico. The data were kindly made available by J.R. Córdoba and M. González. Thanks are also due to a referee whose detailed comments have led to numerous improvements.

Bibliography

Coles, S. G. (2001). *An Introduction to Statistical Modeling of Extreme Values*. London: Springer.

Coles, S. G. and L. Pericchi (2001). Anticipating catastrophes through extreme value modeling. To appear in *Applied Statistics*.

Coles, S. G. and J. A. Tawn (1996). A Bayesian analysis of extreme rainfall data. *Applied Statistics 45*, 463–478.

Cox, D. R. and D. V. Hinkley (1974). *Theoretical Statistics*. London: Chapman & Hall.

Davison, A. C. and R. L. Smith (1990). Models for exceedances over high thresholds (with discussion). *Journal of the Royal Statistical Society, B 52*, 393–442.

Fisher, R. A. and L. H. C. Tippett (1928). Limiting forms of the frequency distributions of the largest or smallest member of a sample. *Proceedings of the Cambridge Philosophical Society 24*, 180–190.

Gilks, W. R., S. Richardson, and D. J. Spiegelhalter (1996). *Markov Chain Monte Carlo in Practice*. London: Chapman & Hall.

Gnedenko, B. V. (1943). Sur la distribution limite du terme maximum d'une série aléatoire. *Annals of Mathematics 44*, 423–453.

Jenkinson, A. F. (1955). The frequency distribution of the annual maximum (or minimum) values of meteorological events. *Quarterly Journal of the Royal Meteorological Society 81*, 158–172.

Larsen, M. C., M. T. V. Conde, and R. A. Clark (2001). Flash-flood related hazards: landslides, with examples from the December, 1999 disaster in Venezuala. In E. Gruntfest and J. Handmer (Eds.), *Coping with Flash Floods*, pp. 259–275. Dordrecht: Kluwer.

Leadbetter, M. R., G. Lindgren, and H. Rootzén (1983). *Extremes and Related Properties of Random Sequences and Series*. New York: Springer Verlag.

Pickands, J. (1975). Statistical inference using extreme order statistics. *Annals of Statistics 3*, 119–131.

Smith, R. L. (1985). Maximum likelihood estimation in a class of nonregular cases. *Biometrika 72*, 67–90.

Smith, R. L. (1989). Extreme value analysis of environmental time series: an application to trend detection in ground-level ozone (with discussion). *Statistical Science 4*, 367–393.

von Mises, R. (1954). La distribution de la plus grande de *n* valeurs. In *Selected Papers, Volume II*, pp. 271–294. Providence, RI: American Mathematical Society.

Weissman, I. (1984). Statistical estimation in extreme value theory. In J. Tiago de Oliveira (Ed.), *Statistical Extremes and Applications*, pp. 109–115. Dordrecht: Reidel.

CHAPTER 3

Risk Management with Extreme Value Theory

Claudia Klüppelberg
Munich University of Technology

Contents

In this chapter we review certain aspects around the Value-at-Risk, which is nowadays the industry benchmark risk measure. As a small quantile (usually 1%) the Value-at-Risk is closely related to extreme value theory, and we explain an estimation method based on this theory. Since the variance of the estimated Value-at-Risk may depend on the dependence structure of the data, we investigate the extreme behaviour of some of the most prominent time series models in finance, continuous as well as discrete time models. We also determine optimal portfolios, when risk is measured by the Value-at-Risk. Again we use realistic financial models, moving away from the traditional Black-Scholes model to the class of Lévy processes.

1-58488-411-8/04/$0.00+$.50

3.1 Introduction

In today's financial world, Value-at-Risk has become the benchmark risk measure. Following the *Basle Accord on Market Risk* (1988, 1995, 1996) every bank in more than 100 countries around the world has to calculate its risk exposure for every individual trading desk. The standard method prescribes: estimate the p-quantile of the profit/loss distribution for the next 10 days and $p = 1\%$ (or $p = 5\%$) based on observations of at least 1 year (220 trading days). Standard model is the normal model. Finally, multiply the estimated quantile by 3. This number is negative and its modulus is called Value-at-Risk (VaR). The factor 3 is supposed to account for certain observed effects, also due to the model risk; it is based on backtesting procedures and can be increased by the regulatory authorities, if the backtesting proves the factor 3 to be insufficient. The importance of VaR is undebated since regulators accept this model as a basis for setting capital requirements for market risk exposure. A textbook treatment of VaR is given in Joriot [50]. Interesting articles on risk management are collected in Embrechts [32].

There were always discussions about the classical risk measure, which has traditionally been the variance, and alternatives have been suggested. They are typically based on the notion of downside risk concepts such as lower partial moments. The lower partial moment of order n is defined as

$$\mathrm{LPM}_n(x) = \int_{-\infty}^{x} (x - r)^n dF(r), \quad x \in \mathbb{R},$$

where F is the distribution function of the portfolio return. Examples can be found in Fishburn [39] or Harlow [47] including the shortfall probability ($n = 0$), which is nothing else but the VaR. An axiomatic approach to risk measures can be found in Artzner et al. [1]; cf. Embrechts [31]. For some discussion see also Rootzén and Klüppelberg [76].

Standard model in the Basle account is the normal distribution which has the property that it is sum stable, i.e., for a dynamic model we obtain

$$\mathrm{VaR}(10 \text{ days}) = \sqrt{10}\,\mathrm{VaR}(1 \text{ day}),$$

and for a multivariate model; i.e., a portfolio with weights w_i for asset i and correlation ρ_{ij} between assets i and j, $i, j = 1, \ldots, q$,

$$\mathrm{VaR}(\text{portfolio}) = \sqrt{\sum_{i,j=1}^{q} \rho_{ij} w_i w_j \mathrm{VaR}_i \mathrm{VaR}_j}.$$

However, the obvious disadvantage of the normal model is that it is wrong and can dangerously underestimate the risk. This is even visible in Figure 3.1 and Figure 3.2.

This is the starting point of the present chapter. Taking also extreme fluctuations of financial data into account, we want to answer the following questions:

1. How does one estimate VaR from financial time series under realistic model assumptions?
2. What is the consequence of VaR as a risk measure based on a low quantile for portfolio optimization?

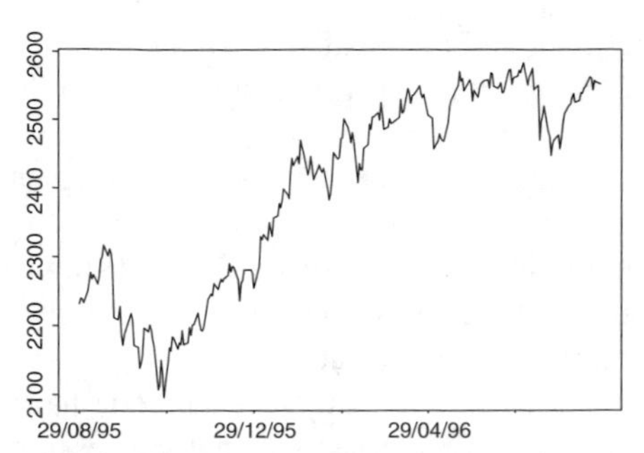

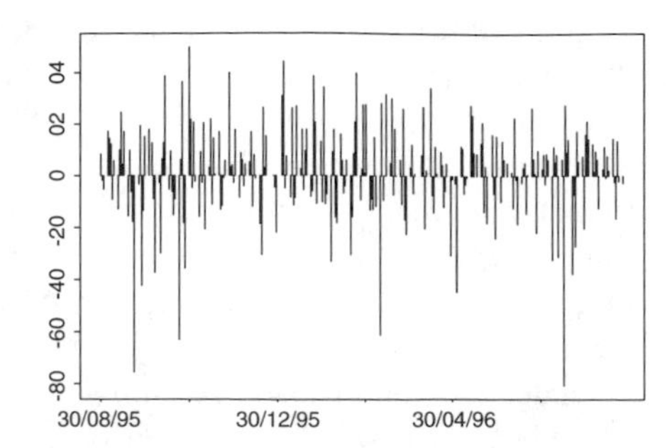

Figure 3.1 *DAX closing prices during 29/8/95 through 26/8/96 (250 data points in total). The corresponding differences, which are the daily price changes (returns), are plotted in the right-hand graph. It is obvious that the returns are not symmetric and that there are more pronounced peaks (in particular negative ones) than one would expect from Gaussian data.*

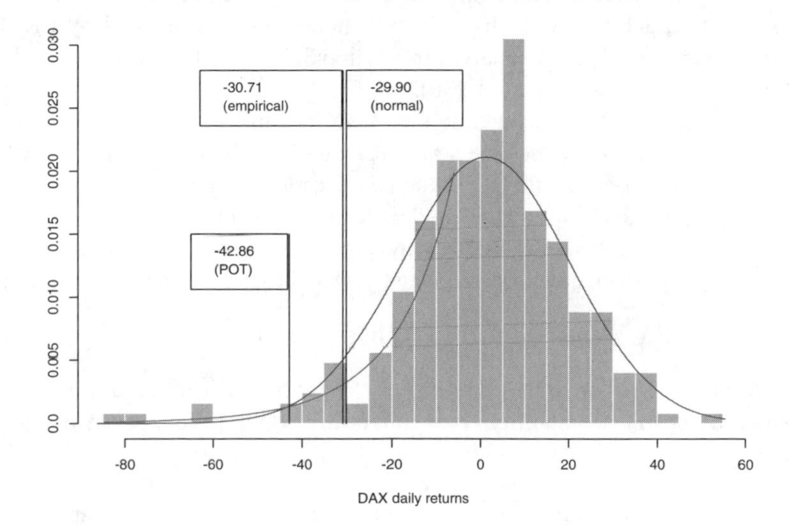

Figure 3.2 *Histogram of the daily price changes of the DAX closing prices with fitted normal distribution. Also fitted is a GPD distribution to the left-hand tail. The corresponding quantiles are estimated by the normal quantile, the GPD quantile, and the empirical quantile.*

Statistical estimation of risk and portfolio optimization are two important issues in risk management, influenced by the choice of risk measure. Pricing of derivatives and hedging of portfolios are other important issues, and the VaR has found its way also to the hedging problem. In incomplete markets, which is the setup for all "realistic" pricing models, the traditional hedge without risk (perfect hedge) has been replaced by a hedge with small remaining risk (so-called quantile-hedging); see Föllmer and Leukert [40] and Cvitanic and Karatzas [24]. This is, however, not a topic for this chapter.

We first turn to the risk estimation problem. In the simplest case, it is assumed that the only source of risk is the price of the portfolio itself, i.e., the risk is modeled in terms of price changes, which are independent and identically distributed (i.i.d.), the underlying planning horizon is $\Delta t = 1$ (1 day), and we estimate just the quantile (without multiplying by 3).

Generally speaking, estimation of a small quantile is not an easy task, as one wants to make inference about the extremal behavior of a portfolio, i.e., in an area of the sample where there is only a very small amount of data. Furthermore (and this is important to note), extrapolation even beyond the range of the data might be wanted, i.e., statements about an area where there are no observations at all.

Under the aphorism *let the tails speak for themselves*, statistical methods have been developed which are based only on that part of the sample that carries the information about the extremal behavior, i.e., only the smallest or largest sample values. This method is not solely based on the data, but includes a probabilistic argument concerning the behavior of the extreme sample values. This leads to a class of semiparametric distributions which can be regarded as plausible.

As a basic reference to modeling and quantifying of extreme events we refer to Embrechts, Klüppelberg, and Mikosch [33], henceforth abbreviated by EKM. The DAX data example, which we analyze in Section 3.2 can be found in greater detail in Emmer, Klüppelberg, and Trüstedt [36].

Unfortunately, most financial time series are not independent, but exhibit some very delicate temporal dependence structure, which is often captured by Markovian volatility models. Consequently, over the last decades a variety of stochastic models have been suggested as appropriate models for financial products.

In a continuous time setting the dynamics of a price or an interest rate process is often modeled as a diffusion process given by a stochastic differential equation (SDE)

$$dX_t = \mu(X_t)dt + \sigma(X_t)dW_t, \quad t > 0, \quad X_0 = x, \tag{3.1}$$

where W is standard Brownian motion, $\mu \in \mathbb{R}$ is the *drift term*, and $\sigma > 0$ is the diffusion coefficient or volatility. Two standard models in finance are of the above form:

1. The Black-Scholes model: $(X_t)_{t \geq 0}$ models the price process of an asset, here $\mu(x) = \mu x$ for $\mu \in \mathbb{R}$ and the volatility $\sigma(x) = \sigma x$ for $\sigma > 0$. The resulting model for the price process is geometric Brownian motion.
2. The Vasicek model: the process $(X_t)_{t \geq 0}$ models an interest rate, the drift term μ is linear, and the volatility $\sigma > 0$ is some constant.

Both models can be considered in the framework of Gaussian models; however, as indicated already, financial data exhibit in general fluctuations which cannot be modeled by Gaussian processes or simple transformations as in the two standard models above. In principle there are two different remedies for the problem.

A first concept sticks to Brownian motion as the driving dynamic of the process, but introduces a path-dependent, time-dependent, or even stochastic volatility into the model. These models are commonly referred to as volatility models, and include diffusions given by the SDE (3.1). We investigate their extremal behavior in Section 3.3.

The second concept replaces the Gaussian driving process in the Black-Scholes or Vasicek model (or any other traditional model) by a process with heavy-tailed marginals, as for instance a Lévy process with nonnormal noise. We consider this approach in Section 3.5 in the context of portfolio optimization.

A discrete time counterpart to (3.1) is the following model

$$X_n = \mu(X_{n-1}) + \sigma(X_{n-1})\,\varepsilon_n, \quad n \in \mathbb{N}, \tag{3.2}$$

where μ is the conditional mean, σ the conditional volatility, and $(\varepsilon_n)_{n\in\mathbb{N}}$ are i.i.d. random variables (rvs), with mean 0 and variance 1. Examples, also Markovian models of higher order, include for instance ARCH and GARCH models, which have been successfully applied in econometrics.

There is one stylized fact in financial data which models of the form (3.2) can capture in contrast to linear diffusion models of the form (3.1). This is the property of persistence in volatility. For many financial time series with high sampling frequency, large changes tend to be followed by large changes, settling down after some time to a more normal behavior. This observation has led to models of the form

$$X_n = \sigma_n\,\varepsilon_n, \quad n \in \mathbb{N}, \tag{3.3}$$

where the innovations ε_n are i.i.d. rvs with mean zero, and the volatility σ_n describes the change of (conditional) variance.

The autoregressive conditionally heteroscedastic (ARCH) models are one of the specifications of (3.3). In this case the conditional variance σ_n^2 is a linear function of the squared past observations. ARCH(p) models introduced by Engle [37] are defined by

$$\sigma_n^2 = \alpha_0 + \sum_{j=1}^{p} \alpha_j X_{n-j}^2, \quad \alpha_0 > 0, \alpha_1, \ldots, \alpha_{p-1} \geq 0, \alpha_p > 0, \quad n \in \mathbb{N}, \tag{3.4}$$

where p is the order of the ARCH process.

There are two natural extensions of this model. Bollerslev [12] proposed the so-called generalized ARCH (GARCH) processes. The conditional variance σ_n^2 is now a linear function of past values of the process X_{n-j}^2, $j = 1, \ldots, p$, and past values of the volatility σ_{n-j}^2, $j = 1, \ldots, q$. An interesting review article is Bollerslev, Chou, and Kroner [13]; a nice collection of some of the most influential articles on ARCH models is Engle [38].

The class of autoregressive (AR) models with ARCH errors introduced by Weiss [88] are another extension; these models are also called SETAR-ARCH models (self-exciting autoregressive). They are defined by

$$X_n = f(X_{n-1}, \ldots, X_{n-k}) + \sigma_n\,\varepsilon_n, \quad n \in \mathbb{N}, \tag{3.5}$$

where f is again a linear function in its arguments, and σ_n is given by (3.4). This model combines the advantages of an AR model, which targets more on the conditional mean of X_n (given the past), and of an ARCH model, which concentrates on the conditional variance of X_n (given the past).

The class of models defined by (3.5) embodies various nonlinear models. In this chapter we focus on the AR(1) process with ARCH(1) errors, i.e.,

$$X_n = \alpha X_{n-1} + \sqrt{\beta + \lambda X_{n-1}^2}\,\varepsilon_n, \quad n \in \mathbb{N},$$

where $\alpha \in \mathbb{R}$, $\beta, \lambda > 0$, and $(\varepsilon_n)_{n\in\mathbb{N}}$ are i.i.d. symmetric rvs with variance one, and X_0 is independent of $(\varepsilon_n)_{n\in\mathbb{N}}$. This Markovian model is analytically tractable and may serve as a prototype for the larger class of models (3.5). Note also that this model is of the form (3.2).

Two early monographs on extreme value theory for stochastic processes are Leadbetter, Lindgren, and Rootzén [62], henceforth abbreviated as LLR, and Berman [9]. They contain all basic results on this topic, and it is this source from which all specific results are derived.

The only models of the form (3.2), whose extremal behavior has been analyzed in detail are the ARCH(1) (by de Haan, Resnick, Rootzén, and de Vries [46], see also EKM [33], Section 8.4); the GARCH(1,1) (by Mikosch and Starica [68]); and the AR(1) model with ARCH(1)-errors (by Borkovec and Klüppelberg [18] and Borkovec [15, 16]). The interesting feature of all these models is that they are able to model heavy-tailedness as well as volatility clustering on a high level.

In Section 3.5 we turn to the second question posed at the beginning. We consider a portfolio optimization problem based on the VaR as a risk measure. Traditional portfolio selection as introduced by Markowitz [65] and Sharpe [81] has been based on the variance as risk measure. In contrast to the variance, the VaR captures the extreme risk. Consequently, it is to be expected that it reacts sensitively to large fluctuations in the data. This is what we investigate here.

We concentrate on the Capital-at-Risk (CaR) as a replacement of the variance in portfolio selection problems. We think of the CaR as the capital reserve in equity to set aside for future risk. The CaR of a portfolio is commonly defined as the difference between the mean of the profit-loss distribution and the VaR. We define the CaR as the difference between the mean wealth of the market (given by the riskless investment) and the VaR of our present portfolio; i.e., we consider the excess loss over the riskless investment.

We aim at closed form solutions and an economic interpretation of our results. This is why we start in a Gaussian world, represented by a Black-Scholes market, where the mean-CaR selection procedure leads to rather explicit solutions for the optimal portfolio. As a first difference to the mean-variance optimization, this approach indeed supports the commonly recommended market strategy that one should always invest in stocks for long-term investment.

As prototypes of models to allow for larger fluctuations than pure Gaussian models, we study Lévy processes, which still have independent and stationary increments, but these increments are no longer normally distributed. Such models have been used as more realistic models for price processes by Barndorff-Nielsen and Shephard [8], Eberlein and his group (see [27] and references therein) and Madan and Seneta [64]; they are meanwhile well understood. The class of normal mixture models supports the observation that volatility changes in time. This is in particular modeled by the normal inverse Gaussian model and the variance gamma model, which have also been recognized and applied in the financial industry. However, as soon as we move away from the Gaussian world, the optimization problem becomes analytically untractable and numerical solutions are called for. We present solutions for the normal mixture models mentioned above.

The data analyses, simulations, and figures presented have been created with the software S-Plus. Most routines for extreme value analysis are contained in the software EVIS written by Alex McNeil and can be downloaded from *http://www.math.ethz.ch/finance/*.

3.2 Starting kit for extreme value analysis

Let $X, X_1, \ldots, X_n$ be i.i.d. rvs, representing financial losses, with distribution function (df) F (we write $X \stackrel{d}{=} F$).

The classical central limit theorem states that for i.i.d. rvs such that $EX = \mu$ and $\mathrm{var}X = \sigma^2 < \infty$ the partial sums $S_n = X_1 + \cdots + X_n$, $n \in \mathbb{N}$, satisfy

$$\lim_{n\to\infty} P\left((S_n - n\mu)/\sqrt{n\sigma^2} \le x\right) = N(x), \quad x \in \mathbb{R},$$

where N is the standard normal df. This result, which holds in a much wider context than just i.i.d. data, supports the normal law for data that can be interpreted as sum or mean of many small effects, whose variance contributions are asymptotically negligible.

Consequently, the normal model is certainly questionable whenever extreme risk has to be quantified. Empirical investigations of financial data show quite clearly that the large values, in particular the large negative values, are much more pronounced than could be explained by a normal model.

In the following, we present the basic notions and ideas of extreme value theory for i.i.d. data. All this and much more can be found in EKM [33]; for more details on the DAX example we refer to Emmer, Klüppelberg, and Trüstedt [36].

3.2.1 *Sample maxima*

The simplest extreme object of a sample is the sample maximum. Define

$$M_1 = X_1, \quad M_n = \max(X_1, \ldots, X_n), \quad n > 1.$$

Then

$$P(M_n \le x) = F^n(x), \quad x \in \mathbb{R},$$

and $M_n \uparrow x_F$ as $n \to \infty$ almost surely, where $x_F = \sup\{x \in \mathbb{R} : F(x) < 1\} \le \infty$ is the right endpoint of F.

In most cases M_n can be normalized such that it converges in distribution to a limit rv, which together with the normalizing constants determines the asymptotic behavior of the sample maxima. The following is the analogue of the CLT for maxima.

Theorem 3.2.1 [Fisher-Tippett theorem]
Suppose we can find sequences of real numbers $a_n > 0$ and $b_n \in \mathbb{R}$ such that

$$\lim_{n\to\infty} P((M_n - b_n)/a_n \leq x) = \lim_{n\to\infty} F^n(a_n x + b_n \leq x) = Q(x), \quad x \in \mathbb{R}, \tag{3.6}$$

for some nondegenerate df Q (we say F is in the maximum domain of attraction of Q and write $F \in \mathrm{MDA}(Q)$). Then Q is one of the following three extreme value dfs:

$$\textit{Fréchet} \quad \Phi_\alpha(x) = \begin{cases} 0, & x \leq 0, \\ \exp(-x^{-\alpha}), & x > 0, \end{cases} \quad \textit{for} \quad \alpha > 0.$$

$$\textit{Gumbel} \quad \Lambda(x) = \exp(-e^{-x}), \quad x \in \mathbb{R}.$$

$$\textit{Weibull} \quad \Psi_\alpha(x) = \begin{cases} \exp(-(-x)^\alpha), & x \leq 0, \\ 1, & x > 0, \end{cases} \quad \textit{for} \quad \alpha > 0.$$

The limit distribution Q is unique up to affine transformations; this whole family is called of the type of Q.

All commonly encountered continuous df are in $\mathrm{MDA}(Q)$ for some extreme value df Q; see EKM [33], pp. 153–157. Here are three examples.

Example 3.2.2

(a) Exponential distribution: $F(x) = 1 - \exp(-\lambda x)$, $x \geq 0$, $\lambda > 0$, is in $\mathrm{MDA}(\Lambda)$ with $a_n = 1/\lambda$, $b_n = \ln n/\lambda$.

(b) Pareto distribution: $F(x) = 1 - \kappa^\alpha/(\kappa + x)^\alpha$, $x \geq 0$, $\kappa\alpha > 0$, is in $\mathrm{MDA}(\Phi_\alpha)$ with $a_n = (n/\kappa)^{1/\alpha}$, $b_n = 0$.

(c) Uniform distribution: $F(x) = x$, $x \in (0, 1)$, is in $\mathrm{MDA}(\Psi_1)$ with $a_n = 1/n$, $b_n = x_F = 1$.

Taking logarithms and invoking a Taylor expansion in (3.6) we obtain

$$F \in \mathrm{MDA}(Q) \Longleftrightarrow \lim_{n\to\infty} n\overline{F}(c_n x + d_n) = -\ln Q(x) =: \tau(x), \quad x \in \mathbb{R}, \tag{3.7}$$

where $\overline{F} = 1 - F$ denotes the tail of F. This MDA condition is a version of Poisson's limit theorem. It can be embedded in the more general theory of point processes as follows.

For i.i.d. rvs $X, X_1, \ldots, X_n$ and threshold u_n we have

$$\mathrm{card}\{i : X_i > u_n, i = 1, \ldots, n\} \stackrel{d}{=} \mathrm{Bin}(n, P(X > u_n)).$$

Define for $n \in \mathbb{N}$

$$N_n(B) = \sum_{i=1}^{n} \varepsilon_{i/n}(B) I\{X_i > u_n\}, \quad B \in \mathcal{B}(0, 1],$$

where $\mathcal{B}(0, 1]$ denotes the Borel σ-algebra on (0,1] and ε the Dirac measure; i.e., $\varepsilon_{i/n}(B) = 1$ if $i/n \in B$ and 0 else. Then N_n is the time normalized point process of exceedances.

The above equivalence (3.7) extends to the following result.

Proposition 3.2.3 *Suppose that $(X_n)_{n\in\mathbb{N}}$ is a sequence of i.i.d. rvs with common df F. Let $(u_n)_{n\in\mathbb{N}}$ be threshold values tending to x_F as $n\to\infty$. Then*

$$\lim_{n\to\infty} nP(X>u_n)=\tau\in(0,\infty) \Longleftrightarrow N_n \overset{d}{\to} N \quad \textit{Poisson process}(\tau), \qquad n\to\infty.$$

From this follows the asymptotic behavior of all upper order statistics, for instance,

$$P(M_n\le u_n)=P(N_n((0,1])=0)\to P(N((0,1])=0)=e^{-\tau}, \qquad n\to\infty.$$

3.2.2 Generalized extreme value distribution (GEV)

For statistical purposes all three extreme value distributions are summarized.

Definition 3.2.4 [Jenkinson-von Mises representation]

$$Q_\xi(x)=\begin{cases}\exp\{-(1+\xi x)^{-1/\xi}\} & \textit{if}\ \xi\ne 0,\\ \exp\{-e^{-x}\} & \textit{if}\ \xi=0,\end{cases}$$

where $1+\xi x>0$ and ξ is the shape parameter.

The GEV represents all three extremal types:

- $\xi>0$ Fréchet $\quad Q_\xi((x-1)/\xi)=\Phi_{1/\xi}(x)$,
- $\xi=0$ Gumbel $\quad Q_0(x)=\Lambda(x)$,
- $\xi<0$ Weibull $\quad Q_\xi(-(x+1)/\xi)=\Psi_{-1/\xi}(x)$.

Additionally, we introduce location and scale parameters $\mu\in\mathbb{R}$ and $\psi>0$ and define $Q_{\xi;\mu,\psi}(x)=Q_\xi(x-\mu)/\psi$. Note that $Q_{\xi;\mu,\psi}$ is of the type of Q_ξ.

This representation is useful for any statistical method that can be based on i.i.d. maxima. These are then modeled by the GEV and the parameters are fitted, leading to tail and quantile estimates; see EKM [33], Section 6.3. The method has its limitations, in particular, if the dependence structure cannot be embedded in an i.i.d. maxima model. Moreover, as for instance the method of annual maxima, it can also be a waste of data material, because it may only use annual maxima, but ignore all other large values of the sample. An excellent remedy, also for non i.i.d. data originates in the hydrology literature and has been developed and very successfully applied by Richard Smith and his collaborators for the last decades; see Smith [83] and references therein.

3.2.3 The POT–method (peaks over threshold)

We explain the POT-method and show it at work for the DAX closing prices of Figure 3.1. A superficial glimpse at the data shows some of the so-called stylized facts of financial data. There are more peaks than can be explained by a normal model and, in particular, the negative peaks are more pronounced than the positive ones. On the other hand, the data are simple in their dependence structure; an analysis of the autocorrelations of the data, their absolute values, and their squares give no indication of dependence. Consequently, we assume that the data are i.i.d. We want to remark, however, that many financial data are not i.i.d., but exhibit a very delicate dependence structure; see Section 3.3 and Section 3.4.

We proceed with a simple exploratory data analysis, which should stand at the beginning of every risk analysis. In a QQ-plot, empirical quantiles are plotted against

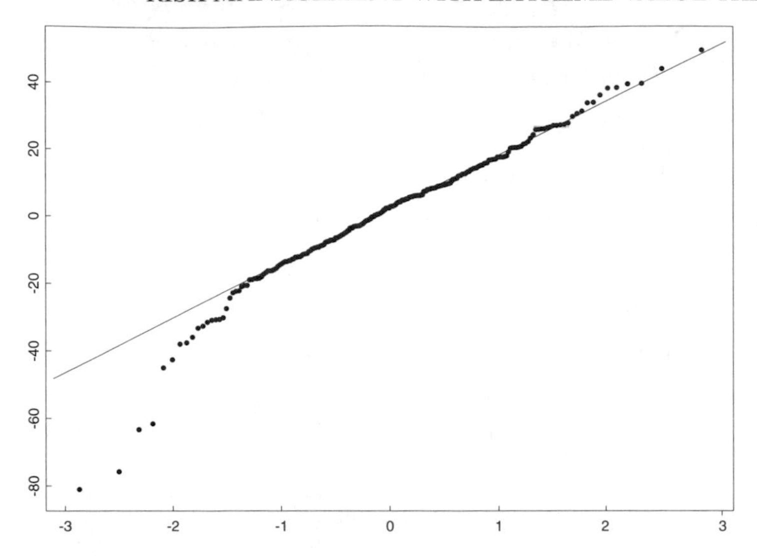

Figure 3.3 *Normal QQ-plot of the daily price changes of the DAX closing prices. The fit is very bad at the left end of the distribution, which is the region of our interest.*

the theoretical quantiles of a given distribution. If the chosen model is correct, nearly all data points will (if the sample size is large enough) lie on the 45-degree line. If the chosen distribution is correct up to its scale and location parameter, the plotted points will still be on a straight line, however with different slope and intersect. Linear regression gives rough estimates for the scale and location parameter, and these are often used as starting values for more sophisticated estimation methods. Figure 3.3 shows a normal QQ-plot of the data. The left end of the plot shows clearly that the left tail of the underlying distribution is much fatter than the left tail of a normal distribution.

Taking the modulus of the negative values of the given sample enables us to apply extreme value theory as introduced above to the left tail of the distribution of daily price changes. This is a sample of size $n = 108$ and will be the basis for the estimation of VaR.

One of the main ingredients of the POT-method is the following result.

Theorem 3.2.5 [Balkema and de Haan [3] and Pickands [70]]

$$F \in \mathrm{MDA}(H_\xi) \Longleftrightarrow \lim_{u \nearrow x_F} \frac{\overline{F}(u + x\beta(u))}{\overline{F}(u)} = \begin{cases} (1+\xi x)^{-1/\xi} & \text{if } \xi \neq 0, \\ e^{-x} & \text{if } \xi = 0, \end{cases}$$

where $1 + \xi x > 0$, *for some (positive measurable) function* $\beta(u)$.

Interpretation. For an rv X with df $F \in \mathrm{MDA}\left(H_\xi\right)$ we have

$$\lim_{u \uparrow x_F} P\left(\frac{X-u}{\beta(u)} > x \,\middle|\, X > u\right) = \begin{cases} (1+\xi x)^{-1/\xi} & \text{if } \xi \neq 0, \\ e^{-x} & \text{if } \xi = 0, \end{cases}$$

i.e., given X exceeds u, the scaled excess converges in distribution.

Definition 3.2.6 [Excess distribution function, mean excess function (MEF)] *Let* $X \stackrel{d}{=} F$ *be an rv with* $x_F \leq \infty$. *For fixed* $u < x_F$ *we call*

$$F_u(x) = P\left(X - u \leq x \mid X > u\right), \quad x + u \leq x_F,$$

the excess df of X or F over the threshold u. *The function*

$$e(u) = E\left[X - u \mid X > u\right] = \int_u^{x_F} \frac{\overline{F}(t)}{\overline{F}(u)} dt, \quad u < x_F, \tag{3.8}$$

is called mean excess function of X *or* F.

It is easy to calculate the mean excess function of an exponential distribution, which is a constant, equal to its parameter. The mean excess function of a distribution with a tail lighter than the tail of an exponential distribution tends to zero as u tends to infinity; for a distribution with tail heavier than exponential, the mean excess function tends to infinity; see Figure 6.2.4 of EKM [33].

Now let $X_1, \ldots, X_n$ denote the sample variables. As usual, $z^+ = \max(z, 0)$ denotes the positive part of z, and $\mathrm{card}A$ is the cardinality of the set A. The empirical function

$$e_n(u) = \frac{1}{\mathrm{card}\{i : X_i > u, i = 1, \ldots, n\}} \sum_{i=1}^{n} (X_i - u)^+, \quad u \geq 0,$$

estimates the mean excess function $e(u)$.

The right-hand side of Figure 3.4 shows the empirical mean excess function of the DAX data corresponding to the left tail. At first, the function is decreasing, but further to the right, it has an upward trend. This shows that in a neighborhood of zero, the data might possibly be modeled by a normal distribution, but this is certainly not the case in the left tail; there, the distribution turns out to have a tail that is clearly heavier than an exponential tail.

Theorem 3.2.5 motivates the following definition.

Definition 3.2.7 [generalized Pareto distribution (GPD)]

$$G_{\xi,\beta}(x) = \begin{cases} 1 - (1 + \xi x/\beta)^{-1/\xi} & \text{if } \xi \neq 0, \\ 1 - e^{-x/\beta} & \text{if } \xi = 0, \end{cases}$$

for $1 + \xi x > 0$. $\xi \in \mathbb{R}$ *is the* shape parameter *and* $\beta > 0$ *is the* scale parameter.

The GPD represents three different limit excess dfs:

- $\xi > 0$ Pareto with support $x \geq 0$,
- $\xi = 0$ exponential with support $x \geq 0$, and
- $\xi < 0$ Pareto with support $0 \leq x \leq -\beta/\xi$.

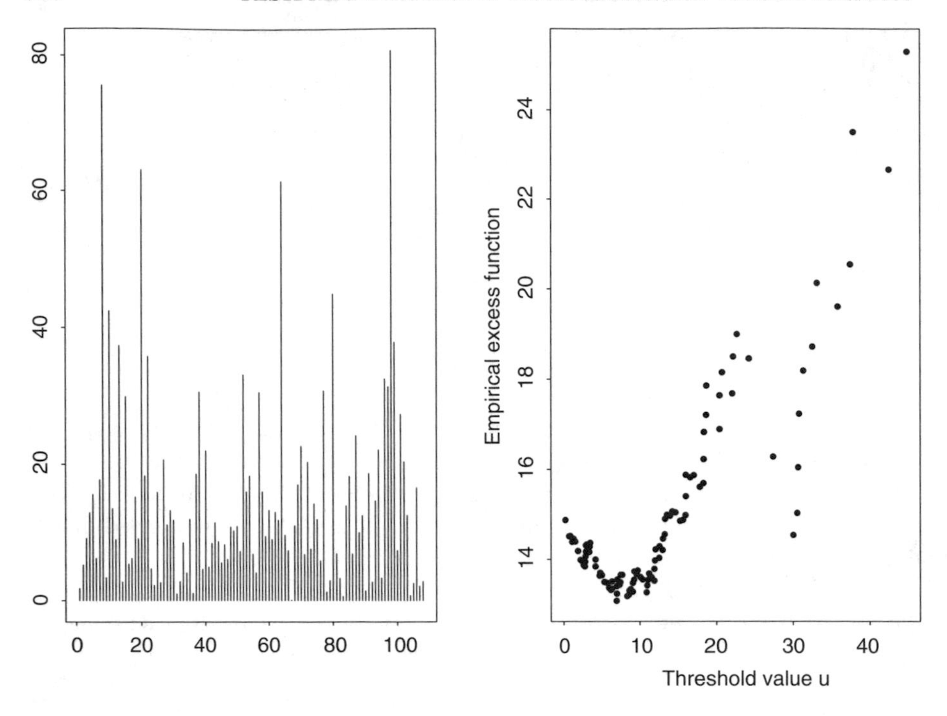

Figure 3.4 *The absolute negative price changes (left) and the corresponding empirical mean excess function (right) of the DAX values.*

These results are applied to model data above a high threshold u as follows:

(1) The point process of exceedances by a Poisson process(τ).

(2) The conditional excesses by a GPD(α, β).

(3) The stochastic quantities of (1) and (2) are independent.

3.2.4 *Estimating tails and quantiles by the POT-method*

Assume that $(X_n)_{n\in\mathbb{N}}$ are i.i.d. and $X_n \stackrel{d}{=} X \stackrel{d}{=} F$. For a high threshold u define

$$N_u = \text{card}\{i : X_i > u, \quad i = 1, \ldots, n\}.$$

Then

$$\overline{F}_u(y) = P(X - u > y \mid X > u) = \frac{\overline{F}(u+y)}{\overline{F}(u)}, \quad y \geq 0,$$

equivalently,

$$\overline{F}(u+y) = \overline{F}(u)\overline{F}_u(y), \quad y \geq 0. \tag{3.9}$$

Estimate $\overline{F}(u)$ and $\overline{F}_u(y)$ by the POT-method:

$$\widehat{\overline{F}(u)} = \frac{1}{n}\sum_{i=1}^{n} I(X_i > u) = \frac{N_u}{n}.$$

Approximate

$$\overline{F}_u(y) \approx \left(1 + \xi \frac{y}{\beta}\right)^{-1/\xi}, \quad y \in \mathbb{R},$$

and estimate ξ and β by $\widehat{\xi}$ and $\widehat{\beta}$ (see below). This results in the following tail and quantile estimates:

Tail estimate

$$\widehat{\overline{F}(u + y)} = \frac{N_u}{n}\left(1 + \widehat{\xi}\frac{y}{\widehat{\beta}}\right)^{-1/\widehat{\xi}}, \quad y \geq 0. \tag{3.10}$$

Quantile estimate

$$\widehat{x}_p = u + \frac{\widehat{\beta}}{\widehat{\xi}}\left(\left(\frac{n}{N_u}(1 - p)\right)^{-\widehat{\xi}} - 1\right), \quad p \in (0, 1). \tag{3.11}$$

A standard method to estimate the parameters ξ and β is maximum likelihood (ML) estimation. It is based on numerically maximizing the likelihood function for the given data, which are the excesses over a threshold u. However, one should bear in mind that the estimation procedure often relies on a very small data set as only the excesses will enter the estimation procedure. For this reason one cannot always rely on the asymptotic optimality properties of the ML-estimators and should therefore possibly use other estimation methods for comparison. For example, the classic Hill estimator could be used as an alternative approach. For a derivation and representation of the Hill estimator as well as a comparison to other tail estimators, see e.g., EKM [33], Chapter 6.

As already mentioned, the ML estimation is based on excess data, hence making it necessary to choose a threshold parameter u. A useful tool here is the plot of the empirical mean excess function in Figure 3.4. Recall that for heavy-tailed distributions the mean excess function in (3.8) tends to infinity. Furthermore it can be shown that for the generalized Pareto distribution, the mean excess function is a linear function (increasing if and only if the parameter ξ is positive). Hence, a possible choice of u is given by the value, above which the empirical mean excess function is approximately linear. Figure 3.4 indicates a reasonable choice of $u = 10$, with corresponding $N_u = 56$. This indicates that the generalized Pareto distribution is not only a good model for the extreme negative daily price changes, but already for about half of them. The ML-estimators are then found to be

$$\widehat{\xi} = 0.186, \qquad \widehat{\beta} = 11.120,$$

which enable us to estimate the lower 5% quantile of the daily price changes.

This estimator leads for the DAX data of Figure 3.1 to the following table.

	Empirical	Normal	GPD
VaR(1day, $p = 0.05$)	−30.654	−29.823	−42.856

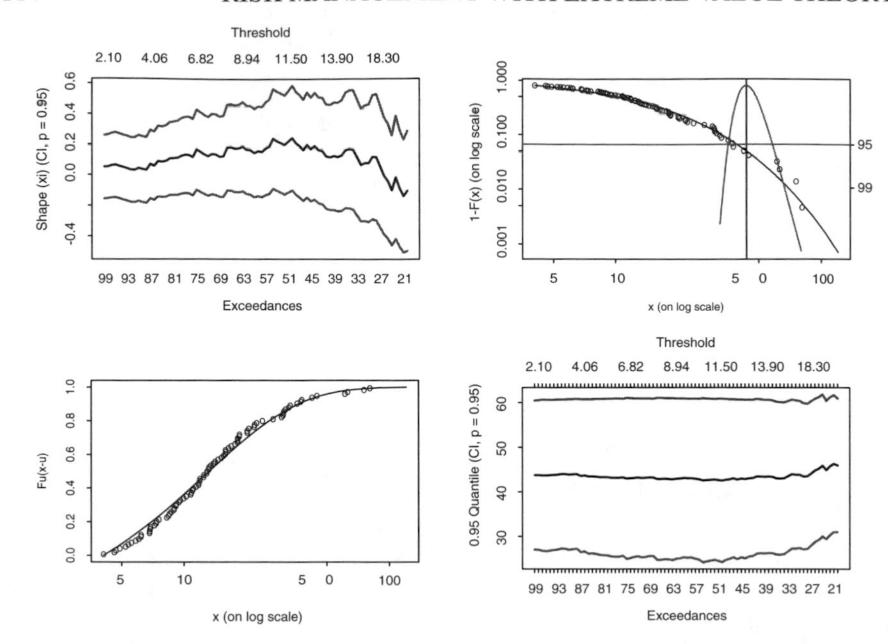

Figure 3.5 *Extreme value analysis of the data. The upper left-hand plot shows the estimated shape parameter* $\widehat{\xi}$ *with pointwise confidence intervals based on the normal asymptotics of the estimator, depending on different threshold values u. The upper right-hand plot shows the fit of the conditional df, and the lower left-hand plot the tail-fit of the DAX daily price changes: the plotted data points are the 56 largest absolute price changes; the solid curves show the estimated df and tail based on these data. In the lower left-hand plot, the vertical line marks the estimated 95% quantile, the curve above is the corresponding profile likelihood. The lower right-hand plot shows the estimated 95% quantile with pointwise confidence bounds depending on the threshold values.*

As is obvious from this table and Figure 3.5, estimation of the quantile by means of extreme value theory results in a much larger risk estimate as for the empirical and normal method.

The estimates fit the given data quite nicely, even in the far end tail. This confirms that the assumption of an underlying heavy tailed distribution is well in line with the data. In this context, the corresponding estimate of the lower 5% quantile of the VaR seems far more plausible than those obtained under the assumption of a normal distribution.

Confidence intervals for the estimated quantile can easily be obtained from the plotted profile likelihood. The 95% confidence interval can be read off the horizontal line. It is the interval [34.37, 60.51], i.e., with probability 0.95 the 95% quantile will lie in the interval [34.37, 60.51]. Not surprisingly, the confidence interval is rather wide, in particular to the right, where very few data are to be found. For the definition and mathematical properties of the profile likelihood we refer to Barndorff-Nielsen and Cox [6].

3.3 Continuous-time diffusion models

In this section, which is based on Borkovec and Klüppelberg [17], we study the extremal behavior of diffusion processes defined by the SDE

$$dX_t = \mu(X_t)dt + \sigma(X_t)dW_t, \quad t > 0, \quad X_0 = x, \tag{3.12}$$

where W is standard Brownian motion, $\mu \in \mathbb{R}$ is the drift term, and $\sigma > 0$ is the diffusion coefficient or volatility.

The stationary distributions of the processes under investigation are often well-known, and one might expect that they influence the extremal behavior of the process in some way. This is however not the case: for any predetermined stationary distribution the process can exhibit quite different behavior in its extremes.

Extremal behavior of a stochastic process $(X_t)_{t \geq 0}$ is as a first step manifested in the asymptotic behavior of the running maxima

$$M_t = \max_{0 \leq s \leq t} X_s, \quad t > 0.$$

The asymptotic distribution of M_t for $t \to \infty$ has been studied by various authors, see Davis [25] for detailed references.

It is remarkable that under quite natural conditions running maxima and minima of $(X_t)_{t \geq 0}$ given by (3.12) are asymptotically independent and have the same behavior as the extremes of i.i.d. rvs. We restrict ourselves to the investigation of maxima, the mathematical treatment for minima being similar.

The diffusion $(X_t)_{t \geq 0}$ given by the SDE (3.12) has state space $(l, r) \subseteq \mathbb{R}$. We only consider the case when the boundaries l and r are inaccessible and $(X_t)_{t \geq 0}$ is recurrent. We require furthermore for all $x \in (l, r)$ that $\sigma^2(x) > 0$ and that there exists some $\varepsilon > 0$ such that $\int_{x-\varepsilon}^{x+\varepsilon} (1 + |\mu(t)|)/\sigma^2(t)dt < \infty$. These two conditions guarantee in particular that the SDE (3.12) has a weak solution which is unique in probability; see Karatzas and Shreve [52], Chapter 5.5.C.

Associated with the diffusion is the scale function s and the speed measure m. The scale function is defined as

$$s(x) = \int_z^x \exp\left(-2 \int_z^y \frac{\mu(t)}{\sigma^2(t)} dt \right) dy, \quad x \in (l, r), \tag{3.13}$$

where z is any interior point of (l, r) whose choice, by the convergence to types theorem, does not affect the extremal behavior. For the speed measure m we know that $m(I) > 0$ for every nonempty open subinterval I of the interior of (l, r). We only consider diffusions with finite speed measure m and denote its total mass by $|m| = m((l, r))$. The speed measure of model (3.12) is absolutely continuous with Lebesgue density

$$m'(x) = \frac{2}{\sigma^2(x)s'(x)}, \quad x \in (l, r),$$

where s' is the Lebesgue density of s. In this situation $(X_t)_{t \geq 0}$ is ergodic, and its stationary distribution is absolutely continuous with Lebesgue density

$$h(x) = m'(x)/|m|, \quad x \in (l, r). \tag{3.14}$$

Notice that the connection between stationary distribution, speed measure, scale function, drift term, and diffusion coefficient (given by (3.13) and (3.14)) allows us to construct diffusions with arbitrary stationary distribution (see Example 3.3.6, Theorem 3.3.4 and Theorem 3.3.5).

Throughout this section, we assume that the diffusion process $(X_t)_{t\geq 0}$ defined in (3.12) satisfies the usual conditions, which guarantee that $(X_t)_{t\geq 0}$ is ergodic with stationary density (3.14):

$$s(r) = -s(l) = \infty \quad \text{and} \quad |m| < \infty. \tag{3.15}$$

For proofs of the above relations and further results on diffusions we refer to the monographs of Karatzas and Shreve [52], Revuz and Yor [75], Rogers and Williams [77], or any other advanced textbook on stochastic processes.

The following formulation can be found in Davis [25].

Proposition 3.3.1 *Let $(X_t)_{t\geq 0}$ satisfy the usual conditions (3.15). Then for any initial value $X_0 = y \in (l, r)$ and any $u_t \uparrow r$,*

$$\lim_{t\to\infty} |P^y(M_t \leq u_t) - F^t(u_t)| = 0,$$

where F is a df, defined for any $z \in (l, r)$ by

$$F(x) = \exp\left(-\frac{1}{|m|s(x)}\right), \quad x \in (z, r). \tag{3.16}$$

The function s and the quantity $|m|$ also depend on the choice of z.

Various proofs of this result exist, and we refer to Davis [25] for further references. Davis' proof is based on a representation of such a diffusion as an Ornstein-Uhlenbeck process after a random time-change. Standard techniques for extremes of Gaussian processes apply, leading to the above result. (The idea is explained in the proof of Theorem 3.3.8.)

As already noted the scale and speed measure of a diffusion $(X_t)_{t\geq 0}$ depend on the choice of z and hence, are not unique. Different scale and speed measures (and therefore different z) lead to different df's F in Proposition 3.3.1. They are, however, all tail-equivalent. This follows immediately by a Taylor expansion from (3.16) and the fact that $s(x) \to \infty$ as $x \uparrow r$.

Corollary 3.3.2 *Under the conditions of Proposition 3.3.1, the tail of the df F in (3.16) satisfies*

$$\overline{F}(x) \sim \left(|m| \int_z^x s'(y)dy\right)^{-1} \sim (|m|s(x))^{-1}, \quad x \uparrow r.$$

In particular, the tail behavior is independent of z.

The extremal behavior (in particular the behavior of the maximum) of an i.i.d. sequence with common df F is determined by the far end of the right tail $\overline{F}$. In our situation the asymptotic behavior of the maxima M_t is determined by the tail of F as in (3.16): if $F \in \text{MDA}(Q)$ with norming constants $a_t > 0$ and $b_t \in \mathbb{R}$, then

$$a_t^{-1}(M_t - b_t) \xrightarrow{d} Q, \quad t \to \infty. \tag{3.17}$$

The notion of regular variation is central in extreme value theory, and we refer to Bingham, Goldie, and Teugels [11], which we henceforth abbreviate by BGT.

Definition 3.3.3 [Regular variation]
A positive measurable function f on $(0, \infty)$ is regularly varying at ∞ with index α (we write $f \in \mathcal{R}(\alpha)$) if

$$\lim_{x\to\infty} \frac{f(tx)}{f(x)} = t^\alpha, \quad t > 0.$$

The following results describe the different behavior of diffusions (3.12) with stationary density h by the df F, which governs the extreme behavior.

Theorem 3.3.4 *Assume that the usual conditions (3.15) hold.*
(a) If $\mu \equiv 0$, then $S = (-\infty, \infty)$ and $\overline{F}(x) \sim cx^{-1}$ as $x \to \infty$ for some $c > 0$.
(b) Let μ and σ be differentiable functions in some left neighborhood of r such that

$$\lim_{x\to r} \frac{d}{dx} \frac{\sigma^2(x)}{\mu(x)} = 0 \quad \text{and} \quad \lim_{x\to r} \frac{\sigma^2(x)}{\mu(x)} \exp\left(-2\int_z^x \frac{\mu(t)}{\sigma^2(t)} dt\right) = -\infty,$$

then

$$\overline{F}(x) \sim |\mu(x)| h(x), \quad x \uparrow r.$$

Theorem 3.3.5 *Assume that the usual conditions (3.15) hold and $r = \infty$.*
(a) If $\sigma^2(x) \sim x^{1-\delta}\ell(x)/h(x)$ as $x \to \infty$ for some $\delta > 0$ and $\ell \in \mathcal{R}(0)$, then

$$\overline{F}(x) \sim \frac{\delta}{2} x^{-\delta}\ell(x), \quad x \to \infty.$$

(b) If $\sigma^2(x) \sim cx^{\delta-1}e^{-\alpha x^\beta}/h(x)$ as $x \to \infty$ for some $\delta \in \mathbb{R}$ and $\alpha, \beta, c > 0$, then

$$\overline{F}(x) \sim \frac{1}{2} c\alpha\beta x^{\delta+\beta-2} \exp(-\alpha x^\beta), \quad x \to \infty.$$

The following example describes the simplest way to construct a diffusion process with prescribed stationary density h.

Example 3.3.6 Define $dX_t = \sigma(X_t)dW_t$, $t > 0$, and $X_0 = x \in (l, r)$ and $\sigma^2(x) = \sigma^2/h(x)$ for $\sigma > 0$ and some density h. Then $\mu(x) = 0$, $s'(x) = 1$, and $(X_t)_{t\geq 0}$ has stationary density h. As a consequence of Theorem 3.3.4(a) this example has a very special extremal behavior, which is — independent of h — the same for all h.

Next we investigate an analogue of the Poisson process approximation for i.i.d. data; see Proposition 3.2.3. Since $(X_t)_{t\geq 0}$ has sample paths with infinite variation, we introduce a discrete skeleton in terms of a point process of so-called ε-upcrossings of a high threshold u by $(X_t)_{t\geq 0}$. For fixed $\varepsilon > 0$ the process has an ε-upcrossing at t if it has remained below u on the interval $(t - \varepsilon, t)$ and is equal to u at t. Under weak conditions, the point process of ε-upcrossings, properly scaled in time and space, converges in distribution to a homogeneous Poisson process, i.e., it behaves again like exceedances of i.i.d. rvs, coming however not from the stationary distribution of $(X_t)_{t\geq 0}$, but from the df F, which describes the growths of the running maxima M_t, $t > 0$ (see Proposition 3.3.1).

Definition 3.3.7 *Let $(X_t)_{t\geq 0}$ be a diffusion satisfying the usual conditions (3.15). Take $\varepsilon > 0$.*
*(a) The process $(X_t)_{t\geq 0}$ is said to have an ε–*upcrossing of the level u at $t_0 > 0$ *if*

$$X_t < u \quad \textit{for} \quad t \in (t_0 - \epsilon, t_0) \quad \textit{and} \quad X_{t_0} = u.$$

(b) For $t > 0$ let $N_{\varepsilon,u_t}(t)$ denote the number of ε-upcrossings of u_t by $(X_s)_{0\leq s\leq t}$. Then

$$N_t^*(B) = N_{\varepsilon,u_t}(tB) = \text{card}\left\{\varepsilon\text{-}upcrossings\ of\ u_t\ by\ (X_s)_{0\leq s\leq t} : \frac{s}{t} \in B\right\},\ B \in \mathcal{B}(0,1],$$

is the time-normalized point process of ε-upcrossings on the Borel sets $\mathcal{B}(0,1]$.

Immediately from the definition ε-upcrossings of a continuous time process, they correspond to exceedances of a discrete time sequence. As we know from Proposition 3.2.3 the point processes of exceedances of i.i.d. data converge weakly to a homogeneous Poisson process. Such results also hold for more general sequences provided the dependence structure is nice enough to prevent clustering of the extremes in the limit.

For diffusions (3.12) the dependence structure of the extremes is such that the point processes of ε-upcrossings converge to a homogeneous Poisson process; however, the intensity is not determined by the stationary df H, but by the df F from Proposition 3.3.1. This means that the ε-upcrossings of $(X_t)_{t\geq 0}$ are likely to behave as the exceedances of i.i.d. rvs with df F. The extra condition (3.18) of the following theorem relates the scale function s and speed measure m of $(X_t)_{t\geq 0}$ to the corresponding quantities s_{ou} and m_{ou} of the standard Ornstein-Uhlenbeck process, defined by

$$s_{ou}(x) = \sqrt{2\pi} \int_0^x e^{t^2/2} dt \quad \text{and} \quad m'_{ou}(x) = 1/s'_{ou}(x), \quad x \in \mathbb{R}.$$

Theorem 3.3.8 *Let $(X_t)_{t\geq 0}$ satisfy the usual conditions (3.15) and $u_t \uparrow r$ such that*

$$\lim_{t\to\infty} t\overline{F}(u_t) = \lim_{t\to\infty} \frac{t}{|m|s(u_t)} = \tau \in (0,\infty).$$

Assume there exists some positive constant c such that

$$\frac{m'_{ou}\left(s_{ou}^{-1}(s(z))\right)}{s'_{ou}\left(s_{ou}^{-1}(s(z))\right)} \frac{s'(z)}{m'(z)} \geq c, \quad \forall z \in (l,r). \tag{3.18}$$

Then for all starting points $y \in (l,r)$ of $(X_t)_{t\geq 0}$ and $\varepsilon > 0$ the time-normalized point processes N_t^* of ε-upcrossings of the levels u_t converge in distribution to N as $t \uparrow \infty$, where N is a homogeneous Poisson process with intensity τ on $(0,1]$.

Proof. The proof invokes a random time change argument. An application of Theorem 12.4.2 of LLR [62] shows that the theorem holds for the standard Ornstein-Uhlenbeck $(O_t)_{t\geq 0}$ process. Denote by

$$Z_t = s_{ou}(O_t), \quad t \geq 0, \quad \text{and} \quad Y_t = s(X_t), \quad t \geq 0,$$

the Ornstein-Uhlenbeck process and our diffusion, both in natural scale. $(Y_t)_{t\geq 0}$ can then be considered as a random time change of the process $(Z_t)_{t\geq 0}$; i.e., for all $t \geq 0$,

$$Y_t = Z_{\tau_t} \quad a.s.$$

for some stochastic process $(\tau_t)_{t\geq 0}$. The random time τ_t has a representation via the local time of the process $(Y_t)_{t\geq 0}$. This is a consequence of the Dambis-Dubins-Schwarz Theorem (Revuz and Yor [75], Theorem 1.6, p. 170), Theorem 47.1 of Rogers and Williams [77], p. 277 and Exercise 2.28 of [75], p. 230. For $z \in (l, r)$ denote by $L_t(z)$ the local time of $(Y_s)_{0\leq s\leq t}$ in z. Then by the occupation time formula (cf. Revuz and Yor [75], p. 209)

$$\tau_t = \int_{-\infty}^{\infty} L_t(z) dm_{ou}\left(s_{ou}^{-1}(z)\right) = \int_0^t \frac{m'_{ou}\left(s_{ou}^{-1}(s(X_s))\right)}{s'_{ou}\left(s_{ou}^{-1}(s(X_s))\right)} \frac{s'(X_s)}{m'(X_s)} ds, \quad t \geq 0.$$

Notice also that τ_t is continuous and strictly increasing in t; i.e., it defines a random time. Under condition (3.18) we obtain

$$\tau_t - \tau_{t-\varepsilon} \geq c\varepsilon, \quad t \geq 0.$$

Moreover, Itô and McKean [49], p. 228 proved the following ergodic theorem

$$\frac{\tau_t}{t} \stackrel{\text{a.s.}}{\rightarrow} \frac{1}{|m|}.$$

The following approximations can be made precise and imply Proposition 3.3.1.

$$\begin{aligned}
P\left(\max_{0\leq s\leq t} X_s > u_t\right) &= P\left(\max_{0\leq s\leq t} Y_s > s(u_t)\right) \\
= P\left(\max_{0\leq s\leq \tau_t} Z_s > s(u_t)\right) &\sim P\left(\max_{0\leq s\leq t/|m|} Z_s > s(u_t)\right) \\
\sim P\left(Z_s > s(u_t)^{t/|m|}\right) &\sim \left(\exp\left(-\frac{1}{s(u_t)}\right)\right)^{t/|m|} \\
= \exp\left(-\frac{t}{s(u_t)|m|}\right), &\qquad t \to \infty, \quad u_t \uparrow r.
\end{aligned}$$

For the point process convergence we use Theorem 4.7 of Kallenberg [51] and prove that for any $y \in (l, r)$

$$\lim_{t\to\infty} P^y\left(N_{\varepsilon,u_t}^X(tU) = 0\right) = P(N(U) = 0),$$

where U is an arbitrary union of semiopen intervals. □

Theorem 3.3.8 describes the asymptotic behavior of the number of ε-upcrossings of a suitably increasing level. In particular, on average there are τ ε-upcrossings of u_t by $(X_s)_{0\leq s\leq t}$ for large t. Notice furthermore, that we get a Poisson process in the limit which is independent of the choice of $\varepsilon > 0$.

The next lemma provides simple sufficient conditions, only on the scale function and the speed measure of $(X_t)_{t\geq 0}$, for (3.18). Notice that by positivity and continuity, (3.18) holds automatically on compact intervals. It remains to check this condition for z in a neighborhood of r and l.

Lemma 3.3.9 *Assume that for* $c_1, c_2 \in (0\,\infty]$

$$\frac{1}{\ln(|s(z)|)s(z)}\left(\frac{s''(z)}{s'(z)m'(z)} - \frac{m''(z)}{(m'(z))^2}\right) \longrightarrow \begin{matrix} c_1 & z\uparrow r, \\ c_2 & z\downarrow l, \end{matrix} \tag{3.19}$$

or (Grigelionis [45]) that for $d_1, d_2 \in (0, \infty]$

$$\frac{s^2(z)h(z)\ln(|s(z)|)}{s'(z)} \longrightarrow \begin{matrix} d_1 & z \uparrow r, \\ d_2 & z \downarrow l, \end{matrix} \tag{3.20}$$

then (3.18) holds.

In the following we investigate some examples that have been prominent in the interest rate modeling. All examples have a linear drift term

$$\mu(x) = c - dx, \quad x \in (l, r), \quad \text{for } c \in \mathbb{R}, d > 0,$$

which implies that the stationary version of $(X_t)_{t \geq 0}$ has mean c/d, provided it exists, and is mean reverting with force d. For financial background we refer to Lamberton and Lapeyre [61] or Merton [66].

Furthermore, $(X_t)_{t \geq 0}$ has state space $\mathbb{R}$ or $\mathbb{R}_+$, hence $F \in \text{MDA}(\Phi_\alpha)$ for some $\alpha > 0$ or $F \in \text{MDA}(\Lambda)$. Note that (3.17) implies that

$$\frac{M_t}{a_t} \xrightarrow{d} \Phi_\alpha \quad \text{if} \quad F \in \text{MDA}(\Phi_\alpha) \tag{3.21}$$

and

$$\frac{M_t - b_t}{a_t} \xrightarrow{d} \Lambda \quad \text{and} \quad \frac{M_t}{b_t} \xrightarrow{P} 1 \quad \text{if} \quad F \in \text{MDA}(\Lambda). \tag{3.22}$$

Figures 3.6 through 3.9 show simulated sample paths of the different models. For simulation methods of solutions of SDEs, see Kloeden and Platen [56]. The solid

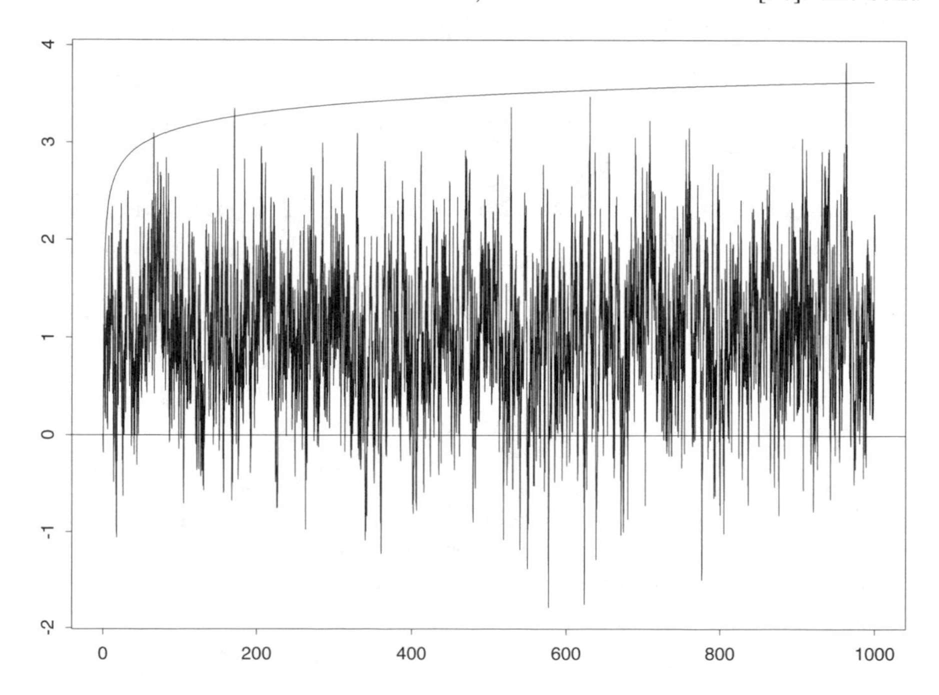

Figure 3.6 *Simulated sample path of the Vasicek model (with parameters $c = d = \sigma = 1$) and corresponding normalizing constants b_t.*

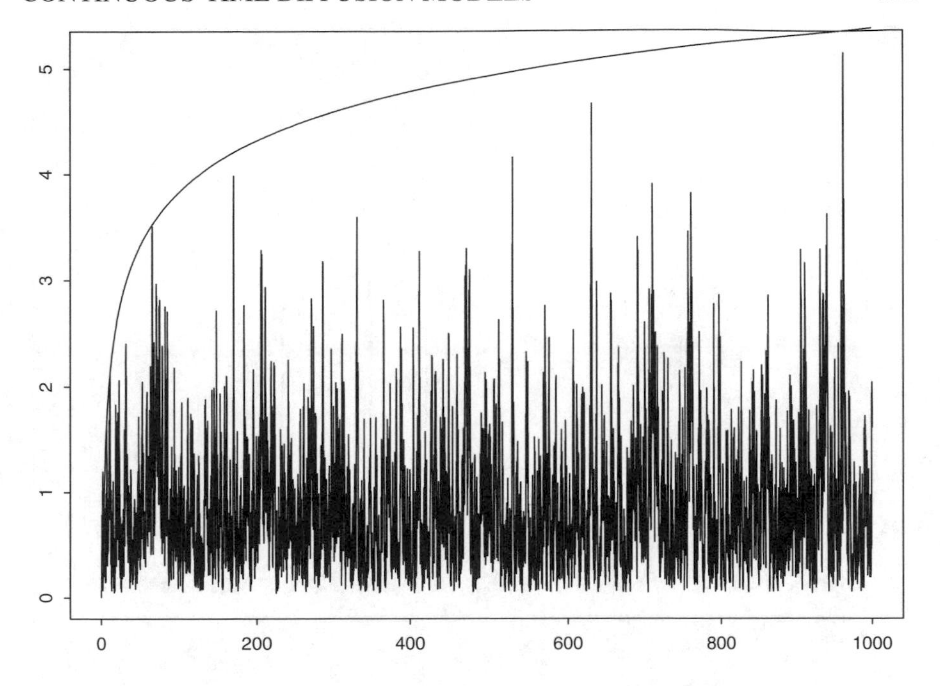

Figure 3.7 *Simulated sample path of the Cox–Ingersoll–Ross model with* $\mu(x) = c - dx$, $c \in \mathbb{R}, d > 0$, *and* $\sigma(x) = \sqrt{x}$. *(The chosen parameters are* $c = d = \sigma = 1$.*) The stationary distribution is a gamma distribution. The solid line shows the corresponding norming constants* b_t.

line indicates those norming constants which describe the increase of M_t for large t, i.e., in MDA(Φ_α) we plot a_t (see (3.21)) and in MDA(Λ) we plot b_t (see (3.22)).

Furthermore, all models in this section except the generalized Cox–Ingersoll–Ross model with $\gamma = 1$ satisfy condition (3.19) of Lemma 3.3.9, hence the Poisson approximation of the ε-upcrossings is also explicitly given for $u_t = a_t x + b_t$ and $\tau = -\ln Q(x)$, where Q is either Φ_α or Λ.

Example 3.3.10 [The Vasicek model (Vasicek [86])]
In this model the diffusion coefficient is $\sigma(x) \equiv \sigma > 0$. The solution of the SDE (3.12) with $X_0 = x$ is given by

$$X_t = \frac{c}{d} + \left(x - \frac{c}{d}\right)e^{-dt} + \sigma \int_0^t e^{-d(t-s)} dW_s, \quad t \geq 0.$$

$(X_t)_{t\geq 0}$ has state space $\mathbb{R}$, mean value, and variance function

$$EX_t = \frac{c}{d} + \left(x - \frac{c}{d}\right)e^{-dt} \to \frac{c}{d}, \quad t \to \infty,$$

$$\text{var}X_t = \frac{\sigma^2}{2d}(1 - e^{-2dt}) \to \frac{\sigma^2}{2d}, \quad t \to \infty.$$

It is well known and easy to calculate from (3.13) and (3.14) that $(X_t)_{t\geq 0}$ has a normal stationary distribution, more precisely, it is $N(\frac{c}{d}, \frac{\sigma^2}{2d})$, where $N(a, b)$ denotes the

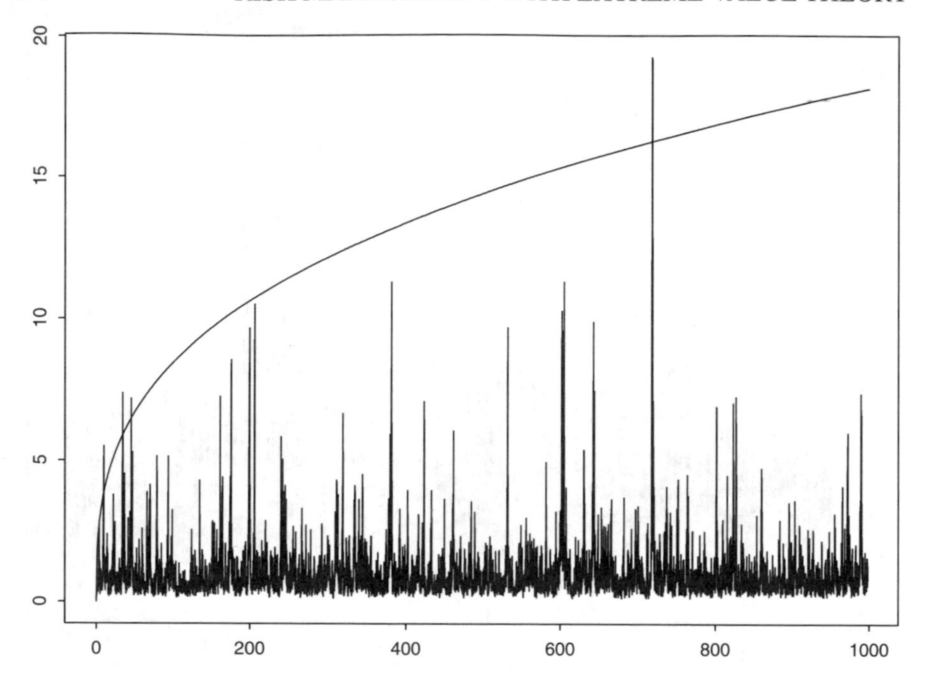

Figure 3.8 *Simulated sample path of the generalized Cox–Ingersoll–Ross model with* $\mu(x) = c - dx$, $c \in \mathbb{R}$, $d > 0$, *and* $\sigma(x) = x^{\gamma}$ *for* $\gamma = 1$. *(The chosen parameters are* $c = d = \sigma = 1$.*) The solid line shows the corresponding norming constants* b_t. *We can calculate* $\overline{F}(x) \sim C\overline{H}(x)$ *as* $x \to \infty$ *for some* $C > 0$.

normal distribution with mean a and variance b. The assumptions of Theorem 3.3.4(b) are satisfied giving

$$\overline{F}(x) \sim \frac{2d^2}{\sigma^2}\left(x - \frac{c}{d}\right)^2 \overline{H}(x), \quad x \to \infty,$$

where $\overline{H}(x)$ is the tail of the stationary normal distribution; hence F has heavier tail than H. It can be shown that $F \in \text{MDA}(\Lambda)$ with norming constants

$$a_t = \frac{\sigma}{2\sqrt{d \ln t}} \quad \text{and} \quad b_t = \frac{\sigma}{\sqrt{d}}\sqrt{\ln t} + \frac{c}{d} + \frac{\sigma}{4\sqrt{d}} \frac{\ln \ln t + \ln(\sigma^2 d/2\pi)}{\sqrt{\ln t}}.$$

Example 3.3.11 [The Cox–Ingersoll–Ross model (Cox, Ingersoll, and Ross [23])] In this model $\sigma(x) = \sigma\sqrt{x}$ for $\sigma > 0$ and $2c \geq \sigma^2$. It has state space $(0, \infty)$, for $X_0 = x$ it has mean value function

$$EX_t = \frac{c}{d} + \left(x - \frac{c}{d}\right)e^{-dt} \to \frac{c}{d}, \quad t \to \infty$$

and variance function

$$\text{var}X_t = \frac{c\sigma^2}{2d^2}\left(1 - \left(1 + \left(x - \frac{c}{d}\right)\frac{2d}{c}\right)e^{-2dt} + \left(x - \frac{c}{d}\right)\frac{2d}{c}e^{-3dt}\right) \to \frac{c\sigma^2}{2d^2}, \quad t \to \infty.$$

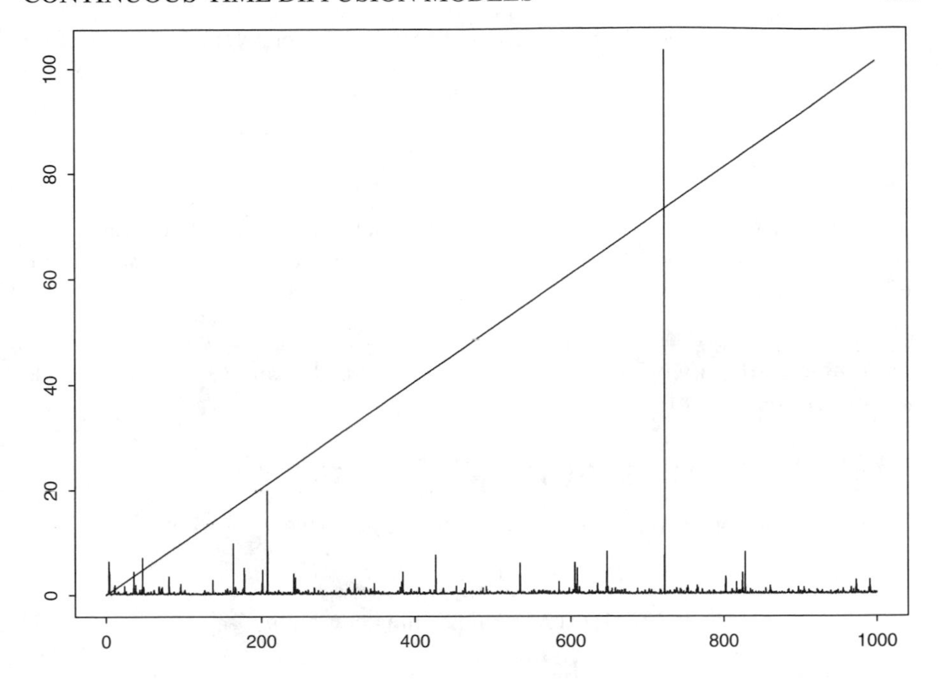

Figure 3.9 *Simulated sample path of the generalized Cox–Ingersoll–Ross model for* $\gamma = 1.5$ *(with parameters* $c = d = \sigma = 1$*) and the corresponding norming constants* a_t.

From (3.13) and (3.14) we obtain that the stationary distribution H is $\Gamma(\frac{2c}{\sigma^2}, \frac{2d}{\sigma^2})$. Theorem 3.3.4(b) applies giving

$$\overline{F}(x) \sim \frac{2cd}{\sigma^2}\overline{G}(x) \sim Ax\overline{H}(x), \quad x \to \infty.$$

where $A > 0$ and $\overline{G}(x)$ is the tail of the $\Gamma(\frac{2c}{\sigma^2}+1, \frac{2d}{\sigma^2})$ distribution. The gamma distributions are in MDA(Λ) and the norming constants for F are

$$a_t = \frac{\sigma^2}{2d} \quad \text{and} \quad b_t = \frac{\sigma^2}{2d}\left(\ln t + \frac{2c}{\sigma^2}\ln\ln t + \ln\left(\frac{d}{\Gamma(2c/\sigma^2)}\right)\right).$$

Example 3.3.12 [The generalized Cox–Ingersoll–Ross model]
In this model $\sigma(x) = \sigma x^\gamma$ for $\gamma \in [\frac{1}{2}, \infty)$. The process is ergodic with state space $(0, \infty)$.

We distinguish the following four cases:

$$\begin{array}{llll} \gamma = 1/2 & : \quad 2c \geq \sigma^2, & d > 0 & \text{(see Example 3.3.11)} \\ 1/2 < \gamma < 1: & \quad c > 0, & d \geq 0 & \\ \gamma = 1 & : \quad c > 0, & d > -\sigma^2/2 & \\ \gamma > 1 & : \quad c > 0,\ d \in \mathbb{R} \text{ or } & c = 0,\ d < 0. & \end{array} \tag{3.23}$$

For $\frac{1}{2} \le \gamma \le 1$ the mean value function of $(X_t)_{t\ge 0}$ is given by

$$EX_t = \begin{cases} \frac{c}{d} + \left(x - \frac{c}{d}\right) e^{-dt} \to \frac{c}{d} & \text{if } d > 0 \\ \frac{c}{d} + \left(x - \frac{c}{d}\right) e^{-dt} \to \infty & \text{if } d < 0 \\ x + ct \qquad\qquad \to \infty & \text{if } d = 0 \end{cases}$$

as $t \to \infty$ where $X_0 = x$. The lack of a first moment indicates already that for certain parameter values the model can capture very large fluctuations in data, which will reflect also in the behavior of the maxima.

$\frac{1}{2} < \gamma < 1$

The stationary density, which can be calculated by (3.13) and (3.14), is for some norming constant $A > 0$

$$h(x) = \frac{2}{A\sigma^2} x^{-2\gamma} \exp\left(-\frac{2}{\sigma^2}\left(\frac{c}{2\gamma-1}x^{-(2\gamma-1)} + \frac{d}{2-2\gamma}x^{2-2\gamma}\right)\right), \quad x > 0.$$

The assumptions of Theorem 3.3.4(b) are satisfied and hence

$$\overline{F}(x) \sim dxh(x) \sim Bx^{2(1-\gamma)}\overline{H}(x), \quad x \to \infty,$$

for some $B > 0$. Then $F \in \text{MDA}(\Lambda)$ with norming constants

$$a_t = \frac{\sigma^2}{2d}\left(\frac{\sigma^2(1-\gamma)}{d}\ln t\right)^{\frac{2\gamma-1}{2-2\gamma}}$$

$$b_t = \left(\frac{\sigma^2(1-\gamma)}{d}\ln t\right)^{\frac{1}{2-2\gamma}}\left(1 - \frac{2\gamma-1}{(2-2\gamma)^2}\frac{\ln\left(\frac{\sigma^2(1-\gamma)}{d}\ln t\right)}{\ln t}\right) + a_t \ln\left(\frac{2d}{A\sigma^2}\right).$$

$\gamma = 1$

In this case the solution of the SDE (3.12) with $X_0 = x$ is explicitly given by

$$X_t = e^{-(d+\frac{\sigma^2}{2})t+\sigma W_t}\left(x + c\int_0^t e^{(d+\frac{\sigma^2}{2})s-\sigma W_s}ds\right), \quad t \ge 0.$$

We obtain from (3.13) and (3.14) that the stationary density is inverse gamma:

$$h(x) = \left(\frac{\sigma^2}{2c}\right)^{-\frac{2d}{\sigma^2}-1}\left(\Gamma\left(\frac{2d}{\sigma^2}+1\right)\right)^{-1} x^{-2d/\sigma^2-2}\exp\left(-\frac{2c}{\sigma^2}x^{-1}\right), \quad x > 0.$$

Notice that $h \in \mathcal{R}(-2d/\sigma^2 - 2)$ and hence by Karamata's theorem (Theorem 1.5.11 of BGT [11]) the tail $\overline{H}$ of the stationary distribution is also regularly varying. This implies that certain moments are infinite:

$$\lim_{t\to\infty} EX_t^\delta = \begin{cases} \left(\frac{2c}{\sigma^2}\right)^\delta \frac{\Gamma\left(\frac{2d}{\sigma^2}+1-\delta\right)}{\Gamma\left(\frac{2d}{\sigma^2}+1\right)} & \text{if } \delta < \frac{2d}{\sigma^2}+1, \\ \infty & \text{if } \delta \ge \frac{2d}{\sigma^2}+1. \end{cases}$$

In particular,

$$\lim_{t\to\infty} \text{var} X_t = \begin{cases} \frac{2c^2}{d(2d-\sigma^2)} < \infty & \text{if } \frac{2d}{\sigma^2} > 1, \\ \infty & \text{if } -1 < \frac{2d}{\sigma^2} \le 1. \end{cases}$$

For the tail of F we obtain by Theorem 3.3.4(b)

$$\overline{F}(x) \sim Bx^{-2d/\sigma^2-1}, \quad x \to \infty,$$

for some $B > 0$. Hence $\overline{F} \in \mathcal{R}(-1-2d/\sigma^2)$, equivalently, $F \in \text{MDA}(\Phi_{1+2d/\sigma^2})$, with norming constants

$$a_t \sim C\, t^{1/(1+2d/\sigma^2)}, \quad t \to \infty,$$

for some $C > 0$.

$\gamma > 1$
Notice first that h is of the same form as in the case $\frac{1}{2} < \gamma < 1$, in particular $\overline{H} \in \mathcal{R}(-2\gamma+1)$ with $1-2\gamma < -1$)). We apply Theorem 3.3.5(a) and obtain for some $A > 0$

$$\overline{F}(x) \sim (Ax)^{-1}, \quad x \to \infty.$$

Hence $F \in \text{MDA}(\Phi_1)$ with norming constants $a_t \sim t/A$. Notice that the order of increase of a_t is always linear. The constant A, which depends on the parameters, decides about the slope.

3.4 The AR(1) model with ARCH(1) errors

In this section we study the extremal behavior of discrete time volatility models of the form

$$X_n = \mu(X_{n-1}) + \sigma(X_{n-1})\,\varepsilon_n, \quad n \in \mathbb{N},$$

where μ is the conditional mean, σ the conditional volatility, and $(\varepsilon_n)_{n\in\mathbb{N}}$ are i.i.d. symmetric rvs with variance 1.

As a prototype model, which can be treated analytically, we focus on the AR(1) process with ARCH(1) errors, i.e.,

$$X_n = \alpha X_{n-1} + \sqrt{\beta + \lambda X_{n-1}^2}\,\varepsilon_n, \quad n \in \mathbb{N}, \tag{3.24}$$

where $\alpha \in \mathbb{R}$, $\beta, \lambda > 0$, and $(\varepsilon_n)_{n\in\mathbb{N}}$ are i.i.d. symmetric rvs with variance 1, and X_0 is independent of $(\varepsilon_n)_{n\in\mathbb{N}}$. This section is based on Borkovec and Klüppelberg [18] and Borkovec [15]; see also [16].

Before we analyze model (3.24) we explain the influence of volatility clusters on a high level within the context of extreme value theory. We also show its consequences for risk management when estimating a high or low quantile.

We start with an introductory example, which we have found useful before.

Example 3.4.1 [EKM [33], Section 4.4, Section 5.5 and Section 8.1]
See Figure 3.10. Let $Y, Y_1, Y_2, \ldots$ be i.i.d. $Y \stackrel{d}{=} \sqrt{F}$, and define $X_n = \max(Y_n, Y_{n+1})$ for $n \in \mathbb{N}$. Then

$$P(X_n \le x) = (P(Y_n \le x))^2 = F(x), \quad x \in \mathbb{R}.$$

Choose u_n such that $nP(X_1 > u_n) \to \tau$ as $n \to \infty$, then $nP(Y_1 > u_n) \to \tau/2$ and

$$\begin{aligned} P(\max(X_1, \ldots, X_n) \le u_n) &= P(\max(Y_1, \ldots, Y_{n+1}) \le u_n) \\ &= P(\max(Y_1, \ldots, Y_n) \le u_n) F(u_n) \to e^{-\tau/2}, \quad n \to \infty. \end{aligned}$$

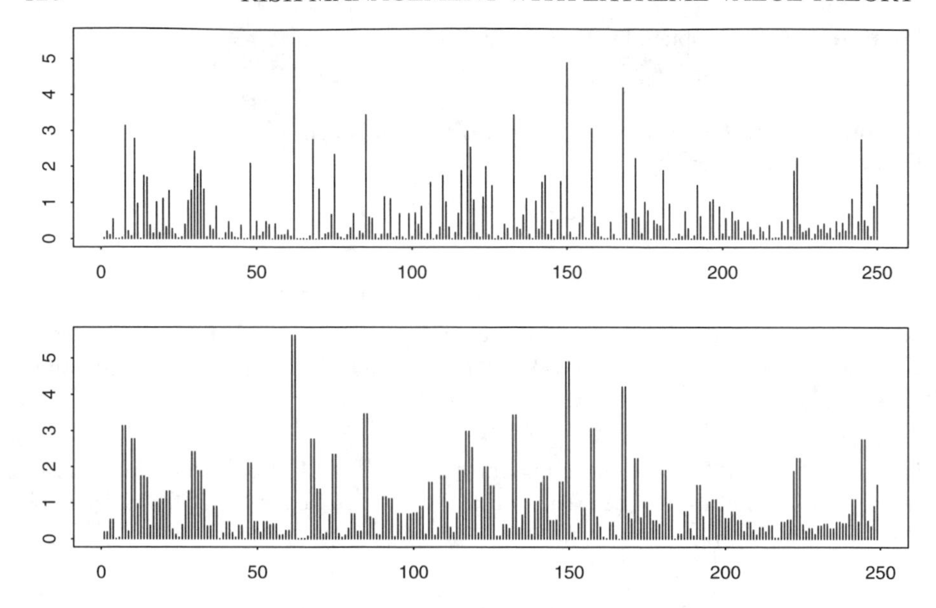

Figure 3.10 *A realization of the sequences* $(Y_n)_{n\in\mathbb{N}}$ *(top) and* $(X_n)_{n\in\mathbb{N}}$ *(bottom) with F standard exponential as discussed in Example 3.4.1.*

Definition 3.4.2 [Extremal index]
Let $(X_n)_{n\in\mathbb{N}}$ *be strictly stationary and define as before*

$$M_1 = X_1, \quad M_n = \max(X_1, \ldots, X_n), \quad n > 1.$$

Assume that for every $\tau > 0$ *there exists a sequence* $(u_n)_{n\in\mathbb{N}}$ *such that*

$$\lim_{n\to\infty} nP(X_1 > u_n) = \tau$$

$$\lim_{n\to\infty} P(M_n \leq u_n) = e^{-\theta\tau}$$

Then $\theta \in [0, 1]$ *is called the extremal index of* $(X_n)_{n\in\mathbb{N}}$.

The extremal index in Example 3.4.1 is $\theta = 1/2$. This indicates already the most intuitive interpretation of the extremal index: $1/\theta$ can be interpreted as the mean cluster size.

In the context of risk management we give an intuitive example.

Example 3.4.3 We want to calculate the VaR = VaR(10 days, $p = 0.05$) of a portfolio; i.e., for daily losses $X_i, i = 1, \ldots, 10$, we want to estimate $P(\max(X_1, \ldots, X_{10}) \leq \text{VaR}) = 0.95$. Assume that we know (1 Mio $= 10^6$)

$$\text{VaR}(1 \text{ day}, p = 0.01) = 10 \text{ Mio and VaR}(1\text{day}, p = 0.005) = 11 \text{ Mio}.$$

For the loss rv X this means that $P(X \leq 10) = 0.99$ and $P(X \leq 11) = 0.995$. Denote by $Z = \max(X_1, \ldots, X_{10})$. If the X_i are i.i.d., then

$$P(Z \leq 11) = P(X \leq 11)^{10} \approx 0.95,$$

whereas for dependent X_i with extremal index $\theta = 0.5$ we obtain

$$P(Z \le 10) = P(X \le 10)^{10/2} \approx 0.95.$$

This means that for i.i.d. data VaR(10 days, $p = 0.05$) is higher than for dependent data.

Using the block maxima method it is easy to compare VaR estimation for independent and dependent stationary financial time series. The data are divided into, say, N blocks, such that the corresponding block maxima can be considered as independent. Moreover, if the sample variables are in MDA(Q) for some extreme value distribution Q, then the block maxima, we call them $Z_1, \dots, Z_N$, can be viewed as an i.i.d. sample of rvs with df Q. Consequently, we assume that $Z_1, \dots, Z_N$ are i.i.d. GEV distributed; i.e., introducing a location parameter $\mu \in \mathbb{R}$ and a scale parameter $\psi > 0$,

$$P(Z \le \mathrm{VaR}(p)) \approx \exp\left(- \left(1 + \xi \frac{\mathrm{VaR}(p) - \mu}{\psi} \right)^{-1/\xi} \right). \tag{3.25}$$

Defining for given $p \in (0, 1)$ the VaR(p) by $1 - p = P(Z \le -\mathrm{VaR}(p))$, we obtain by inversion

$$\mathrm{VaR}(p) = \mu + \frac{\psi}{\xi}((- \ln(1 - p))^{-\xi} - 1).$$

By Definition 3.4.2 dependence introduces an additional factor θ in the exponent of (3.25) giving

$$\mathrm{VaR}(p) = \mu + \frac{\psi}{\xi}\left(\left(-\frac{1}{\theta} \ln(1 - p) \right)^{-\xi} - 1 \right).$$

In the context of risk management we expect $\xi \ge 0$ and for $\xi = 0$ we take the limit

$$\mathrm{VaR}(p) = \mu - \psi \ln \left(-\frac{1}{\theta} \ln(1 - p) \right).$$

A different method is a dependent version of the POT-method; i.e., the quantile estimate (3.11). Starting again with (3.9), the estimation of the tail in (3.10) changes, when $\overline{F}(u)$ is estimated. The empirical estimator N_u/n for i.i.d. data is replaced by $N_u^b/(n\widehat{\theta}_u)$, where N_u^b is the number of block maxima exceeding u and $\widehat{\theta}_u$ is the estimated extremal index; see EKM [33], Section 8.1 and references therein. For the quantile estimate (3.11) this means that

$$\widehat{x}_p = u + \frac{\widehat{\beta}}{\widehat{\xi}}\left(\left(\frac{n\widehat{\theta}_u}{N_u^b}(1 - p) \right)^{-\widehat{\xi}} - 1 \right), \quad p \in (0, 1).$$

3.4.1 Stationarity and tail behavior

In this section we present an extreme value analysis of the AR(1) process with ARCH(1) errors as given by (3.24). As a prerequisite we first need to know whether we are dealing with a stationary model and what the tail of the stationary distribution looks like.

For $\lambda = 0$ the process is an AR(1) process whose stationary distribution is determined by the innovations $(\varepsilon_n)_{n \in \mathbb{N}}$ and stationarity is guaranteed for $|\alpha| < 1$. In the

ARCH(1) case (the case when $\alpha = 0$) the process is geometric ergodic provided that $\beta > 0$ and $0 < \lambda < 2e^\gamma$, where γ is Euler's constant. The tail of the stationary distribution is known to be Pareto-like (see e.g., Goldie [41] or EKM [33], Section 8.4). This result was obtained by considering the square ARCH(1) process leading to a stochastic recurrence equation that fits into the setting of Kesten [53, 54] and Vervaat [87]; see also Diaconis and Freedman [26] for an interesting overview and Brandt, Franken, and Lisek [19]. Goldie and Maller [42] give necessary and sufficient conditions for stationarity of stochastic processes, which are solutions of stochastic recurrence equations.

For the general case we follow the standard procedure as, for instance, in the case $\alpha = 0$ to find the parameter region of stationarity of the process. For the tail behavior, however, we apply a technique, which differs completely from Kesten's renewal type arguments, by invoking the Drasin-Shea Tauberian theorem. This approach has the drawback that it ensures regular variation of the stationary tail, but gives no information on the slowly varying function. However, the method does apply to processes that do not fit into the framework of Kesten [53]. Moreover, the Tauberian approach does not depend on additional assumptions that are often very hard to check (as e.g., the existence of certain moments of the stationary distribution). Combining the Tauberian method with results in Goldie [41], we finally specify the slowly varying function as a constant.

We shall need the following assumptions on the noise variables. Denote by ε a generic rv with the same df G as ε_1. Throughout this section the following general conditions are in force:

- ε Is symmetric with variance 1.
- ε Is absolutely continuous with respect to Lebesgue measure with density g, which is positive on the whole of $\mathbb{R}$ and decreasing on $\mathbb{R}_+$. (3.26)

We summarize in Theorem 3.4.6 some properties of the process $(X_n)_{n\in\mathbb{N}}$. In particular, geometric ergodicity guarantees the existence and uniqueness of a stationary distribution. For an introduction to Markov chain terminology we refer to Tweedie [85] or Meyn and Tweedie [67].

The next proposition follows easily from well-known properties of moment generating functions (one can follow the proof of the case $\alpha = 0$; see e.g., Lemma 8.4.6 of EKM [33]).

Proposition 3.4.4 *Let ε be a rv with probability density g satisfying the general conditions (3.26). Define $h_{\alpha,\lambda} : [0,\infty) \to [0,\infty]$ for $\alpha \in \mathbb{R}$ and $\lambda > 0$ by*

$$h_{\alpha,\lambda}(u) := E[|\alpha + \sqrt{\lambda}\,\varepsilon|^u], \quad u \geq 0. \tag{3.27}$$

(a) The function $h_{\alpha,\lambda}(\cdot)$ is strictly convex in $[0, T)$, where

$$T := \inf\{u \geq 0 \mid E[|\sqrt{\lambda}\,\varepsilon|^u] = \infty\}.$$

(b) If furthermore the parameters α and λ are chosen such that

$$h'_{\alpha,\lambda}(0) = E[\ln|\alpha + \sqrt{\lambda}\,\varepsilon|] < 0, \tag{3.28}$$

then there exists a unique solution $\kappa = \kappa(\alpha, \lambda) > 0$ *to the equation* $h_{\alpha,\lambda}(u) = 1$. *Moreover, under* $h'_{\alpha,\lambda}(0) < 0$,

$$\kappa(\alpha, \lambda) \begin{cases} >2, & \text{if } \alpha^2 + \lambda E[\varepsilon^2] < 1, \\ =2, & \text{if } \alpha^2 + \lambda E[\varepsilon^2] = 1, \\ <2, & \text{if } \alpha^2 + \lambda E[\varepsilon^2] > 1. \end{cases}$$

Remark 3.4.5

(a) By Jensen's inequality $\alpha^2 + \lambda\, E[\varepsilon^2] < 1$ implies $h'_{\alpha,\lambda}(0) < 0$.

(b) Proposition 3.4.4 holds in particular for a standard normal rv ε. In this case $T = \infty$.

(c) In general, it is not possible to determine explicitly which parameters α and λ satisfy (3.28). If $\alpha = 0$ (i.e., in the ARCH(1)-case) and $\varepsilon \stackrel{d}{=} N(0, 1)$ (3.28) is satisfied if and only if $\lambda \in (0, 2e^{\gamma})$, where γ is Euler's constant (see e.g., EKM [33], Section 8.4). For $\alpha \neq 0$, Table 3.1 through Table 3.3 show numerical domains of α and λ; see Kiefersbeck [55] for more examples.

(d) Note that κ is a function of α and λ. Since ε is symmetric κ does not depend on the sign of α. For $\varepsilon \stackrel{d}{=} N(0, 1)$ we can show that for fixed λ the function κ is decreasing in $|\alpha|$. See also Table 3.1. □

Theorem 3.4.6 *Consider the process* $(X_n)_{n\in\mathbb{N}}$ *in (3.24) with* $(\varepsilon_n)_{n\in\mathbb{N}}$ *satisfying the general conditions (3.26) and with parameters* α *and* λ *satisfying (3.28). Then the following assertions hold:*

(a) Let ν *be the normalized Lebesgue-measure on the interval* $[-M, M] \subset \mathbb{R}$*; i.e.,* $\nu(\cdot) := \lambda(\cdot \cap [-M, M])/\lambda([-M, M])$*. Then* $(X_n)_{n\in\mathbb{N}}$ *is an aperiodic positive* ν*-recurrent Harris chain with regeneration set* $[-M, M]$ *for* M *large enough.*

(b) $(X_n)_{n\in\mathbb{N}}$ *is geometric ergodic. In particular,* $(X_n)_{n\in\mathbb{N}}$ *has a unique stationary distribution and satisfies the strong mixing condition with geometric rate of convergence. The stationary distribution is continuous and symmetric.*

Table 3.1 *Range of stationarity of the AR(1)+ARCH(1) model with parameters* α *and* λ*. The matrix components contain the estimated tail index* κ *for standard normal noise. There is no estimate given if the estimated* κ *is less than* 10^{-2} *or (3.28) is not satisfied.*

	λ											
$\lvert\alpha\rvert$	0.2	0.4	0.6	0.8	1.0	1.4	1.8	2.2	2.6	3.0	3.4	3.5
0.0	12.89	6.09	3.82	2.68	2.00	1.21	0.77	0.49	0.29	0.14	0.03	0.01
0.2	11.00	5.50	3.54	2.52	1.89	1.16	0.74	0.47	0.28	0.13	0.02	—
0.4	8.14	4.30	2.88	2.11	1.61	1.00	0.64	0.40	0.23	0.10	—	—
0.6	5.45	3.03	2.12	1.60	1.24	0.79	0.50	0.30	0.15	0.04	—	—
0.8	3.02	1.85	1.37	1.07	0.85	0.55	0.33	0.18	0.06	—	—	—
1.0	0.96	0.83	0.70	0.57	0.47	0.29	0.15	0.04	—	—	—	—
1.1	0.12	0.39	0.40	0.35	0.29	0.17	0.07	—	—	—	—	—
1.2	—	0.01	0.12	0.14	0.12	0.05	—	—	—	—	—	—

Table 3.2 *Range of stationarity of the AR(1)+ARCH(1) model with parameters α and λ. The matrix components contain the estimated tail index κ for student-t noise with 5 degrees of freedom. The range of stationarity has shrunk compared to the normal noise. Moreover, the corresponding tails are heavier than for normal noise (cf. Table 3.1).*

	λ									
$\lvert\alpha\rvert$	0.2	0.4	0.6	0.8	1.0	1.4	1.8	2.2	2.6	2.8
0.0	4.00	2.76	2.00	1.50	1.14	0.69	0.41	0.21	0.07	0.02
0.2	3.93	2.68	1.92	1.44	1.10	0.66	0.39	0.20	0.06	0.01
0.4	3.70	2.41	1.70	1.27	0.97	0.58	0.32	0.15	0.03	—
0.6	3.14	1.93	1.36	1.01	0.77	0.44	0.23	0.08	—	—
0.8	2.10	1.29	0.92	0.68	0.51	0.28	0.11	—	—	—
1.0	0.78	0.60	0.45	0.34	0.24	0.09	—	—	—	—
1.1	0.19	0.27	0.22	0.16	0.10	—	—	—	—	—
1.2	—	—	0.02	0.01	—	—	—	—	—	—

Table 3.3 *Range of stationarity of the AR(1)+ARCH(1) model with parameters α and λ. The matrix components contain the estimated and tail index for student-t noise with 3 degrees of freedom. The range of stationarity has further decreased and the tails have become very heavy indeed; a third moment does not exist (cf. Tables 3.1 and 3.2).*

	λ								
$\lvert\alpha\rvert$	0.2	0.4	0.6	0.8	1.0	1.4	1.8	2.2	2.4
0.0	2.43	1.80	1.35	1.02	0.78	0.45	0.23	0.08	0.01
0.2	2.41	1.76	1.31	0.99	0.75	0.43	0.21	0.06	0.01
0.4	2.31	1.62	1.18	0.88	0.66	0.36	0.16	0.02	—
0.6	2.06	1.35	0.96	0.70	0.51	0.26	0.09	—	—
0.8	1.50	0.93	0.64	0.45	0.32	0.12	—	—	—
1.0	0.59	0.41	0.28	0.18	0.10	—	—	—	—
1.1	0.13	0.15	0.10	0.04	—	—	—	—	—

Remark 3.4.7 When we study the stationary distribution of $(X_n)_{n\in\mathbb{N}}$ we may wlog assume that $\alpha \geq 0$. For a justification, consider the process $(\widetilde{X}_n)_{n\in\mathbb{N}} = ((-1)^n X_n)_{n\in\mathbb{N}}$ which solves the stochastic difference equation

$$\widetilde{X}_n = -\alpha \widetilde{X}_{n-1} + \sqrt{\beta + \lambda \widetilde{X}_{n-1}^2}\, \varepsilon_n, \quad n \in \mathbb{N},$$

where $(\varepsilon_n)_{n\in\mathbb{N}}$ are the same rvs as in (3.24) and $\widetilde{X}_0 = X_0$. If $\alpha < 0$, because of the symmetry of the stationary distribution, we may hence study the new process $(\widetilde{X}_n)_{n\in\mathbb{N}}$. □

In order to determine the tail of the stationary distribution F we need some additional technical assumptions on g and $\overline{G} = 1 - G$, the density and the distribution tail of ε:

D$_1$ The lower and upper Matuszewska indices of $\overline{H}$ are equal and satisfy in particular

$$-\infty \leq \gamma := \lim_{\nu\to\infty} \frac{\ln \limsup_{x\to\infty} \overline{H}(\nu x)/\overline{H}(x)}{\ln \nu}$$
$$= \lim_{\nu\to\infty} \frac{\ln \liminf_{x\to\infty} \overline{H}(\nu x)/\overline{H}(x)}{\ln \nu} \leq 0.$$

D$_2$ If $\gamma = -\infty$ then for all $\delta > 0$ there exist constants $q \in (0, 1)$ and $x_0 > 0$ such that for all $x > x_0$ and $t > x^q$

$$g\left(\frac{x \pm \alpha t}{\sqrt{\lambda t^2}}\right) \geq (1 - \delta)\, g\left(\frac{x \pm \alpha t}{\sqrt{\beta + \lambda t^2}}\right). \tag{3.29}$$

If $\gamma > -\infty$ then for all $\delta > 0$ there exist constants $x_0 > 0$ and $T > 0$ such that for all $x > x_0$ and $t > T$ the inequality (3.29) holds anyway.

The definition of the lower and upper Matuszewska indices can be found, e.g., in BGT [11], p. 68; for the above representation we used Theorem 2.1.5 and Corollary 2.1.6. The case $\gamma = -\infty$ corresponds to a tail which is exponentially decreasing. For $\gamma \in (-\infty, 0]$ condition D_1 is equivalent to the existence of constants $0 \leq c \leq C < \infty$ such that for all $\Lambda > 1$, uniformly in $\nu \in [1, \Lambda]$,

$$c(1 + o(1))\nu^\gamma \leq \frac{\overline{G}(\nu x)}{\overline{G}(x)} \leq C(1 + o(1))\nu^\gamma, \quad x \to \infty. \tag{3.30}$$

In particular, a distribution with a regularly varying tail satisfies D_1; the value γ is then the tail index. Due to the equality of the Matuszewska indices and the monotonicity of g we obtain easily some asymptotic properties of $\overline{G}$ and of g, respectively.

Proposition 3.4.8 *Suppose the general conditions (3.26) and $D_1 - D_2$ hold. Then the following holds:*

(a) $\lim_{x\to\infty} x^m \overline{G}(x) = 0$ *and* $E[|\varepsilon|^m] < \infty$ *for all* $m < -\gamma$.

(b) $\lim_{x\to\infty} x^m \overline{G}(x) = \infty$ *and* $E[|\varepsilon|^m] = \infty$ *for all* $m > -\gamma$.

(c) $\lim_{x\to\infty} x^{m+1} g(x) = 0$ *for all* $m < -\gamma$.

(d) If $\gamma > -\infty$, *there exist constants* $0 < c \leq C < \infty$ *such that*

$$c \leq \liminf_{x\to\infty} \frac{x\, g(x)}{\overline{G}(x)} \leq \limsup_{x\to\infty} \frac{x\, g(x)}{\overline{G}(x)} \leq C.$$

Moreover, there exist constants $0 \leq d \leq D < \infty$ *such that for all* $\Lambda > 1$, *uniformly in* $\nu \in [1, \Lambda]$,

$$d(1 + o(1))\nu^{\gamma-1} \leq \frac{g(\nu x)}{g(x)} \leq D(1 + o(1))\nu^{\gamma-1}, \quad x \to \infty. \tag{3.31}$$

Furthermore, in this case (3.31) is equivalent to (3.30) or D_1.

The general conditions (3.26) are fairly simple and can be checked easily, whereas D_1 and in particular D_2 seem to be quite technical and intractable. Nevertheless, numerous densities satisfy these assumptions.

Example 3.4.9 The following two families of densities satisfy the general conditions (3.26) and $D_1 - D_2$.

(a) $g_{\rho,\theta}(x) \propto \exp(-\theta^{-1}|x|^\rho)$, $x \in \mathbb{R}$, for $\rho, \theta > 0$.

Note that this family includes the Laplace (double exponential for $\rho = 1$) and the normal density with mean 0 ($\rho = 2$).

(b) $g_{a,\rho,\theta}(x) \propto (1+x^2/\theta)^{-(\rho+1)/2}(1+a \sin(2\pi \ln(1+x^2/\theta)))$, $x \in \mathbb{R}$, for parameters $\rho > 2$, $\theta > 0$ and $a \in [0, (\rho+1)/(\rho+1+4\pi))$.

This family includes, e.g., the Student-t distribution with parameter ρ (set $a = 0$ and $\theta = \rho$).

The following modification of the Drasin-Shea Theorem (BGT [11], Theorem 5.2.3, p. 273) is the key to our result.

Theorem 3.4.10 *Let $k : [0, \infty) \to [0, \infty)$ be an integrable function and let (a, b) be the maximal open interval (where $a < 0$) such that*

$$\widehat{k}(z) = \int_{(0,\infty)} t^{-z} k(t) \frac{dt}{t} < \infty, \quad \text{for } z \in (a, b).$$

If $a > -\infty$, assume $\lim_{\delta \downarrow 0} \widehat{k}(a+\delta) = \infty$, if $b < \infty$, assume $\lim_{\delta \downarrow 0} \widehat{k}(b-\delta) = \infty$. Let H be a df on $\mathbb{R}_+$ with tail $\overline{H}$. If

$$\lim_{x\to\infty} \int_{(0,\infty)} k\left(\frac{x}{t}\right) \frac{\overline{H}(t)}{\overline{H}(x)} \frac{dt}{t} = c > 0,$$

then

$$c = \widehat{k}(\rho) \text{ for some } \rho \in (a, b) \text{ and } \overline{H}(x) \sim x^\rho l(x), \quad x \to \infty,$$

where $l \in \mathcal{R}(0)$.

The following is the main theorem of this section.

Theorem 3.4.11 *Suppose $(X_n)_{n\in\mathbb{N}}$ is given by equation (3.24) with $(\varepsilon_n)_{n\in\mathbb{N}}$ satisfying the general conditions (3.26) and $D_1 - D_2$ and with parameters α and λ satisfying (3.28). Let $\overline{F}(x) = P(X > x)$, $x \ge 0$, be the right tail of the stationary distribution. Then*

$$\overline{F}(x) \sim c\, x^{-\kappa}, \quad x \to \infty, \tag{3.32}$$

where

$$c = \frac{1}{2\kappa} \frac{E[|\alpha|X| + \sqrt{\beta + \lambda X^2}\varepsilon|^\kappa - |(\alpha + \sqrt{\lambda}\varepsilon)|X||^\kappa]}{E[|\alpha + \sqrt{\lambda}\varepsilon|^\kappa \ln|\alpha + \sqrt{\lambda}\varepsilon|]}$$

and κ is given as the unique positive solution to

$$E[|\alpha + \sqrt{\lambda}\varepsilon|^\kappa] = 1. \tag{3.33}$$

Remark 3.4.12 (a) Let $E[|\alpha + \sqrt{\lambda}\varepsilon|^\kappa] = h_{\alpha,\lambda}(\kappa)$ be as in Lemma 3.4.4. Recall that for $\varepsilon \stackrel{d}{=} N(0,1)$ and fixed λ, the exponent κ is decreasing in $|\alpha|$. This means that the distribution of X gets heavier tails. In particular, the AR(1) process with ARCH(1) errors has for $\alpha \neq 0$ heavier tails than the ARCH(1) process (see also Table 3.1), and (b) Theorem 3.4.11, together with Proposition 3.4.4, implies that the second moment of the stationary distribution exists if and only if $\alpha^2 + \lambda\, E[\varepsilon^2] < 1$. □

Idea of Proof. Recall that $P(\varepsilon > x) = \overline{G}(x)$ with density g.

$$\begin{aligned}
\overline{F}(x) &= \int_{-\infty}^{\infty} P(\alpha t + \sqrt{\beta + \lambda t^2}\varepsilon > x) dF(t) \\
&= \int_0^{\infty} \left(\overline{G}\left(\frac{x + \alpha t}{\sqrt{\beta + \lambda t^2}} \right) + \overline{G}\left(\frac{x - \alpha t}{\sqrt{\beta + \lambda t^2}} \right) \right) dF(t) \\
&\sim \int_0^{\infty} \left(g\left(\frac{x + \alpha t}{\sqrt{\beta + \lambda t^2}} \right) + g\left(\frac{x - \alpha t}{\sqrt{\beta + \lambda t^2}} \right) \right) x\overline{F}(t)\frac{dt}{t} \\
&= \int_0^{\infty} k\left(\frac{x}{t}\right)\overline{F}(t)\frac{dt}{t},
\end{aligned}$$

where

$$k(x) = x\left(g\left(\frac{x+\alpha}{\sqrt{\lambda}}\right) + g\left(\frac{x-\alpha}{\sqrt{\lambda}}\right) \right), \quad x > 0,$$

then

$$\lim_{x\to\infty} \frac{1}{\overline{F}(x)} \int_0^{\infty} k\left(\frac{x}{t}\right)\overline{F}(t)\,\frac{dt}{t} = 1. \tag{3.34}$$

Define the transform

$$\begin{aligned}
\hat{k}(z) = \int_0^{\infty} t^{-z}k(t)\frac{dt}{t} &= \int_0^{\infty} t^{-z}\left(g\left(\frac{t+\alpha}{\sqrt{\lambda}}\right) + g\left(\frac{t-\alpha}{\sqrt{\lambda}}\right) \right) dt \\
&= E[|\,\alpha + \sqrt{\lambda}\varepsilon\,|^{-z}] < \infty, \quad z \in (-\infty, 1).
\end{aligned}$$

Since (3.34) holds, the conditions of the Drasin-Shea Tauberian theorem are satisfied. Hence there exists some $\rho \in (-\infty, 1)$ such that $\widehat{k}(z) = 1$ and

$$\overline{F}(x) \sim x^{\rho}\ell(x), \quad x \to \infty.$$

But $\widehat{k}(z) = E[|\alpha + \sqrt{\lambda}\varepsilon|^{-z}] = 1$ for $\rho = -\kappa$ and hence

$$\overline{F}(x) \sim x^{-\kappa}\ell(x), \quad x \to \infty. \tag{3.35}$$

We apply now Corollary 2.4 of Goldie [41] to the process $(Y_n)_{n\in\mathbb{N}}$ given by the stochastic recurrence equation

$$Y_n = \left|\alpha Y_{n-1} + \sqrt{\beta + \lambda Y_{n-1}^2}\right|, \quad n \in \mathbb{N}, \quad \text{and} \quad Y_0 = |X_0| \text{ a.s.},$$

which satisfies $(Y_n) \stackrel{d}{=} (|X_n|)$. By (3.35) $EY^{\kappa-1} < \infty$ and hence the moment condition

$$E|(|\alpha Y + \sqrt{\beta + \lambda Y^2}\varepsilon|)^{\kappa} - (|\alpha + \sqrt{\lambda}\varepsilon|Y)^{\kappa}| < \infty$$

requested in Goldie [41] is satisfied. By symmetry of X we conclude finally

$$\ell(x) = c = \frac{E[|\alpha|X| + \sqrt{\beta + \lambda X^2}\varepsilon|^{\kappa} - |(\alpha + \sqrt{\lambda}\varepsilon)|X||^{\kappa}]}{2\kappa E[|\alpha + \sqrt{\lambda}\varepsilon|^{\kappa} \ln|\alpha + \sqrt{\lambda}\varepsilon|]}.$$

□

3.4.2 Extreme value analysis

Theorems 3.4.6 and 3.4.11 are crucial for investigating the extremal behavior of $(X_n)_{n\in\mathbb{N}}$. The strong mixing property implies automatically that the sequence $(X_n)_{n\in\mathbb{N}}$ satisfies the conditions $D(u_n)$ and $\Delta(u_n)$. These conditions are frequently used mixing conditions in extreme value theory, which, as we do not need them explicitly, we will not define; instead we refer to Hsing, Hüsler, and Leadbetter [48] or Perfekt [69] for precise definitions. Loosely speaking, $D(u_n)$ and $\Delta(u_n)$ give the "degree of independence" of extremes situated far apart from each other. This property together with (3.32) implies that the maximum of the process $(X_n)_{n\in\mathbb{N}}$ belongs to the domain of attraction of a Fréchet distribution Φ_κ, where κ is given as solution to (3.33).

In the following denote by P^μ the probability law for $(X_n)_{n\in\mathbb{N}}$ when X_0 starts with distribution μ and π is the stationary distribution.

Theorem 3.4.13 [Borkovec [15]]
Let $(X_n)_{n\in\mathbb{N}}$ be the AR(1) *process with* ARCH(1) *errors (3.24) with noise satisfying the usual conditions and $D_1 - D_2$. Let $X_0 \stackrel{d}{=} \mu$, then*

$$\lim_{n\to\infty} P^\mu\left(n^{-1/\kappa}\max_{1\le j\le n} X_j \le x\right) = \exp(-c\theta x^{-\kappa}), \quad x \ge 0,$$

where κ solves the equation (3.33), c is the constant in the tail of the stationary distribution (3.32) and

$$\theta = \kappa \int_1^\infty P\left(\sup_{n\in\mathbb{N}} \prod_{i=1}^{n}(\alpha + \sqrt{\lambda}\varepsilon_i) \le y^{-1}\right) y^{-\kappa-1} dy.$$

For $x \in \mathbb{R}$ and $n \in \mathbb{N}$ let N_n be the point process of exceedances of the threshold $u_n = n^{1/\kappa}x$ by $X_1, \ldots, X_n$. Then

$$N_n \stackrel{d}{\to} N, \quad n \to \infty,$$

where N is a compound Poisson process with intensity $c\theta x^{-\kappa}$ and cluster probabilities

$$\pi_k = \frac{\theta_k - \theta_{k+1}}{\theta}, \quad k \in \mathbb{N},$$

with

$$\theta_k = \kappa \int_1^\infty P\left(\operatorname{card}\left\{n \in \mathbb{N} : \prod_{i=1}^{n}(\alpha + \sqrt{\lambda}\varepsilon_i) > y^{-1}\right\} = k-1\right) y^{-\kappa-1} dy.$$

In particular, $\theta_1 = \theta$.

We want to explain the idea of the proof: recall first from Theorem 3.4.6 that $(X_n)_{n\in\mathbb{N}}$ is Harris recurrent with regeneration set $[-e^{a/2}, e^{a/2}]$ for a large enough. Thus there exists a renewal point process $(T_n)_{n\ge 0}$ (e.g., the successive entrance times in $[-e^{a/2}, e^{a/2}]$), which describes the regenerative structure of $(X_n)_{n\in\mathbb{N}}$. This process $(T_n)_{n\ge 0}$ is aperiodic and has finite mean recurrence times.

Hence we can apply a coupling argument giving for any probability measure μ, the stationary distribution π and any sequence $(u_n)_{n\in\mathbb{N}}$

$$\left| P^{\mu}\left(\max_{1\le k\le n} X_k \le u_n\right) - P^{\pi}\left(\max_{1\le k\le n} X_k \le u_n\right)\right| \to 0, \quad n\to\infty.$$

Consequently, we suppose in the following that $(X_n)_{n\in\mathbb{N}}$ is stationary.

On a high level, the process $(X_n)_{n\in\mathbb{N}}$ can be linked to some random walk as follows. Define

$$S_0 = 0, \quad S_n = \sum_{i=1}^{n} \ln(\alpha + \sqrt{\lambda}\varepsilon_i), \quad n\in\mathbb{N}.$$

Although it is not as natural as for pure volatility models we consider besides $(X_n)_{n\in\mathbb{N}}$ also $(X_n^2)_{n\in\mathbb{N}}$. Define the auxiliary process $(Z_n)_{n\in\mathbb{N}} := (\ln(X_n^2))_{n\in\mathbb{N}}$, which satisfies the stochastic difference equation

$$Z_n = Z_{n-1} + \ln((\alpha + \sqrt{\beta\, e^{-Z_{n-1}} + \lambda}\,\varepsilon_n)^2), \quad n\in\mathbb{N}, \quad Z_0 = \ln\left(X_0^2\right) \quad \text{a.s.}$$

Note that, since strong mixing is a property of the underlying σ-algebra of the process, $(X_n^2)_{n\in\mathbb{N}}$ and $(Z_n)_{n\in\mathbb{N}}$ are also strong mixing. Since ε is symmetric the process $(Z_n)_{n\in\mathbb{N}}$ is independent of the sign of the parameter α. Hence we may wlog in the following assume that $\alpha \ge 0$.

We show that $(Z_n)_{n\in\mathbb{N}}$ can be bounded by two random walks $(S_n^{l,a})_{n\in\mathbb{N}}$ and $(S_n^{u,a})_{n\in\mathbb{N}}$ from below and above, respectively. For the construction of the two random walks $(S_n^{l,a})_{n\in\mathbb{N}}$ and $(S_n^{u,a})_{n\in\mathbb{N}}$ we define with the same notation as before

$$\begin{aligned}
A_a &:= \left\{ \frac{-\alpha}{\sqrt{\beta\, e^{-a} + \lambda} - \sqrt{\beta\, e^{-a/2}}} \le \varepsilon \le \frac{-\alpha}{\sqrt{\beta\, e^{-a} + \lambda} + \sqrt{\beta\, e^{-a/2}}} \right\}, \\
p(a,\varepsilon) &:= \ln((\alpha + \sqrt{\beta\, e^{-a} + \lambda\varepsilon})^2), \\
q(a,\varepsilon) &:= \ln\left(1 - \frac{2\alpha\sqrt{\beta} e^{-a/2}\varepsilon}{(\alpha + \sqrt{\beta\, e^{-a} + \lambda}\,\varepsilon)^2} 1_{\{\varepsilon<0\}}\right), \\
r(a,\varepsilon) &:= \ln\left(1 - \frac{\beta\varepsilon^2 e^{-a}}{(\alpha + \sqrt{\beta\, e^{-a} + \lambda}\,\varepsilon)^2} 1_{\{\varepsilon<0\}}\right).
\end{aligned}$$

Note that $q(a)$ and $r(a)$ both converge to 0 a.s. as $a\to\infty$. Define the lower and upper random walks

$$S_n^{l,a} := \sum_{j=1}^{n} U_j^a \quad \text{and} \quad S_n^{u,a} := \sum_{j=1}^{n} V_j^a, \quad n\in\mathbb{N}, \tag{3.36}$$

where for each $j = 1,\dots,n$

$$U_j^a := -\infty\cdot 1_{A_a} + (p(a,\varepsilon_j) + r(a,\varepsilon_j))1_{A_a^c\cap\{\varepsilon_j<0\}} \ln(\alpha + \sqrt{\lambda}\varepsilon)^2)\, 1_{\{\varepsilon_j\ge 0\}}, \tag{3.37}$$

$$V_j^a := p(a,\varepsilon_j) + q(a,\varepsilon_j). \tag{3.38}$$

The following lemma summarizes some properties of the random walks defined in (3.36) through (3.38).

Lemma 3.4.14 *Let a be large enough, $Z_0 > a$ and $N_a := \inf\{j \geq 1 \mid Z_j \leq a\}$. Then*

(a) $Z_0 + S_k^{l,a} \leq Z_k \leq Z_0 + S_k^{u,a}$ for all $k \leq N_a$ a.s.

(b) $(S_n^{u,a})_{n\in\mathbb{N}}$ and $(S_n^{l,a})_{n\in\mathbb{N}}$ are random walks with negative drift.

(c) Define $S_0 = 0$ and $S_k = \sum_{j=1}^{k} \ln((\alpha + \sqrt{\lambda}\varepsilon_j)^2)$ for $k \in \mathbb{N}$.

Then

$$S_k^{l,a} \xrightarrow{P} S_k \quad and \quad S_k^{u,a} \xrightarrow{a.s.} S_k, \quad a \uparrow \infty, \ and$$

(d) $\sup_{k\geq 1} S_k^{l,a} \xrightarrow{d} \sup_{k\geq 1} S_k$ *and* $\sup_{k\geq 1} S_k^{u,a} \xrightarrow{a.s.} \sup_{k\geq 1} S_k$ *as $a \uparrow \infty$.*

Lemma 3.4.14 characterizes the behavior of the process $(Z_n)_{n\in\mathbb{N}}$ above a high treshold a and hence also the behavior of $(X_n^2)_{n\in\mathbb{N}}$. This is the key to what follows: the process $(S_n)_{n\in\mathbb{N}}$ will completely determine the extremal behavior of $(X_n^2)_{n\in\mathbb{N}}$.

We shall assume strong mixing with a special rate of convergence. This will become clear in the next lemma. For a stochastic process $(X_n)_{n\in\mathbb{N}}$, for any $k, m \in \mathbb{N}$, suppose that for $A \in \sigma(X_l, l \leq m)$ and $B \in \sigma(X_l, l \geq m + k)$,

$$P(A \cap B) - P(A)P(b)| \leq \gamma(k).$$

Then γ is called a *mixing function* of $(X_n)_{n\in\mathbb{N}}$.

Lemma 3.4.15 *Let γ be the mixing function of $(X_n)_{n\in\mathbb{N}}$ and $(p_n)_{n\in\mathbb{N}}$ an increasing sequence such that*

$$\frac{p_n}{n} \to 0 \quad and \quad \frac{n\gamma(\sqrt{p_n})}{p_n} \to 0 \quad as\ n \to \infty. \tag{3.39}$$

Then for $u_n = n^{2/\kappa}x$, $x > 0$,

$$\lim_{p\to\infty} \limsup_{n\to\infty} P\left(\max_{p\leq j\leq p_n} X_j^2 > u_n \mid X_0^2 > u_n \right) = 0, \tag{3.40}$$

and for $u_n = n^{1/\kappa}x$, $x > 0$,

$$\lim_{p\to\infty} \limsup_{n\to\infty} P\left(\max_{p\leq j\leq p_n} X_j > u_n \mid X_0 > u_n \right) = 0. \tag{3.41}$$

Proof. The proof of (3.40) is very technical, and we refer to Borkovec [15] for details. It is, however, easy to see that (3.40) implies (3.41):

$$\begin{aligned} P\left(\max_{p\leq j\leq p_n} X_j^2 > u_n^2 \mid X_0^2 > u_n^2 \right) &= \frac{P\left(\max_{p\leq j\leq p_n} X_j^2 > u_n^2,\ X_0^2 > u_n^2 \right)}{P\left(X_0^2 > u_n^2\right)} \\ &\geq \frac{P(\max_{p\leq j\leq p_n} X_j > u_n,\ X_0 > u_n)}{P(X_0 > u_n) + P(X_0 < -u_n)} \\ &= \frac{1}{2} P\left(\max_{p\leq j\leq p_n} X_j > u_n \mid X_0 > u_n \right). \end{aligned}$$

□

Remark 3.4.16 (a) Since $(X_n)_{n\in\mathbb{N}}$ is geometric ergodic, the mixing function γ decreases exponentially fast, hence it is not difficult to find a sequence $(p_n)_{n\in\mathbb{N}}$ to satisfy

(3.39), (b) As mentioned already, the strong mixing condition is a property of the underlying σ-field of a process. Hence γ is also the mixing function of $(X_n^2)_{n\in\mathbb{N}}$, and $(Z_n)_{n\in\mathbb{N}}$, and we may work for all these processes with the same sequence $(p_n)_{n\in\mathbb{N}}$, and (c) In the case of a strong mixing process, conditions (3.39) are sufficient to guarantee that $(p_n)_{n\in\mathbb{N}}$ is a $\Delta(u_n)$-separating sequence. It describes somehow the interval length needed to accomplish asymptotic independence of extremal events over a high level u_n in separate intervals. For a definition see Perfekt [69]. Note that $(p_n)_{n\in\mathbb{N}}$ is in the case of a strong mixing process independent of $(u_n)_{n\in\mathbb{N}}$. □

The following theorem is an extension of Theorem 3.2 of Perfekt [69], p. 543 adapted to our situation.

Theorem 3.4.17 *Suppose $(X_n)_{n\in\mathbb{N}}$ is a strongly mixing stationary Markov chain whose stationary df F is symmetric with tail $\overline{F} \in \mathcal{R}(-\kappa)$ on $\mathbb{R}_+$. Suppose furthermore that*

$$\lim_{u\to\infty} P(X_1 \le xu \mid X_0 = u) = H(x), \quad x \in \mathbb{R},$$

for some df H. Let $(A_n)_{n\in\mathbb{N}}$ be an i.i.d. sequence with df H and define $Y_n = A_n Y_{n-1}$ for $n \in \mathbb{N}$ with Y_0 independent of $(A_n)_{n\in\mathbb{N}}$ and $Y_0 \stackrel{d}{=} \mu$ given by $\mu(dx) := \kappa^{-1}x^{-1/\kappa-1}dx$, for $x > 1$. For every $\tau > 0$ let $(u_n(\tau))_{n\in\mathbb{N}}$ be a sequence satisfying

$$\lim_{n\to\infty} n\overline{F}(u_n(\tau)) = \tau.$$

Then $(X_n)_{n\in\mathbb{N}}$ has extremal index θ given by

$$\theta = P^{\mu}(\text{card}\{n \in \mathbb{N} : Y_n > 1\} = 0).$$

Moreover, for $n \in \mathbb{N}$ the time normalized point process of exceedances

$$N_n^{\tau}(B) := \sum_{i=1}^{n} \varepsilon_{i/n}(\cdot) I\{X_k > u_n(\tau)\} \stackrel{d}{\to} N(B), \quad B \in \mathcal{B}(0, 1],$$

where N is a compound Poisson process with intensity $\theta\tau$ and jump probabilities $(\pi_k)_{k\in\mathbb{N}}$ given by

$$\pi_k = \frac{\theta_k - \theta_{k+1}}{\theta}, \quad k \in \mathbb{N},$$

where

$$\theta_k = P^{\mu}(\text{card}\{n \in \mathbb{N} : Y_n > 1\} = k - 1), \quad k \in \mathbb{N}.$$

Proof of Theorem 3.4.13 The proof is an application of Theorem 3.4.17. As stated already we may assume wlog that $(X_n)_{n\in\mathbb{N}}$ is stationary. Let $x \in \mathbb{R}$ be arbitrary. Note that by (3.32)

$$\lim_{u\to\infty} P(X_1 \le ux \mid X_0 = u) = P(\alpha + \sqrt{\lambda}\,\varepsilon \le x), \quad x \in \mathbb{R}.$$

$(X_n)_{n\in\mathbb{N}}$ satisfies all assumptions of Theorem 3.4.17 and we have the extremal index

$$\theta = \int_1^\infty P\left(\text{card}\left\{n \in \mathbb{N} : \left(\prod_{i=1}^n (\alpha + \sqrt{\lambda}\,\varepsilon_i)\right) Y_0 > 1\right\} = 0 \mid Y_0 = y\right) \kappa y^{-\kappa-1} dy$$

$$= \kappa \int_1^\infty P\left(\sup_{n\geq 1}\left(\prod_{i=1}^n (\alpha + \sqrt{\lambda}\,\varepsilon_i\right) \leq y^{-1}\right) y^{-\kappa-1} dy.$$

The cluster probabilities can be determined in the same way and hence the statement follows. □

Remark 3.4.18 (i) Notice that for the squared process the extremal index and the cluster probabilities can be described by the random walk $(S_n)_{n\in\mathbb{N}}$, namely

$$\theta_k^{(2)} = \frac{\kappa}{2}\int_0^\infty P(\text{card}\{n \in \mathbb{N} \mid S_n > -x\} = k-1)\, e^{-\frac{\kappa}{2}x}\, dx, \quad k \in \mathbb{N}.$$

The description of the extremal behavior of $(X_n^2)_{n\in\mathbb{N}}$ by the random walk $(S_n)_{n\in\mathbb{N}}$ is to be expected since by Lemma 3.4.14 the process $(Z_n)_{n\in\mathbb{N}} = (\ln(X_n^2))_{n\in\mathbb{N}}$ behaves

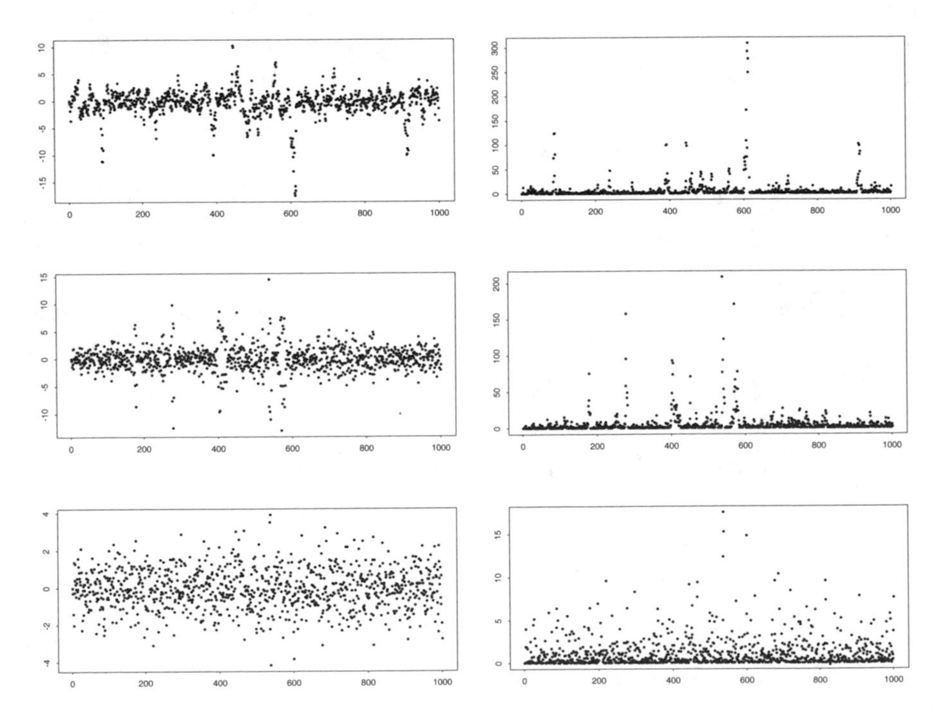

Figure 3.11 *Simulated sample path of $(X_n)_{n\in\mathbb{N}}$ with parameters $(\alpha, \beta, \lambda) = (0.8, 1, 0.2)$ (top, left), of $(X_n^2)_{n\in\mathbb{N}}$ with the same parameters (top, right), of $(X_n)_{n\in\mathbb{N}}$ with parameters $(\alpha, \beta, \lambda) = (-0.8, 1, 0.2)$ (middle, left), of $(X_n^2)_{n\in\mathbb{N}}$ with the same parameters (middle, right), of $(X_n)_{n\in\mathbb{N}}$ with parameters $(\alpha, \beta, \lambda) = (0, 1, 0.2)$ (bottom, left) and of $(X_n^2)_{n\in\mathbb{N}}$ with the same parameters (bottom, right) in the case $\varepsilon \stackrel{d}{=} N(0, 1)$. All simulations are based on the same simulated noise sequence $(\varepsilon_n)_{n\in\mathbb{N}}$.*

above a high threshold asymptotically like $(S_n)_{n\in\mathbb{N}}$. Unfortunately, this link fails for $(X_n)_{n\in\mathbb{N}}$.
(ii) Analogous to de Haan et al. [46] we may construct "estimators" for the extremal indices $\theta^{(2)}$ and $\theta_k^{(2)}$ of $(X_n^2)_{n\in\mathbb{N}}$, respectively, by

$$\widehat{\theta}^{(2)} = \frac{1}{N}\sum_{i=1}^{N} 1\left\{ \sup_{1\le j\le m} S_j^{(i)} \le -E_\kappa^{(i)} \right\}$$

and

$$\widehat{\theta}_k^{(2)} = \frac{1}{N}\sum_{i=1}^{N} 1\left\{ \sum_{j=1}^{m} 1\{S_j^{(i)} > -E_\kappa^{(i)}\} = k-1 \right\}, \quad \text{for } k \in \mathbb{N},$$

where N denotes the number of independent simulated sample paths of $(S_n)_{n\in\mathbb{N}}$, $E_\kappa^{(i)}$ are i.i.d. exponential rvs with rate κ, and m is chosen large enough. These estimators can be studied as in the case $\alpha = 0$ and $\varepsilon \stackrel{d}{=} N(0,1)$ in de Haan et al. [46]. In particular,

$$\sqrt{N}\frac{\theta^{(2)} - \widehat{\theta}^{(2)}}{(\theta^{(2)}(1-\theta^{(2)}))^{1/2}} \stackrel{d}{\to} N(0,1), \quad N, m \to \infty.$$

(iii) The approach chosen in (ii) is not possible for $(X_n)_{n\in\mathbb{N}}$, because $\prod_{l=1}^{j}(\alpha + \sqrt{\lambda}\,\varepsilon_l)$ may be negative. In a similar spirit we choose as "estimators" for θ and θ_k for $(X_n)_{n\in\mathbb{N}}$

$$\widehat{\theta} = \frac{1}{N}\sum_{i=1}^{N} 1\left\{ \sup_{1\le j\le m} \prod_{l=1}^{j}(\alpha + \sqrt{\lambda}\,\varepsilon_l) \le 1/P_\kappa^{(i)} \right\}$$

and

$$\widehat{\theta}_k = \frac{1}{N}\sum_{i=1}^{N} 1\left\{ \sum_{j=1}^{m} 1\left\{ \prod_{l=1}^{j}(\alpha + \sqrt{\lambda}\,\varepsilon_l) > 1/P_\kappa^{(i)} \right\} = k-1 \right\}, \quad \text{for } k \in \mathbb{N},$$

where N denotes the number of simulated paths of $(\prod_{l=1}^{n}(\alpha + \sqrt{\lambda}\,\varepsilon_l))_{n\in\mathbb{N}}$, $P_\kappa^{(i)}$ are i.i.d. Pareto rvs with shape parameter κ, i.e., with distribution function $G(x) = 1 - x^{-\kappa}, x \ge 1$, and m is large enough. These are suggestive estimators since $\prod_{l=1}^{n}(\alpha + \sqrt{\lambda}\,\varepsilon_l) \to 0$ a.s. as $n \to \infty$ because of assumption (3.27).
(iv) Note that the extremal index θ of $(X_n)_{n\in\mathbb{N}}$ is not symmetric in α (see Table 3.4). This is not surprising since the clustering is for $\alpha > 0$ stronger by the autoregressive part than for $\alpha < 0$. □

Remark 3.4.19 (i) Model (3.24) has a natural extension to higher order: the autoregressive model of order q with ARCH(q)-errors has been investigated in Klüppelberg and Pergamenchtchikov [58, 59]. It is also shown there that for Gaussian error variables this model is in distribution equivalent to a random coefficient model.
(ii) Such models also lead to interesting statistical theory, some can be found in econometric textbooks; see, e.g., Campbell, Lo and MacKinley [21], Gouriéroux [44], Shephard [82], or Taylor [84]. In Klüppelberg et al. [57] tests for models including (3.24) are suggested. A pseudo-likelihood ratio test for the hypotheses that the model

Table 3.4 *"Estimated" extremal index θ of $(X_n)_{n\in\mathbb{N}}$ in the case $\varepsilon \stackrel{d}{=} N(0,1)$. We chose $N = m = 2\,000$. Note that the extremal index decreases as $|\alpha|$ increases and that we have no symmetry in α.*

α	λ										
	0.2	0.4	0.6	0.8	1.0	1.2	1.5	2.0	2.5	3.0	3.5
−1.2	—	0.001	0.001	0.003	0.004	0.001	0.000	—	—	—	—
−1	0.15	0.19	0.19	0.16	0.13	0.09	0.05	0.01	—	—	—
−0.8	0.56	0.47	0.41	0.34	0.26	0.21	0.13	0.05	0.01	—	—
−0.6	0.86	0.71	0.61	0.50	0.41	0.33	0.22	0.10	0.03	0.00	—
−0.4	0.96	0.85	0.71	0.60	0.50	0.40	0.30	0.14	0.06	0.01	—
−0.2	0.98	0.89	0.77	0.65	0.56	0.47	0.33	0.18	0.07	0.02	0.00
0	0.98	0.89	0.78	0.65	0.55	0.45	0.33	0.18	0.08	0.02	0.00
0.2	0.94	0.82	0.72	0.61	0.52	0.43	0.32	0.18	0.07	0.02	0.00
0.4	0.85	0.72	0.63	0.53	0.45	0.37	0.28	0.13	0.06	0.01	—
0.6	0.68	0.55	0.48	0.41	0.35	0.29	0.21	0.10	0.03	0.00	—
0.8	0.39	0.34	0.32	0.27	0.22	0.19	0.12	0.05	0.01	—	—
1.0	0.09	0.14	0.13	0.13	0.11	0.08	0.04	0.01	—	—	—
1.2	—	0.000	0.001	0.003	0.004	0.001	0.000	—	—	—	—

Table 3.5 *"Estimated" extremal index θ and cluster probabilities $(\pi_k)_{1\le k\le 6}$ of $(X_n)_{n\in\mathbb{N}}$ dependent on α and λ in the case $\varepsilon \stackrel{d}{=} N(0,1)$. We chose $N = m = 2000$. Note that the extremal index for $\alpha > 0$ is much larger than for $\alpha < 0$.*

α	λ	θ	π_1	π_2	π_3	π_4	π_5	π_6
0	0.2	0.974	0.973	0.027	0.000	0.000	0.000	0.000
0	0.6	0.781	0.799	0.147	0.036	0.012	0.005	0.001
0	1	0.549	0.607	0.188	0.107	0.036	0.034	0.017
−0.4	0.2	0.962	0.962	0.037	0.001	0.000	0.000	0.000
0.4	0.2	0.853	0.867	0.103	0.026	0.002	0.002	0.000
−0.4	0.6	0.715	0.747	0.168	0.048	0.026	0.006	0.002
0.4	0.6	0.624	0.676	0.182	0.066	0.040	0.019	0.012
−0.4	1	0.497	0.540	0.210	0.115	0.075	0.040	0.004
0.4	1	0.445	0.533	0.185	0.080	0.109	0.032	0.017
−0.8	0.2	0.572	0.626	0.185	0.111	0.026	0.033	0.001
0.8	0.2	0.386	0.470	0.172	0.148	0.062	0.068	0.006
−0.8	0.6	0.414	0.520	0.159	0.134	0.072	0.043	0.016
0.8	0.6	0.314	0.443	0.156	0.110	0.087	0.073	0.041
−0.8	1	0.273	0.429	0.137	0.126	0.106	0.016	0.012
0.8	1	0.224	0.346	0.132	0.114	0.129	0.045	0.004

reduces to random walk or i.i.d. data is investigated and the distributional limit of the test statistic is derived. □

3.5 Optimal portfolios with bounded Value-at-Risk (VaR)

In this section we investigate the influence of large fluctuations and the Value-at-Risk as a risk measure to portfolio optimization. The section is based on Emmer, Klüppelberg, and Korn [35] and Emmer and Klüppelberg [34].

Starting with the traditional Black-Scholes model, where stock prices follow a geometric Brownian motion we first study the difference between the classical risk measure, i.e., the variance, and the VaR.

Since the variance of Brownian motion increases linearly, the use of the variance as a risk measure of an investment leads to a decreasing proportion of risky assets in a portfolio, when the planning horizon increases. This is not true for the Capital-at-Risk which—as a function of the planning horizon—increases first, but decreases, when the planning horizon becomes larger. We show for the CaR that, as seems to be common wisdom in asset management, long term stock investment leads to an almost sure gain over locally riskless bond investments. In the long run stock indices are growing faster than riskless rates, despite the repeated occurrence of stock market declines. The CaR therefore supports the portfolio manager's advice that the more distant the planning horizon, the greater should be one's wealth in risky assets. Interestingly, the VaR as risk measure supports the empirical observation above and hence resolves the contradiction between theory and empirical facts.

Then we study the optimal portfolio problem for more realistic price processes, i.e., Lévy processes that model also large fluctuations. Here, as is to be expected, the VaR reacts to exactly those and, consequently, so does the CaR. We investigate, in particular, the normal inverse Gaussian and variance gamma Lévy processes.

3.5.1 The Black-Scholes model

In this section, we consider a standard Black-Scholes type market consisting of one *riskless bond* and several *risky stocks*. Their respective prices $(P_0(t))_{t\geq 0}$ and $(P_i(t))_{t\geq 0}$ for $i = 1, \dots, d$ evolve according to the equations

$$P_0(t) = e^{rt}, \quad t \geq 0,$$

$$P_i(t) = p_i \exp\left(\left(b_i - \frac{1}{2}\sum_{j=1}^{d}\sigma_{ij}^2\right)t + \sum_{j=1}^{d}\sigma_{ij}W_j(t)\right), \quad t \geq 0.$$

Here $W(t) = (W_1(t), \dots, W_d(t))'$ is a standard d-dimensional Brownian motion, $r \in \mathbb{R}$ is the riskless interest rate, $b = (b_1, \dots, b_d)'$ the vector of stock-appreciation rates, and $\sigma = (\sigma_{ij})_{1\leq i,j\leq d}$ is the matrix of stock-volatilities. For simplicity, we assume that σ is invertible and that $b_i \geq r$ for $i = 1, \dots, d$. Since the assets are traded on the same market, they show some correlation structure which we model by a linear combination of the same Brownian motions $W_1, \dots, W_d$ for each traded asset. Throughout this chapter we denote by $\mathbb{R}^d$ the d-dimensional Euclidean space.

Its elements are column vectors and for $x \in \mathbb{R}^d$ we denote by x' the transposed vector; analogously, for a matrix β we denote by β' its transposed matrix. We further denote by $|x| = (\sum_{i=1}^d x_i^2)^{1/2}$ the Euclidean norm of $x \in \mathbb{R}^d$.

We need the SDE corresponding to the price processes above.

$$\begin{aligned} dP_0(t) &= P_0(t)r\,dt, & P_0(0) &= 1, \\ dP_i(t) &= P_i(t)\Big(b_i\,dt + \sum_{j=1}^d \sigma_{ij}dW_j(t)\Big), & P_i(0) &= p_i, \quad i = 1, \ldots, d. \end{aligned} \tag{3.42}$$

Let $\pi(t) = (\pi_1(t), \ldots, \pi_d(t))' \in \mathbb{R}^d$ be an admissible portfolio process, i.e., $\pi_i(t)$ is the fraction of the wealth $X^\pi(t)$, which is invested in asset i (see Korn [60], Section 2.1 for relevant definitions). Denoting by $(X^\pi(t))_{t\geq 0}$ the wealth process, it follows the dynamic

$$dX^\pi(t) = X^\pi(t)\{((1 - \pi(t)'\underline{1})r + \pi(t)'b)dt + \pi(t)'\sigma dW(t)\}, \quad X^\pi(0) = x,$$

where $x \in \mathbb{R}$ denotes the initial capital of the investor and $\underline{1} = (1, \ldots, 1)'$ denotes the vector (of appropriate dimension) having unit components. The fraction of the investment in the bond is $\pi_0(t) = 1 - \pi(t)'\underline{1}$. Throughout the chapter, we restrict ourselves to constant portfolios $\pi(t) = \pi = (\pi_1, \ldots, \pi_d)$ for all $t \in [0, T]$. This means that the fractions in the different stocks and the bond remain constant on $[0, T]$. The advantage of this is twofold: first we obtain, at least in a Gaussian setting, explicit results, and, furthermore, the economic interpretation of the mathematical results is comparatively easy. It is also important to point out that following a constant portfolio process does not mean that there is no trading. As the stock prices evolve randomly one has to trade at every time instant to keep the fractions of wealth invested in the different securities constant. Thus, following a constant portfolio process still means one must follow a dynamic trading strategy.

Standard Itô integration and the fact that $Ee^{sW(1)} = e^{s^2/2}$, $s \in \mathbb{R}$, yield the following explicit formulae for the wealth process for all $t \in [0, T]$.

$$X^\pi(t) = x\exp((\pi'(b - r\underline{1}) + r - |\pi'\sigma|^2/2)t + \pi'\sigma W(t)), \tag{3.43}$$

$$E[X^\pi(t)] = x\exp((\pi'(b - r\underline{1}) + r)t), \tag{3.44}$$

$$\mathrm{var}(X^\pi(t)) = x^2\exp(2(\pi'(b - r\underline{1}) + r)t)(\exp(|\pi'\sigma|^2 t) - 1). \tag{3.45}$$

Definition 3.5.1 [Capital-at-Risk]
Let x be the initial capital and T a given planning horizon. Let z_α be the α-quantile of the standard normal distribution. For some portfolio $\pi \in \mathbb{R}^d$ and the corresponding terminal wealth $X^\pi(T)$, the VaR *of $X^\pi(T)$ is given by*

$$\begin{aligned} \mathrm{VaR}(x, \pi, T) &= \inf\{z \in \mathbb{R} : P(X^\pi(T) \leq z) \geq \alpha\} \\ &= x\exp((\pi'(b - r\underline{1}) + r - |\pi'\sigma|^2/2)T + z_\alpha|\pi'\sigma|\sqrt{T}). \end{aligned}$$

Then we define

$$\begin{aligned}\mathrm{CaR}(x, \pi, T) &= x\exp(rT) - \mathrm{VaR}(x, \pi, T)\\ &= x\exp(rT)\\ &\quad\times(1-\exp((\pi'(b-r\underline{1}) - |\pi'\sigma|^2/2)T + z_\alpha|\pi'\sigma|\sqrt{T}))\end{aligned} \tag{3.46}$$

the Capital-at-Risk of the portfolio π (with initial capital x and planning horizon T).

To avoid (nonrelevant) subcases in some of the following results we always assume $\alpha < 0.5$ which leads to $z_\alpha < 0$.

Remark 3.5.2 (i) Our definition of the Capital-at-Risk limits the possibility of excess losses over the riskless investment.
(ii) We typically want to have a positive CaR (although it can be negative in our definition as the examples below will show) as the upper bound for the "likely losses." Further, we concentrate on the actual amount of losses appearing at the planning horizon T. This is in line with the mean-variance selection procedure enabling us to directly compare the results of the two approaches; see below. □

In the following it will be convenient to introduce the function $f(\pi)$ for the exponent in (3.46), that is

$$f(\pi) := z_\alpha|\pi'\sigma|\sqrt{T} - |\pi'\sigma|^2 T/2 + \pi'(b-r\underline{1})T, \quad \pi\in\mathbb{R}^d. \tag{3.47}$$

By the obvious fact that $f(\pi)\to-\infty$ as $|\pi'\sigma|\to\infty$ we have the natural upper bound $\sup_{\pi\in\mathbb{R}^d}\mathrm{CaR}(x,\pi,T) = x\exp(rT)$; i.e., the use of extremely risky strategies (in the sense of a high norm $|\pi'\sigma|$) can lead to a CaR which is close to the total capital. The computation of the minimal CaR is done in the following proposition.

Proposition 3.5.3 *Let* $\theta = |\sigma^{-1}(b-r1)|$.
(a) If $b_i = r$ for all $i = 1,\ldots,d$, then $f(\pi)$ attains its maximum for $\pi^* = 0$ leading to a minimum Capital-at-Risk of $\mathrm{CaR}(x,\pi^*,T) = 0$.
(b) If $b_i \neq r$ for some $i\in\{1,\ldots,d\}$ and $\theta\sqrt{T} < |z_\alpha|$, then again the minimal CaR *equals zero and is only attained for the pure bond strategy $\pi^* = 0$.*
(c) If $b_i \neq r$ for some $i\in\{1,\ldots,d\}$ and $\theta\sqrt{T}\geq|z_\alpha|$, then the minimal CaR *is attained for*

$$\pi^* = \left(\theta - \frac{|z_\alpha|}{\sqrt{T}}\right)\frac{(\sigma\sigma)^{-1}(b-r\underline{1})}{|\sigma^{-1}(b-r\underline{1})|} \tag{3.48}$$

with

$$\mathrm{CaR}(x,\pi^*,T) = x\exp(rT)\left(1-\exp\left(\frac{1}{2}(\sqrt{T}\theta - |z_\alpha|)^2\right)\right) < 0. \tag{3.49}$$

Proof. (a) follows directly from the explicit form of $f(\pi)$ under the assumption of $b_i = r$ for all $i = 1,\ldots,d$ and the fact that σ is invertible.
(b), (c) Consider the problem of maximizing $f(\pi)$ over all π which satisfy

$$|\pi'\sigma| = \varepsilon \tag{3.50}$$

for a fixed positive ε. Over the (boundary of the) ellipsoid defined by (3.50) $f(\pi)$ equals

$$f(\pi) = z_\alpha \varepsilon \sqrt{T} - \varepsilon^2 T/2 + \pi'(b - r\underline{1})T.$$

Thus, the problem is reduced to maximizing a linear function (in π) over the boundary of an ellipsoid. Such a problem has the explicit solution

$$\pi_\varepsilon^* = \varepsilon \frac{(\sigma\sigma')^{-1}(b - r\underline{1})}{|\sigma^{-1}(b - r\underline{1})|} \tag{3.51}$$

with

$$f(\pi_\varepsilon^*) = -\varepsilon^2 T/2 + \varepsilon(\theta T - |z_\alpha|\sqrt{T}). \tag{3.52}$$

As every $\pi \in \mathbb{R}^d$ satisfies relation (3.50) with a suitable value of ε (due to the fact that σ is regular), we obtain the minimum CaR strategy π^* by maximizing $f(\pi_\varepsilon^*)$ over all nonnegative ε. Due to the form of $f(\pi_\varepsilon^*)$ the optimal ε is positive if and only if the multiplier of ε in representation (3.52) is positive. Thus, in the situation of Proposition 3.5.3(b) the assertion holds. In the situation of Proposition 3.5.3(c) the optimal ε is given as

$$\varepsilon = \theta - \frac{|z_\alpha|}{\sqrt{T}}.$$

Inserting this into equations (3.51) and (3.52) yields the assertions (3.48) and (3.49) (with the help of equations (3.46) and (3.47)). □

Remark 3.5.4 (i) Part (a) of Proposition 3.5.3 states that in a risk-neutral market the CaR of every strategy containing stock investment is bigger than the CaR of the pure bond strategy.
(ii) Part (c) states the (at first sight surprising) fact that the existence of at least one stock with a mean rate of return different from the riskless rate implies the existence of a stock and bond strategy with a negative CaR as soon as the planning horizon T is large. Thus, even if the CaR would be the only criterion to judge an investment strategy the pure bond investment would not be optimal if the planning horizon is far away. On one hand this fact is in line with empirical results on stock and bond markets. On the other hand this shows a remarkable difference between the behavior of the CaR and the variance as risk measures. Independent of the planning horizon and the market coefficients, pure bond investment would always be optimal with respect to the variance of the corresponding wealth process. □

We now turn to a Markowitz mean-variance type optimization problem where we replace the variance constraint by a constraint on the CaR of the terminal wealth. More precisely, we solve the following problem:

$$\max_{\pi \in \mathbb{R}^d} E[X^\pi(T)] \quad \text{subject to} \quad \text{CaR}(x, \pi, T) \leq C, \tag{3.53}$$

where C is a given constant of which we assume that it satisfies $C \leq x \exp(rT)$.

Due to the explicit representations (3.45), (3.46) and a variant of the decomposition method as applied in the proof of Proposition 3.5.3 we can solve problem (3.53) explicitly.

Proposition 3.5.5 *Let* $\theta = |\sigma^{-1}(b - r\underline{1})|$ *and assume that* $b_i \neq r$ *for at least one* $i \in \{1, \ldots, d\}$. *Assume furthermore that* C *satisfies*

$$0 \leq C \leq x \exp(rT) \quad \text{if } \theta\sqrt{T} < |z_\alpha|, \tag{3.54}$$

$$x \exp(rT)\left(1 - \exp\left(\frac{1}{2}(\sqrt{T}\theta - |z_\alpha|)^2\right)\right) \leq C \leq x \exp(rT) \quad \text{if } \theta\sqrt{T} \geq |z_\alpha|. \tag{3.55}$$

Then problem (3.53) has solution

$$\pi^* = \varepsilon^* \frac{(\sigma\sigma')^{-1}(b - r\underline{1})}{|\sigma^{-1}(b - r\underline{1})|}$$

with

$$\varepsilon^* = (\theta + z_\alpha/\sqrt{T}) + \sqrt{(\theta + z_\alpha/\sqrt{T})^2 - 2c/T},$$

where $c = \ln(1 - \frac{C}{x}\exp(-rT))$. *The corresponding maximal expected terminal wealth under the* CaR *constraint equals*

$$E[X^{\pi^*}(T)] = x \exp((r + \varepsilon^*|\sigma^{-1}(b - r\underline{1})|)T). \tag{3.56}$$

Proof. The requirements (3.54) and (3.55) on C ensure that the CaR constraint in problem (3.53) cannot be ignored: in both cases C lies between the minimum and the maximum value that CaR can attain (see also Proposition 3.5.3). Every admissible π for problem (3.53) with $|\pi'\sigma| = \varepsilon$ satisfies the relation

$$(b - r\underline{1})'\pi T \geq c + \frac{1}{2}\varepsilon^2 T - z_\alpha \varepsilon \sqrt{T} \tag{3.57}$$

which is in this case equivalent to the CaR constraint in (3.53). But again, on the set given by $|\pi'\sigma| = \varepsilon$ the linear function $(b - r\underline{1})'\pi T$ is maximized by

$$\pi_\varepsilon = \varepsilon \frac{(\sigma\sigma')^{-1}(b - r\underline{1})}{|\sigma^{-1}(b - r\underline{1})|}. \tag{3.58}$$

Hence, if there is an admissible π for problem (3.53) with $|\pi'\sigma| = \varepsilon$ then π_ε must also be admissible. Further, due to the explicit form (3.44) of the expected terminal wealth, π_ε also maximizes the expected terminal wealth over the ellipsoid. Consequently, to obtain π for problem (3.53) it suffices to consider all vectors of the form π_ε for all positive ε such that requirement (3.56) is satisfied. Inserting (3.58) into the left-hand side of inequality (3.57) results in

$$(b - r\underline{1})'\pi_\varepsilon T = \varepsilon|\sigma^{-1}(b - r\underline{1})|T, \tag{3.59}$$

which is an increasing linear function in ε equalling zero in $\varepsilon = 0$. Therefore, we obtain the solution of problem (3.53) by determining the biggest positive ε such that (3.57) is still valid. But the right-hand side of (3.59) stays above the right-hand side of (3.57) until their largest positive point of intersection which is given by

$$\varepsilon^* = (\theta + z_\alpha/\sqrt{T}) + \sqrt{(\theta + z_\alpha/\sqrt{T})^2 - 2c/T}.$$

The remaining assertion (3.56) can be verified by inserting π^* into equation (3.44).

□

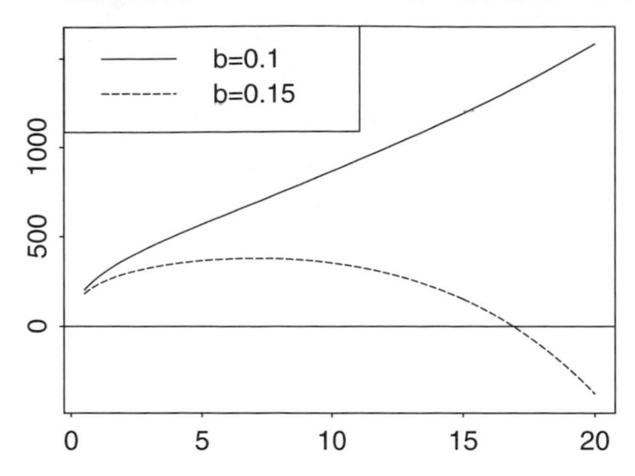

Figure 3.12 CaR(1 000, 1, T) *of the pure stock portfolio (one risky asset only) for different appreciation rates as a function of the planning horizon* T*;* $0 < T \leq 20$*. The volatility is* $\sigma = 0.2$*. The riskless rate is* $r = 0.05$*.*

Remark 3.5.6 Note that the optimal expected value only depends on the stocks via the norm $|\sigma^{-1}(b - r\underline{1})|$. There is no explicit dependence on the number of different stocks. We therefore interpret Proposition 3.5.3 as a kind of mutual fund theorem as there is no difference between investment in our multistock market and a market consisting of the bond and just one stock with appropriate market coefficients b and σ. □

Example 3.5.7 Figure 3.12 shows the dependence of CaR on the planning horizon T illustrated by CaR(1 000,1,T). Note that the CaR first increases and then decreases with time, a behavior which was already indicated by Proposition 3.5.3. It differs substantially from the behavior of the variance of the pure stock strategy, which increases with T. Figure 3.13 illustrates the behavior of the optimal expected terminal wealth with varying planning horizon corresponding to the pure bond strategy and the pure stock strategy as functions of the planning horizon T. For a planning horizon of up to 5 years the expected terminal wealth of the optimal portfolio even exceeds the pure stock investment. The reason for this becomes clear if we look at the corresponding portfolios. The optimal portfolio always contains a short position in the bond as long as this is tolerated by the CaR constraint. This is shown in Figure 3.14 where we have plotted the optimal portfolio together with the pure stock portfolio as function of the planning horizon. For $b = 0.15$ the optimal portfolio always contains a short position in the bond. For $b = 0.1$ and $T > 5$ the optimal portfolio (with the same CaR constraint as in Figure 3.13) again contains a long position in both bond and stock (with decreasing tendency of π as time increases). This is an immediate consequence of the increasing CaR of the stock price. For the smaller appreciation rate of the stock it is simply not attractive enough to take the risk of a large stock investment. Figure 3.14 shows the mean-CaR efficient frontier for the above parameters with $b = 0.1$ and fixed planning horizon $T = 5$. As expected it has a similar form as a typical mean-variance efficient frontier.

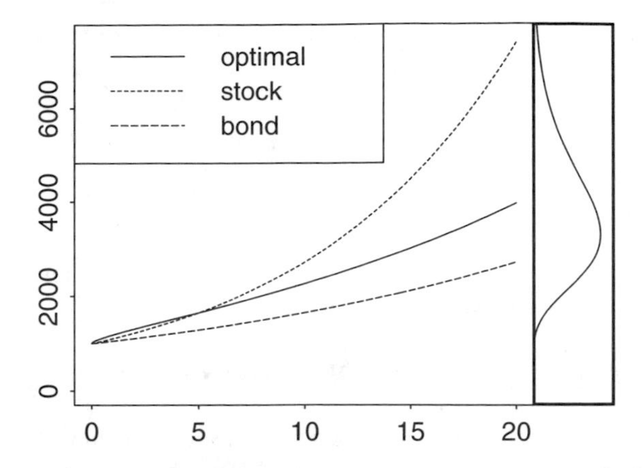

Figure 3.13 *Expected terminal wealth of different investment strategies depending on the planning horizon T, $0 \leq T \leq 20$. The parameters are $d = 1$, $r = 0.05$, $b = 0.1$, $\sigma = 0.2$, and $\alpha = 0.05$. As the upper bound C of the* CaR *we used* CaR$(1\,000, 1, 5)$, *the* CaR *of the pure stock strategy with planning horizon $T = 5$. On the right border we have plotted the density function of the wealth for the optimal portfolio.*

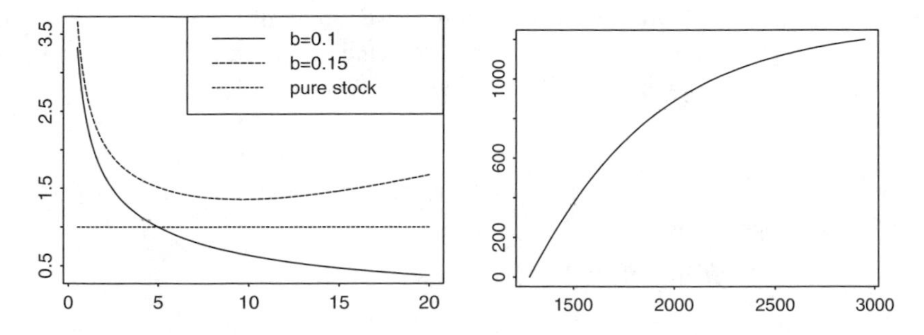

Figure 3.14 *For different appreciation rates the left-hand figure shows the optimal portfolio π and the pure stock portfolio. An optimal portfolio of $\pi > 1$ corresponds to short selling. The right-hand figure shows the mean-*CaR *efficient frontier with the mean on the horizontal axis and the* CaR *on the vertical axis. The parameters are the same as in Figure 3.13.*

We compare now the behavior of the optimal portfolios for the mean-CaR with solutions of a corresponding mean-variance problem. To this end we consider the following simpler optimization problem:

$$\max_{\pi \in \mathbb{R}^d} E[X^\pi(T)] \quad \text{subject to} \quad \text{var}(X^\pi(T)) \leq C\,. \tag{3.60}$$

Proposition 3.5.8 *If $b_i \neq r$ for at least one $i \in \{1, \ldots, d\}$, then the optimal solution of the mean-variance problem (3.60) is given by*

$$\widehat{\pi} = \widehat{\varepsilon}\,\frac{(\sigma\sigma')^{-1}(b - r\underline{1})}{|\sigma^{-1}(b - r\underline{1})|}\,,$$

where $\widehat{\varepsilon}$ is the unique positive solution of the nonlinear equation

$$rT + |\sigma^{-1}(b - r\underline{1})|\varepsilon T - \frac{1}{2}\ln\left(\frac{C}{x^2}\right) + \frac{1}{2}\ln(\exp(\varepsilon^2 T) - 1) = 0.$$

The corresponding maximal expected terminal wealth under the variance constraint equals

$$E[X^{\widehat{\pi}}(T)] = x\exp((r + \widehat{\varepsilon}\,|\sigma^{-1}(b - r\underline{1})|)T).$$

Proof. By using the explicit form (3.45) of the variance of the terminal wealth, we can rewrite the variance constraint in problem (3.60) as

$$(b - r\underline{1})'\pi T \le \frac{1}{2}\ln\left(\frac{C}{x^2}\right) - \frac{1}{2}\ln(\exp(\varepsilon^2 T) - 1)) - rT =: h(\varepsilon), \quad |\pi'\sigma| = \varepsilon \quad (3.61)$$

for $\varepsilon > 0$. More precisely, if $\pi \in \mathbb{R}^d$ satisfies the constraints in (3.61) for one $\varepsilon > 0$ then it also satisfies the variance constraint in (3.60) and vice versa. Noting that $h(\varepsilon)$ is strictly decreasing in $\varepsilon > 0$ with

$$\lim_{\varepsilon \downarrow 0} h(\varepsilon) = \infty \quad \text{and} \quad \lim_{\varepsilon \to \infty} h(\varepsilon) = -\infty$$

we see that the left-hand side of (3.61) must be smaller than the right-hand side for small values of $\varepsilon > 0$ if we plug in π_ε as given by equation (3.58). Recall that this was the portfolio with the highest expected terminal wealth of all portfolios π satisfying $|\pi'\sigma| = \varepsilon$. It even maximizes $(b - r\underline{1})'\pi T$ over the set given by $|\pi'\sigma| \le \varepsilon$. If we have equality

$$(b - r\underline{1})'\pi_{\widehat{\varepsilon}}\, T = h(\widehat{\varepsilon}) \quad (3.62)$$

for the first time with increasing $\varepsilon > 0$, then this determines the optimal $\widehat{\varepsilon} > 0$. To see this, note that we have

$$E[X^\pi(T)] \le E[X^{\pi_{\widehat{\varepsilon}}}(T)] \quad \text{for all } \pi \text{ with} \quad |\pi'\sigma| \le \widehat{\varepsilon}\,,$$

and for all admissible π with $\varepsilon = |\pi'\sigma| > \widehat{\varepsilon}$ we obtain

$$(b - r\underline{1})'\pi T \le h(\varepsilon) < h(\widehat{\varepsilon}) = (b - r\underline{1})'\pi_{\widehat{\varepsilon}}\, T\,.$$

By solving the nonlinear equation (3.62) for $\widehat{\varepsilon}$ we have thus completely determined the solution of problem (3.60). □

Example 3.5.9 Figure 3.15 compares the behavior of $\widehat{\varepsilon}$ and ε^* as functions of the planning horizon T. We have used the same data as in Example 3.5.7. To make the solutions of problems (3.53) and (3.60) comparable, we have chosen C differently for the variance and the CaR risk measures in such a way that $\widehat{\varepsilon}$ and ε^* coincide for $T = 5$. Notice that C for the variance problem is roughly the square of C for the CaR problem taking into account that the variance measures an L^2-distance, whereas CaR measures an L^1-distance. The (of course expected) bottom line of Figure 3.15 is that with increasing time the variance constraint demands a smaller fraction of risky securities in the portfolio. This is also true for the CaR constraint for small time horizons. For larger planning horizon T ($T \ge 20$) ε^* increases again due to the fact that the CaR decreases. In contrast to that, $\widehat{\varepsilon}$ decreases to 0, since the variance increases.

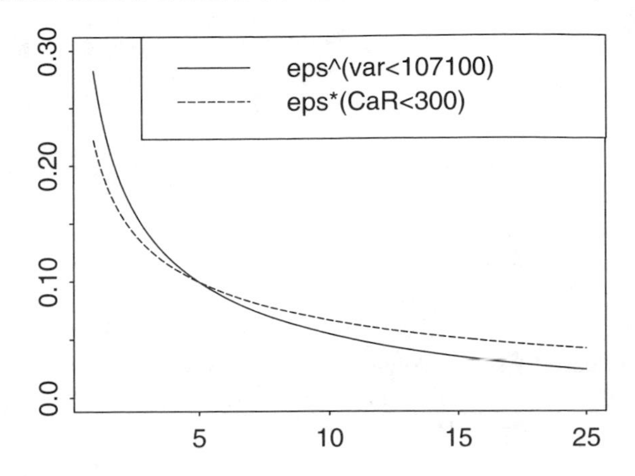

Figure 3.15 $\widehat{\varepsilon}$ *and* ε^* *as functions of the planning horizon;* $0 < T \leq 20$. *The parameters are the same as in Figure 3.13.*

3.5.2 The exponential Lévy model

As in Section 3.5.1 we consider a standard Black-Scholes type market consisting of a riskless bond and several risky stocks; however, we assume now that their prices follow exponential Lévy processes. This is a large class of models, including besides the geometric Brownian motion also much more realistic price models. The respective prices $(P_0(t))_{t\geq 0}$ and $(P_i(t))_{t\geq 0}$ for $i = 1, \dots, d$ evolve according to the equations

$$P_0(t) = e^{rt} \quad \text{and} \quad P_i(t) = p_i \exp\left(b_i t + \sum_{j=1}^{d} \sigma_{ij} L_j(t)\right), \quad t \geq 0. \quad (3.63)$$

Here $r \in \mathbb{R}$ is the riskless interest rate, $b \in \mathbb{R}^d$ and $\sigma = (\sigma_{ij})_{1\leq i,j\leq d}$ is an invertible matrix. $(L(t))_{t\geq 0} = (L_1(t), \dots, L_d(t))_{t\geq 0}$ is a d-dimensional Lévy process with independent components. Hence we assume that each $(L_i(t))_{t\geq 0}$ for $i = 1, \dots, d$ has stationary independent increments with cadlag sample paths. We define this model analogously to the Black-Scholes model in Section 3.5.1, but replace the Brownian motion by a general Lévy process L.

Before we specify this model further we summarize some results on Lévy processes. For relevant background we refer to Bertoin [10], Protter [72] and, in particular, Sato [80]. A very interesting collection of research articles is Barndorff-Nielsen, Mikosch and Resnick [7].

Each infinitely divisible df F on $\mathbb{R}^d$ generates a Lévy process L by choosing F as df of the d-dimensional vector $L(1)$. This can be seen immediately, since the characteristic function is for each $t > 0$ given by

$$E \exp(is'L(t)) = \exp(t\Psi(s)), \quad s \in \mathbb{R}^d,$$

where Ψ has Lévy-Khintchine representation

$$\Psi(s) = is'a - \frac{s'\beta'\beta s}{2} + \int_{\mathbb{R}^d} (e^{is'x} - 1 - is'xI(|x| \leq 1))\nu(dx), \quad s \in \mathbb{R}^d. \quad (3.64)$$

Here $a \in \mathbb{R}^d$, $\beta'\beta$ is a nonnegative definite symmetric $d \times d$-matrix, and ν is a measure on $\mathbb{R}^d$ satisfying $\nu(\{0\}) = 0$ and $\int_{\mathbb{R}^d}(|x|^2 \wedge 1)\nu(dx) < \infty$, called the Lévy measure of the process L. The term corresponding to $xI(|x| \leq 1)$ represents a centering, without which the integral (3.64) may not converge. The characteristic triplet $(a, \beta'\beta, \nu)$ characterizes the Lévy process L.

According to Sato [80], Chapter 4 (see Theorem 19.2), the following holds. For each ω in the probability space, define $\Delta L(t, \omega) = (\Delta L_1(t, \omega), \ldots, \Delta L_d(t, \omega))$ with $\Delta L_j(t, \omega) = L_j(t, \omega) - L_j(t-, \omega)$ for $j = 1, \ldots, d$. For each Borel set $B \subset [0, \infty) \times \mathbb{R}^{d*}$ ($\mathbb{R}^{d*} = \mathbb{R}^d \setminus \{0\}$) set

$$M(B, \omega) = \text{card}\{t \geq 0 \,:\, (t, \Delta L(t, \omega)) \in B\}.$$

Lévy's theory says that M is a Poisson random measure with intensity

$$m(dt, dx) = dt\nu(dx),$$

where ν is the Lévy measure of the process L. Notice that m is σ-finite and $M(B, \cdot) = \infty$ a.s. when $m(B) = \infty$.

With this notation, representation (3.64) corresponds to the Lévy-Itô representation

$$L(t) = at + \beta W(t) + \sum_{0<s\leq t} \Delta L(s) I(|\Delta L(s)| > 1) + \int_0^t \int_{|x|\leq 1} x(M(dx, ds) - \nu(dx)ds), \quad t \geq 0. \quad (3.65)$$

This means that $L(t)$ is the independent sum of a drift term, a Brownian component $\beta W(t)$ and a pure jump part with Lévy measure ν, having the interpretation that a jump of size x occurs at rate $\nu(dx)$. To ensure finiteness of the integral (3.64), the small jumps are compensated by their expectation. If the sample paths of L have almost surely finite variation on $(0, t]$ for any $t > 0$, then this representation reduces to

$$L(t) = \gamma t + \beta W(t) + \sum_{0<s\leq t} \Delta L(s), \quad t \geq 0, \quad (3.66)$$

where $\gamma = a - \int_{|x|\leq 1} x\nu(dx)$.

We return to model (3.63) with L having characteristic triplet $(a, \beta'\beta, \nu)$, where a is a d-dimensional vector, $\beta = \text{diag}(\beta_1, \ldots, \beta_d)$ is a d-dimensional diagonal matrix and ν is the Lévy measure, which corresponds to the product measure of the independent components of L on $\mathbb{R}^d$. This means that, e.g., for $d = 2$ and a rectangle $A = (a, b] \times (c, d] \subset \mathbb{R}^2$ the Lévy measure $\nu(A) = \nu_1((a, b]) + \nu_2((c, d])$, where ν_i is the Lévy measure of L_i for $i = 1, 2$. The diagonal matrix β means that the d-dimensional Wiener process W has independent components with different variances possible. This allows for different scaling factors in the Wiener processes and the nonGaussian components; moreover, if some $\beta_i = 0$ the model allows for Lévy processes without Gaussian component as asset price models.

In order to derive the wealth process of a portfolio we need the corresponding SDE. By Itô's formula (see, e.g., Protter [72]), $P_i, i = 1, \ldots, d$, is the solution to the SDE

$$\begin{aligned} dP_i(t) &= P_i(t-)(b_i dt + d\widehat{L}_i(t)) \\ &= P_i(t-)\Bigg(\Bigg(b_i + \frac{1}{2}\sum_{j=1}^{d}(\sigma_{ij}\beta_j)^2\Bigg)dt + \sum_{j=1}^{d}\sigma_{ij}(dL_j(t) - \Delta L_j(t)) \\ &\quad + \exp\Bigg(\sum_{j=1}^{d}\sigma_{ij}\Delta L_j(t)\Bigg) - 1\Bigg), \quad t > 0, \quad P_i(0) = p_i. \end{aligned} \tag{3.67}$$

Remark 3.5.10 (i) Note the similarity but also the difference to the geometric Brownian motion model (3.42). Again the Wiener process introduces an Itô term in the drift component of the SDE. However, there is a main change in the jumps of the Lévy processes. First note that, because of the independence, jumps of the different processes $L_1, \ldots, L_d$ occur at different times (see Sato [80], Exercise E12.10 on p. 67). Then every jump of one of the original processes is replaced: a jump of size $\sum_{j=1}^{d}\sigma_{ij}\Delta L_j$ is replaced by a jump of size $\exp(\sum_{j=1}^{d}\sigma_{ij}\Delta L_j) - 1$.
(ii) Note also that $\widehat{L}_i$ is such that

$$\exp\Bigg(\sum_{j=1}^{d}\sigma_{ij}L_j\Bigg) = \mathcal{E}(\widehat{L}_i), \quad i = 1, \ldots, d,$$

where $\mathcal{E}$ denotes the stochastic exponential of a process. □

We shall use the following lemma which relates the characteristic triplet of an exponential Lévy process and its stochastic exponential in $\mathbb{R}$.

Lemma 3.5.11 (Goll and Kallsen [43])
If L is a real-valued Lévy process with characteristic triplet (a, β, ν), then also $\widehat{L}$ defined by $e^L = \mathcal{E}(\widehat{L})$ is a Lévy process with characteristic triplet $(\widehat{a}, \widehat{\beta}, \widehat{\nu})$ given by

$$\begin{aligned} \widehat{a} - a &= \frac{1}{2}\beta^2 + \int((e^x - 1)1_{\{(|e^x-1|<1\}} - x1_{\{|x|<1\}})\nu(dx) \\ \widehat{\beta} &= \beta \\ \widehat{\nu}(\Lambda) &= \nu(\{x \in \mathbb{R} : e^x - 1 \in \Lambda\}) \textit{ for any Borel set } \Lambda \subset \mathbb{R}^*. \end{aligned}$$

As in the Black-Scholes model before, we restrict ourselves to constant portfolios; i.e., $\pi(t) = \pi, t \in [0, T]$, for some fixed planning horizon T. In order to avoid negative wealth we require that $\pi \in [0, 1]^d$ and $\pi'\underline{1} \leq 1$. Denoting by $(X^\pi(t))_{t\geq 0}$ the *wealth process*, it follows the dynamic

$$dX^\pi(t) = X^\pi(t-)(((1 - \pi'\underline{1})r + \pi'b)dt + \pi'd\widehat{L}(t)), \quad t > 0, \quad X^\pi(0) = x,$$

where $x \in \mathbb{R}$ denotes the initial capital of the investor.
Using Itô's formula, this SDE has the solution

$$X^\pi(t) = x\exp((r + \pi'(b - r\underline{1}))t)\mathcal{E}(\pi'\widehat{L}(t)), \quad t \geq 0. \tag{3.68}$$

One important consequence of this representation is the fact that a jump $\Delta L(t)$ is transformed into a jump $\Delta X^\pi(t) = \ln(1 + \pi'(e^{\sigma\Delta L(t)} - 1)) > \ln(1 - \pi'\underline{1})$ and hence we also require for the portfolio that $\pi'\underline{1} \le 1$. (Note that $\sigma\Delta L(t)$ is a vector and $e^{\sigma\Delta L(t)}$ is defined componentwise.)

From (3.67) it is clear that $(X^\pi(t))_{t\ge 0}$ cannot have a nice and simple representation as in the case of geometric Brownian motion; see (3.43). In any case, $(X^\pi(t))_{t\ge 0}$ is again an exponential Lévy process and we calculate the characteristic triplet of its logarithm.

Lemma 3.5.12 *Consider model (3.63) with Lévy process L and characteristic triplet (a, β, ν). Define for the $d \times d$-matrix $\sigma\beta$ the vector $[\sigma\beta]^2$ with components*

$$[\sigma\beta]_i^2 = \sum_{j=1}^{d} (\sigma_{ij}\beta_j)^2, \quad i = 1, \dots, d.$$

The process $(\ln X^\pi(t))_{t\ge 0}$ is a Lévy process with triplet (a_X, β_X, ν_X) given by

$$\begin{aligned} a_X &= r + \pi'(b - r\underline{1} + [\sigma\beta]^2/2 + \sigma a) - |\pi'\sigma\beta|^2/2 \\ &\quad + \int_{\mathbb{R}^d} \left(\ln(1 + \pi'(e^{\sigma x} - \underline{1}))1_{\{|\ln(1+\pi'(e^{\sigma x}-1))|\le 1\}} - \pi'\sigma x 1_{\{|x|\le 1\}} \right) \nu(dx), \\ \beta_X &= |\pi'\sigma\beta|, \\ \nu_X(A) &= \nu(\{x \in \mathbb{R}^d : \ln(1 + \pi'(e^{\sigma x} - 1)) \in A\}) \text{ for any Borel set } A \subset \mathbb{R}^*. \end{aligned}$$

In the finite variation case we obtain

$$\begin{aligned} \ln \mathcal{E}(\pi'\widehat{L}(t)) &= \gamma_X t + \pi'\sigma\beta W(t) + \sum_{0<s\le t} \ln\left(1 + \sum_{i=1}^{d} \pi_i \Delta\widehat{L}_i(s)\right) \\ &= \gamma_X t + \pi'\sigma\beta W(t) \\ &\quad + \sum_{0<s\le t} \ln\left(1 + \sum_{i=1}^{d} \pi_i \left(\exp\left(\sum_{j=1}^{d} \sigma_{ij}\Delta L_j(s)\right) - 1\right)\right), \quad t \ge 0, \end{aligned}$$

where

$$\gamma_X = \pi'(\sigma\gamma + [\sigma\beta]^2/2) - |\pi'\sigma\beta|^2/2,$$

and $\gamma = a - \int_{|x|\le 1} x\nu(dx)$ as in (3.66).

By Lemma 3.5.12 $\ln X^\pi(t)$ has characteristic function $E\exp(is\ln X^\pi(t)) = \exp(t\psi_X(s))$, $s \in \mathbb{R}$. If it can be analytically extended around $s = 0$ in $\mathbb{C}$, then by Theorem 25.17 in Sato [80] we obtain for all $k \in \mathbb{N}$

$$E[(X^\pi(t))^k] = \exp(t\Psi_X(-ik)), \quad t \ge 0. \tag{3.69}$$

In particular, $E\exp(s\ln X^\pi(t)) = E[(X^\pi(t))^s] < \infty$ for one and hence all $t > 0$ if and only if $\int_{|x|>1} e^{sx}\nu_X(dx) < \infty$.

Proposition 3.5.13 *Let $L = (L_1, \dots, L_d)$ be a Lévy process with independent components and assume that for all $j = 1, \dots, d$ the rv $L_j(1)$ has finite moment generating* *function $\widehat{f}_j$ such that $\widehat{f}_j(\sigma_{ij}) = E\exp(\sigma_{ij}L_j(1)) < \infty$ for $i, j = 1, \ldots, d$. Denote*

$$\widehat{f}(\sigma) := E\exp(\sigma L(1)) = \left(E\exp\left(\sum_{j=1}^{d}\sigma_{1j}L_j(1)\right), \ldots, E\exp\left(\sum_{j=1}^{d}\sigma_{dj}L_j(1)\right)\right). \tag{3.70}$$

Let $X^\pi(t)$ be as in equation (3.68). Then

$$E[X^\pi(t)] = x\exp(t(r + \pi'(b - r\underline{1} + \ln\widehat{f}(\sigma)))),$$

$$\operatorname{var}(X^\pi(t)) = x^2\exp(2t(r + \pi'(b - r1 + \ln\widehat{f}(\sigma))))(\exp(t\pi' A\pi) - 1),$$

where A is a $d \times d$-matrix with components

$$A_{ij} = E\exp\left(\sum_{l=1}^{d}(\sigma_{il} + \sigma_{jl})L_l(1)\right) - E\exp\left(\sum_{l=1}^{d}\sigma_{il}L_l(1)\right)$$

$$- E\exp\left(\sum_{l=1}^{d}\sigma_{jl}L_l(1)\right), \quad 1 \le i, j \le d.$$

Proof. Formula (3.69) reduces for $k = 1$ and $k = 2$ somewhat, giving together with the expression of ν_X in terms of ν of Lemma 3.5.12,

$$E[X^\pi(t)] = x\exp\left(t\left(r + \pi'\left(b - r\underline{1} + \frac{1}{2}[\sigma\beta]^2 + \sigma a\right.\right.\right.$$

$$\left.\left.\left. + \int_{\mathbb{R}^d}(e^{\sigma x} - 1 - \sigma x 1_{\{|x|<1\}})\nu(dx)\right)\right)\right),$$

$$\operatorname{var}(X^\pi(t)) = x^2\exp\left(2t\left(r + \pi'\left(b - r\underline{1} + \frac{1}{2}[\sigma\beta]^2 + \sigma a\right.\right.\right.$$

$$\left.\left.\left. + \int_{\mathbb{R}^d}(e^{\sigma x} - 1 - \sigma x 1_{\{|x|<1\}})\nu(dx)\right)\right)\right)$$

$$\times\left(\exp\left(t\left(|\pi'\sigma\beta|^2 + \int_{\mathbb{R}^d}(\pi'(e^{\sigma x} - \underline{1}))^2\nu(dx)\right)\right) - 1\right).$$

For $i = 1, \ldots, d$ denote by e_i the i-th unit vector in $\mathbb{R}^d$. Then the i-th component of (3.70) is obtained by

$$E\exp\left(\sum_{l=1}^{d}\sigma_{il}L_l(1)\right) = \exp\left(\left(\sigma a + [\sigma\beta]^2/2 + \int(e^{\sigma x} - 1 - \sigma x 1_{\{|x|<1\}})\nu(dx)\right)_i\right).$$

which corresponds to the i-th component of $\ln(E\exp(\sigma L(1)))$. The formula for the variance is obtained analogously. □

Remark 3.5.14 Note that for $l = 1, \ldots, d$ ($i = \sqrt{-1}$)

$$\ln\left(E\exp\left(\sum_{j=1}^{d}\sigma_{lj}L_j(1)\right)\right) = \sum_{j=1}^{d}\ln\widehat{f}_j(\sigma_{lj}) = \ln E[\mathcal{E}(\widehat{L}_l)(1)] = \sum_{j=1}^{d}\Psi(-i\sigma_{lj}).$$

This implies in particular $E\mathcal{E}(\pi'\widehat{L}(t)) = (\prod_{l=1}^{d}(E[\mathcal{E}(\widehat{L}_l(t))])^{\pi_l})$, □

3.5.3 Portfolio optimization

We consider now the portfolio optimization problem using the Capital-at-Risk as risk measure in the more general setting of Lévy processes. The definition of the CaR from Definition 3.5.1 adapted to the more general situation reads as follows.

Definition 3.5.15 [Capital-at-Risk]
Let x be the initial capital and T a given planning horizon. Let z_α be the α-quantile of $\mathcal{E}(\pi'\widehat{L}(T))$ for some portfolio $\pi \in \mathbb{R}^d$ and $X^\pi(T)$ the corresponding terminal wealth. Then the VaR *of $X^\pi(T)$ is given by*

$$\begin{aligned}\mathrm{VaR}(x,\pi,T) &= \inf\{z \in \mathbb{R} : P(X^\pi(T) \le z) \ge \alpha\} \\ &= x z_\alpha \exp((\pi'(b - r\underline{1}) + r)T)\end{aligned}$$

and we define

$$\begin{aligned}\mathrm{CaR}(x,\pi,T) &= x\exp(rT) - \mathrm{VaR}(x,\pi,T) \\ &= x\exp(rT)(1 - z_\alpha \exp(\pi'(b - r\underline{1})T))\end{aligned}$$

the Capital-at-Risk of the portfolio π (with initial capital x and planning horizon T).

We consider now the following optimization problem.

$$\max_{\pi\in[0,1]^d,\,\pi'\underline{1}\le 1} E[X^\pi(T)] \quad \text{subject to} \quad \mathrm{CaR}(x,\pi,T) \le C.$$

In general, quantiles of Lévy processes cannot be calculated explicitly. Usually, the df of $X^\pi(T)$ is not known explicitly. At first sight there are various possibilities for approximations, and we discuss their applicability for quantile estimation below.

For simplicity we restrict ourselves to $d = 1$, i.e., the portfolio consists of the bond and one risky asset, which is modeled by the exponential Lévy process

$$P(t) = p\exp(bt + L(t)) \quad t \ge 0,$$

where L has characteristic function $Ee^{isL(t)} = e^{t\Psi(s)}$, $s \in \mathbb{R}$. We set $\pi_1 = \pi$ and $X^\pi(t)$ reduces to

$$X^\pi(t) = x\exp((r + \pi(b - r))t)\mathcal{E}(\pi\widehat{L}(t)), \quad t \ge 0, \quad X^\pi(0) = x,$$

where $(\ln \mathcal{E}(\pi\widehat{L}(t)))_{t\ge 0}$ is a Lévy process with characteristic triplet (a_X, β_X, ν_X) given by

$$\begin{aligned}a_X &= \pi\left(a - \frac{1}{2}(1-\pi)\beta^2\right) + \int (\ln(1 + \pi(e^x - 1))1(|\ln(1 + \pi(e^x - 1))| \le 1) \\ &\quad - \pi x 1(|x| \le 1))\nu(dx), \\ \beta_X &= \pi\beta, \\ \nu_X(A) &= \nu(\{x \in \mathbb{R} : \ln(1 + \pi(e^x - 1)) \in A\}) \quad \text{for any Borel set } A \subset \mathbb{R}^*.\end{aligned}$$

Setting $\Psi(-si) = \ln Ee^{sL(1)}$ for $s \in \mathbb{R}$ such that the moment generating function is finite, also the existing moments reduce for $t \geq 0$ to

$$E[X^\pi(t)] = x \exp(t(r + \pi(b - r + \Psi(-i))),$$

$$\operatorname{var}(X^\pi(t)) = x \exp(2t((r + \pi(b - r + \Psi(-i)))(\exp(\pi^2 t(\Psi(-2i) - 2\Psi(-i))) - 1).$$

We obtain in the case of a jump part of finite variation for $t \geq 0$,

$$E[X^\pi(t)] = x \exp\left(\left(r + \pi\left(b - r + \frac{1}{2}\beta^2 + \gamma + \widehat{\mu}\right)\right)t\right), \tag{3.71}$$

$$\operatorname{var}(X^\pi(t)) = x^2 \exp\left(2t\left(r + \pi(b - r + \gamma + \widehat{\mu} + \frac{1}{2}\beta^2\right)\right)$$
$$\times(\exp(\pi^2 t(\beta^2 + \widehat{\mu}_2 - 2\widehat{\mu})) - 1), \tag{3.72}$$

where $\widehat{\mu} = \int(e^x - 1)\nu(dx)$, $\widehat{\mu}_2 = \int(e^{2x} - 1)\nu(dx)$, and $\gamma = a - \int_{|x|<1} x\nu(dx)$.

In the following we discuss some estimation methods for the CaR, which means that we have to estimate a small quantile of $\mathcal{E}(\pi'\widehat{L}(T))$; see Definition 3.5.15.

Simulation methods of Lévy processes are often based on infinite series representations; see Rosinski [78] and references therein. In principle, such methods can be applied here to simulate independent copies of $X^\pi(T)$ and estimate the quantile by its empirical counterpart. Such methods are based on the Lévy measure ν_X, which we derived in Lemma 3.5.12. There are, however, two serious drawbacks. The first is that low and high quantiles are even in straightforward models not well estimated by their empirical counterparts; the second is that the infinite series has to be truncated, which obviously is another source of inaccuracy.

We invoke instead an idea used for instance by Bondesson [14] and Rydberg [79] for simulation purposes and made mathematically precise by Asmussen and Rosinski [2]. Before we apply their result to approximate a low quantile as the VaR above we first explain the idea. The intuition behind it is that small jumps ($< \varepsilon$) may be approximated by Brownian motion, whereas large ones ($\geq \varepsilon$) constitute a compound Poisson process N^ε. This normal approximation works for various, but not for all models. In particular, it fails for the exponential variance-gamma model, which has become an important model also in practice. We formulate therefore a more general result.

For a Lévy process with representation (3.65) we write for small $\varepsilon > 0$,

$$L(t) = \mu(\varepsilon)t + \beta W(t) + N^\varepsilon(t) + \int_0^t \int_{|x|<\varepsilon} x(M(ds, dx) - ds\nu(dx))$$
$$\approx \mu(\varepsilon)t + \beta W(t) + N^\varepsilon(t) + \sigma(\varepsilon)V(t), \quad t \geq 0, \tag{3.73}$$

where V is some (hopefully simple) Lévy process and

$$\sigma^2(\varepsilon) = \int_{|x|<\varepsilon} x^2\nu(dx), \tag{3.74}$$

$$\mu(\varepsilon) = a - \int_{\varepsilon\leq|x|\leq 1} x\nu(dx), \tag{3.75}$$

$$N^\varepsilon(t) = \sum_{s\leq t} \Delta L(s) 1_{\{|\Delta L(s)|\geq\varepsilon\}}. \tag{3.76}$$

The approximation (3.73) can be made precise. Under certain conditions it is a consequence of a functional central limit theorem, provided that for $\varepsilon \to 0$ and V standard Brownian motion,

$$\sigma(\varepsilon)^{-1} \int_0^t \int_{|x|<\varepsilon} x(M(ds, dx) - ds\nu(dx)) = \sigma(\varepsilon)^{-1}(L(t) - L_\varepsilon(t)) \xrightarrow{d} V(t), \quad t \geq 0, \quad (3.77)$$

where

$$L_\varepsilon(t) = \mu(\varepsilon)t + \beta W(t) + N_\varepsilon(t), \quad t \geq 0. \quad (3.78)$$

We denote by $\xrightarrow{d}$ weak convergence in $D[0, \infty)$ with the supremum norm uniformly on compacta; see Pollard [71].

Since the Brownian component and the jump component of a Lévy process are independent, (3.77) justifies approximation in distribution (3.73).

We want to invoke this result to approximate quantiles of $\mathcal{E}(\pi'\widehat{L}(T))$. We do this in two steps: firstly, we approximate $\mathcal{E}(\pi'\widehat{L}(T))$; secondly, we use that convergence of dfs implies also convergence of their generalized inverses. This gives the approximation of the quantiles.

Theorem 3.5.16 [Emmer and Klüppelberg [34]]
Let Y be any Lévy process with Lévy measure ν. Let $\mathcal{E}^{\leftarrow}(\exp(Y(\cdot)) = Z(\cdot)$ be such that $\mathcal{E}Z(\cdot) = \exp Y(\cdot)$ with characteristic triplets given in Lemma 3.5.11. Let furthermore, $\sigma(\cdot)$ be defined as in (3.74), and Y_ε and Z_ε as L_ε in (3.78), respectively.
Let V be a Lévy process. Equivalent are for $\varepsilon \to 0$

$$\sigma(\varepsilon)^{-1}(Y(t) - Y_\varepsilon(t)) \xrightarrow{d} V(t), \quad t \geq 0, \quad (3.79)$$

$$(\pi\sigma(\varepsilon))^{-1} (\ln \mathcal{E}(\pi Z(t)) - \ln \mathcal{E}(\pi Z_\varepsilon(t))) \xrightarrow{d} V(t), \quad t \geq 0. \quad (3.80)$$

For the proof we need the following theorem.

Theorem 3.5.17 *Let Z^ε, $\varepsilon > 0$, be Lévy processes without Brownian component and $Y^\varepsilon = \ln \mathcal{E}(Z^\varepsilon)$ their logarithmic stochastic exponentials with characteristic triplets (a_Z, β_Z, ν_Z) and (a_Y, β_Y, ν_Y) as defined in Lemma 3.5.11. Let $g : \mathbb{R} \to \mathbb{R}_+$ with $g(\varepsilon) \to 0$ as $\varepsilon \to 0$ and V some Lévy process. Then equivalent are as $\varepsilon \to 0$,*

$$\frac{Z^\varepsilon(t)}{g(\varepsilon)} \xrightarrow{d} V(t), \quad t \geq 0,$$

$$\frac{Y^\varepsilon(t)}{g(\varepsilon)} \xrightarrow{d} V(t), \quad t \geq 0.$$

Proof of (3.79) ⇔ (3.80). Setting $g(\varepsilon) := \sigma(\varepsilon)$ and $Y^\varepsilon := Y - Y_\varepsilon$ in Theorem 3.5.17 we obtain that (3.79) holds if and only if

$$\sigma(\varepsilon)^{-1}\mathcal{E}^{\leftarrow}(\exp(Y(t) - Y_\varepsilon(t))) \xrightarrow{d} W(t), \quad t \geq 0. \quad (3.81)$$

Applying Theorem 3.5.17 to $g(\varepsilon) := \pi\sigma(\varepsilon)$ and $Z_\varepsilon := \pi\mathcal{E}^{\leftarrow}(\exp(Y(t) - Y_\varepsilon(t)))$ leads to the equivalence of (3.81) and

$$(\pi\sigma(\varepsilon))^{-1}\ln\mathcal{E}(\pi\mathcal{E}^{\leftarrow}(\exp(Y(t) - Y_\varepsilon(t)))) \xrightarrow{d} W(t), \quad t \geq 0.$$

The identity

$$\begin{aligned}&\ln\mathcal{E}(\pi\mathcal{E}^{\leftarrow}(\exp(Y(t) - Y_\varepsilon(t)))) \\ &\quad = \ln\mathcal{E}(\pi\mathcal{E}^{\leftarrow}(\exp(Y(t)))) - \ln\mathcal{E}(\pi\mathcal{E}^{\leftarrow}(\exp Y_\varepsilon(t)))), \quad t \geq 0, \quad (3.82)\end{aligned}$$

which can be proven by calculating all three logarithmic exponentials by Itô's formula (see Emmer and Klüppelberg [34]), leads to the equivalence with (3.80). □

In the finite variation case (3.77), i.e., (3.79) can be rewritten to

$$\sigma(\varepsilon)^{-1}\left(\sum_{0<s\leq t}\Delta L(s)I(|\Delta L(s)| < \varepsilon) - E\left[\sum_{0<s\leq t}\Delta L(s)I(|\Delta L(s)| < \varepsilon)\right]\right) \xrightarrow{d} V(t),$$

as $t \Rightarrow \infty$, which shows immediately the connection to the classical central limit theorem.

We apply (3.80) and (3.82) to approximate $\ln\mathcal{E}(\pi\widehat{L})$ for $\pi \in (0, 1]$ as follows.

$$\ln\mathcal{E}(\pi\mathcal{E}^{\leftarrow}(\exp(L(t)))) \approx \ln\mathcal{E}(\pi\mathcal{E}^{\leftarrow}(L_\varepsilon(t)))) + \pi\sigma(\varepsilon)V(t), \quad t \geq 0,$$

and hence we obtain

$$\ln\mathcal{E}(\pi\mathcal{E}^{\leftarrow}(\exp(L(t)))) \approx \gamma_\pi^\varepsilon t + \pi\beta W(t) + M_\pi^\varepsilon(t) + \pi\sigma(\varepsilon)V(t), \quad t \geq 0,$$

where

$$\begin{aligned}\gamma_\pi^\varepsilon &= \pi(\mu(\varepsilon) + (1-\pi)\beta^2/2), \\ M_\pi^\varepsilon(t) &= \sum_{s\leq t}\ln(1 + \pi(\exp(\Delta L(s)1_{\{|\Delta L(s)|>\varepsilon\}}) - 1)),\end{aligned}$$

i.e., M_π^ε is a compound Poisson process with jump measure

$$\nu_{M_\pi^\varepsilon}(\Lambda) = \nu_L(\{x : \ln(1 + \pi(e^x - 1)) \in \Lambda\}\backslash(-\varepsilon, \varepsilon))$$

for any Borel set $\Lambda \subset \mathbb{R}^*$. □

By Proposition 0.1 of Resnick [74] we obtain the corresponding approximation for the α-quantile z_α of $\mathcal{E}(\pi\widehat{L}(T)$.

Proposition 3.5.18 *With the quantities as defined above we obtain*

$$z_\alpha \approx z_\alpha^\varepsilon(\pi) = \inf\{z \in \mathbb{R} : P\left(\gamma_\pi^\varepsilon T + \pi\beta W(T) + M_\pi^\varepsilon(T) + \pi\sigma_L(\varepsilon)V(T) \leq \ln z\right) \geq \alpha\},$$

giving the following approximations

$$\begin{aligned}\text{VaR}(x, \pi, T) &\approx x z_\alpha^\varepsilon(\pi)\exp((\pi(b-r) + r)T), \\ \text{CaR}(x, \pi, T) &\approx x e^{rT}\left(1 - z_\alpha^\varepsilon(\pi)e^{\pi(b-r)T}\right).\end{aligned}$$

The following corollary characterizes, the normal approximation.

Corollary 3.5.19 [Asmussen and Rosinski [2]]
(a) V is standard Brownian motion if and only if

$$\sigma(h\sigma(\varepsilon) \wedge \varepsilon) \sim \sigma(\varepsilon) \text{ for each } h > 0. \tag{3.83}$$

(b) Condition (3.83) holds if $\lim_{\varepsilon\downarrow 0} \sigma(\varepsilon)/\varepsilon = \infty$.
(c) If the Lévy measure has no atoms in a neighborhood of 0, then condition (3.83) is equivalent to $\lim_{\varepsilon\downarrow 0} \sigma(\varepsilon)/\varepsilon = \infty$.

Provided the above condition is satisfied, we have reduced the problem of estimating a quantile of a complicated Lévy process to the estimation of a quantile of the sum of the compound Poisson rv $M_\pi^\varepsilon(T)$ and the normal rv $\widetilde{W}(T) := \pi'(\beta^2+\sigma_L^2(\varepsilon))^{1/2} W(T)$. We calculate the density of $M_\pi^\varepsilon(T)+\widetilde{W}(T)$ using the Fast Fourier Transform method, henceforth abbreviated as FFT. By independence, we have for the characteristic function of $M_\pi^\varepsilon(T) + \widetilde{W}(T)$

$$\phi(u) = \phi_{M_\pi^\varepsilon(T)}(u)\phi_{\widetilde{W}(T)}(u), \quad u \in \mathbb{R}. \tag{3.84}$$

Denote by $h_{M_\pi^\varepsilon}$ the Lévy density of M_π^ε, which we assume to exist, then we obtain

$$\phi_{M_\pi^\varepsilon(T)}(u) = \exp(T\nu_{M_\pi^\varepsilon}(\mathbb{R})(\phi_Y(u) - 1)), \quad u \in \mathbb{R},$$

where

$$\phi_Y(u) = \frac{1}{\nu_{M_\pi^\varepsilon}(\mathbb{R})} \int e^{iux} h_{M_\pi^\varepsilon}(x)dx, \quad u \in \mathbb{R}. \tag{3.85}$$

Furthermore, by normality,

$$\phi_{\widetilde{W}(T)}(u) = \exp\big(-Tu^2\pi^2\big(\beta^2 + \sigma_L^2(\varepsilon)\big)\big), \quad u \in \mathbb{R}.$$

We approximate the integral in (3.85) by the trapezoid rule, choosing a number n (a power of 2) of intervals and a step size Δx. Set $g = h_{M_\pi^\varepsilon}/\nu_{M_\pi^\varepsilon}(\mathbb{R})$. We truncate the integral ϕ_Y and obtain

$$\begin{aligned}
\int_{-\infty}^{\infty} e^{iux} g(x)dx &\approx \int_{-(n/2)\Delta x}^{(n/2-1)\Delta x} e^{iux} g(x)dx \\
&\approx \sum_{k=-(n/2)}^{n/2-1)} e^{iuk\Delta x} g((k\Delta x)\Delta x \\
&= \sum_{k=0}^{n-1} e^{iu(k-n/2)\Delta x} g((k-n/2)\Delta x)\Delta x \\
&= \Delta x e^{-iun\Delta x/2} \sum_{k=0}^{n-1} e^{iuk\Delta x} g((k-n/2)\Delta x).
\end{aligned}$$

For $g_k := g((k - n/2)\Delta x)$, $k = 0, \ldots, n-1$, the sum is the discrete Fourier transform of the complex numbers g_k and can be calculated by the FFT algorithm for $u_k = 2\pi k/(n\Delta x)$, $k = 0, \ldots, n-1$, simultaneously (see, e.g., Brigham [20],

Chapter 10). This results in an approximation for ϕ in (3.84). By the inverse FFT we obtain the density of $M_\pi^\varepsilon(T) + W(T)$.

Example 3.5.20 *[Exponential Brownian motion with jumps]*
Here the Lévy process is the sum of a Brownian motion with drift $(\beta W(t)+\gamma t)_{t\geq 0}$, and a compound Poisson process $(L(t))_{t\geq 0}$, with Poisson intensity c and p as distribution of the jump heights $(Y_i)_{i\in\mathbb{N}}$. For illustrative purpose we restrict this example to one compound Poisson process, we could as well take several different ones, see, e.g., [35]. The drift $\gamma = -\frac{1}{2}\beta^2 - \widehat{\mu}$ is chosen such that it compensates the jumps. The Lévy measure is $\nu(dx) = cp(dx)$ and hence $\widehat{\mu} = c(\widehat{g}(1) - 1)$ and $\widehat{\mu}_2 = c(\widehat{g}(2) - 1)$, where g is the moment generating function of Y_1, which we assume to exist at the required points. By (3.71) and (3.72) we obtain for $t \geq 0$

$$\begin{aligned} X^\pi(t) &= x \exp\left(t\left(r + \pi(b-r) - \pi\widehat{\mu} - \frac{1}{2}\pi^2\beta^2\right) + \pi\beta W(t)\right) \\ &\quad \times \prod_{i=1}^{N(t)} (1 + \pi(e^{Y_i} - 1)), \\ E[X^\pi(t)] &= x \exp(t(r + \pi(b-r))), \end{aligned}$$

$$\mathrm{var}(X^\pi(t)) = x^2 \exp(2t(r + \pi(b-r))) \times (\exp(\pi^2 t(\beta^2 + c(\widehat{g}(2) - 2\widehat{g}(1) + 1))) - 1).$$

Note that for $c = 0$ the model reduces to exponential Brownian motion; i.e.

$$X^\pi(t) = x \exp\left(t\left(r + \pi(b-r) - \frac{1}{2}\pi^2\beta^2\right) + \pi\beta W(t)\right).$$

On the other hand, if $\beta = 0$ the model reduces to exponential compound Poisson process; i.e.,

$$X^\pi(t) = x \exp(t(r + \pi(b-r) - \pi\widehat{\mu})) \prod_{i=1}^{N(t)} (1 + \pi(e^{Y_i} - 1)).$$

Example 3.5.21 *[Exponential normal inverse Gaussian (NIG) Lévy process]*
This normal mixture model has been suggested by Barndorff-Nielsen [5, 4]; see also Eberlein and collaborators [28, 29, 30]. It has the representation

$$L(t) = \rho t + \lambda\sigma^2(t) + W(\sigma^2(t)), \quad t \geq 0,$$

where $\rho, \lambda \in \mathbb{R}$, W is standard Brownian motion and $(\sigma^2(t))_{t\geq 0}$ has inverse Gaussian increments. The process $(L(t))_{t\geq 0}$ is uniquely determined by the distribution of the increment $L(1)$, which is NIG (see Barndorff-Nielsen [5]). This means that $L(1) \stackrel{d}{=} N(\rho + \lambda Z, Z)$, where $N(a, b)$ denotes a normal rv with mean a and variance b and Z is inverse Gauss distributed; more precisely, the density of $L(1)$ is given by

$$nig(x, \alpha, \lambda, \rho, \delta) := \frac{\alpha}{\pi} \exp(\delta\sqrt{\alpha^2 - \lambda^2} + \lambda(x-\rho)) \frac{K_1(\delta\alpha g(x-\rho))}{g(x-\rho)}, \quad x \in \mathbb{R},$$

where $\alpha > |\lambda| \geq 0$, $\delta > 0$, $\rho \in \mathbb{R}$, $g(x) = \sqrt{\delta^2 + x^2}$ and

$$K_1(s) = \frac{1}{2}\int_0^\infty \exp\Big(-\frac{1}{2}s(x + x^{-1})\Big)dx$$

is the modified Bessel function of the third kind. The parameter α is a steepness parameter, i.e., for larger α we get less large and small jumps and more jumps of middle height, δ is a scale parameter, λ is a symmetry parameter and ρ a location parameter. For $\rho = \lambda = 0$ (symmetry around 0) the characteristic triplet of a NIG Lévy process is given by $(0, 0, \nu)$ with

$$\nu(dx) = \frac{\delta\alpha}{\pi}|x|^{-1}K_1(\alpha|x|)dx, \quad x \in \mathbb{R}^*.$$

We can calculate $\widehat{\mu}$ and $\widehat{\mu}_2$ via the moment generating function of $L(1)$, which is for the NIG distribution given by

$$E\exp(sL(1)) = \exp(\delta(\alpha - \sqrt{\alpha^2 - s^2})), \quad |s| < \alpha,$$

(see, e.g., Raible [73], Example 1.6) and hence because of symmetry,

$$\widehat{\mu} = \delta(\alpha - \sqrt{\alpha^2 - 1}),$$
$$\widehat{\mu}_2 = \delta(\alpha - \sqrt{\alpha^2 - 4}).$$

Plugging these results into (3.71) and (3.72), and choosing $b_{nig} = b_{BS} - \delta(\alpha - \sqrt{\alpha^2 - 1})$ (b_{BS} is the quantity b from Example 3.5.20), such that the expectation of an asset in the NIG model is the same as for the exponential Brownian motion, we obtain for $t \geq 0$,

$$\begin{aligned} X^\pi(t) &= x\exp(t(r + \pi(b_{BS} - r - \delta(\alpha - \sqrt{\alpha^2 - 1})))) \\ &\quad \times \prod_{0<s\leq t}\left(1 + \pi(e^{\Delta L(s)} - 1)\right), \\ E[X^\pi(t)] &= x\exp(t(r + \pi(b_{BS} - r)), \\ \operatorname{var}(X^\pi(t)) &= x^2\exp(2t(\pi(b_{BS} - r)) + r)) \\ &\quad \times(\exp(\delta\pi^2 t(2\sqrt{\alpha^2 - 1} - \alpha - \sqrt{\alpha^2 - 4})) - 1). \end{aligned}$$

By Corollary 3.5.19, for the exponential normal inverse Gaussian Lévy process the normal approximation for small jumps is allowed since $\sigma^2(\varepsilon) \sim (2\delta/\pi)\varepsilon$ as $\varepsilon \to 0$. For an estimate of the α-quantile we invoke Proposition 3.5.18 and use FFT. Figure 3.16 shows sample paths of a geometric NIG-Lévy process with certain parameter values.

Example 3.5.22 *[Exponential variance gamma (VG) model]*
This normal mixture model has been suggested by Madan and Seneta [64], its nonsymmetric version can be found in Madan, Carr and Chang [63]. An interesting empirical investigation has been conducted by Carr et al. [22]. The nonsymmetric model is

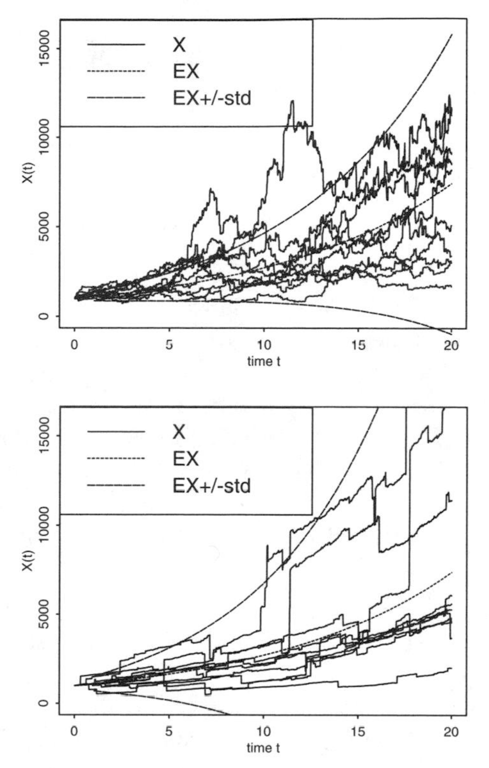

Figure 3.16 *Ten sample paths of the exponential NIG Lévy process with* $\alpha = 8$ *and* $\delta = 0.32$ *(top) and with* $\alpha = 2$ *and* $\delta = 0.08$ *(bottom), its expectation* $\mathcal{E}(L(T))$ *and expectation* $\pm$ *standard deviation for* $x = 1\,000$, $b = 0.1$, *and* $r = 0.05$.

defined as follows.

$$L(t) = \rho t + \lambda\sigma^2(t) + W(\sigma^2(t)), \quad t \geq 0,$$

where $\rho, \lambda \in \mathbb{R}$, W is standard Brownian motion and $(\sigma^2(t))_{t\geq 0}$ has gamma increments, more precisely, $\sigma^2(s) \stackrel{d}{=} \Gamma(\alpha s, \theta)$ for parameters $\alpha > 0$ and $\theta > 0$; i.e., it has density

$$h(x) = \frac{x^{\alpha s - 1}}{\Gamma(\alpha s)\theta^{\alpha s}} e^{-x/\theta}, \quad x > 0.$$

The characteristic function of $L(1)$ is given by

$$E\exp(isL(1)) = \frac{\exp(is\rho t)}{(1 - is\theta\lambda + s^2\theta/2)^{\alpha t}} \quad s \in \mathbb{R}.$$

The Lévy process L is a pure jump process with Lévy density

$$\nu(dx) = \frac{\alpha}{|x|} \exp\left(-\sqrt{\frac{2}{\theta} + \lambda^2}\, |x| + \lambda x \right) dx, \quad x \in \mathbb{R}^*.$$

We obtain as before

$$X^\pi(t) = x\exp(t(r + \pi(b - r + \rho)))\prod_{s\le t}(1 + \pi(e^{\Delta L(s)} - 1)),$$

$$E[X^\pi(t)] = x(1 - \theta\lambda - \theta/2)^{-\alpha\pi t}\exp(t(r + \pi(b - r + \rho))),$$

$$\begin{aligned}\operatorname{var}(X^\pi(t)) &= x^2(1 - \theta\lambda - \theta/2)^{-2\alpha\pi t}\exp(2t(r + \pi(b - r + \rho)))\\ &\quad\times\left(\left(\frac{(1 - \theta\lambda - \theta/2)^2}{1 - 2\theta\lambda - 2\theta}\right)^{\alpha\pi^2 t} - 1\right).\end{aligned}$$

For our figures we choose $\rho = \alpha\ln(1 + \theta\lambda - \theta/2)$ and $b = b_{BS}$ such that $E[X^\pi(t)] = x\exp((r + \pi(b_{BS} - r))t)$. In order to find an approximation for the VaR we calculate $\sigma^2(\varepsilon) \sim \alpha\varepsilon^2$ as $\varepsilon \to 0$. Since its Lévy measure has no atoms in a neighborhood of 0, by Corollary 3.5.19, the normal approximation for small jumps is not allowed.

However, there is another limit process to allow for approximation of the small jumps: for $\varepsilon \to 0$

$$\sigma(\varepsilon)^{-1}(L(t) - L_\varepsilon(t)) \stackrel{d}{\to} V(t), \quad t \ge 0,$$

where V is a Lévy process with characteristic triplet $(0, 0, \nu_V)$ where the Lévy measure ν_V has density $\nu_V(dv) = (\alpha/v)1_{(-1/\sqrt{\alpha},1/\sqrt{\alpha})}(v)dv$. This means that the following approximation is valid

$$z_\alpha \approx z_\alpha^\varepsilon(\pi) = \inf\{z \in \mathbb{R} : P(\gamma_\pi^\varepsilon T + M_\pi^\varepsilon + \pi\sigma(\varepsilon)V(T) \le \ln z) \ge \alpha\},$$

giving again approximations as in Proposition 3.5.18.

Remark 3.5.23 When we want to perform a portfolio optimization for the different exponential Lévy models as price processes, then certain structures can be exploited. Note, e.g., that the expected wealth process is increasing in π; hence the optimal portfolio is always the largest π such that the risk bound is satisfied. For Lévy processes additionally $\pi \le 1$ has to be satisfied. Such π can always easily be found by a simple numerical iteration procedure.

Next note that to make results comparable we have chosen all mean portfolio processes equal.

(a) *Mean-variance optimization:* Since NIG and VG models have so many parameters we can always choose them so that all variances are equal in the different examples. Then, of course, the mean-variance optimization problem always leads to the same result.

(b) *Mean-CaR optimization:* Here the shape of the distribution in the left tail enters; see Figure 3.17. The heavier the tail at the corresponding α-quantile, the higher the risk, i.e., the more cautious the investment π into the risky stock.

For more details see Emmer and Klüppelberg [34]. □

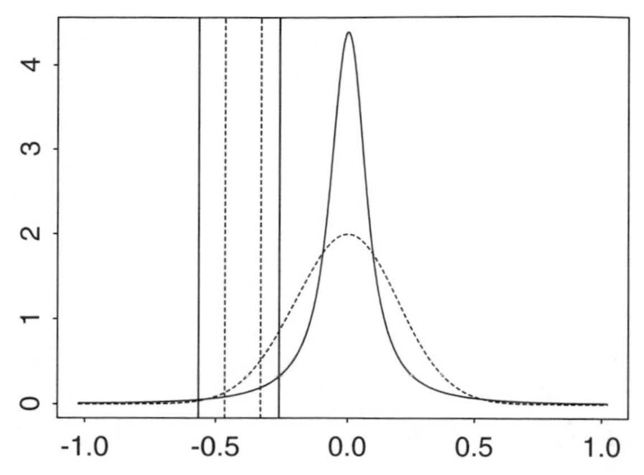

Figure 3.17 *Density of $L(1)$ for the NIG model with parameters $\alpha = 2$, $\delta = 0.08$, $\lambda = \rho = 0$, $x = 1\,000$, $b = 0.1$, and $r = 0.05$. The normal density with the same variance 0.04 is plotted for comparison. Moreover, the respective 1%-quantiles (left vertical lines) and 5%-quantiles (right vertical lines) are plotted. All solid lines correspond to the NIG model, all dotted ones to the normal model.*

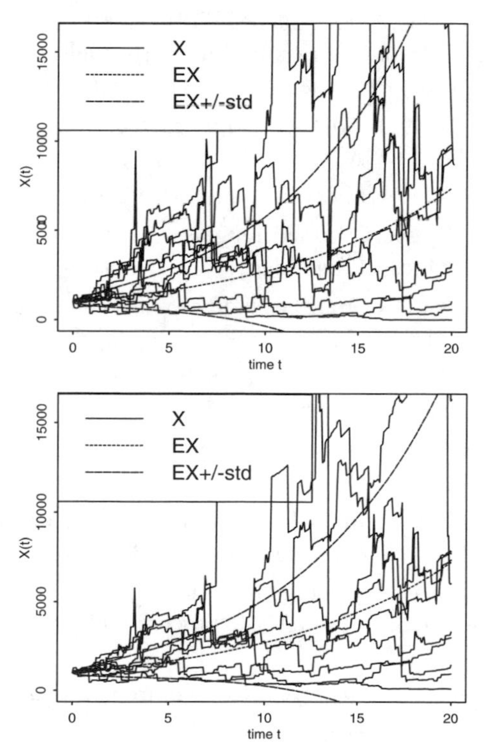

Figure 3.18 *Ten sample paths of the exponential VG Lévy process with $\alpha = 0.1$, $\delta = 0$, $\theta = 0.35$, and $\mu = -0.019$ (top) and with $\alpha = 0.2$, $\delta = 0$, $\theta = 0.2$, and $\mu = -0.022$ (bottom), its expectation $E[L(T)]$ and expectation $\pm$ standard deviation for $x = 1\,000$, $b = 0.1$, and $r = 0.05$.*

Bibliography

[1] Artzner, P., Delbaen, F., Eber, J.M. and Heath, D. (1999) Coherent measures of risk. *Math. Finance* 9, 203–228.

[2] Asmussen, S. and Rosinski, J. (2000) Approximation of small jumps of Lévy processes with a view towards simulation. *J. Appl. Probab.* 38, 482–493.

[3] Balkema, A.A. and Haan, L. de (1974) Residual lifetime at great age. *Ann. Probab.* 2, 792–804.

[4] Barndorff-Nielsen, O.E. (1997) Normal inverse Gaussian distributions and stochastic volatility modeling. *Scand. J. Statist.* 24, 1–14.

[5] Barndorff-Nielsen, O.E. (1998) Processes of normal inverse Gaussian type. *Finance and Stochastics* 2, 41–68.

[6] Barndorff-Nielsen, O.E. and Cox, D. (1994) *Inference and Asymptotics.* Chapman & Hall, London.

[7] Barndorff-Nielsen, O.E., Mikosch, T. and Resnick, S.I. (Eds.) (2001) *Lévy Processes — Theory and Applications.* Birkhäuser, Boston.

[8] Barndorff-Nielsen, O.E. and Shephard, N. (2001) Modelling by Lévy processes for financial econometrics. In: Barndorff-Nielsen, O.E., Mikosch, T. and Resnick, S. (eds) *Lévy Processes –Theory and Applications.* Birkhäuser, Boston.

[9] Berman, S.M. (1992) *Sojourns and Extremes of Stochastic Processes.* Wadsworth, Belmond, CA.

[10] Bertoin, J. (1996) *Lévy Processes.* Cambridge University Press, Cambridge.

[11] Bingham, N.H., Goldie, C.M. and Teugels, J.L. (1987) *Regular Variation.* Cambridge University Press, Cambridge.

[12] Bollerslev, T. (1986) Generalized autoregressive conditional heteroscedasticity. *J. Econometrics* 31, 307–327.

[13] Bollerslev, T., Chou, R.Y. and Kroner, K.F. (1992) ARCH modeling in finance: a review of the theory and empirical evidence. *J. Econometrics* 52, 5–59.

[14] Bondesson, L. (1982) On simulation from infinitely divisible distributions. *Adv. Appl. Probab.* 14, 855–869.

[15] Borkovec, M. (2000) Extremal behaviour of the autoregressive process with ARCH(1) errors. *Stoch. Proc. Appl.* 85, 189–207.

[16] Borkovec, M. (2001) Asymptotic behaviour of the sample autocovariance and autocorrelation function of the AR(1) process with ARCH(1) errors. *Bernoulli* 7, 847–872.

[17] Borkovec, M. and Klüppelberg, C. (1998) Extremal behaviour of diffusion models in finance. *Extremes* 1, 47–80.

[18] Borkovec, M. and Klüppelberg, C. (2001) The tail of the stationary distribution of an autoregressive process with ARCH(1) errors. *Ann. Appl. Probab.* 11, 1220–1241.

[19] Brandt, A., Franken, P. and Lisek, B. (1990) *Stationary Stochastic Models.* Wiley, Chichester.

[20] Brigham, E.O. (1974) *The Fast Fourier Transform.* Prentice-Hall, Englewood Cliffs, NJ.

[21] Campbell, H.Y., Lo, A.W. and MacKinley, A.C. (1997) *The Econometrics of Financial Markets.* Princeton University Press, Princeton, NJ.

[22] Carr, P., Geman, H., Madan, D.B. and Yor, M. (2001) The fine structure of asset returns: an empirical investigation. *Journal of Business.*

[23] Cox, J.C., Ingersoll, J.E. and Ross, S.A. (1985) A theory of the term structure of interest rates. *Econometrica* 53, 385–408.

[24] Cvitanic, J. and Karatzas, I. (1999) On dynamic measures of risk. *Finance and Stochastics* 3, 451–482.

[25] Davis, R.A. (1982) Maximum and minimum of one-dimensional diffusions. *Stoch. Proc. Appl.* 13, 1–9.

[26] Diaconis, P. and Freedman, D. (1999) Iterated random functions. *SIAM Review* 41, 45–76.

[27] Eberlein, E. (2001) Application of generalized hyperbolic Lévy motion to finance. In: Barndorff-Nielsen, O.E., Mikosch, T. and Resnick, S. (eds) *Lévy Processes — Theory and Applications.* Birkhäuser, Boston.

[28] Eberlein, E. and Keller, U. (1995) Hyperbolic distributions in finance. *Bernoulli* 1, 281–299.

[29] Eberlein, E., Keller, U. and Prause, K. (1998) New insights into smile, mispricing and value at risk: the hyperbolic model. *Journal of Business* 71, 371–405.

[30] Eberlein, E. and Raible, S. (1999) Term structure models driven by general Lévy processes. *Math. Finance.* 9, 31–53.

[31] Embrechts. P. (2004) Extremes in economics and the economics of extremes, in *Extreme Values in Finance, Telecommunications, and the Environment*, Finkenstädt, B. and Rootźen, H., Eds., CRC/Chapman & Hall, Boca Raton, Ch. 4.

[32] Embrechts. P. (Ed.) (2000) *Extremes and Integrated Risk Management.* UBS Warburg and Risk Books.

[33] Embrechts, P., Klüppelberg, C. and Mikosch, T. (1997) *Modelling Extremal Events for Insurance and Finance.* Springer, Berlin.

[34] Emmer, S. and Klüppelberg, C. (2001) Optimal portfolios when stock prices follow an exponential Lévy process. *Finance and Stochastics*, to appear.

[35] Emmer, S., Klüppelberg, C. and Korn, R. (2000) Optimal portfolios with bounded Capital-at-Risk. *Math. Finance* 11, 365–384.

[36] Emmer, S., Klüppelberg, C., Trüstedt, M. (1998) VaR - ein Mass für das extreme Risiko. *Solutions* 2, 53–63. English version available at *http://www.ma.tum.de/stat/*

[37] Engle, R. F. (1982) Autoregressive conditional heteroscedasticity with estimates of the variance of U.K. inflation. *Econometrica* 50, 987–1007.

[38] Engle, R.F. (Ed.) (1995) *ARCH. Selected Readings.* Oxford University Press, Oxford.

[39] Fishburn, P.C. (1977) Mean-risk analysis with risk associated with below-target returns. *American Economic Review* 67, 116–126.

[40] Föllmer, H. and Leukert, P. (1999): Quantile hedging. *Finance and Stochastics* 3, 251–273.

[41] Goldie, C.M. (1991) Implicit renewal theory and tails of solutions of random equations. *Ann. Appl. Prob.* 1, 126–166.

[42] Goldie, C.M. and Maller, R. (2000) Stability of perpetuities. *Ann. Probab.* 28, 1195–1218.

[43] Goll, T. and Kallsen, J. (2000) Optimal portfolios for logarithmic utility. *Stoch. Proc. Appl.* 89, 31–48.

[44] Gouriéroux, C. (1997) *ARCH Models and Financial Applications.* Springer, New York.

[45] Grigelionis, B. (2001) On point measures of ε-upcrossings for stationary diffusions. Preprint. Vilnius University.

[46] Haan, L. de, Resnick, S.I., Rootzén, H. and Vries, C.G. de (1989) Extremal behaviour of solutions to a stochastic difference equation, with applications to ARCH processes. *Stoch. Proc. Appl.* 32, 213–224.

[47] Harlow, W. (1991) Asset allocation in a downside-risk framework. *Financial Analyst Journal*, 28–40.

[48] Hsing, T., Hüsler, J. and Leadbetter, M.R. (1988) On the exceedance point process for a stationary sequence. *Probab. Theory Related Fields* 78, 97–112.

[49] Itô, K.I. and McKean, H.P. (1974) *Diffusion Processes and Their Sample Paths.* Springer, Berlin.

[50] Joriot, P. (1997) *Value at Risk: the New Benchmark for Controlling Market Risk.* Irwin, Chicago.

[51] Kallenberg, O. (1983) *Random Measures.* Akademie Verlag, Berlin.

[52] Karatzas, I. and Shreve, S.E. (1988) *Brownian Motion and Stochastic Calculus.* Springer, New York.

[53] Kesten, H. (1973) Random difference equations and renewal theory for products of random matrices. *Acta Math.* 131, 207–248.

[54] Kesten, H. (1974) Renewal theory for functional of a Markov chain with general state space. *Ann. Probab.* 2, 355–386.

[55] Kiefersbeck, N. (1999) Stationarität und Tailindex bei zeitdiskreten Volatilitätsmodellen. Diploma Thesis. Munich University of Technology. *www.ma.tum.de/stat/*

[56] Kloeden, P.E. and Platen, E. (1992) *Numerical Solutions of Stochastic Differential Equations.* Springer, Berlin.

[57] Klüppelberg, C., Maller R.A., Van De Vyver M. and Wee, D. (2000) Testing for reduction to random walk in autoregressive conditional heteroscedasticity models. *Econometrics Journal* 5, 387–416.

[58] Klüppelberg, C. and Pergamenchtchikov, S. (2001) Renewal theory for functionals of a Markov chain with compact state space. *Ann. Probab.*, to appear.

[59] Klüppelberg, C. and Pergamenchtchikov, S. (2001) The tail of the stationary distribution of a random coefficient AR(q) process with applications to an ARCH(q) process. Submitted for publication. *www.ma.tum.de/stat/*

[60] Korn, R. (1997) *Optimal Portfolios.* World Scientific, Singapore.

[61] Lamberton, D. and Lapeyre, B. (1991) *Introduction au calcul stochastique appliqué à la finance.* Edition Marketing S.A., Ellipses, SMAI, Paris. English translation by Rabeau, N. and Mantion, F. (1996) *Introduction to Stochastic Calculus Applied to Finance.* Chapman & Hall, London.

[62] Leadbetter, M.R., Lindgren, G. and Rootzén, H. (1983) *Extremes and Related Properties of Random Sequences and Processes.* Springer, Berlin.

[63] Madan, D.B., Carr, P. and Chang, E. (1998) The variance gamma process and option pricing. *European Finance Review*, 2, 79–105.

[64] Madan, D.B. and Seneta, E. (1990) The variance gamma (VG) model for share market returns. *Journal of Business* 63, 511–524.

[65] Markowitz, H. (1959) *Portfolio Selection — Efficient Diversification of Investments.* New York, Wiley.

[66] Merton, R.C. (1994) *Continuous–Time Finance*, revised ed., Blackwell, Cambridge, MA.

[67] Meyn, S. and Tweedie, R. (1993) *Markov Chains and Stochastic Stability.* Springer, New York.

[68] Mikosch, T. and Starica, C. (2000) Limit theory for the sample autocorrelations and extremes of a GARCH(1,1) process. *Ann. Statist.* 28, 1427–1451.

[69] Perfekt, R. (1994) Extremal behaviour of stationary Markov chains with applications. *Ann. Appl. Probab.* 4, 529–548.

[70] Pickands, J. III (1975) Statistical inference using extreme order statistics. *Ann. Statist.* 3, 119–131.

[71] Pollard, D. (1984) *Convergence of Stochastic Processes.* Springer, New York.

[72] Protter, P. (1990) *Stochastic Integrals and Differential Equations.* Springer, New York.

[73] Raible, S. (2000) *Lévy Processes in Finance: Theory, Numerics, and Empirical Facts.* PhD thesis, University of Freiburg. *http://www.freidok.uni-freiburg.de/volltexte/51*

[74] Resnick, S.I. (1987) *Extreme Values, Regular Variation and Point Processes.* Springer, New York.

[75] Revuz, D. and Yor, M. (1991) *Continuous Martingales and Brownian Motion.* Springer, Berlin.

[76] Rootzén, H. and Klüppelberg, C. (1999) A single number can't hedge against economic catastrophes. *Ambio* 28, 6, 550–555. Royal Swedish Academy of Sciences.

[77] Rogers, L.C.G. and Williams, D. (2000) *Diffusions, Markov Processes, and Martingales, Vol. 2, 2nd Ed.* Cambridge University Press, Cambridge.

[78] Rosinski, J. (2001) Series representations of Lévy processes from the perspective of point processes. In: Barndorff-Nielsen, O.E., Mikosch, T. and Resnick, S. (eds) *Lévy Processes — Theory and Applications.* Birkhäuser, Boston.

[79] Rydberg, T. (1997) The normal inverse Gaussian Lévy process: simulation and approximation. *Stoch. Models* 13, 887–910.

[80] Sato, K. I. (1999) *Lévy Processes and Infinitely Divisible Distributions.* Cambridge University Press, Cambridge.

[81] Sharpe, W. (1964) A theory of market equilibrium under condition of risk. *Journal of Finance* 19, 1536–1575.

[82] Shephard, N. (1996) Statistical aspects of ARCH and stochastic volatility. In: Cox, D.R., Hinkley, D.V. and Barndorff-Nielsen, O.E. (Eds.) *Likelihood, Time Series with Econometric and other Applications.* Chapman and Hall, London.

[83] Smith, R. (2004) Statistics of Extremes with applications in environment, insurance, and finance, in *Extreme Values in Finance, Telecommunications, and the Environment*, Finkenstädt, B. and Rootźen, H., Eds., CRC/Chapman & Hall, Boca Raton, Ch. 1.

[84] Taylor, S. (1995) *Modelling Financial Time Series.* Wiley, Chichester.

[85] Tweedie, R.L. (1976) Criteria for classifying general Markov chains. *Adv. Appl. Prob.* 8, 737–771.

[86] Vasicek, O.A. (1977) An equilibrium characterization of the term structure. *J. Fin. Econ.* 5, 177–188.

[87] Vervaat, W. (1979) On a stochastic difference equation and a representation of nonnegative infinitely divisible random variables. *Adv. Appl. Prob.* 11, 750–783.

[88] Weiss, A.A. (1984) ARMA models with ARCH errors. *J. Time Ser. Analysis* 3, 129–143.

CHAPTER 4

Extremes in Economics and the Economics of Extremes

Paul Embrechts
Department of Mathematics, ETH Zürich

Contents

4.1 About the title

Within econometrics, probability theory and statistics, an enormous literature exists on the topic of extremes in economics. See, for instance, Mikosch [47] and Klüppelberg [36] in this volume, and the numerous references listed in those contributions. For a long time, econometric research has shown that, for instance, logarithmic returns of financial data are nonnormal. Extremal moves up or down do occur much more regularly than standard (normal based) models make us believe. Extreme value theory (EVT) has become a standard toolkit within quantitative finance useful for describing these nonnormal phenomena. Statistically exploring and stochastically modelling such extremes in financial data is, however, a rather different task from answering the question: "Given a financial market where such extremes occur, how are they to be handled from an economic point of view?" Perhaps the most striking answer to the question came in the late eighties, early nineties through the emergence of risk management (RM) regulations. In this chapter, I want to highlight some of these developments, stress the interplay between EVT and RM and hint at possible areas

1-58488-411-8/04/$0.00+$.50

of research where the focus is more on the second part of the title, "the Economics of Extremes." Several references will guide the reader to related publications.

4.2 On the history of financial risk management

In his excellent text Steinherr [51], the author, states, "Risk Management: one of the most important innovations of the 20th century." This statement summarizes the revolution we witnessed on financial markets during the second half of the 20th century. Some key dates that emphasize this revolution are (without striving for completeness):

1933 (four years after the 1929 crash): the Glass–Steagall Act was passed in the U.S. in the aftermath of the Depression, prohibiting commercial banks from underwriting insurance and most kind of securities. From that act emerged a new trend of financial institutions: the investment bank. Many of the limitations embedded in the Glass–Steagall Act were gradually softened, leading to its abolishment and reformulation through the 1999 Financial Services Act repealing many key provisions of Glass–Steagall. As a consequence, bank holding companies will continue to expand the range of their financial services, and further convergence of finance and insurance is likely. For a critical discussion on the latter, see Cummins [14].

Around the early fifties, the foundation of modern portfolio theory was laid, for instance through the seminal work of Harry Markowitz. Risk (measured through standard deviation) entered as an extra dimension next to (excess) return. The risk-return diagram with its efficient frontier became the bread and butter of any portfolio manager.

In 1970, the Bretton Woods system of fixed exchange rates was abolished, leading overnight to increased exchange rate volatility. The seventies also saw several oil crises making energy an unpredictable, highly volatile market component. Investors were looking for instruments that would enable them to hedge this increased riskiness on financial markets.

Through the work of Fisher Black, Myron Scholes, and Robert Merton (1972), this search for hedging instruments got a scientific response in that financial derivatives could be rationally priced and hedged. Advanced mathematics and finance joined forces in order to come up with what we now call the Black-Scholes pricing formula (framework) for options; see Black and Scholes [7]. The floodgates opened starting with the opening in 1973 of the Chicago Board of Options Exchange (CBOE). In those days, it seemed that the sky was the limit.

An enormous growth in both volume and complexity of instruments traded on financial markets resulted. For example, on the New York Stock Exchange, the 3.5 million shares traded daily in 1970 grew to a volume of 40 million in 1990. The nominal value of the so-called over-the-counter (OTC) derivatives increased over the period from 1995 until 1998 from $13 trillion to $18 trillion for forex contracts, $26 to $50 trillion for interest rate contracts and across all types from \$47 to \$80 trillion. To be sure, \$1 trillion $= 1 \times 10^{12}$; an enormous figure indeed! For details on these statistics, and much more, see Crouchy et al. [12]. At this point, it needs stressing that all these developments were made possible by an unprecedented growth in IT technology.

In the late eighties and early nineties, first attempts were made, both industry-internally as well as from a regulatory point of view, to get these so-called off-balance instruments (derivatives) under control. Again, Crouchy et al. [12] has the full story. For the purpose of this chapter, it suffices to recall the work of the Basel Committee on Banking Supervision. No doubt this regulatory framework came very much into being due to some spectacular losses including Orange County, Metallgesellschaft, Barings, and indeed more recently LTCM and Enron. A lot has been written on these spectacular losses, so much that Professor Stephen Ross (MIT) has been going around starting his talk with "I am like a financial pathologist, I dissect financial corpses." Under the title "Disasters: Divine Results Rocked by Human Recklessness," Boyle and Boyle [9] have written an excellent account on Orange County, Barings, and LTCM. See also Jorion [34] for an interesting discussion on LTCM. Specific contributions on RM for alternative investments (hedge funds, private equity, alternative risk transfer, ...) are Jaeger [32] and Lane [38].

It is essentially the regulatory framework that contributes strongly to laying down the rules for the economics of extremes, hence in the next section, we will have a closer look at these rules.

4.3 Basel I and II

When I refer to extremes in economics, I refer to the modeling and analysis of extremes in econometric data. As an example, look at the recent paper by Longin [43], one of the pioneers of EVT in finance. In that paper, the author, for instance, estimates the probability of exceedance and waiting time period for the ten largest daily return price movements in the U.S. equity market (S&P 500) over the period July 1962 to December 1999. This "hit parade" ranges from -18.35% on October 19, 1987 to -3.29% on October 9, 1979. Other, perhaps less well known applied econometric work on extremes in economics concerns spill-over events; see for instance Hartmann et al. [31]. In this case, EVT in a multidimensional setup appears.

The second part of my title, "the Economics of Extremes," concentrates on the crucial question: given the econometric evidence on quantifiable extremal events in finance (and insurance), how can we handle these extremes from an economic point of view. Some concrete questions could be:

How can one devise prudent regulatory rules aiming at market stability? Here the Basel Committee enters; see below for more details.

Measure and (more importantly) price the time-dimension of system-wide risk. For these questions, see, for instance, Crockett [11] and Borio et al. [8]. An interesting review on systemic risk, an area where EVT as a quantitative tool has a lot to offer, is De Brandt and Hartmann [17]. Important in these problems is finding typically macroeconomic structures that help the economy/market to dampen (hopefully avoid) the more negative consequences of extremal events.

For most of the more mathematically minded extreme value theorists, working in risk management is equivalent to estimating Value-at-Risk (VaR) for ever more complicated stochastic models. In their (our) terminology, VaR is "just" a quantile of some underlying process or distribution. However, VaR is to finance what body temperature is to a patient; an indicator of bad health but not an instrument telling us what is wrong and far less a clue on how to get the patient (system) healthy again. Let us look at some of the main issues in risk management (RM) from the perspective of the regulator as personified by the famous Basel Committee. Details underlying the summary below are to be found on the homepage www.bis.org of the Bank of International Settlements in Basel.

The Basel Committee was established by the Central-Bank Governors of the Group of Ten at the end of 1974. The committee does not possess any formal supranational supervisory authority, and hence its conclusions do not have legal force. Rather, it formulates broad supervisory standards and guidelines and recommends statements of best practice in the expectation that individual authorities will take steps to implement them through detailed arrangements — statutory or otherwise — that are best suited to their own national system. In 1988, the committee introduced a capital measurement system, commonly referred to as the Basel Capital Accord (also called Basel I). This system provided for the implementation of a credit risk measurement framework with a minimum capital standard of 8% (a so-called haircut) by the end of 1992. From the start, banks criticized the lack of risk sensitivity in this approach. On the credit risk side, this led to the New Capital Adequacy (so-called Basel II) framework of June 1999. The latter is now under discussion with the industry and is planned to become operational by the end of 2006. Besides these key developments within the credit risk area, already around 1994 we saw various amendments to Basel I catering to market risk, in particular for derivative positions. The 1996 report on the amendment to the Capital Accord to incorporate market risks opened the floodgates for the VaR–modelers. Through this amendment, a direct link between the quantitative VaR measure for market risk and regulatory capital was established. The exact form of the link very much depends on the statistical qualities of the underlying market risk models through backtesting. For banks opting for the so-called internal modelling approach, the following formula yields the capital charge C_t at time t:

$$C_t = \max\left\{ \mathrm{VaR}_{t-1} + d_t\ \mathrm{ASR}_{t-1}^{\mathrm{VaR}},\ M_t \frac{1}{60}\sum_{j=1}^{60} \mathrm{VaR}_{t-j} + d_t \frac{1}{60}\sum_{j=1}^{60} \mathrm{ASR}_{t-j} \right\}$$

where

VaR_{t-i} is the 99%, 10-day VaR at day $t-i$;

M_t is the multiplier for day t, $M_t \geq 3$, mainly depending on the statistical qualities of the model, in particular, depending on backtesting results;

ASR^{VaR} is the extra VaR-based charge derived from specific portfolio risk for equity and interest rate instruments (using the CAPM language), and

d_t is a $\{0, 1\}$-indicator function which for day t possibly includes specific risk.

For details on the formula, and its related economic interpretation, see, for instance, Jovic [35]. There are also various regulatory rules on the allowable size of C_t in function to the banks' so-called Tier 1 and 2 capital, as well as bank internal allocation rules of C_t to subunits and the safeguarding of limits spoken based on VaR measures. Moreover, the independent calculation/supervision/verification of VaR and ASR^{VaR} poses a major problem, implying that there is much, much more to the calculation of VaR than just saying that "we are estimating a quantile." I find it very important that EVT specialists, especially those participants to this SemStat meeting with an interest in finance, take a deeper interest in these underlying economic and more detailed computational issues. As already stated above, the BIS website is a good place to start. J.P. Morgan's RiskMetrics is a further source of more applied reading, especially its more recent updated technical document, Mina and Xiao [48], which makes a nice link between current EVT research and its impact on Market RM.

Although I have already used various examples of risk classes, in order to move more in detail to Basel II it may be useful to give a brief classification of financial risks as referred to in the Basel documents:

Market risk (MR): the risk associated with fluctuations in the value of traded assets.

Credit risk (CR): the risk associated with the uncertainty that debtors will honor their financial obligations.

Operational risk (OR): the risk of loss resulting from inadequate or failed internal processes, people, and systems or from external events.

Liquidity risk (LR): the risk that positions cannot be unwound quickly enough at critical times.

Other risks: business, reputational.

A modern financial institution will have to map the above zoology of risks with indications of relevance, size, organizational issues, qualitative, and quantitative assessment. See, for instance, Litterman [41], [42] for a discussion of some of these issues. An important point concerns aggregation across risk classes and allocation of resulting risk capital to the various relevant layers within the institution. In all these fundamental steps, mathematical techniques enter at more or less prominent levels.

The key improvement within Basel II concerns an increased risk sensitivity for credit risk internal models. This can be achieved through various analytical models that go under the names of contingent claim, actuarial, and reduced form approaches; see Crouchy et al. [12] for details. Gordy [29] gives an excellent review, Frey and McNeil [27] unify the above models from a statistical (latent variable) point of view, whereas a definitive textbook may well become Duffie and Singleton [21]. See also Arvanitis and Gregory [5] for a guide to pricing, hedging, and risk management of credit positions. Within CR management, extremes play a role through the typical skewness of the loss distributions, but more importantly through the nonGaussian dependence between credit loss events. As shown by Frey and McNeil [27], the EVT based modelling of default correlation is of key importance to any well functioning CR management system. See also the latter paper for further references on this. A critical discussion on the use of EVT to CR modeling is Lucas et al. [44].

An important economic consequence of the risk sensitiveness improvements made for CR within Basel II is an anticipated reduction in total regulatory capital. At the same time, however, the Basel Committee introduced within Basel II operational risk (OR) as a new risk class. Although the consultation with industry is still ongoing, it is to be expected that the decrease in CR capital charge will be (approximately) offset by the new OR charge. Given that there will be a new OR capital charge forthcoming, EVT in combination with standard actuarial modeling will be called for in a fundamental way. In order to see this, consider the following OR setup. A stylized OR database will look as follows:

$$\{Y_k^{i,j}, i, j, k\}$$

where

$$\begin{aligned} i &= 1, 2, \ldots, T && \text{(years, say, e.g., } T = 10\text{);} \\ j &= 1, 2, \ldots, s && \text{(\# claim types, e.g., } s = 6\text{), and} \\ k &= 1, 2, \ldots, N^{i,j} && \text{(\# claims of type } j \text{ in year } i\text{).} \end{aligned}$$

Note that typically $Y_k^{i,j}$ is censored from below, i.e.,

$$Y_k^{i,j} = \left(\widetilde{Y}_k^{i,j} - d^{i,j}\right)^+$$

for the full (ground up) claims ($\widetilde{Y}_k^{i,j}$) and some company specific lower thresholds ($d^{i,j}$). As a result, the total yearly OR loss amounts across all s types are

$$\left\{ \sum_{j=1}^{s} \sum_{k=1}^{N_{i,j}} Y_k^{i,j}, \quad i = 1, \ldots, T \right\}.$$

Because of Basel II, banks using an internal OR modelling approach will have to come up with an estimate of the $100(1-\alpha)\%$ quantile (OR-VaR) with α small ($\alpha = 0.0005$, say) of the distribution function of next year's total loss

$$\sum_{j=1}^{s} \sum_{k=1}^{N_{T+1,j}} Y_k^{T+1,j}.$$

Of the few facts available for real OR losses, one is very clear: losses are heavy tailed. Hence from an extremes in economics point of view, actuarial total loss modelling under a heavy tailed regime is natural; for some publications along these lines, see Medova [46] and Cruz [13]. Embrechts and Samorodnitsky [26] contains some advanced ruin theoretic results motivated by OR. At present, the more important issue falls under the economics of extremes heading: why introduce an OR capital charge in the first place? The already quoted BIS website (www.bis.org) contains under "Basel Committee: Comments Received" several discussion papers on this topic.

As an example, see Daníelsson et al. [16] where some of the more fundamental economic issues underlying quantitative risk management regulations à la Basel II are critically assessed.

In the next section, some current mathematical research is summarized originating from the above discussions on risk management in general and Basel I and II in particular. The choice of topics is rather subjective. An attempt is made (mainly from an economics of extremes point of view) to complement other EVT applications within finance and insurance discussed elsewhere in this volume.

4.4 Some current research

4.4.1 Coherent risk measurement

In a sequel of fundamental papers, Artzner, Delbaen, Eber, and Heath ([3], [4], [18], [19]) posed and answered the following questions:

(a) What economic properties should a "good" risk measure have?

(b) Characterize all "good" risk measures.

(c) Is VaR "good"?

(d) If the answer to (c) is no, suggest improvements.

In a one-period setup, a risk X is a bounded random variable ($X \in L^o(\Omega, \mathcal{F}, P)$) denoting the profit-and-loss of a financial position that we hold today for a fixed future period, 10 days, say. Suppose the risk free interest over this one period is $r \geq 1$. In the above publications, a "good" risk measure

$$\rho : L^o(\Omega, \mathcal{F}, P) \to \mathbb{R}$$

is termed coherent and has to satisfy the following axioms:

(C1) (translation invariance)
$$\forall X \in L^o, \alpha \in \mathbb{R} : \rho(X + \alpha r) = \rho(X) - \alpha.$$

(C2) (subadditivity)
$$\forall X, Y \in L^o : \rho(X + Y) \leq \rho(X) + \rho(Y).$$

(C3) (positive homogeneity)
$$\forall X \in L^o, \lambda \geq 0 : \rho(\lambda X) = \lambda \rho(X).$$

(C4) (monotinicity)
$$\forall X, Y \in L^o, X \leq Y, \text{ we have } \rho(X) \geq \rho(Y).$$

In Artzner et al. [4], the link to economics is made through the notion of acceptance set associated with a coherent risk measure ρ:

$$\mathcal{A}_\rho = \left\{ X \in L^o, \quad \rho(X) \leq 0 \right\}.$$

Hence $\mathcal{A}_\rho$ contains those financial positions for which, using the risk measure ρ, no further capital charge ($\rho(X) = 0$) is necessary, or even ($\rho(X) < 0$) capital can be redrawn. A further result from this paper is the following representation theorem: a risk measure ρ is coherent if and only if there exists a set $\mathcal{P}$ of probability measures on $(\Omega, \mathcal{F})$, such that

$$\rho(X) = \sup\{E^Q(-X/r), \quad Q \in \mathcal{P}\},$$

i.e., ρ is a so-called generalized scenario through which Q2 is answered. It is not difficult to see that in general VaR is not coherent. Indeed typically for nonelliptically distributed portfolios, VaR fails to satisfy the (for economic purposes) important subadditive property (C2). The following easy example goes back to Claudio Albanese; a similar example is to be found in Artzner et al. [3].

Example 4.1 (VaR is not necessarily coherent) Suppose $X_1, \ldots, X_{100}$ correspond to the profit-and-loss (P&L) positions of 100 defaultable (one year) bonds, each with face value \$100, default probability 1% and 2% yearly coupon. Hence, for $i = 1, \ldots, 100$,

$$X_i = \begin{cases} 2 & \text{with probability } 99\% \\ -100 & \text{with probability } 1\%. \end{cases}$$

Surely, the more diversified position $\sum_{i=1}^{100} X_i$ should have a lower capital charge as the "all eggs in one basket" position $100X_1$. When we take $\rho = \text{VaR}_{95\%}$ however, it is easy to check that

$$\rho(100X_1) = \sum_{i=1}^{100} \rho(X_i) < 0 < \rho\left(\sum_{i=1}^{100} X_i\right).$$

The sign convention, $\text{VaR}_{95\%}$ = minus the 5% left quantile of the P&L distribution, corresponds to usage in practice to report VaR positively and the definition of coherence used above where losses are in the left tail of the P&L distribution. Indeed (C3 and C4) yield that a risky position ($X \leq 0$) becomes a positive net regulatory capital charge ($\rho(X) \geq 0$). This convention is not material for the example.

The main reason that VaR fails the subadditivity property is the high skewness of the positions X_i; these so-called "spike-the-firm" positions (terminology coined by Dilip Madan) do however occur in practice, especially in markets where high severity–low probability events occur. Embrechts et al. [25] discuss the relevance of the above for portfolio theory and stress that one other reason why VaR may lack subadditivity in more general situations is the typical nonGaussian dependence structure of financial data. We will come back to this point in Section 4.4.2.

VaR has a further, very obvious shortcoming in that it only yields a frequency estimate of a high loss; it does not give information on the severity for when that (rare) loss happens. For instance, saying that a 99%, 10-day VaR equals \$1 Mio means that with probability 1%, by the end of a 10-day period, our present portfolio (held fixed) will incur a loss of \$1 Mio or more. VaR does not give any information on this crucial "or more." Going now to question (d), an obvious risk measure stressing the "or more" would be the conditional VaR measure

$$\rho_{\text{CV}}(X) = E(-X/r \mid -X/r > \text{VaR}(X))$$

for some given VaR. Under weak conditions, see Delbaen [19] and Acerbi and Tasche [1], ρ_{CV} is coherent. Clearly $\rho_{\text{CV}} \geq \text{VaR}$ and in some extreme (though realistic) cases

$\rho_{CV} \gg$ VaR. When we would move to ρ_{CV} as a measure determining regulatory capital, the immediate economic question arises of how to handle in practice (given the present regulatory environment) the difference $\rho_{CV} -$ VaR. One would have to come up with a fully, economically sound regulatory capital framework based on ρ_{CV}; this task is definitely doable but needs combined input from academia, regulators, and industry. I refer to Daníelsson et al. [16] for a discussion of the relevant economic pitfalls underlying such a task. At this point, I would like to stress that so far we do not have a full theory for coherent multiperiod risk measurement. First attempts to come up with such a theory, already showing that ρ_{CV} is also problematic, are under discussion (private communication with Freddy Delbaen and Philippe Artzner).

4.4.2 Allocation and aggregation of risk

As soon as one has reached a consensus between regulators and industry on how to quantify and measure risk, one immediately faces the question: how to use this technology to improve capital allocation. As in the previous section, it would be useful to come up with an axiomatic definition of (risk) capital allocation. Denault [20] has worked out the details of such a coherent risk capital allocation, based on the work on coherent risk measures (Section 4.1), game theory and related results from the actuarial literature. As before, after understanding what rules are scientifically sound, the next step is to work out their actual implementation in practice. On the latter, Matten [45] yields a good introduction.

A related (in some sense, reverse) question concerns the aggregation of risk measures. One often faces the problem that risk measures $\rho(X_i)$, $i = 1, \ldots, d$ have been calculated for separate risk classes; how can we estimate the risk measure $\rho(\Psi(\underset{\sim}{X}))$ of a global position $\Psi(\underset{\sim}{X})$ on $\underset{\sim}{X} = (X_1, \ldots, X_d)'$. Typical examples for $X_1, \ldots, X_d$ are one-period risks within certain risk classes (market, credit, operational), but also across different classes. At the highest level, one could think of X_1 standing for market, X_2 for credit and X_3 for operational risk of a particular financial institution over a comparable fixed time period. Another example would correspond to d lines of business in a multiline insurance contract. Depending on the context, for Ψ one could think of examples like:

$$\Psi(\underset{\sim}{X}) = \sum_{i=1}^{d} X_i = S_d,$$

$$\Psi(\underset{\sim}{X}) = \max_{i=1,\ldots,d} X_i = M_d,$$

$$\Psi(\underset{\sim}{X}) = \sum_{i=1}^{d} (X_i - k)^+ ,$$

$$\Psi(\underset{\sim}{X}) = \left(\sum_{i=1}^{d} X_i - k \right)^+ ,$$

$$\Psi(\underset{\sim}{X}) = M_d I_{\{S_d > q_\alpha\}}, \quad \text{etc.}$$

For the risk measure ρ one could restrict attention to the class of coherent risk measures. The more interesting case however is that of noncoherence like VaR. In general, one could even take $\rho(\Psi(\underset{\sim}{X})) = F_{\Psi(\underset{\sim}{X})}$, the distribution of the financial position $\Psi(\underset{\sim}{X})$ or some functional of $\Psi(\underset{\sim}{X})$ as for instance moments $E(\Psi(\underset{\sim}{X})^k)$. Often in practice one is given only the marginal distribution functions $F_1, \ldots, F_d$ of $X_1, \ldots, X_d$ respectively, together with some notion of dependence between $X_1, \ldots, X_d$. The crucial point is that most often one does not have full (even usable statistical) information on $F_{\underset{\sim}{X}}$. How can one construct optimal bounds

$$\rho_L(\Psi(\underset{\sim}{X})) \leq \rho(\Psi(\underset{\sim}{X})) \leq \rho_U(\Psi(\underset{\sim}{X})) \tag{4.1}$$

in agreement with the above assumptions? A full discussion of this problem, with several examples, is to be found in Embrechts et al. [23]. The notion of dependence is defined using the language of copulas; suppose $F_1, \ldots, F_d$ are continuous, then there exists a unique function

$$C : [0, 1]^d \to [0, 1]$$

which is a distribution function with standard uniform marginals so that

$$P(X_1 \leq x_1, \ldots, X_d \leq x_d) = C(F_1(x_1), \ldots, F_d(x_d)).$$

The function C is called copula as it couples the marginal laws $F_1, \ldots, F_d$ to the joint distribution of $\underset{\sim}{X}$. A typical dependence condition on the unknown copula C of $\underset{\sim}{X}$ could be $C \geq C_o$ for some known copula C_o, e.g., the independence copula $C_o(\underset{\sim}{u}) = \prod_{i=1}^d u_i$. From Embrechts et al. [23] take for instance $d = 2$, $\Psi(\underset{\sim}{X}) = X_1 + X_2$ and $\rho = \text{VaR}_{95\%}$ (here we only look at the right tail of the distribution function so that $\text{VaR}_{95\%}$ corresponds to the 95th percentile). If we assume for example that $F_i = \Gamma(3, 1)$, $i = 1, 2$, then $\rho(X_i) = 6.3$, $i = 1, 2$. The unconstrained range of possible values for $\rho(X_1 + X_2)$ in (4.1) is [6.47, 14.44]. If we assume X_1 and X_2 to be independent, then $\rho(X_1 + X_2) = 10.52$. Whenever X_1 and X_2 are comonotone, i.e., there exist increasing functions f_1, f_2 and a random variable Z so that $X_i = f_i(Z)$, $i = 1, 2$, then ρ becomes additive so that $\rho(X_1 + X_2) = \rho(X_1) + \rho(X_2) = 12.60$. In case $C \geq C_o$, the independent copula, then the possible range becomes [8.17, 14.41]. The crucial observation stems from a comparison of the (attainable) upper bound for the unconstrained case (14.44) and the value of $\rho(X_1 + X_2) = \rho(X_1) + \rho(X_2)$ for comonotonic risks (12.60). The gap [12.60, 14.44] corresponds to dependence structures (copulas) on (F_1, F_2) which yield, for the corresponding bivariate model for (X_1, X_2), a nonsubadditive risk measure $\rho = \text{VaR}_{95\%}$. The key issue here is not the shape of F_i (we could also have taken $F_i = N(0, 1)$) but rather the damage nonGaussian dependence structures can cause on risk management systems. These issues, and their economic implications, need further investigation. See Embrechts et al. [25] for a start.

4.4.3 Portfolio management under general constraints

Using the one-period setup so far, given $X_o, X_1, \ldots, X_d$ where X_o corresponds to a riskless investment and $X_1, \ldots, X_d$ correspond to risky positions, the basic problem of portfolio analysis concerns the following. Given some risk measure ρ, find the portfolio weights $a_0^*, a_1^*, \ldots, a_d^*$ so that

$$\underset{\sim}{a}^* = \underset{\underset{\sim}{a} \in \mathbb{R}^{d+1}}{\arg\min}\, \rho(\underset{\sim}{a}'\underset{\sim}{X})$$

$$\text{so that } r(\underset{\sim}{a}'\underset{\sim}{X}) = r_0, \text{ fixed.}$$

Here, $r(Y)$ stands for the one-period excess return on the investment Y. The case $\rho = \sigma$ (standard deviation) corresponds to the classical Markowitz problem, leading to the notion of efficient frontier/portfolios. Numerous authors have considered this problem for a variety of risk measures ρ. For instance, going from $\rho = \sigma$ to $\rho =$ VaR seems a very natural thing to do; however, from an economic (stability) point of view, such optimization can readily lead to dangerous situations and should be treated with care. For this particular case, a detailed discussion is to be found in Basak and Shapiro [6]. See also Krokhmal et al. [37] and Rockafeller and Uryasev [49]. I also would like to stress that already in the early days of portfolio theory, optimization with respect to alternative risk measures was considered; see for example Lemus et al. [39] for a review.

4.4.4 Dynamic models catering for extremes

Most of the work within the realm of extremes in economics has centered around either static or discrete time modelling. However, given that extremal moves do occur and are important, one has to take the logical step following on from this and come up with dynamic models for derivative pricing and hedging replacing the Black-Scholes-Merton framework based on geometric Brownian motion:

$$dS(t) = S(t)(\mu\, dt + \sigma\, dW(t)).$$

One of the key models already in use in practice is the one where in the above SDE, standard Brownian motion is replaced by a more general Lévy process $\{L(t), t \geq 0\}$. Replacing $(W(t))$ in such a way immediately leads to an incomplete market where there is no unique pricing martingale measure. Fairly recently, several authors have worked out a possible framework. Readers interested in this area of research could, for instance, consult Eberlein [22], Carr et al. [10], Geman et al. [28], and Levin and Tchernitser [40]. The latter paper can be downloaded via www.gloriamundi.org/var/wps.html, a website containing numerous working papers on VaR.

4.5 Final comments

In the above discussion, I have tried to stress the need for more economic thinking/modelling in the interplay between EVT and risk management. Especially now when, through Basel II, new basic guidelines for quantitative risk management are under discussion, extreme value theorists with an interest in applying their techniques to finance have to take a closer look at the underlying economic fundamentals. That extremes in finance matter is clear. Looking back at the LTCM case in 1998 where extreme market movements resulted from the Russian moratorium on government bonds, it is interesting to see that the key player in LTCM's up and down, John Meriwether on 21/8/2000 (*Wall Street Journal*), talking about his new business JWM Partners, was quoted as follows: "With globalization increasing, you'll see more crises. Our whole focus is on the extremes now — what's the worst that can happen to you in any situation — because we never want to go through that again." Already in the introduction to our book Embrechts et al. [24], we stated that "Though not providing a risk manager in a bank with the final product he or she can use for monitoring financial risk on a global scale, we (i.e., EVT models) will provide that manager with stochastic methodology needed for the construction of various components of such a global tool." By now, EVT has provided RM for banking and insurance with a useful set of techniques for looking more realistically at extremes. The main emphasis of the present paper is for EVT researchers to take the step beyond and look at the economic implications of their research. The following papers may offer some further guidance along this road:

Daníelsson [15] offers a critical assessment of the use of statistical, EVT based techniques in RM. With respect to VaR based RM, the author states that "For regulatory use, the VaR measure may give misleading information about risk, and in some cases may actually increase both idiosyncratic and systemic risk." This paper also formed the basis of the earlier quoted Basel II response Daníelsson et al. [16]. Also the following papers offer a useful introduction to the main issues at hand: Jorgensen et al. [33] and Zigrand and Daníelsson [52].

In [50], Myron Scholes rediscusses some of the basic issues underlying the collapse of LTCM, stressing the crucial importance of market liquidity. Concerning VaR, Scholes concludes the following: "Over the last number of years, regulators have encouraged financial entities to use portfolio theory to produce dynamic measures of risk. VaR, the product of portfolio theory, is used for short-run day-to-day profit and loss exposures. Now is the time to encourage the BIS and other regulatory bodies to support studies on stress test and concentration methodologies. Planning for crises is more important than VaR analysis. And such new methodologies are the correct response to recent crises in the financial industry."

The discussions above are mainly restricted to EVT issues in finance. Similarly, we could have discussed more specifically examples in insurance. At present one witnesses an increasing collaboration between insurance and banking regulators. On the website of the Canadian Institute of Actuaries (www.actuaries.ca), one finds the following vision statement: "for actuaries to be recognized as the leading professionals in the financial modelling and management of risk and contingent events." On that same website, a presentation by Allen Brender on capital requirements and stochastic methods can be found where he concludes "We are only in the early days of a new actuarial age." The publication Hancock et al. [30] gives a very readable introduction to this "new actuarial age."

Bibliography

[1] Acerbi, C. and Tasche, D. (2002) Expected shortfall: a natural coherent alternative to Value at Risk. *Economic Notes* 31(2), 379–388.

[2] Acerbi, C. and Tasche, D. (2001) On the coherence of expected shortfall. *Journal of Banking and Finance* 26(7), 1487–1503.

[3] Artzner, P., Delbaen, F., Eber, J.M. and Heath, D. (1997) Thinking coherently. *Risk*, November 1997, 68–71.

[4] Artzner, P., Delbaen, F., Eber, J.M. and Heath, D. (1999) Coherent measures of risk. *Math. Finance* 9, 203–228.

[5] Arvanitis, A. and Gregory, J. (2001) *Credit: A Complete Guide to Pricing, Hedging and Risk Management*. Risk Waters Group, London.

[6] Basak, S. and Shapiro, A. (2001) Value-at-Risk based risk management: optimal policies and asset prices. *Review of Financial Studies* 14, 371–405.

[7] Black, F. and Scholes, M. (1973) The pricing of options and corporate liabilities. *J. Political Economy* 81, 637–654.

[8] Borio, C., Furline, C. and Lowe, P. (2001) Procyclicality of the financial system and financial stability; issues and policy options. Working Paper, BIS, Basel.

[9] Boyle, P. and Boyle, F. (2001) *Derivatives. The Tools that Changed Finance.* Risk Waters Group, London.

[10] Carr, P., Geman, H., Madan, D. and Yor, M. (2000) The fine structure of asset returns: an empirical investigation. Preprint, University of Maryland.

[11] Crockett, A. (2000) Marrying the micro- and macro-prudential dimensions of financial stability. Working Paper, BIS, Basel.

[12] Crouchy, M., Galai, D. and Mark, R. (2001) *Risk Management.* McGraw-Hill, New York.

[13] Cruz, M.G. (2002) *Modeling, Measuring and Hedging Operational Risk. A Quantitative Approach.* Wiley, New York.

[14] Cummins, D. (2002) Convergence of banking and insurance: opportunities in the wholesale financial services. IFCI Geneva Research Paper No. 9 (www.riskinstitute.ch).

[15] Daníelsson, J. (2002) The emperor has no clothes: limits to risk modelling. *Journal of Banking and Finance* 26(7), 1273–1296.

[16] Daníelsson, J., Embrechts, P., Goodhart, C., Keating, C., Muennich, F., Renault, O. and Shin, H.S. (2001) An academic response to Basel II. Special Paper No. 130, Financial Markets Group, LSE.

[17] De Brandt, O. and Hartmann, P. (2000) Systemic risk: a survey. Working Paper No. 35, European Central Bank, Frankfurt.

[18] Delbaen, F. (2000) Coherent risk measures on general probability spaces. Preprint, ETH Zurich.

[19] Delbaen, F. (2002) *Coherent Risk Measures.* Lecture Notes, Cattedra Galileiana 2000, Scuola Normale, Pisa.

[20] Denault, M. (2001) Coherent allocation of risk capital. Preprint, RiskLab, ETH Zurich.

[21] Duffie, D. and Singleton, K. (2003) *Credit Risk Modeling for Financial Institutions*. Princeton University Press.

[22] Eberlein, E. (2001) Application of generalized hyperbolic Lévy motions to finance, in *Lévy Processes: Theory and Applications*, O.E. Barndorff-Nielsen, T. Mikosch and S. Resnick (Eds.), Birkhäuser, 319–337.

[23] Embrechts, P., Hoeing, A. and Juri, A. (2003) Using copulae to bound the Value-at-Risk for functions of dependent risks. *Finance and Stochastics* 7(2), 145–167.

[24] Embrechts, P., Klüppelberg, C. and Mikosch, T. (1997) *Modelling Extremal Events for Insurance and Finance*, Springer, Berlin.

[25] Embrechts, P., McNeil, A. and Straumann, D. (2002) Correlation and dependence in risk management: properties and pitfalls. In: *Risk Management: Value at Risk and Beyond*, M. Dempster (Ed.) Cambridge University Press, p. 176–223.

[26] Embrechts, P. and Samorodnitsky, G. (2003) Ruin problem and how fast stochastic processes mix. *Annals of Applied Probability* 13, 1–36.

[27] Frey, R. and McNeil, A. (2000) Modelling dependent defaults. Preprint, ETH Zürich.

[28] Geman, H., Madan, D. and Yor. M. (2001) Time changes for Lévy processes. *Math. Finance* 11, 79–96.

[29] Gordy, M. (2000) A comparative anatomy of credit risk models. *Journal of Banking and Finance* 24, 119–149.

[30] Hancock, J., Huber, P. and Koch, P. (2002) The economics of insurance. How insurers create value for shareholders. Swiss Reinsurance Company, Zurich (www.swissre.com).

[31] Hartmann, P., Straetmans, S. and de Vries, C.G. (2001) Asset market linkages in crisis periods. Working Paper, No. 71, European Central Bank, Frankfurt.

[32] Jaeger, L. (2002) *Managing Risk in Alternative Investment Strategies.* Financial Times, Prentice Hall, London.

[33] Jorgensen, B., de Vries, C. and Daníelsson, J. (2002) Incentives for effective risk management. *Journal of Banking and Finance* 26(7), 1407–1425.

[34] Jorion, P. (2000) Risk management lessons from Long-Term Capital Management. *European Financial Management* 6, 277–300.

[35] Jovic, D. (1999) *Risikoorientierte Eigenkapitalallokation und Performencemessung für Banken.* Verlag Paul Haupt, Bern.

[36] Klüppelberg, C. (2004) Risk management with extreme value theory. In: *Extreme Values in Finance, Telecommunications, and the Environment*, Finkenstädt, B. and Rootzén, H., Eds., CRC/Chapman & Hall, Boca Raton, Ch. 3.

[37] Krokhmal, P., Palmquist, J. and Uryasev, S. (2002) Portfolio optimization with conditional Value-at-Risk objective and constraints. *The Journal of Risk* 4(2).

[38] Lane, M. (Ed.) (2002) *Alternative Risk Strategies*, Risk Waters Group, London.

[39] Lemus, G., Samarov, A. and Welsch, R. (2001) Optimal portfolio selection based on alternative risk measures. Preprint, MIT, Sloan School of Management.

[40] Levin, A. and Tchernitser, A. (2001) Multifactor stochastic variance models in risk management: maximum entropy approach and Lévy processes. Working paper, Bank of Montreal.

[41] Litterman, R. (1997) Hot spots and hedges I. *Risk* 10(3), 42–45.

[42] Litterman, R. (1997) Hot spots and hedges II. *Risk* 10(5), 38–42.

[43] Longin, F. (2001) Stock market crashes: Some quantitative results based on extreme value theory. *Derivatives Use, Trading and Regulation*, 7, 197–205.

[44] Lucas, A., Klaassen, P., Spreij, P. and Straetmans, S. (2002) Tail behavior of credit loss distributions for general latent factor models. Paper presented at the Third Joint Central Bank Research Conference on Risk Measurement and Systemic Risk (www.bis.org/cgfs/cgfsconf2002prog.htm).

[45] Matten, C. (2000) *Managing Bank Capital: Capital Allocation and Performance Measurement*, 2nd Ed. Wiley, New York.

[46] Medova, E. (2000) Measuring risk by extreme values. *Risk*, November 2000, S20–S26.

[47] Mikosch, T. (2004) Modeling dependence and tails of financial time series. In: *Extreme Values in Finance, Telecommunications, and the Environment*, Finkenstädt, B. and Rootzén, H., Eds., CRC/Chapman & Hall, Boca Raton, Ch. 5.

[48] Mina, J. and Xiao, J.Y. (2001) *Return to RiskMetrics. The Evolution of a Standard.* RiskMetrics Group, New York (www.riskmetrics.com).

[49] Rockafellar, R.T. and Uryasev, S. (2000) Optimization of conditional Value-at-Risk. *The Journal of Risk* 2, 21–41.

[50] Scholes, M.S. (2000) Crisis and risk management. *American Economic Review*, May 2000, 17–22.

[51] Steinherr, A. (1998) *Derivatives. The Wild Beast of Finance.* Wiley, New York.

[52] Zigrand, J.-P. and Daníelsson, J. (2003) What happens when you regulate risk? Evidence from a simple equilibrium model. Preprint, LSE.

CHAPTER 5

Modeling Dependence and Tails of Financial Time Series

Thomas Mikosch
University of Copenhagen

Contents

1-58488-411-8/04/$0.00+$.50

The aim of this chapter is to discuss the interplay between the tail behavior and the dependence structure of financial data.

We start by discovering some of the "stylized facts" of real-life returns: long range dependence effects that affect the sample autocorrelation behavior of the absolute log-returns and erratic behavior that results in heavy-tailed marginal distributions. In practice one often observes that financial log-return series can have infinite 3rd, 4th, 5th, ... moments.

Then we look at some of the standard econometric models that try to capture the empirical behavior. Those include GARCH (generalized autoregressive conditionally heteroscedastic) and stochastic volatility processes. The GARCH case turns out to be a very complex one. We embed these processes in finite-dimensional stochastic recurrence equations. Following classical work of Kesten [88], we show that the finite-dimensional distributions of these processes are multivariate regularly varying, thus they have infinite power moments of a certain degree. Heavy tails cause the classical limit theory for the sample autocorrelations to break down. Therefore one has to modify this asymptotic theory: unfamiliar infinite variance limit distributions occur and rates of convergence can be extremely slow. Point process techniques turn out to be important in this context.

The long range dependence effect of real-life data cannot be explained by GARCH models. A possible explanation for this effect is nonstationarity of the underlying time series. We will learn how statistical tools behave under nonstationarity and how they can fool us to see things that are not there.

5.1 "Stylized facts" about financial data

Over the last few years financial mathematics has been one of the mathematical success stories that has attracted the attention not only of mathematicians but also of various groups of people outside these circles of specialists. Among them are nonmathematicians such as economists, econometricians, physicists, psychologists, and many more. Mathematicians clearly get excited because difficult parts of their theory are applicable in an area that is far away from mathematics, whereas the finance industry thinks that this part of mathematics might be useful to make money in one form or the other. Certain people find it intellectually stimulating that, eventually, there is a formula that explains a difficult procedure such as the pricing of an option very much in the same way as Newton's law gives a quantitative description of gravitation.

It is mainly that part of financial mathematics related to the pricing of derivatives such as options and futures that has been considered as most useful. The Black-Scholes option pricing formula, which, by now, is taught in any good course on stochastic integration and martingales, stands as a synonym for this theory. It has been pushed forward since 1973 when the two fundamental papers of Black-Scholes and Merton appeared. Brownian motion and Itô stochastic integral, Girsanov transformation and change of measure are the basic notions in this framework. A deep knowledge of the theory of stochastic processes is the basis for anybody who wants to conduct serious research in this area.

Parallel to the pricing theory of derivatives, econometricians have modeled financial time series. Time series are discrete time processes, and the link to continuous time processes is not obvious and in general difficult to establish, i.e., the embedding of a discrete time process in a continuous time one is by no means an easy matter. However, the aims of time series analysis are different from those of pricing. Whereas the latter requires a continuous time framework in order to make martingale and stochastic integration techniques applicable, time series analysis has always been directed towards the understanding of the mechanism that drives a given series of data with the aim of possibly predicting future values in the series. This does not mean that stochastic differential equations which are commonly used to model price movements for derivative pricing do not describe a certain physical mechanism of the price evolution. This approach, however, does not primarily aim at the most realistic model for prices. Its basic goal is to get a reasonable model that is mathematically tractable and can be understood or interpreted by financial practitioners.

Thus, financial time series analysis focuses on the truth behind the data, meaning that one is interested in finding physical models that explain, at least to some extent, the empirically observed features of real-life data. Because there are many financial data, it might be difficult to say anything about some common properties. Surprisingly, a large variety of financial data which we denote consistently by $P_t, t = 0, 1, 2, \ldots$, (t can be minutes, hours, days, etc.) exhibits similar properties after the transformation

$$X_t = \log P_t - \log P_{t-1} = \log\left(1 + \frac{P_t - P_{t-1}}{P_{t-1}}\right),$$

at least if one focuses on share prices (of Microsoft, say), stock indices (DAX, Nikkei, Dow Jones, etc.) or foreign exchange rates (such as USD/JPY, USD/DEM, or JPY/DEM). As a matter of fact, these similar properties depend on the time scale chosen. Depending on whether the time unit is a second, half an hour, or a day, a month, or a year, qualitative differences in the time series can be expected, and different models need to be introduced. For example, if the time scale is too small (minutes, seconds, etc.) P_t lives on a grid and varies little; one often observes that P_t does not change over longer (relative to the time unit) periods. This would imply that the distribution of X_t has an atom at zero. This would be an unacceptable assumption if X_t was calculated on a daily basis, in which case one often requires that X_t has a density. Since we do not want to get into too large or small time scales; in what follows we think of t in units of hours, days, or weeks.

The resulting (X_t) is the time series of log-returns which, by a Taylor series argument, is close to the relative returns series $((P_t - P_{t-1})/P_{t-1})$. The relative returns give some more intuition on the log-returns (the relative returns are very small and therefore almost indistinguishable from the log-returns), i.e., they describe the relative change over time of the price process. In what follows, we often refer, for short, to returns instead of log-returns. See Figure 5.1.

The log-differences X_t have the advantage that they are free of any unit, therefore, comparable among each other. Moreover, there is an important mathematical issue as well: it is believed that the time series (X_t) can be modeled by a stationary (in the strict or wide senses) stochastic process, i.e., this transformation yields one realization of a stationary process. In turn, stationarity is a basic assumption for any kind of time series analysis. In the beginning, we will accept stationarity of (X_t) as a working hypothesis. Later, in Section 5.11, we will criticize it.

5.1.1 Distribution and tails

Samples of returns $X_1, \ldots, X_n$ have the following stylized facts in common:

- The sample mean of the data is close to zero; the sample variance is of the order 10^{-4} and smaller.

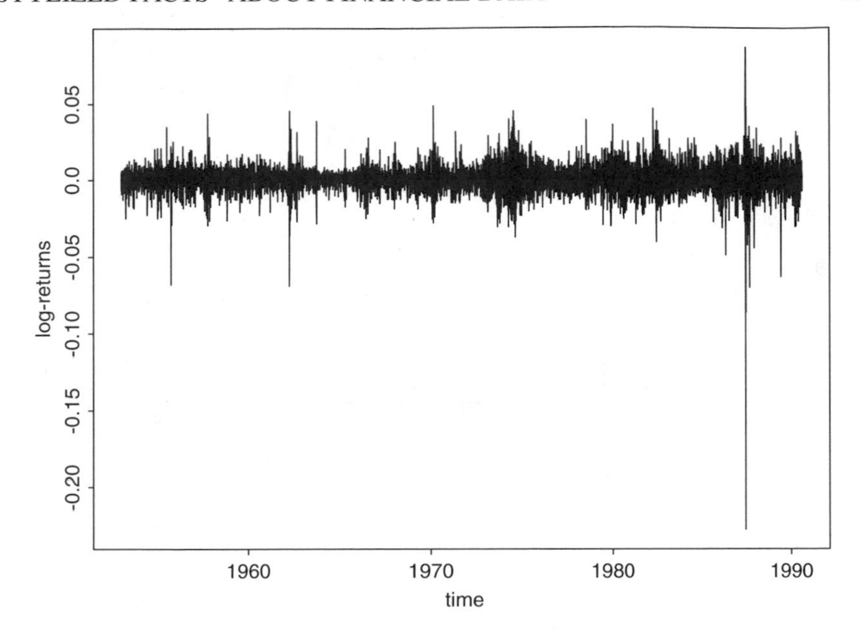

Figure 5.1 *Plot of 9558* S&P500 *daily log-returns from January* 2, 1953, *to December* 31, 1990. *The year marks indicate the beginning of the calendar year.*

This is not surprising because price changes are in general very small; a daily change of more than 2 to 3% is unusual.

- A density plot of the data shows that the distribution of the data is roughly symmetric in its center, sharply peaked around zero with heavy tails on both sides.

See the top graph of Figure 5.2 for an illustration in the S&P500 series case.

The shape of the density indicates that the normal distribution is perhaps not the most appropriate model for returns. This is also seen by a QQ-plot of the data against the normal distribution (see the bottom graph of Figure 5.2), i.e., the empirical quantiles are plotted against the quantiles of the normal distribution with the same mean and variance as the data. The deviation from normality is in contrast to the Black-Scholes model, which is most frequently used for stock prices. Indeed, the latter assumes that prices are modeled by the continuous time stochastic process

$$P_t = P_0 \, \mathrm{e}^{\,ct+\sigma B_t}, \quad t \geq 0,$$

with constants c, σ, and standard Brownian motion B, i.e., (P_t/P_0) is geometric Brownian motion. Choosing t at equidistant instants of time, the structure of P implies that the log-return series (X_t) constitutes an iid normal sequence. As we have seen by empirical means, this is not a realistic assumption. But it is convenient for the purposes of derivative pricing.

The very large and very small values in the density plot and the shape of the normal QQ-plot (which curves down at the right and curves up at the left) are indications for the next stylized fact:

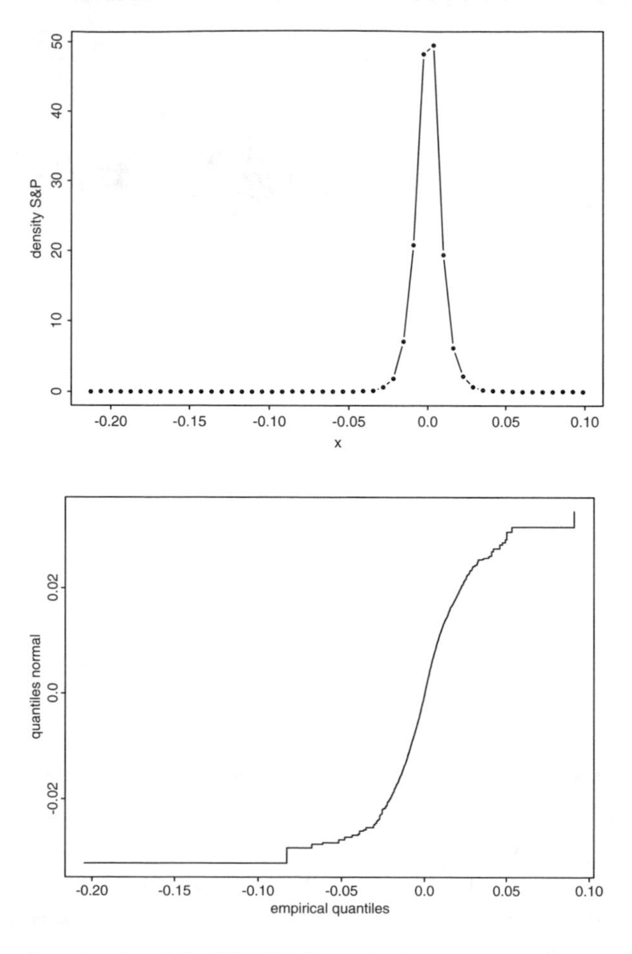

Figure 5.2 Top: *Density plot of the* S&P500 *data. The limits on the x-axis indicate the range of the data.* Bottom: *QQ-plot of the* S&P500 *data against the normal distribution whose mean and variance are estimated from the* S&P500 *data.*

- Log-return data have heavy-tailed marginal distributions.

Later, in Section 5.5, we will see that the tails of log-returns are conveniently modeled by distributions with power law tails, i.e., for large x and some positive number α (the tail index),

$$P(X_t > x) \approx x^{-\alpha}.$$

We use $\approx$ in a heuristic way meaning that $P(X_t > x)$ is roughly of the order $x^{-\alpha}$ for large x; in Section 5.4 we introduce the notion of regularly varying tail in order to make these heuristics precise. Examples of distributions satisfying the latter right tail condition are the Pareto distribution (with an exact power law) and the t-distribution with α degrees of freedom. Thus, for the moment, it is convenient to think of log-returns as t-distributed random variables. This is closer to the truth than the normal

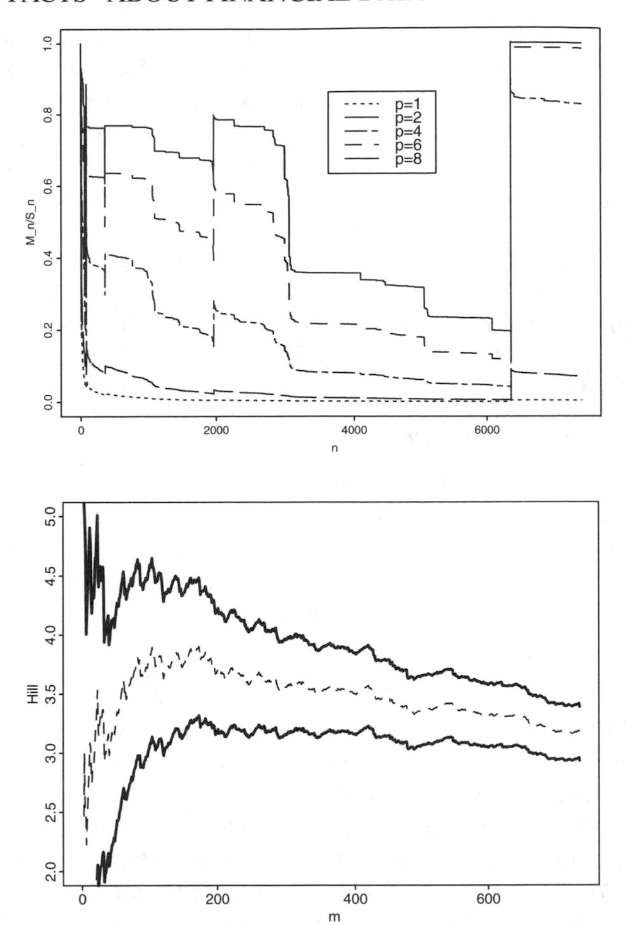

Figure 5.3 Top: *Plot of the ratio* $\max_{i=1,\dots,n} |X_i|^p/(|X_1|^p + \cdots + |X_n|^p)$ *against n for the* S&P500 *data* (X_t) *for various values of p. If* $E|X_1|^p < \infty$ *and the data came from a stationary ergodic model, the ratio should converge to zero a.s., by virtue of the strong law of large numbers.* Bottom: *Hill plot (dotted line) for the* S&P500 data *with 95% asymptotic confidence bounds. The Hill estimator estimates the tail index* α *in the model* $P(X_1 > x) \sim cx^{-\alpha}$ *as a function of the m upper order statistics in the return sample. The estimate is taken in the m-region where the plot does not change much;* $\alpha = 3.7$ *would be reasonable. See Embrechts et al. [58], Chapter* 6, *for details and more tools on extreme value statistics, or Resnick [121]* (*this volume*) *for an extensive discussion of Hill plots.*

distribution as can be seen in Figure 5.3, where some procedures for determining the tail index α of the distribution of the S&P500 are exercised.

Since the 1960s there has been discussion about whether or not log-returns can be modeled by an infinite variance α-stable distribution. This discussion was started by Benoit Mandelbrot [99]; see Mandelbrot [100, 101] for some recent thoughts on this hypothesis. The definition of an α-stable distribution is given in Section 5.7, but for our present purposes we only need to know that the α-stable hypothesis implies the tail behavior $P(X_t > x) \sim c\,x^{-\alpha}$ as $x \to \infty$, for some $\alpha \in (0, 2)$. (Here and in what follows, $f(x) \sim g(x)$ for two functions f and g means that their ratio converges to 1 as $x \to \infty$.) Thus we are in the framework of power law tails for log-returns. Although stable distributions (more generally, stable processes) have very desirable scaling properties, there is quite general agreement in the literature that log-return series cannot be reasonably modeled by stable distributions. The author has never found convincing statistical evidence for any log-return series to be in agreement with the infinite variance marginal distribution hypothesis. The contrary position is defended, for example, in the monograph by Mittnik and Rachev [111].

5.1.2 Dependence, autocorrelations, and clusters of extremes

A further deviation from the Black-Scholes model is that returns are dependent.

In standard time series analysis such as presented in the excellent books of Brockwell and Davis [29, 30] the second order structure of a stationary time series (X_t) is fundamental for any studies. In particular, the autocovariances $\gamma_X(h)$ and autocorrelations $\rho_X(h)$ *at lag* $h \in \mathbb{Z}$, given by

$$\gamma_X(h) = \mathrm{cov}(X_0, X_h) \quad \text{and} \quad \rho_X(h) = \mathrm{corr}(X_0, X_h), \tag{5.1.1}$$

are the main objects of interest. The functions γ_X and ρ_X will be referred to as the autocovariance function (ACVF) and the autocorrelation function (ACF) of the time series (X_t). For a stationary Gaussian time series, the mean and the ACVF completely determine the dependence structure and the finite-dimensional distributions of the sequence (X_t). This, in general, is not the case for a nonGaussian time series. Nevertheless, the ACVF and in particular its normed version, the ACF, are considered as important indicators for the dependence structure in a time series. The ACVF and ACF are essential ingredients for parameter estimation, model testing, and prediction in classical time series analysis, which is the analysis of linear processes, in particular, ARMA (autoregressive moving average) processes.

Since the ACVF and ACF are not known for a real-life data set, they have to be estimated by statistical means. Standard estimators are given by the sample autocovariances $\gamma_{n,X}(h)$ and sample autocorrelations $\rho_{n,X}(h)$ at lag $h \in \mathbb{Z}$:

$$\gamma_{n,X}(h) = \frac{1}{n} \sum_{t=1}^{n-|h|} (X_t - \overline{X}_n)(X_{t+h} - \overline{X}_n) \quad \text{and} \quad \rho_{n,X}(h) = \frac{\gamma_{n,X}(h)}{\gamma_{n,X}(0)}, \tag{5.1.2}$$

where we interpret $\gamma_{n,X}(h) = \rho_{n,X}(h) = 0$ for $|h| \geq n$, and, as usual, $\overline{X}_n$ stands for the sample mean. A straightforward application of the ergodic theorem yields that for every $h \in \mathbb{Z}$,

$$\gamma_{n,X}(h) \stackrel{\text{a.s.}}{\to} \gamma_X(h) \quad \text{and} \quad \rho_{n,X}(h) \stackrel{\text{a.s.}}{\to} \rho_X(h), \tag{5.1.3}$$

provided that (X_t) is stationary ergodic and $\mathrm{var}(X_t) < \infty$. For this reason, the sample ACVF and sample ACF are usually taken as surrogates for the ACVF and ACF, respectively, although common sense tells us that we may get in trouble with this procedure if n is too small or h too large with respect to n. We will see later, in Section 5.9.1, that the sample ACF may also lead to misinterpretations if the distribution of the X_t's is very heavy-tailed (in the sense that $EX_t^4 = \infty$) since the rate of convergence in (5.1.3) can then be extremely slow and, hence, asymptotic confidence bands for the ACF can be unusually wide. But let us assume for the moment that the sample ACF is what it is supposed to be: a reasonable estimator for the ACF. Then the following properties of log-return series (X_t) are commonly observed.

- The sample ACF $\rho_{n,X}$ is negligible at all lags. (A possible exception is the first lag. However, the estimated value is usually small in absolute value as well.)
- The sample ACFs $\rho_{n,|X|}$ of the absolute values $|X_t|$ (further referred to as absolute log-returns) and ρ_{n,X^2} of the squares X_t^2 are different from zero for a large number of lags and stay almost constant and positive for large lags.

An illustration of these empirically observed facts for the S&P500 data of Figure 5.1 is presented in Figure 5.4.

The slow decay of the sample ACF for the absolute log-returns we observe in Figure 5.4 is very typical for longer time series. Since the autocorrelation $\rho_{|X|}(h)$ describes the degree of dependence in the process $(|X_t|)$ over a period of h units of time, and $\rho_{n,|X|}(h)$ is not negligible even for lags 200, 250 (one business year), 300, etc., one often refers in this context to the long memory or long range dependence of absolute returns and their squares. We will discuss this issue later in Section 5.11.1, and try to find an explanation for this effect.

At this point it is worthwhile explaining why we focus on the time series of the absolute values and squares of the log-returns. Empirical evidence shows that the sequence of the signs of a log-return series has similar statistical properties as a sequence of i.i.d. symmetric Bernoulli random variables. Therefore many models are of the form $X_t = |X_t| \,\mathrm{sign}(X_t)$, where the sequence $(\mathrm{sign}(X_t))$ consists of i.i.d. symmetric Bernoulli random variables, and so one is left to model absolute returns. To be more precise, as we will see in Section 5.2.2, various popular models are of multiplicative type: $X_t = \sigma_t Z_t$, where (Z_t) is an i.i.d. symmetric sequence, the volatility sequence (σ_t) is stationary and nonnegative, and σ_t and Z_t are independent for every fixed t. Thus the sequence $(\mathrm{sign}(X_t)) = (\mathrm{sign}(Z_t))$ is i.i.d. symmetric Bernoulli, as desired. The volatility of a log-return is of major interest but, by construction, it cannot be observed. Therefore the observable quantities $|X_t|$ and X_t^2 are often considered as surrogate values or estimators of σ_t and σ_t^2, respectively. To some extent this may explain the interest in modeling the dependence of the absolute log-returns and their squares.

We focus in this chapter on extremes, so it is worthwhile mentioning that autocorrelations are not good tools for explaining large and small values in a time series. Indeed, covariances and correlations are moments, hence they are integrated characteristics

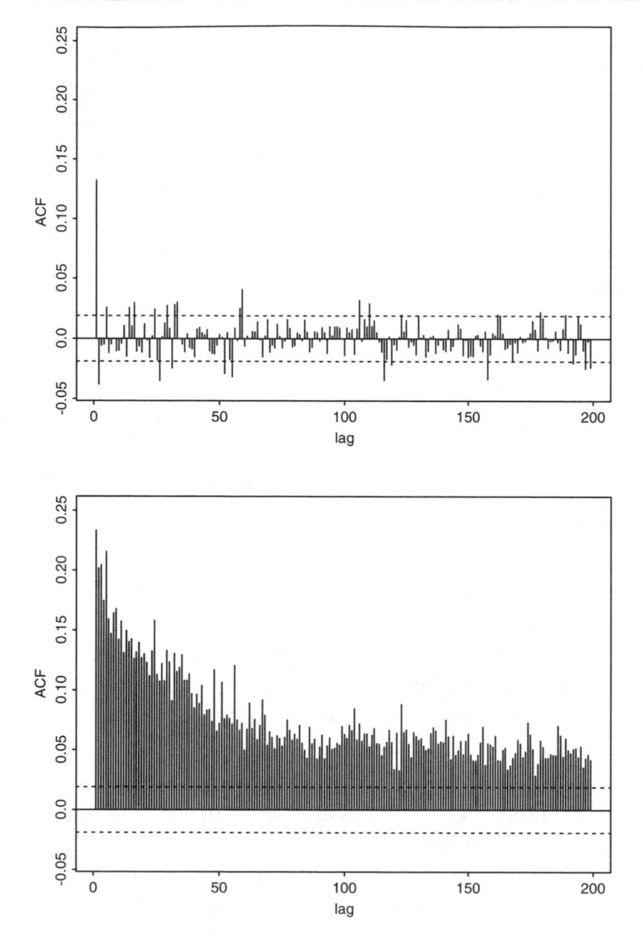

Figure 5.4 *Sample ACFs for the log-returns* (top) *and absolute log-returns* (bottom) *of the* S&P500. *Here and in what follows, the horizontal lines in graphs displaying sample ACFs are set as the* 95% *confidence bands* ($\pm 1.96/\sqrt{n}$) *corresponding to the ACF of iid Gaussian noise.*

of the distribution of the underlying time series; the contribution of the probabilities in the tails of the distributions is averaged out and disappears.

The dependence of extremal return values is obvious if one looks, for example, at pairs $|X_t|$, $|X_{t+1}|$ exceeding a high threshold; see Figure 5.5. For comparison we include the graph of the joint exceedances of the same threshold for the successive values $|X_t|$, $|X_{t+1}|$ in an i.i.d. sequence (X_t) with a student distribution with four degrees of freedom which often fits returns nicely. It is obvious that joint pairwise exceedances of a high threshold occur in clumps for log-return data due to the dependence of the extreme values. We refer to dependence in the tails. For an i.i.d. sequence, exceedances of high positive or low negative thresholds (relatively to the size of the data) occur

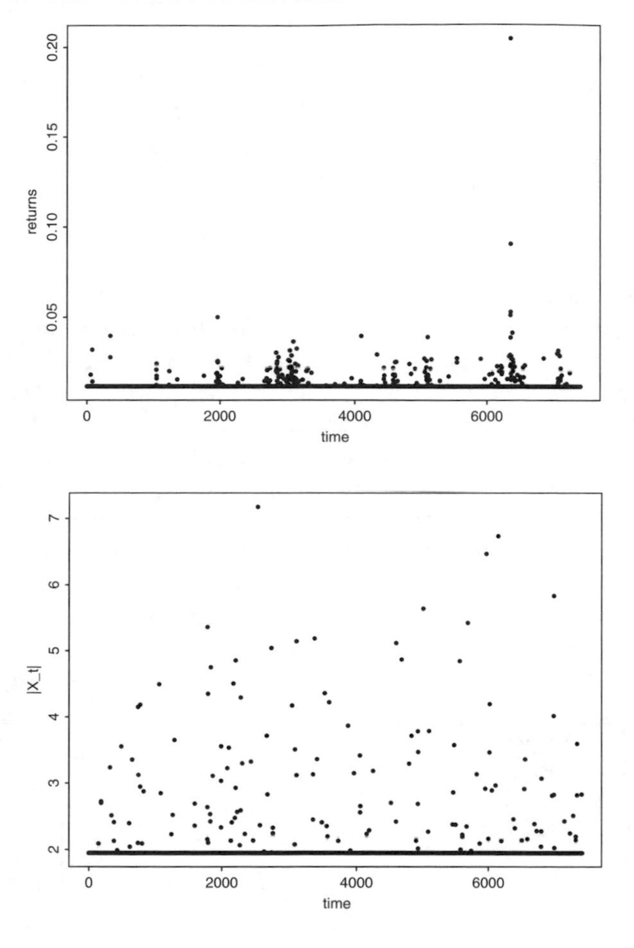

Figure 5.5 Top: *Absolute returns* $|X_t|$ *of the* S&P500 *series for which both* $|X_t|$ *and* $|X_{t+1}|$ *exceed the 87% quantile of the data. The latter is indicated by the bottom line.* Bottom: *The same kind of plot for an i.i.d. sequence from a student distribution with 4 degrees of freedom. In the former case pairwise exceedances occur in clusters, in the latter case exceedances appear uniformly scattered over time.*

separated over time, roughly according to a homogeneous Poisson process. For returns, this is not true.

This leads one to the next stylized fact:

- The large and small values in the log-return sample occur in clusters. There is dependence in the tails.

Since autocorrelations are not appropriate for describing the dependence of large and small X_t-values, further tools have been considered. One of them we want to explain now. For an i.i.d. sequence (X_t) we know that

$$P(M_n \leq x) = [P(X_1 \leq x)]^n, \quad n = 1, 2, \dots,$$

where

$$M_n = \max(X_1, \ldots, X_n).$$

For large classes of strictly stationary sequences (X_t) one can show the existence of a number $\theta \in [0, 1]$ such that

$$P(M_n \leq x_n) = [P(X_1 \leq x_n)]^{\theta n} + o(1),$$

where (x_n) is a suitable sequence converging to the right endpoint of the distribution of X_1. This number θ is the *extremal index* of (X_t). It can be estimated from the data by statistical methods; see Figure 5.6 for an illustration.

Due to dependence, the serial indices of the data can be separated into blocks in such a way that the size of the data is comparable within such a group which, for obvious reasons, is referred to as a cluster. We avoid giving a more precise definition. (This is a difficult task with various possible answers.) The value $1/\theta$ has interpretation as the expected value or the mean cluster size of high level exceedances of the data, and so an estimated value $\theta < 1$ for returns gives some indication on the clustering behavior for very large values of the returns. For more information on modeling of clusters and estimation of θ, see [58], Section 8.1, and the references therein, or the classic in extreme value theory, Leadbetter et al. [94].

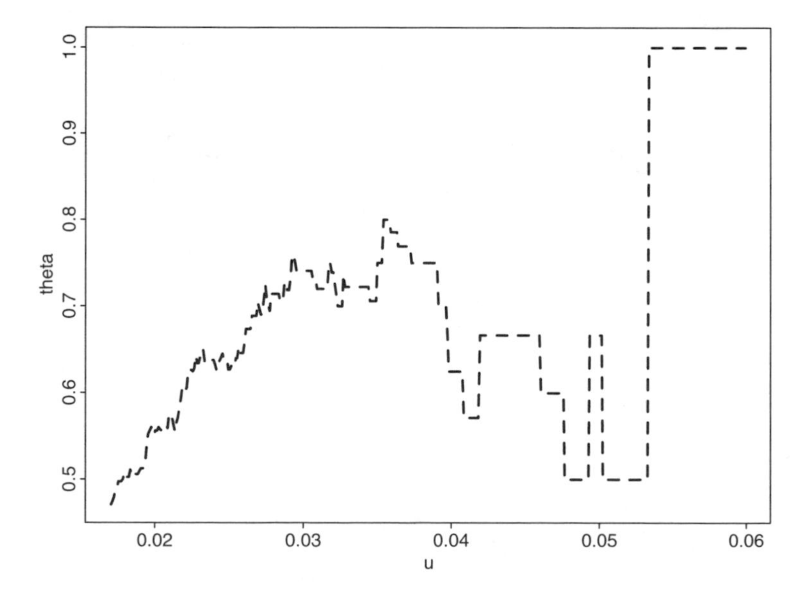

Figure 5.6 *Point estimation of the extremal index for the* S&P500 *data. The estimators are based on the upper order statistics exceeding the threshold u. The smallest u is the 97% quantile of the data. The estimator estimates θ in a u-region where the plot does not change much (between 0.03 and 0.04) resulting in an extremal index of about 0.7. If u is too high (above 0.04 say), the estimator is based on too few order statistics and not reliable.*

The extremal index gives a rough description of the clustering behavior of high level exceedances. In Section 5.4.2 we will learn about more sophisticated measures of dependence in the tails.

5.1.3 Aggregational Gaussianity

The list of stylized facts for returns is by far not complete. In the literature one finds more properties that are typical for a large variety of return series. Among them is

- the aggregational Gaussianity, i.e., the distribution of log-returns over longer periods of time (such as a month, half a year, a year) is closer to the normal distribution than for hourly or daily log-returns.

This means that density plots of returns are sharply peaked about zero and exhibit heavier tails than consistent with normal behavior (see Figure 5.2), but that under aggregation exhibit more of a bell shape characteristic of the normal density. Alternatively, this is an indication that daily or hourly returns satisfy the central limit theorem (CLT). In turn, because the condition of finite variance is close to necessity for the CLT with normal limit of a strictly stationary mixing series (cf. Ibragimov, Linnik [84] or Doukhan [52]), this indicates that the variance of the distribution of the data is finite. Aggregational Gaussianity is another piece of empirical evidence against Mandelbrot's hypothesis on infinite variance stable distributions; see the discussion in Section 5.1.1.

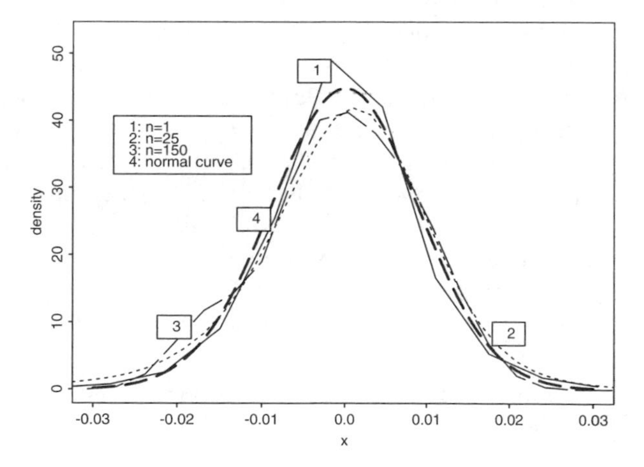

Figure 5.7 *Estimated probability density for the scaled (with $1/\sqrt{n}$) aggregated log-returns of the* S&P500 *from disjoint periods of* $n = 1, 25, 150$ *days and the normal density curve with the same mean and variance as the complete* S&P500 *series. Density plots are sharply peaked about zero and exhibit heavier tails than consistent with normal behavior, but under aggregation they exhibit more of a bell shape characteristic of the normal. If the aggregation period increases the sample size used for the density estimation becomes smaller and, therefore, the density curves are less smooth than for a lower aggregation level. (For periods of* $n = 1, 25, 150$ *days, the sample sizes used for density estimation are 9558,382, and 63, respectively.)*

In the literature one can often find the statement that log-returns over longer periods of time have tails less heavy than compared to those over shorter periods. This is frequently explained by the aggregational Gaussianity. This statement is incorrect if the data come from a distribution with a power law tail and the dependence structure of the data is specified in a suitable way; see Example 5.4.9 for more details. Indeed, in this case it can be shown for large classes of strictly stationary time series (X_t) that the sums $X_t + \cdots + X_{t+h}$ have very much the same asymptotic power law tails as X_t itself. In particular, they have the same tail index.

We finally mention a property of log-return data that is often referred to as leverage effect. We claimed before that the distribution of log-returns is roughly symmetric in the center of the distribution. However, there is a tendency for the data to be negative in periods of higher activity compared to periods of lower activity. This means that the market is prone to react differently to positive as opposed to negative returns. This dynamic version of asymmetry (Barndorff-Nielsen and Shephard [5]) will not be addressed in this chapter where we mainly focus on the issue of modeling the dependence in the tails.

5.2 Standard models

5.2.1 *Linear and ARMA processes*

The stylized facts of returns indicate the difficult task of modeling time series with such a complicated structure. We start by asking what answers *classical time series analysis* would yield to the mentioned problems. Classical (or *standard*) time series analysis is motivated by the theory of Gaussian stationary processes where the second order structure of the time series tells us everything about the finite-dimensional distributions and, hence, the dependence in the process. Although this statement is not correct for an arbitrary (strictly) stationary process, the detection of the second order structure of the time series remains the basic goal with the aims of fitting models to real-life data and predicting future values out of the data. For returns X_t the second order structure of the time series (all autocorrelations are negligible) does not help one much to understand its structure, and therefore it is useful to consider alternative means, in particular, when one is interested in the extremal behavior of the return series.

In classical time series analysis as presented in Brockwell and Davis [29], linear processes constitute the basic model:

$$X_t = \sum_{j=0}^{\infty} \psi_j Z_{t-j}, \quad t \in \mathbb{Z}, \tag{5.2.1}$$

where (Z_t) is an i.i.d. noise sequence, and (ψ_j) is a sequence of real numbers satisfying some mild summability condition depending on the tails of the distribution of Z_t. The role of linear processes is justified by the fact that any reasonable stationary (i.e., nondeterministic) Gaussian process has a representation as Gaussian linear process; see Brockwell and Davis [29], Section 5.7. *ARMA processes* (autoregressive moving average processes) of a certain order (p, q) are particular linear processes (under certain conditions on their parameters). An ARMA(p, q) process (X_t) is given as the solution to the difference equations

$$\varphi(B)X_t = \theta(B)Z_t, \quad t \in \mathbb{Z}. \tag{5.2.2}$$

Here

$$\varphi(z) = 1 - \sum_{i=1}^{p} \varphi_i z^i \quad \text{and} \quad \theta(z) = 1 + \sum_{j=1}^{q} \theta_j z^j,$$

are polynomials with real coefficients φ_i, θ_j and

$$B^k A_t = B^{k-1}(BA_t) = B^{k-1}A_{t-1} = A_{t-k}$$

is the kth power of the backshift operator B satisfying $BA_t = A_{t-1}$ for any sequence of random variables (A_t). Also recall that (X_t) is said to be an AR(p) process (autoregressive process of order p) if $q = 0$ and an MA(q) process (moving average process of order q) if $p = 0$.

The heavy tails of the distribution of X_t can easily be explained by assuming power law behavior for the distribution of the Z_ts; see Section 5.4.3 for more details. However, the only causal[1] ARMA processes driven by i.i.d. noise whose autocorrelation structures are in agreement with the sample ACF of the returns X_t would be an i.i.d. finite variance noise sequence or an MA(1) process with a small parameter θ_1. Clearly, the latter models cannot explain the dependence in the data, in particular in the absolute and squared returns. Thus one is forced to model return data by a nonlinear process.[2]

5.2.2 *Multiplicative models*

Most models for return data used in practice are of a multiplicative form

$$X_t = \mu + \sigma_t Z_t, \quad t \in \mathbb{Z}, \tag{5.2.3}$$

where (Z_t) is an i.i.d. sequence, (σ_t) is a stochastic process with nonnegative values and such that σ_t and Z_t are independent for fixed t. Mostly, the volatility process (σ_t) and (X_t) are assumed to be strictly stationary. In order to model the symmetry of the return distribution, it is often assumed that Z_t is symmetric or, at least, has mean zero. In what follows, we will always assume that Z_t is symmetric and has unit variance. The latter assumption does not restrict generality because a scaling trade-off between σ_t and Z_t is always possible. The former assumption is an idealization — there is no perfect symmetry in real-life return data. This condition can be abandoned in most parts of the theory presented, but we stick to it in what follows, because it often leads to an easier presentation. Moreover, we suppose that μ can be estimated from the data and therefore it will be convenient to assume $\mu = 0$ as well.

[1] This means that (X_t) has representation (5.2.1).

[2] The time series models considered below are white noise models, i.e., they are stationary with zero ACF. In the terminology of classical time series analysis, these models would be called linear processes in the sense of the Wold decomposition; see Brockwell and Davis [29], Section 5.7, for details. We call these processes nonlinear since they do not have representation as a linear process (5.2.1) with i.i.d. innovations Z_t.

There are various reasons for the particular choice of the simple multiplicative model.

- First of all, the direction of price changes is modeled only by the sign of Z_t, independently of the order of magnitude of this change, which is directed by the volatility σ_t. This is in agreement with the empirical observation that it is difficult or even impossible to predict the sign of price changes.
- Secondly, since σ_t and Z_t are independent and Z_t is assumed to have mean zero and variance 1, σ_t^2 is then the conditional variance of X_t given σ_t. Some models assume that σ_t is a function of the past, i.e., $X_{t-1}, X_{t-2}, \ldots$ and $\sigma_{t-1}, \sigma_{t-2}, \ldots$, therefore, it is in principle known at time t. For this reason, σ_t in connection with the distribution of Z_t can be used to give a distributional forecast of X_t.

For example, assume Z_t to be standard normal. Then, conditionally on the past values $X_{t-1}, \sigma_{t-1}, X_{t-2}, \sigma_{t-2} \ldots$, today's return X_t has an $N(0, \sigma_t^2)$ distribution. This allows one to give conditional confidence bands for X_t. For example, it is an easy matter to determine the conditional VaR (Value at Risk) of the sequence (X_t), i.e., the 1% or 5% quantiles of the distribution of X_t. The determination of VaR is one of the main requirements of regulators of banks: based on their own historical data, banks have to prove that they have enough reserves to cover essential future losses at the 1% or 5% levels of the distribution of the returns X_t. Multiplicative models of type (5.2.3) are therefore attractive models, in particular when paired with Gaussian noise (Z_t). The above lines should not give the impression that it is an easy task to determine VaR. The latter is an extremely difficult statistical problem — because of the sparseness of data and the fact that in reality one has to face portfolios of assets, i.e., one has to deal with linearly combined returns from various (typically dependent) assets. For a recent account on the VaR methodology we refer to Crouhy et al. [33], the website www.gloriamundi.org on recent developments on VaR, or the RiskMetrics group's website www.riskmetrics.com.

It is an easy exercise to see that the multiplicative model (5.2.3) with σ_t being a function of the past values of the σs and Xs has ACF

$$\rho_X(h) = \operatorname{corr}(X_t, X_{t+h}) = 0, \quad t \in \mathbb{Z},$$

for all lags $h \neq 0$, provided the latter expressions are well defined, i.e., when the variance of X_t is finite. At least this stylized fact of real-life returns is captured, given that the autocorrelation $\rho_X(h)$ at lag h is well approximated by its sample analogue $\rho_{n,X}(h)$, which is not significantly different from zero as we have seen in Section 5.1.2.

5.2.3 The ARCH family

Definition and relation to ARMA processes

Engle [59] suggested the following simple model for the volatility σ_t:

$$\sigma_t^2 = \alpha_0 + \sum_{i=1}^{p} \alpha_i X_{t-i}^2, \quad t \in \mathbb{Z}. \tag{5.2.4}$$

Here α_i are nonnegative constants with $\alpha_p > 0$ if $p \geq 1$. The model (5.2.3) with the specification (5.2.4) for σ_t^2 is called an ARCH(p) (autoregressive conditionally heteroscedastic model of order p).

The autoregressive structure can be seen by the following argument. Writing

$$\nu_t = X_t^2 - \sigma_t^2 = \sigma_t^2 \left(Z_t^2 - 1\right),$$

with the help of (5.2.4) one obtains

$$\varphi(B) X_t^2 = \alpha_0 + \nu_t, \quad t \in \mathbb{Z}, \tag{5.2.5}$$

where

$$\varphi(z) = 1 - \sum_{i=1}^{p} \alpha_i z^i,$$

and $B A_t = A_{t-1}$ is the backshift operator. If (Z_t) is an i.i.d. sequence with mean zero and unit variance, and (X_t) is second-order stationary, then (ν_t) constitutes a white noise sequence. Therefore we observe a formal analogy with an AR(p) process given in (5.2.2) with $\theta(z) \equiv 1$. But we also observe one crucial difference: the right-hand side of (5.2.2) depends on the i.i.d. sequence (Z_t), whereas the right-hand side of (5.2.5) depends on the sequence (ν_t) which consists of dependent random variables.

Since ARCH(p) processes do not fit log-returns very well unless one chooses the order p quite large (which is not desirable when the sample is small), various people have thought about improvements. Because (5.2.5) bears some resemblance with an AR structure, the similarity with (5.2.2) suggests that we introduce an ARMA structure for squared returns:

$$\varphi(B)\, X_t^2 = \alpha_0 + \beta(B)\, \nu_t, \quad t \in \mathbb{Z}, \tag{5.2.6}$$

where $\varphi(B)$ and $\beta(B)$ are polynomials in the backshift operator B with coefficients φ_i, β_j. More precisely, let α_i, $i = 0, \ldots, p$, and β_j, $j = 1, \ldots, q$, be nonnegative coefficients with $\alpha_p > 0$ if $p \geq 1$ and $\beta_q > 0$ if $q \geq 1$, then

$$\varphi(z) = 1 - \sum_{i=1}^{p} \alpha_i z^i - \sum_{j=1}^{q} \beta_j z^j \quad \text{and} \quad \beta(z) = 1 - \sum_{j=1}^{q} \beta_j z^j.$$

This construction leads to the GARCH(p, q) process (generalized ARCH process of order (p, q)), which was independently introduced by Bollerslev [15] and Taylor [132]. The latter process, with its ramifications and modifications, has become the model for returns used most frequently in applications. It is more conveniently written as the multiplicative model (5.2.3) with specification:

$$\sigma_t^2 = \alpha_0 + \sum_{i=1}^{p} \alpha_i X_{t-i}^2 + \sum_{j=1}^{q} \beta_j \sigma_{t-j}^2, \quad t \in \mathbb{Z}. \tag{5.2.7}$$

Here α_i and β_j are nonnegative constants. To understand the motivation behind the ARCH processes it pays to read some of the original articles of which some were mentioned; see Engle [60] for a good collection.

Why GARCH?

The popularity of the GARCH model can be explained by various arguments.

- Its relation to ARMA processes suggests that the theory behind it might be closely related to ARMA process theory, which is well studied, widely known, and seemingly easy.

This opinon is, however, wishful thinking. The difference to standard ARMA processes is due to the fact that the noise sequence (ν_t) in (5.2.6) depends on the X_ts themselves, so a complicated nonlinear relationship of the X_ts builds up. For example, in order to show that a stationary version of (X_t^2) exists, one would have to iterate equation (5.2.6), hoping that X_t^2 becomes an explicit expression only of the sequence (ν_t), which expression one might take as the solution to the difference equation (5.2.6). For an i.i.d. noise sequence (ν_t) this recipe is known to work; see Brockwell and Davis [29], Chapter 3, who study conditions for the validity of this approach. However, the noise $\nu_t = X_t^2 - \sigma_t^2$ itself depends on the stationary sequence (X_t) to be constructed, and so one has basically gained nothing by this approach.

Conditions for the existence of the stationary version of a general GARCH process are not easily given. An exception is the GARCH(1,1) process for which necessary and sufficient conditions for the existence of a stationary version of (X_t) in terms of $\alpha_0, \alpha_1, \beta_1$, and the distribution of Z_1 are known. More arguments on this issue will be given in Section 5.3.2.

If one knows that (X_t) is a well defined stationary process, the relation with ARMA processes can be useful. For example, one can derive formulae for the moments of X_t^2 by using the moments of an ARMA process in terms of the ARMA parameters and the moments of the underlying noise sequence (ν_t).

A second argument for the use of GARCH models is that,

- even for a GARCH(1,1) model with three parameters one often gets a reasonable fit to real-life financial data, provided that the sample has not been chosen from a too long period making the stationarity assumption questionable. Tests for the residuals of GARCH(1,1) models with estimated parameters $\alpha_0, \alpha_1, \beta_1$ give the impression that the residuals very much behave like an i.i.d. sequence.

Some evidence on this issue can be found in the paper of Mikosch and Stărică [107]. A third, and perhaps the most powerful argument in favor of GARCH models, from an applied point of view, is the fact that

- the statistical estimation of the parameters of a GARCH process is rather uncomplicated.

This attractive property has led S+ to provide us with a module for the statistical inference and simulations of GARCH models.

Estimation: Gaussian quasi-maximum likelihood

The estimation technique used most frequently in applications is a Gaussian quasi-maximum likelihood procedure that we want to explain briefly. Assume for the moment that the noise (Z_t) in an ARCH(p) model of a given order p is i.i.d. standard normal. Then X_t is Gaussian $N(0, \sigma_t^2)$ given the whole past $X_{t-1}, X_{t-2}, \dots$, and a conditioning argument yields the density function $f_{X_p,\dots,X_n}$ of $X_p, \dots, X_n$ through the conditional Gaussian densities of the X_ts given $X_1 = x_1, \dots, X_n = x_n$:

$$\begin{aligned}
&f_{X_p,\dots,X_n}(x_p, \dots, x_n) \\
&\quad = f_{X_n}(x_n \mid X_{n-1} = x_{n-1}, \dots, X_1 = x_1) f_{X_{n-1}}(x_{n-1} \mid X_{n-2} = x_{n-2}, \dots, X_1 = x_1) \cdots \\
&\qquad f_{X_{p+1}}(x_{p+1} \mid X_p = x_p, \dots, X_1 = x_1) f_{X_p}(x_p) \\
&\quad = (2\pi)^{-(n-p)/2} \prod_{t=p+1}^{n} \sigma_t^{-1} \mathrm{e}^{-x_t^2/(2\sigma_t^2)} f_{X_p}(x_p).
\end{aligned} \tag{5.2.8}$$

Ignoring the density f_{X_p} and replacing $t = p + 1$ by $t = 1$ in (5.2.8), the Gaussian log-likelihood of $X_1, \dots, X_n$ is given by

$$L_n(\alpha_0, \alpha_1, \dots, \alpha_p, \beta_1, \dots, \beta_q) = -\frac{1}{2n} \sum_{t=1}^{n} \left[2 \log \sigma_t + \sigma_t^{-2} X_t^2\right]. \tag{5.2.9}$$

The latter quantity is also formally defined for general GARCH(p, q) processes, and it can be maximized as a function of the α_is and β_js involved. The resulting value in the parameter space is the Gaussian quasi-maximum likelihood estimator (MLE) of the parameters of a GARCH(p, q) process.

There are obvious problems with this estimation procedure. For example, one might be surprised about the assumption of Gaussian noise (Z_t). Although this is not the most realistic assumption,[3] theoretical work (see the references below) shows that asymptotic properties such as $\sqrt{n}$-*consistency* (i.e., consistency and asymptotic normality with $\sqrt{n}$-rate) of the Gaussian quasi-MLE remain valid for large classes of noise distributions. This observation is similar to other estimation procedures in time series analysis where one does not maximize the true maximum likelihood function of the underlying data, but rather assumes Gaussianity of the data and maximizes a corresponding score function. This approach works for the Gaussian maximum likelihood of ARMA processes (see Brockwell and Davis [29], Section 10.8) and in more general situations, see Heyde [82].

Attempts to replace the Gaussian densities in L_n by a more realistic density of the Z_ts (for example, a t-density) can lead to nonconsistency of the MLE. Consistency of the estimators can be achieved if one knows the exact density underlying Z_t, but when dealing with data one can never rely on this assumption. Even if one tries to

[3] Empirical evidence indicates that the Z_ts are much better modeled by a t-distribution; see Section 5.10.1.

estimate the parameters of the density of Z_t together with the GARCH parameters (for example, some professional software offers to estimate the degrees of freedom of t-distributed Z_ts from the data), the MLE based on these densities can lead to nonconsistent estimators.[4]

The careful reader might also have observed that the derivation of the maximum likelihood function (5.2.8) is not directly applicable if the model deviates from an ARCH(p) process. Indeed, that formula requires calculating the unobservable values σ_t, $t = 1, \ldots, n$, from the observed sample $X_1, \ldots, X_n$. A glance at the defining formula (5.2.7) convinces one that this is not possible in the general GARCH(p, q) case. Indeed, an iteration of (5.2.7) yields that one would have to know all values $X_{n-1}, \ldots, X_0, X_{-1}, \ldots$ for the calculation of $\sigma_1, \ldots, \sigma_n$. Alternatively, one needs to know finitely many values of the unobservable values $X_0, X_{-1}, \ldots$ and $\sigma_0, \sigma_{-1}, \ldots$. Therefore practitioners (and software packages) have to choose a finite number of such initial values in order to make the iteration for the σs run. The choice of deterministic initial values implies that the calculated $\sigma_1, \ldots, \sigma_n$ cannot be considered as a realization of a stationary sequence. One may, however, hope that the dependence on the initial values disappears for large values of n in a way similar to a Markov chain with arbitrary initial value whose distribution becomes closer to the stationary distribution. We will see in Section 5.3.2 that the σ_t^2s can be embedded in a multivariate Markov chain, and, therefore, this reasoning has some foundation. However, the Gaussian quasi-MLE is a complicated functional of the Xs and σs and, therefore, the theoretical properties of this estimator are not easy to derive. Alternatively, the MLE with fixed initial values can be viewed as conditional MLE.

There exist various papers dealing with the asymptotic properties of the quasi-MLE. The first one to be mentioned in this context is Weiss [135] who treated the ARCH(p) case. Notice that in this case it is possible to calculate $\sigma_{p+1}, \ldots, \sigma_n$ from $X_1, \ldots, X_n$. Then, working with the sample $X_{p+1}, \ldots, X_n$, one can avoid the mentioned theoretical problems. Lee and Hansen [96] addressed the GARCH(1, 1) case, assuming the stationarity of the sequence (σ_n), thus avoiding the problem of the initial values. They showed $\sqrt{n}$-consistency of the quasi-MLE. A similar result, but under different conditions, was given by Lumsdaine [98] in the GARCH(1, 1) case. He first shows the result for the stationary sequence (σ_t), then he addresses the problem of deterministic initial values and shows that the result proved for the stationary sequence remains valid. The paper of Berkes et al. [11] contains an alternative proof of the $\sqrt{n}$-consistency in the general GARCH(p, q) case under weak assumptions; see also Comte and Lieberman [32] for the multivariate case.

Simulation results, see Figure 5.8, indicate that the Gaussian quasi-MLE does not work too well for small sample sizes of a couple of hundred values. On the other hand, it is not very realistic to fit a particular GARCH model to several years of daily log-returns — the data do not behave like a stationary process over such long periods of time; see the discussion in Section 5.11. (In the course of this chapter the author will deviate from this principle and fit GARCH models to data worth many years or returns.) The accuracy of the estimation procedure based on one business year of data

[4] Daniel Straumann, personal communication.

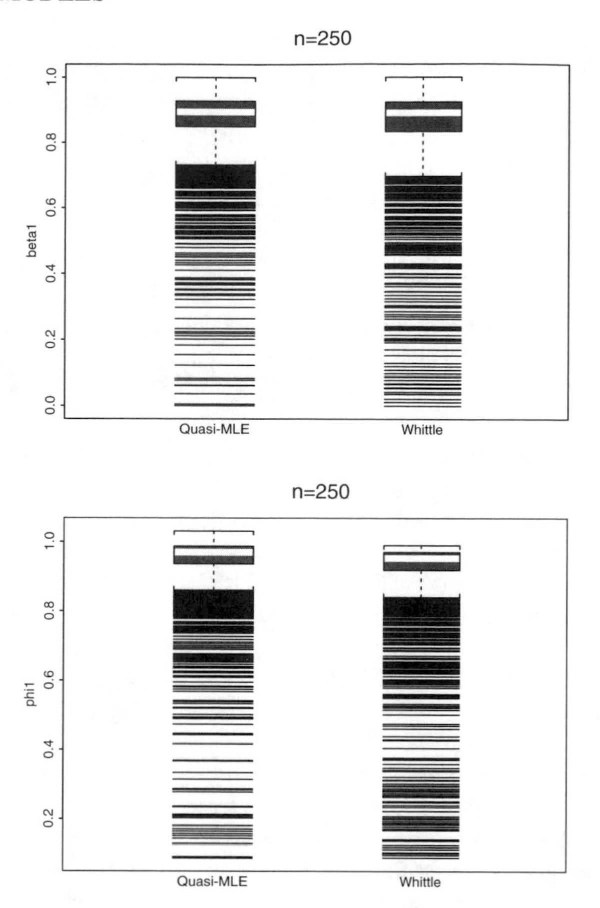

Figure 5.8 *A boxplot comparison of the distributions of the Gaussian quasi-MLE and the Whittle estimator for* β_1 (top) *and* $\varphi_1 = \alpha_1 + \beta_1$ (bottom) *in a* GARCH(1, 1) *model with parameters* $\alpha_0 = 8.58 \times 10^{-6}$, $\alpha_1 = 0.072$, *and* $\beta_1 = 0.92$. *The sample size is* $n = 250$. *The boxplots are based on 1000 independent repetitions of the parameter estimation procedures.*

(250 days) is rather nonsatisfactory. The poor behavior of the quasi-MLE for GARCH models seems to be due to the fact that the log-likelihood function L_n in (5.2.9) is rather flat in the parameter space and therefore it is difficult to find its maximum. This fact is well known as the author has learned in various conversations, but he did not find a reference where theoretical reasons are given.

Whittle estimation

The emphasis in the literature on estimation in GARCH models is on the Gaussian quasi-MLE. Other estimation techniques have often been ignored. As an alternative to quasi-maximum likelihood one might think of using classical estimation techniques developed for the ARMA case. We have learned in Section 5.2.1 that the squares

X_t^2 of a GARCH process satisfy the system of ARMA equations (5.2.6) with noise sequence (ν_t), which is uncorrelated if $EX_t^4 < \infty$. This idea was taken up by Giraitis and Robinson [67] who showed in the context of parametric ARCH(∞) models, which include the GARCH(p, q) case, that the so-called Whittle estimator of the α_is and β_js is $\sqrt{n}$-consistent provided the 8th moment of the data is finite. They pointed out that the variance of the Whittle estimator is larger as compared to the ARMA case with the same coefficients and i.i.d. noise with the same variance as ν_t.

We mention here that Whittle estimation is one of the three classical estimation techniques for ARMA processes with i.i.d. noise. It is based on the idea of approximating the spectral density of the underlying process and can be shown to be asymptotically equivalent to Gaussian quasi-maximum likelihood and least squares estimation in the sense that all three methods are $\sqrt{n}$-consistent with the same asymptotic variance. See Brockwell and Davis [29], Section 10.8, for a comparison of these methods. We also mention that the Whittle estimator coincides with the least squares estimator and the celebrated Yule-Walker estimator for AR(p) processes. The Whittle estimator for ARMA processes even works if the noise has infinite variance and its rate of convergence compares favorably to $\sqrt{n}$-rates; see Mikosch et al. [104] for details and Section 7.5 in Embrechts et al. [58] for related results for various other estimators.

We mentioned in Section 5.1.1 that log-return data tend to be heavy-tailed with 3rd, 4th, 5th, etc., moments possibly infinite. It is worthwhile studying the Whittle estimator for the squared GARCH process under the moment assumption $EX_t^8 = \infty$. The asymptotic results, however, differ very much from the ARMA case with i.i.d. noise. As shown in Mikosch and Straumann [110], two cases can occur. If $EX_t^4 < \infty$ the rate of convergence for the Whittle estimator based on the squares of a GARCH(1, 1) process is slower the fewer moments of X_t exist. Even worse, if $EX_t^4 = \infty$ the Whittle estimator converges in distribution to a nondegenerate limit. Thus the quasi-MLE seems to outperform the Whittle estimator as regards basic consistency questions and convergence rates. This is very much in contrast to the classical ARMA case. This is also illustrated by some simulation evidence in Figure 5.8 and Figure 5.9.

For small sample sizes such as one business year of data, both estimators behave very much in the same way for parameter values that ensure the existence of the 4th moment of the X_ts. For larger sample sizes the Whittle estimator is clearly inferior to the MLE, and although no theoretical results seem to be available for this case, the MLE improves its performance when the tails of the distribution of X_t become heavier. It is difficult to compare the asymptotic covariance matrices for the Whittle estimator (they depend only on the variance of the ν_ts and the parameters) and for the quasi-MLE (they depend on the distribution of the noise Z_t and the parameters, are rather unattractive and need to be evaluated by using simulations) in the cases when it makes sense to compare them, i.e., when $EX_t^8 < \infty$. However, the problems noted above for the MLE estimation show that the MLE is not an ideal estimator that one should always use without seeking alternatives. Many theoretical problems in this context remain.

Finally, the question arises as to whether the GARCH model captures any of the stylized facts of real-life returns (heavy-tailed distribution, dependence in the tails, long memory effects of the absolute, and squared returns) mentioned in Section 5.1.

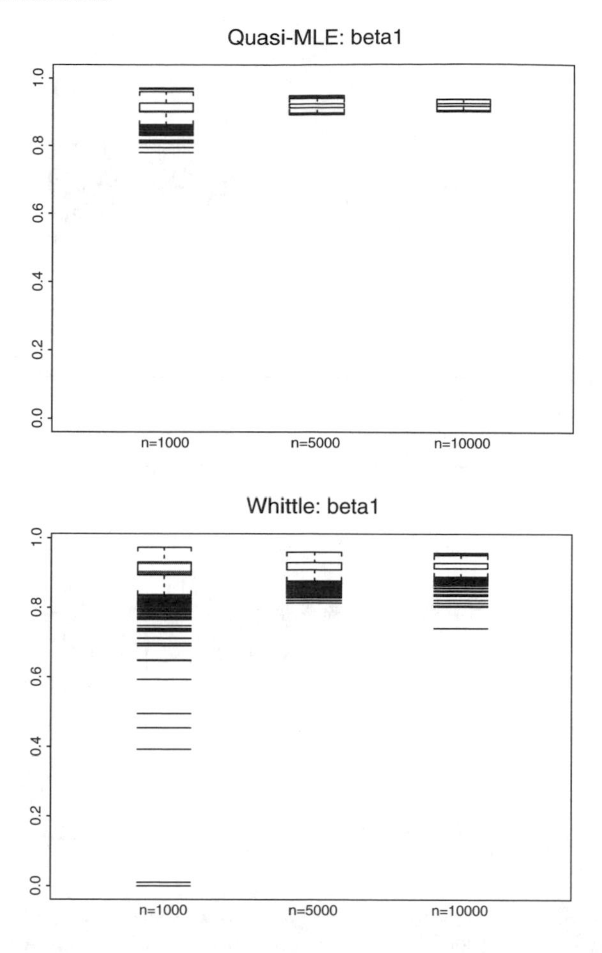

Figure 5.9 *A boxplot comparison of the distributions of the Gaussian quasi-MLE and the Whittle estimator for* $\beta_1 = 0.92$ *in the* GARCH(1, 1) *model of Figure 5.8 for different sample sizes n. The boxplots are based on 1000 independent repetitions of the estimation procedures. Gaussian quasi-MLE outperforms the Whittle estimator, but one needs samples of size about* $n = 5000$ *before the asymptotic normality assumption starts working. All estimators are negatively biased.*

These problems will be addressed in detail starting in Section 5.3 and summarized in Section 5.10.1.

5.2.4 *The stochastic volatility model*

The GARCH model is explicitly defined by the difference equation (5.2.7) for σ_t^2. However, its probabilistic properties (existence of stationary solution, dependence structure, tails, etc.) are by no means easy to derive and not in all cases well understood. We will consider another multiplicative model, the stochastic volatility process $X_t = \sigma_t Z_t$, where (σ_t) is a strictly stationary process, independent of the i.i.d.

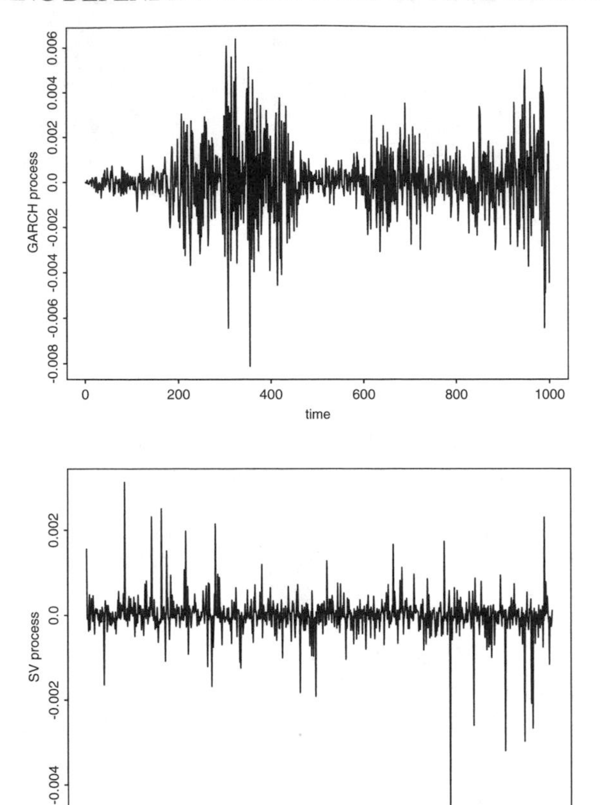

Figure 5.10 Top: *A simulated path of the* GARCH(1, 1) *time series* $X_t = (0.0001 + 0.1X_{t-1}^2 + 0.9\sigma_{t-1}^2)^{0.5} Z_t$, $t = 1, \ldots, 1000$, *for i.i.d. standard normal* (Z_t). Bottom: *A simulated path of the stochastic volatility time series* $X_t = e^{Y_t} Z_t$, $t = 1, \ldots, 1000$, *for i.i.d. student noise* (Z_t) *with 4 degrees of freedom,* $Y_t = 0.5Y_{t-1} + 0.3\eta_{t-1} + \eta_t$ *is an* ARMA(*1,1*) *process with i.i.d. standard normal noise* (η_t). *An eyeball inspection gives the impression that the characteristics of a GARCH time series are more in agreement with return series than those of a stochastic volatility time series. In particular, the clusters of high and low volatility observed in real-life data are more appropriately mimicked in the top than in the bottom graph. Compare with Figure 5.1 and Figure 5.23.*

symmetric noise process (Z_t). In contrast to GARCH-type processes, there is no feedback between the noise (Z_t) and the volatility process (σ_t), i.e., there are two independent sources of randomness. For this reason, the stochastic volatility models are sometimes considered as unnatural.[5] However, the probabilistic properties of a stochastic volatility process are much better understood than the properties of

[5] A referee of this chapter pointed out that the independence of (σ_t) and (Z_t) is not universally assumed for stochastic volatility models; some econometricians allow for dependence between (σ_t) and (Z_t) in order to model leverage.

GARCH processes: the mutual independence of (σ_t) and (Z_t) allows one to model dependency only via the volatilities σ_t, whereas the heavy tails of the distribution of X_t can be regulated via the interplay between the tails of σ_t and Z_t.

Usually, (σ_t) is given by a parametric model such as a Gaussian ARMA process for $(\log \sigma_t)$. By first squaring X_t and then taking logarithms, one can see that the ARMA process $2 \log \sigma_t$ gets perturbed by the extra noise $2 \log |Z_t|$ which makes estimation more complicated because it is impossible to give an explicit expression for the likelihood function. In comparison to ARCH, the lack of estimation procedures for stochastic volatility models made them less attractive for modeling purposes than the GARCH family. In recent years the attitude towards stochastic volatility has changed and a variety of estimation techniques (such as GMM and quasi-MLE) has been developed as well; see, for example, Section 5 in Ghysels et al. [65] or Shephard [129] for some surveys. Often one needs to resort to simulation based methods to calculate efficient estimates. Quasi-MLE methods are discussed in Harvey et al. [81], Breidt and Carriquiry [26], and Breidt et al. [27]. Simulation based methods are used in Jacquier et al. [85] and Durbin and Koopman [56].

As for GARCH processes, the question arises as to which of the stylized facts of real-life returns are actually captured by stochastic volatility models. This issue will be discussed in detail in the following sections and summarized in Section 5.10.2.

5.3 The stationarity issue

The assumption of stationarity of the time series (X_t) is basic for the statistical analysis of the data. In this section we address the question under which conditions strictly stationary versions of the GARCH and stochastic volatility processes exist.

5.3.1 The stochastic volatility case

For a stochastic volatility model, (σ_t) and (Z_t) are independent, and (Z_t) is i.i.d. Hence (X_t) is strictly stationary if (σ_t) has this property. In what follows, we assume the following representation for the log-volatilities:

$$\log \sigma_t = \sum_{j=0}^{\infty} \psi_j \eta_{t-j}, \quad t \in \mathbb{Z}, \tag{5.3.1}$$

where (η_t) is an i.i.d. sequence with mean zero and finite variance and the real coefficients ψ_j are square summable. Under these conditions, $(\log \sigma_t)$, hence (σ_t), is a strictly stationary process; see Brockwell and Davis [29], Chapter 3. We will often assume that (η_t) is i.i.d. Gaussian. In this case we know that σ_t is log-normal.

5.3.2 The GARCH case

The stationarity issue for a GARCH process is a difficult one and requires some sophisticated arguments that we want to discuss now.

In what follows, we assume that the noise (Z_t) is i.i.d. with mean zero and unit variance, i.e., the symmetry assumption on Z_t is not needed here. As for the stochastic volatility model, $X_t = \sigma_t Z_t$ constitutes a strictly stationary sequence if and only if

(σ_t^2) is strictly stationary. Indeed, by iterating the defining equation of the σ_t^2-process, it is not difficult to see that (σ_t^2, Z_t) is strictly stationary if and only if (σ_t^2) is strictly stationary, and so are processes given by $(f(\sigma_t^2, Z_t))$ for some measurable f. Hence we will focus on showing that a stationary version of (σ_t^2) exists.

For the sake of argument, we first focus on the GARCH(1,1) case. Writing

$$A_t = \alpha_1 Z_{t-1}^2 + \beta_1, \quad B_t = \alpha_0 \quad \text{and} \quad Y_t = \sigma_t^2,$$

we see that

$$\sigma_t^2 = \alpha_0 + \alpha_1 X_{t-1}^2 + \beta_1 \sigma_{t-1}^2 = \alpha_0 + \left(\alpha_1 Z_{t-1}^2 + \beta_1\right) \sigma_{t-1}^2$$

or

$$Y_t = A_t Y_{t-1} + B_t. \tag{5.3.2}$$

Notice that A_t and Y_{t-1} are independent, and (A_t, B_t) constitute an i.i.d. sequence. We refer to (5.3.2) as a stochastic recurrence equation (SRE) or random coefficient autoregressive model. Indeed, if $A_t = \varphi$ were a constant, (5.3.2) would describe an AR(1) process with parameter φ. For an AR(1) process with i.i.d. noise (B_t) we know that a unique stationary causal[6] solution of (5.3.2) exists if $|\varphi| < 1$ and B_t satisfies a mild moment condition. This solution is obtained by iterating the defining SRE, and one can proceed in a similar way for (5.3.2). Iterating r times back and noticing that $B_t = B_1$, we obtain:

$$Y_t = A_t \cdots A_{t-r} Y_{t-r-1} + \sum_{i=t-r}^{t} A_t \cdots A_{i+1} B_1.$$

Now, letting r go to infinity, we hope that the first term on the right-hand side will disappear, and the second one will converge. Notice that

$$\sum_{i=-\infty}^{t} A_t \cdots A_{i+1} = 1 + \sum_{i=-\infty}^{t-1} \exp\left\{(t-i)\left[\frac{1}{t-i} \sum_{j=i+1}^{t} \log A_j\right]\right\}. \tag{5.3.3}$$

For fixed t, the strong law of large numbers tells us that as $i \to -\infty$,

$$\frac{1}{t-i} \sum_{j=i+1}^{t-1} \log A_j \overset{\text{a.s.}}{\to} E \log A_1 ,$$

provided that $E \log A_1$ is defined, finite or infinite. Hence, under the moment condition $-\infty \leq E \log A_1 < 0$, the infinite series (5.3.3) converges a.s. for every fixed t. Then the sequence

$$\widetilde{Y}_t = \sum_{i=-\infty}^{t} A_t \cdots A_{i+1} B_1, \quad t \in \mathbb{Z},$$

[6] This means the solution depends only on past and present values of the noise (B_t).

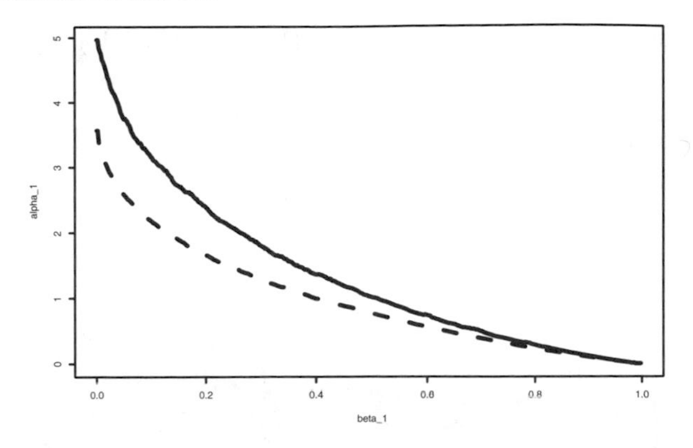

Figure 5.11 *The* (α_1, β_1)*-areas below the two curves guarantee the existence of a stationary GARCH(1, 1) process.* Solid line: *IID student noise with 4 degrees of freedom with variance 1.* Dotted line: *i.i.d. standard normal noise. The regions were determined by checking the condition* $E\log(\alpha_1 Z_1^2 + \beta_1) < 0$ *through simulating* $n = 1{,}000{,}000$ *i.i.d.* Z_t*s and then calculating* $n^{-1}\sum_{i=1}^{n}\log(\alpha_1 Z_t^2 + \beta_1)$*. According to the strong law of large numbers, the latter quantity converges a.s. to* $E\log(\alpha_1 Z_1^2 + \beta_1)$*. If* Z_1 *has a density, an alternative way to verify the stationarity assumption is to calculate* $E\log(\alpha_1 Z_1^2 + \beta_1)$ *by numerical integration.*

constitutes a strictly stationary solution to equation (5.3.2). If there is another stationary solution $(\widehat{Y}_t)$ we have by iterating (5.3.2),

$$|\widetilde{Y}_t - \widehat{Y}_t| = A_t \cdots A_{t-r}\, |\widetilde{Y}_{t-r-1} - \widehat{Y}_{t-r-1}|, \tag{5.3.4}$$

and since $A_t \cdots A_{t-r}$ and $|\widetilde{Y}_{t-r-1} - \widehat{Y}_{t-r-1}|$ are independent, the weak law of large numbers and $E\log A_1 < 0$ imply that the right-hand side in (5.3.4) converges to zero in probability as $r \to \infty$. Therefore $\widetilde{Y}_t = \widehat{Y}_t$ for every t with probability 1.

More sophisticated arguments show that $E\log A_1 < 0$ is also necessary for the existence of a nontrivial stationary solution.

Theorem 5.3.1 (Nelson [112], Bougerol and Picard [21]) *There exists an a.s. unique nonvanishing strictly stationary causal*[7] *solution of the SRE* (5.3.2) *with* $B_t \equiv \alpha_0$ *if and only if* $\alpha_0 > 0$ *and* $E\log(\alpha_1 Z_1^2 + \beta_1) < 0$.

Remark 5.3.2 The assumption $\alpha_0 > 0$ is crucial; otherwise $X_t \equiv 0$ a.s. would be a solution to (5.3.2). Since

$$\log(\beta_1) \le E\log(\alpha_1 Z_1^2 + \beta_1) < 0,$$

$0 \le \beta_1 < 1$ is a necessary condition on β_1. □

Remark 5.3.3 The condition $EA_1 = E\log(\alpha_1 Z_1^2 + \beta_1) < 0$ can be interpreted in the context of the random coefficient AR(1) model $Y_t = A_t Y_{t-1} + B_t$. For a constant coefficient AR(1) model $Y_t = \varphi Y_{t-1} + B_t$ the condition $|\varphi| < 1$ is necessary for the

[7] This means it depends only on past and present values of the Zs.

existence of a stationary causal solution (Y_t). The condition $E \log A_1 < 0$ means in a sense that A_t is less than 1 on average. □

A general GARCH(p, q) process can be embedded in a SRE in a similar way. Indeed, writing

$$\mathbf{Y}_t = \left(\sigma_{t+1}^2, \ldots, \sigma_{t-q+2}^2, X_t^2, \ldots, X_{t-p+2}^2\right)', \tag{5.3.5}$$

$$\mathbf{A}_t = \begin{pmatrix} \alpha_1 Z_t^2 + \beta_1 & \beta_2 & \cdots & \beta_{q-1} & \beta_q & \alpha_2 & \alpha_3 & \cdots & \alpha_p \\ 1 & 0 & \cdots & 0 & 0 & 0 & 0 & \cdots & 0 \\ 0 & 1 & \cdots & 0 & 0 & 0 & 0 & \cdots & 0 \\ \vdots & \vdots & \ddots & \vdots & \vdots & \vdots & \vdots & \ddots & \vdots \\ 0 & 0 & \cdots & 1 & 0 & 0 & 0 & \cdots & 0 \\ Z_t^2 & 0 & \cdots & 0 & 0 & 0 & 0 & \cdots & 0 \\ 0 & 0 & \cdots & 0 & 0 & 1 & 0 & \cdots & 0 \\ \vdots & \vdots & \ddots & \vdots & \vdots & \vdots & \vdots & \ddots & \vdots \\ 0 & 0 & \cdots & 0 & 0 & 0 & \cdots & 1 & 0 \end{pmatrix}, \tag{5.3.6}$$

$$\mathbf{B}_t = (\alpha_0, 0, \ldots, 0)', \tag{5.3.7}$$

$((\mathbf{A}_t, \mathbf{B}_t))$ is an i.i.d. sequence, $\mathbf{Y}_{t-1}$ and $(\mathbf{A}_t, \mathbf{B}_t)$ are independent, where the $\mathbf{A}_t$s are i.i.d. random $(p + q - 1) \times (p + q - 1)$ matrices and the $\mathbf{B}_t$s i.i.d. $(p + q - 1)$-dimensional random vectors. Moreover, $(\mathbf{Y}_t)$ satisfies the following vector SRE which is analogous to the scalar SRE (5.3.2):

$$\mathbf{Y}_t = \mathbf{A}_t \mathbf{Y}_{t-1} + \mathbf{B}_t, \quad t \in \mathbb{Z}. \tag{5.3.8}$$

Proceeding in a way similar to the one above, one can show the existence of a strictly stationary solution to (5.3.8). The condition $E \log A_1 < 0$ has then to be replaced by negativity of the so-called top Lyapunov exponent γ:

$$\gamma = \inf_n \left\{ n^{-1} \, E \log \|\mathbf{A}_n \cdots \mathbf{A}_1\| \right\} < 0,$$

where $\|\cdot\|$ is the operator norm corresponding to a given norm in $\mathbb{R}^{p+q-1}$. In general, the value γ cannot be calculated explicitly. According to the subadditive ergodic theorem (cf. Kingman [89], Theorem 6), it follows that

$$n^{-1} \log \|\mathbf{A}_n \cdots \mathbf{A}_1\| \stackrel{\text{a.s.}}{\to} \gamma, \tag{5.3.9}$$

which relation offers a Monte-Carlo procedure for calculating γ. Asymptotic confidence bands for γ can be obtained by using central limit theory due to Goldsheid [72].

The GARCH(1,1) case is a real exception:

$$A_n \cdots A_1 = \prod_{t=1}^{n} \left(\alpha_1 Z_t^2 + \beta_1\right),$$

and so we immediately have $\gamma = E \log(\alpha_1 Z_1^2 + \beta_1)$ provided the latter expectation makes sense.

Theorem 5.3.4 (Bougerol and Picard [21]) *The* GARCH(p, q) *SRE specified in* (5.3.5) *through* (5.3.8) *has the a.s. unique strictly stationary nonvanishing causal solution*

$$\mathbf{Y}_t = \sum_{i=-\infty}^{t} \mathbf{A}_t \cdots \mathbf{A}_{i+1} \mathbf{B}_i = \mathbf{B}_t + \sum_{i=-\infty}^{t-1} \mathbf{A}_t \cdots \mathbf{A}_{i+1} \mathbf{B}_i, \quad t \in \mathbb{Z},$$

if and only if $\alpha_0 > 0$ *and* $\gamma < 0$. *Moreover, this solution is also ergodic.*

Remark 5.3.5 Similar to the GARCH(1, 1) case (see Remark 5.3.2), $\alpha_0 > 0$ ensures that $X_t \equiv 0$ a.s. is excluded as a solution to the SRE (5.2.7) and

$$\sum_{j=1}^{q} \beta_j < 1$$

is a necessary condition for $\gamma < 0$; see Bougerol and Picard [21], Corollary 2.3. A sufficient condition for $\gamma < 0$ is given by

$$\sum_{i=1}^{p} \alpha_i + \sum_{j=1}^{q} \beta_j < 1,$$

provided that $EZ_1^2 = 1$ and $EZ_1 = 0$; see remark on p. 122 of [21]. □

Above we learned that the existence of a strictly stationary volatility sequence (σ_t) is equivalent to strict stationarity of $X_t = \sigma_t Z_t$. So we can conclude the following:

Corollary 5.3.6 *The* GARCH(p, q) *process* $X_t = \sigma_t Z_t$, $t \in \mathbb{Z}$, *with the specification* (5.2.7) *for* σ_t^2 *and an i.i.d. noise sequence* (Z_t) *with mean zero and unit variance has a nonvanishing strictly stationary causal version if and only if* $\alpha_0 > 0$ *and* $\gamma < 0$.

Example 5.3.7 (The IGARCH process)
In financial practice one often observes that the sum of the estimated parameters $\sum_i \alpha_i + \sum_j \beta_j$ is close to one; see Figure 5.12 for an illustration. For the parameter setting

$$\sum_{i=1}^{p} \alpha_i + \sum_{j=1}^{q} \beta_j = 1,$$

Engle and Bollerslev [61], in analogy to integrated ARMA (ARIMA) processes, coined the name integrated GARCH(p, q) or IGARCH(p, q) process. The integrated refers to the fact that there might be a unit root problem (as for ARIMA processes) which could lead to the nonexistence of a stationary version of (X_t). However, this is not the case for the IGARCH under the conditions of Corollary 5.3.6, as shown by Nelson [112] for the IGARCH(1, 1) case and by Bougerol and Picard [21], Corollary 2.2, for the general IGARCH(p, q) case, assuming some mild additional assumptions. As we will see in Example 5.5.8, the IGARCH process has infinite variance, a property that is not exhibited in real-life log-returns. Later, in Section 5.11.2, we will give another possible explanation for the *IGARCH* effect: nonstationarity in the data.

Figure 5.12 *The estimated values of* $\alpha_1 + \beta_1$, *using quasi-MLE, for an increasing sample of the* S&P500 *log-returns from Figure 5.1. An initial* GARCH(1,1) *model is fitted to the first* 1500 *observations* (6 *business years*)*. Then* $k * 100$, $k = 1, 2, \ldots$, *data points are successively added to the sample and* α_1 *and* β_1 *are reestimated on these samples. The labels on the time axis indicate the date of the latest observation used for the estimation procedure.*

The author hopes that he has convinced the reader that the question about a stationary version of a GARCH(p, q) process is not a trivial one. In general, given a set of parameter values α_i, β_j and the distribution of Z_t, we cannot decide whether a stationary version of the corresponding GARCH(p, q) process exists or not. Remark 5.3.5, however, indicates that the situation is not totally hopeless since $\sum_i \alpha_i + \sum_j \beta_j < 1$ is a sufficient condition for the existence of a stationary solution provided Z_t has mean zero and unit variance. As we will see in Example 5.5.8, the remaining cases $\sum_i \alpha_i + \sum_j \beta_j \geq 1$ correspond to infinite variance GARCH processes.

The stationarity region of the α- and β-values clearly depends on the distribution of Z_t: one has to check where $\gamma < 0$. In the GARCH(1, 1) case this is more transparent since $\gamma = E \log(\alpha_1 Z_1^2 + \beta_1)$. But even in this case one depends on Monte Carlo or numerical integration techniques in order to determine those α_1 and β_1 for which $\gamma < 0$; see Figure 5.11 for an illustration. The ARCH(1) case with i.i.d. standard normal Z_ts has been studied in detail, and the α_1-values ensuring stationarity were tabulated; see de Haan et al. [79] or Section 8.4 in Embrechts et al. [58]. It turns out that $\alpha_1 < \mathrm{e}^E \approx 3.56856$, where E is Euler's constant, is the correct α_1-area of stationarity. As for the IGARCH case, we see that parameter values greater or equal to 1 are possible, in sharp contrast to a stationary constant coefficient AR(1) model with i.i.d. noise where the AR-parameter cannot exceed 1; see also Remark 5.3.3.

The literature on SREs of type $\mathbf{Y}_t = \mathbf{A}_t \mathbf{Y}_{t-1} + \mathbf{B}_t$ for i.i.d. and stationary ergodic sequences $((\mathbf{A}_t, \mathbf{B}_t))$ is vast. We refer to pioneering work of Kesten [88] (a paper which will be essential for the study of the tails of GARCH processes), Goldie [70], Le Page [97] for extensions and refinements of Kesten's work, Brandt [24] for the existence of a stationary version $(\mathbf{Y}_t)$ for dependent $(\mathbf{A}_t, \mathbf{B}_t)$s, Brandt et al. [25] for a general theory of SREs, the papers of Bougerol and Picard [20, 21] and Babillot et al. [1] who give necessary and sufficient conditions for the existence of stationary $(\mathbf{Y}_t)$ or conditions close to the optimal ones. The model $\mathbf{Y}_t = \mathbf{A}_t \mathbf{Y}_{t-1} + \mathbf{B}_t$ with independent $(\mathbf{A}_t, \mathbf{B}_t)$ is a simple example of a Markov chain or random iterated mapping. The book by Meyn and Tweedie [103] on Markov chains yields various results on the dependence structure and the distribution for this particular model. The random iterated mapping aspect is covered by the recent overview paper of Diaconis and Freedman [48]. The paper by Vervaat [134] gives an accessible introduction to random autoregressive models.

5.4 An excursion to regular variation

In what follows, we will need the notion of regularly varying random variable and regularly varying random vector. Regular variation is one of the basic notions for describing heavy-tailed distributions and dependence in the tails. Regular variation of random vectors or random variables is used to describe domain of attraction conditions for stable distributions in the summation theory of i.i.d. random vectors and maximum domain of attraction conditions for max-stable distributions in the limit theory for maxima of i.i.d. random vectors. We refer to Feller [63] for the classical summation theory of i.i.d. random variables, Rvačeva [127] for the summation theory of i.i.d. random vectors, Embrechts et al. [58] for the limit theory of partial maxima of i.i.d. random variables and Resnick [119] for the corresponding theory for component-wise maxima of random vectors. An encyclopedic treatment of univariate regular variation and its applications is given in the monograph of Bingham et al. [14]. A more recent treatment, including multivariate regular variation, is Meerschaert and Scheffler [102].

5.4.1 Univariate regular variation

Recall that a measurable function L on $(0, \infty)$ is said to be slowly varying if it is positive for large x and satisfies the relation $L(cx)/L(x) \to 1$ as $x \to \infty$ for all positive c. Simple examples are the functions $L(x) = c$ for some $c > 0$, $L(x) = \log^\alpha x$ for any real α, iterated logarithms, etc.

We say that the random variable X and its distribution F are regularly varying[8] with index $\alpha \geq 0$ if there exist $p, q \geq 0$, $p + q = 1$, and a slowly varying function L such that as $x \to \infty$,

$$\begin{cases} \overline{F}(x) = P(X > x) = p\,(1 + o(1))\,\frac{L(x)}{x^\alpha}, \\ F(-x) = P(X \leq -x) = q\,(1 + o(1))\,\frac{L(x)}{x^\alpha}. \end{cases} \tag{5.4.1}$$

Condition (5.4.1) is often referred to as the tail balance condition. It is a semiparametric assumption about the tails: they are close to power laws, but we do not specify the function L. Its behavior in any neighborhood of the origin is not of interest.

[8] It is common use to call the function $g(x) = x^p\, L(x)$ regularly varying with index $p \in \mathbb{R}$. If we followed this convention we should say that $F(-x)$ and $1 - F(x)$ are regularly varying with index $-\alpha$. In this chapter, we prefer to assign a nonnegative index α to the distribution F.

Relation (5.4.1) is equivalent to the following: for any $t > 0$,

$$\lim_{x\to\infty} \frac{P(X > tx)}{P(|X| > x)} = p\,t^{-\alpha} \quad \text{and} \quad \lim_{x\to\infty} \frac{P(X \le -tx)}{P(|X| > x)} = q\,t^{-\alpha}. \tag{5.4.2}$$

The limiting expressions determine a measure μ on $\overline{\mathbb{R}}\backslash\{0\}$ given by

$$\mu(dt) = p\,\alpha\,t^{-\alpha-1}\,I_{(0,\infty)}(t)dt + q\alpha\,|t|^{-\alpha-1}\,I_{(-\infty,0)}(t)dt.$$

Using the measure μ, relation (5.4.2) can be shown to be equivalent to

$$\lim_{x\to\infty} \frac{P(X \in xA)}{P(|X| > x)} = \mu(A), \tag{5.4.3}$$

for every Borel set $A \subset \overline{\mathbb{R}}\backslash\{0\}$ bounded away from zero whose boundary ∂A satisfies $\mu(\partial A) = 0$. For such A we also have

$$\mu(tA) = \lim_{x\to\infty} \frac{P(X \in txA)}{P(|X| > tx)} \frac{P(|X| > tx)}{P(|X| > x)} = t^{-\alpha}\,\mu(A).$$

Hence μ satisfies the homogeneity property $\mu(tA) = t^{-\alpha}\mu(A)$ for every $t > 0$. It is not difficult to see that relation (5.4.2) can be rewritten in spherical coordinates: let $\theta \in \{-1, 1\}$ be a random variable with distribution $P(\theta = 1) = p$ and $P(\theta = -1) = q$. Then for any $t > 0$ and set $S \subset \{-1, 1\}$,

$$\lim_{x\to\infty} \frac{P\left(|X| > tx,\, X/|X| \in S\right)}{P(|X| > x)} = t^{-\alpha}\,P(\theta \in S). \tag{5.4.4}$$

The limit relations (5.4.3) and (5.4.4) look very much like the ordinary notion of weak convergence of probability measures. However, the limiting measure is infinite and therefore one refers to the convergences in (5.4.3) and (5.4.4) as vague convergence. Vague convergence plays an important role in the weak convergence theory of point processes; see Kallenberg [87] or Resnick [119].

Example 5.4.1 (Some univariate regularly varying distributions)
Various well known univariate distributions are regularly varying. Among them are the following distributions F whose right tail $\overline{F}$ or density f we give here:

Pareto	$\overline{F}(x) = \dfrac{\kappa^\alpha}{(x+\kappa)^\alpha}, \quad x \ge 0, \quad \kappa > 0, \quad \alpha > 0,$
student with n degrees of freedom	$f(x) = \dfrac{\Gamma((n+1)/2)}{\Gamma(n/2)\sqrt{\pi n}}\,(1 + x^2/n)^{-(n+1)/2}, \quad x \in \mathbb{R},$
log-gamma	$f(x) = \dfrac{\alpha^\beta}{\Gamma(\beta)}\,(\log x)^{\beta-1}\,x^{-\alpha-1}, \quad x \ge 1, \quad \alpha, \beta > 0.$

Remark 5.4.2 The notion of regular variation is central in the weak limit theory for sums and maxima of i.i.d. random variables. To make this statement more transparent, let (X_i) be a sequence of i.i.d. random variables with common distribution F, and denote their partial maxima and partial sums respectively by

$$M_n = \max(X_1, \ldots, X_n) \quad \text{and} \quad S_n = X_1 + \cdots + X_n, \quad n \geq 1.$$

It is well known that there exist only three location-scale families of extreme value distributions H for which one can find constants $c_n > 0$ and $d_n \in \mathbb{R}$

$$\lim_{n\to\infty} P(c_n^{-1}(M_n - d_n) \leq x) = H(x), \quad x \in \mathbb{R},$$

and the distribution F is then said to be in the maximum domain of attraction of H ($F \in \text{MDA}(H)$). One of these classes consists of the scale-location family defined by the Fréchet distribution Φ_α of order $\alpha > 0$. Its distribution function is given by $\Phi_\alpha(x) = \exp\{-x^{-\alpha}\}$ for $x > 0$. It is well known that $F \in \text{MDA}(\Phi_\alpha)$ if and only if the right tail $\overline{F} = 1 - F$ is a regularly varying function with index $-\alpha$, and then one can choose c_n according to $\overline{F}(c_n) \sim n^{-1}$ and $d_n = 0$; see Embrechts et al. [58], Section 3.3.1, for the corresponding theory.

A similar theory holds for the partial sums. It is well known that the only limiting distributions H for standardized sums of i.i.d. random variables are the stable distributions; see Section 5.7 for a short introduction. This means that there exist $c_n > 0$ and $d_n \in \mathbb{R}$ such that

$$\lim_{n\to\infty} P(c_n^{-1}(S_n - d_n) \leq x) = H(x), \quad x \in \mathbb{R},$$

and the distribution F is then said to be in the domain of attraction of the stable distribution H ($F \in \text{DA}(H)$). Clearly, the normal distribution is stable. The remaining stable distributions are less familiar; the Cauchy distribution is perhaps its best known representative. All nonnormal stable distributions have infinite variance, in particular they are regularly varying with index $\alpha < 2$, and therefore we denote a representative of the class of α-stable distributions by H_α. Moreover, $F \in \text{DA}(H_\alpha)$ for some $\alpha < 2$ if and only if X_1 is regularly varying with index α; see Gnedenko and Kolmogorov [69], Ibragimov and Linnik [84], or Feller [63] for extensive discussions of summation theory. □

5.4.2 Multivariate regular variation

Definition

Since we have already reformulated univariate regular variation in terms of vague convergence, we can immediately take over relations (5.4.3) and (5.4.4) to define multivariate regular variation: the random vector $\mathbf{X}$ with values in $\mathbb{R}^d$ is regularly varying with index $\alpha \geq 0$ if there exists a random vector $\mathbf{\Theta}$ with values in the unit sphere $\mathbb{S}^{d-1}$ of $\mathbb{R}^d$ such that for any $t > 0$ and any Borel set $S \subset \mathbb{S}^{d-1}$ with $P(\boldsymbol{\Theta} \in \partial S) = 0$,

$$\lim_{x\to\infty} \frac{P(|\mathbf{X}| > tx, \widetilde{\mathbf{X}} \in S)}{P(|\mathbf{X}| > x)} = t^{-\alpha}\, P(\boldsymbol{\Theta} \in S), \tag{5.4.5}$$

where for any vector $\mathbf{x} \neq \mathbf{0}$,

$$\widetilde{\mathbf{x}} = \frac{\mathbf{x}}{|\mathbf{x}|}.$$

The distribution of $\boldsymbol{\Theta}$ is called the spectral measure of $\mathbf{X}$. Here and in what follows, we write $|\cdot|$ for any fixed norm.

Relation (5.4.5) holds if and only if the limit relations

$$\lim_{x\to\infty} \frac{P(|\mathbf{X}| > tx)}{P(|\mathbf{X}| > x)} = t^{-\alpha} \quad \text{for all } t > 0, \tag{5.4.6}$$

$$\lim_{x\to\infty} P(\widetilde{\mathbf{X}} \in S \mid |\mathbf{X}| > x) = P(\boldsymbol{\Theta} \in S), \tag{5.4.7}$$

are satisfied. Indeed, choosing $S = \mathbb{S}^{d-1}$ in (5.4.5), we see that $|\mathbf{X}|$ is regularly varying with index α, hence (5.4.6) holds. Therefore (5.4.5) implies

$$\lim_{x\to\infty} \frac{P(|\mathbf{X}| > tx, \widetilde{\mathbf{X}} \in S)}{P(|\mathbf{X}| > x)} = \lim_{x\to\infty} P(\widetilde{\mathbf{X}} \in S \mid |\mathbf{X}| > tx) \lim_{x\to\infty} \frac{P(|\mathbf{X}| > tx)}{P(|\mathbf{X}| > x)}, \tag{5.4.8}$$

hence (5.4.7) follows. On the other hand, if (5.4.6) and (5.4.7) hold, (5.4.5) follows from (5.4.8). Relations (5.4.6) and (5.4.7) illustrate the meaning of regular variation of the vector $\mathbf{X}$: its radial and spherical parts become independent in an asymptotic sense if $|\mathbf{X}|$ is sufficiently large.

The right-hand side of (5.4.5) can be interpreted as the μ-measure of the Borel set

$$A(t, S) = \left\{\mathbf{x} : |\mathbf{x}| > t,\ \widetilde{\mathbf{x}} \in S\right\} \subset \overline{\mathbb{R}}^d \backslash \{\mathbf{0}\}, \tag{5.4.9}$$

and (5.4.5) can be written as

$$\begin{aligned} \lim_{x\to\infty} \frac{P(\mathbf{X} \in xA(t, S))}{P(|\mathbf{X}| > x)} &= \lim_{x\to\infty} \frac{P(\mathbf{X} \in xA(t, S))}{P(\mathbf{X} \in xA(1, \mathbb{S}^{d-1}))} \\ &= \mu(A(t, S)) = t^{-\alpha}\, \mu(A(1, S)). \end{aligned}$$

The totality of the values $\mu(A(t, S))$ determines a Radon measure μ (i.e., finite on compact sets) on $\overline{\mathbb{R}}^d \backslash \{\mathbf{0}\}$. It can be shown that relation (5.4.5) is equivalent to

$$\lim_{x\to\infty} \frac{P(\mathbf{X} \in xA)}{P(|\mathbf{X}| > x)} = \mu(A), \tag{5.4.10}$$

for every Borel set $A \subset \overline{\mathbb{R}}^d \backslash \{\mathbf{0}\}$ bounded away from the origin satisfying $\mu(\partial A) = 0$, where μ is a measure on the Borel sets of $\overline{\mathbb{R}}^d \backslash \{\mathbf{0}\}$ satisfying the homogeneity assumption $\mu(tA) = t^{-\alpha}\mu(A)$.

Both (5.4.5) and (5.4.10) define vague convergence of the left-hand measures to the measure μ; we again refer to Kallenberg [87] and Resnick [118, 119] for an exact description of vague convergence and its relation to weak convergence of point processes. In what follows we write, for example,

$$\frac{P(\mathbf{X} \in x\, \cdot)}{P(|\mathbf{X}| > x)} \stackrel{v}{\to} \mu(\cdot)$$

for the totality of the relations (5.4.10). Note that this vague convergence is defined on the σ-field of the Borel sets of $\overline{\mathbb{R}}^d \backslash \{\mathbf{0}\}$.

In relation (5.4.10) the case $\mu(A) = 0$ for certain sets A is not excluded. This means in particular that the vector $\mathbf{X}$ may contain components whose tails decay to zero faster than other components. For example, the vector $\mathbf{X} = (X, 0, \ldots, 0) \in \mathbb{R}^d$ is regularly varying if X is a regularly varying random variable, and then $\mu(A_1 \times A_2) = 0$ if $A_2 \subset \mathbb{R}^{d-1}$ does not contain the vector $\mathbf{0} \in \mathbb{R}^{\mathbf{d-1}}$.

In theoretical work one often assumes that the nondegenerate components of the vector $\mathbf{X}$ are transformed in such a way that they are regularly varying with the same index; see, for example, Resnick [119], Chapter 5, or de Haan and de Ronde [80] for details. In statistical applications this fact creates problems because the transformation of all components to the same order of magnitude requires knowledge of all marginal distributions. We refer for example to de Haan and de Ronde [80], Draisma [54], or Einmahl et al. [57] for solutions in various statistical applications. An alternative approach has been advocated for many years: use operator normalizations for the components, i.e., choose for each component a normalization adjusted to the tails of the components. We refer to de Haan and Resnick [76, 77] and Greenwood and Resnick [74] for some early work; see also Meerschaert and Scheffler [102] for a recent account.

Regular variation of $\mathbf{X}$ implies regular variation of $|\mathbf{X}|$ with the same index. This is immediate from (5.4.5) with $S = \mathbb{S}^{d-1}$. Then we can choose a sequence (x_n) of positive numbers satisfying the asymptotic relation

$$P(|\mathbf{X}| > x_n) \sim n^{-1} ,$$

and this sequence is regularly varying with index $1/\alpha$, i.e., there exists a slowly varying function ℓ such that $x_n = n^{1/\alpha} \ell(n)$. This is a well known fact from the theory of regular variation; see, e.g., Bingham et al. [14]. Replacing x by x_n in (5.4.5) and (5.4.10), we arrive at the relations

$$\begin{aligned} &n P(\mathbf{X} \in x_n \cdot) \stackrel{v}{\to} \mu(\cdot) \quad \text{and} \\ &\quad n P(|\mathbf{X}| > t x_n \,, \widetilde{\mathbf{X}} \in \cdot) \stackrel{v}{\to} t^{-\alpha}\, P(\mathbf{\Theta} \in \cdot), \quad t > 0 \,. \end{aligned} \tag{5.4.11}$$

In the former case, $\stackrel{v}{\to}$ refers to vague convergence on the Borel σ-field of $\overline{\mathbb{R}}^d \backslash \{\mathbf{0}\}$, in the latter case to vague convergence on the Borel σ-field of $\mathbb{S}^{d-1}$. If one requires that $\mu(tA) = t^{-\alpha} \mu(A)$, any of the convergence relations with x_n as chosen above is equivalent to the regular variation of $\mathbf{X}$. We refer to any of the relations (5.4.11) as the sequential definition of regular variation.

Example 5.4.3 (Some multivariate regularly varying distributions)
Various well known multivariate distributions are regularly varying. Those include the following ones whose precise description can be found in Johnson and Kotz [86] or Tong [133]: multivariate student, multivariate F, and multivariate Cauchy.

Resnick [119], Proposition 5.20 and examples following it, discusses various absolutely continuous regularly varying distributions, including the ones above. In particular, he gives conditions on the densities implying regular variation of the distribution.

Analogously to the univariate case (see Remark 5.4.2) multivariate regular variation is used to characterize maximum domains of attractions for partial componentwise maxima of i.i.d. random vectors attracted by a multivariate extreme value distribution with Fréchet marginals; see Resnick [119] and the discussion at the end of Section 5.4.2. Moreover, as in the univariate case, multivariate regular variation is used as domain of attraction condition for partial sums of i.i.d. random vectors attracted by α-stable distributions; see Rvačeva [127] for the summation theory and Section 5.7 for a brief introduction to stable random vectors.

Multivariate regular variation is by no means an easy notion and is far from being completely understood. Indeed, many results for univariate regular variation (such as Abelian and Tauberian results) do not exist for multivariate regular variation or are only known in some special cases; see Bingham et al. [14], p. 426, for some existing results. See also Section 5.4.3 for another open problem related to regular variation. A recent account on regular variation and its applications can be found in Meerschaert and Scheffler [102].

Examples and intuitive interpretation of the spectral measure

In this section we first consider some examples of regularly varying random vectors and try to determine their spectral measures. For the purpose of illustration we focus on two-dimensional vectors.

Example 5.4.4 (Total independence)
Consider a vector $\mathbf{X} = (X_1, X_2)$ of i.i.d. components X_1, X_2 with common distribution function F, and assume for simplicity that $X_i > 0$ a.s. Suppose that both are regularly varying with index α. In this context it will be convenient to choose the max-norm $|\mathbf{x}| = \max(x_1, x_2)$ and the sequential characterization of regular variation introduced in Section 5.4.2. Thus we have to study the limits, for $t > 0$, of the quantities

$$nP\left(\{\max(X_1, X_2) > t x_n\} \cap A(S)\right), \quad \text{where} \quad A(S) = \left\{ \frac{(X_1, X_2)}{\max(X_1, X_2)} \in S \right\}.$$

Notice that

$$\begin{aligned} nP(|\mathbf{X}| > x_n) &= n\,(1 - P(\max(X_1, X_2) \le x_n)) \\ &= [n\overline{F}(x_n)]\,(1 + F(x_n)) \sim 2\,[n\overline{F}(x_n)] \sim 1 \end{aligned}$$

and therefore

$$\overline{F}(x_n) \sim 0.5\, n^{-1}. \tag{5.4.12}$$

Choose $\delta \in (0, 1)$. Then

$$\{\max(X_1, X_2) > tx_n\} = \{X_1 > tx_n, X_2 > \delta tx_n\} \cup \{X_1 > tx_n, X_2 \leq \delta tx_n\}$$
$$\cup \{X_2 > tx_n, X_1 > \delta tx_n\} \cup \{X_2 > tx_n, X_1 \leq \delta tx_n\}.$$

Hence by the independence of X_1 and X_2,

$$\begin{aligned} &nP(\{\max(X_1, X_2) > tx_n\} \cap A(S)) \\ &= O(n[\overline{F}(\delta tx_n)]^2) + nP(X_1 > tx_n, X_2 \leq \delta tx_n, (1, X_2/X_1) \in S) \\ &+ nP(X_2 > tx_n, X_1 \leq \delta tx_n, (X_1/X_2, 1) \in S). \end{aligned}$$

Because $\delta \in (0, 1)$ can be chosen arbitrarily small, the right-hand expressions make nonnegligible contributions to the limit if and only if $(1, 0) \in S$ or $(0, 1) \in S$. (We have to assume that neither $(0, 1)$ nor $(1, 0)$ are at the boundary of S.) In the latter case, by fine tuning the convergences as $n \to \infty$ and $\delta \downarrow 0$ and taking account of (5.4.12), the quantity $nP(X_2 > tx_n, X_1 \leq \delta tx_n, (X_1/X_2, 1) \in S)$ can be shown to be of the asymptotic order

$$nP(X_2 > tx_n)\, I_S((0, 1)) \sim 0.5\, t^{-\alpha}\, I_S((0, 1)).$$

Using analogous arguments, it finally turns out that

$$P(\mathbf{\Theta} \in S) = 0.5\, I_S((1, 0)) + 0.5\, I_S((0, 1)),$$

or, in words, the spectral measure is concentrated on the intersection between the axes and the unit circle. As a matter of fact, a more precise argument shows that this statement remains valid for any norm in $\mathbb{R}^2$.

An intuitive interpretation of this fact is that for an i.i.d. sequence $\mathbf{X}_1, \ldots, \mathbf{X}_n$ with the same distribution as $\mathbf{X}$, extremes, i.e., $\mathbf{X}_i$ with large $|\mathbf{X}_i|$, are caused with high probability by $\mathbf{X}_i$ with one component significantly larger than the other one. Thus, for a plot of the two-dimensional points $\mathbf{X}_1, \ldots, \mathbf{X}_n$ we expect that the points $\mathbf{X}_i$ which are far away from the origin are concentrated along the axes. See Figure 5.13 for an illustration of this fact.

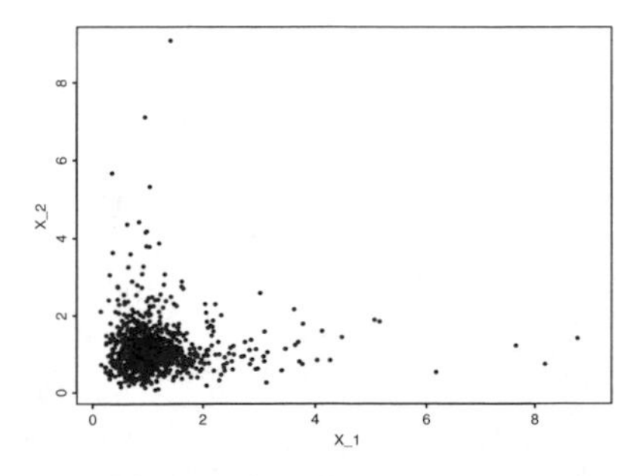

Figure 5.13 *Plot of 1000 realizations from an i.i.d. sample* $(\mathbf{X}_i)$ *with i.i.d. components satisfying* $P(X_1 > x) \sim x^{-3}$. *The* $\mathbf{X}_i$*s with large distance from the origin are concentrated along the axes.*

Example 5.4.5 (Total dependence)
We assume that $\mathbf{X} = (X, X)$ for some regularly varying X with index $\alpha > 0$ and again choose the max-norm $|\mathbf{x}| = \max(x_1, x_2)$. In this case, with $nP(|X| > x_n) \sim 1$,

$$nP(|\mathbf{X}| > x_n, \widetilde{\mathbf{X}} \in S) = nP(|X| > x_n, (1, 1) \in S) \sim I_S((1, 1)),$$

where we assume that $(1, 1)$ is not at the boundary of S. The spectral measure is degenerate and concentrated on the intersection of the unit sphere with the line $x = y$. The same remark applies to any norm in $\mathbb{R}^2$.

For an i.i.d. sequence $\mathbf{X}_1, \ldots, \mathbf{X}_n$ with the same distribution as $\mathbf{X}$, it is clear from the dependence structure of the components that an $\mathbf{X}_i$ far away from the origin occurs if both components in the vector are large at the same time.

In contrast to the previous artificial example, for a real-life time series $\mathbf{X}_1, \ldots, \mathbf{X}_n$ with dependent nonidentical components we do not expect that all these vectors lie on the line $x = y$. However, when $\mathbf{X}_i$s of large modulus stay away strongly from the axes, we have an indication of asymptotic dependence.

The observed features can be generalized to vectors with values in $\mathbb{R}^d$. Again, the spectral measure of vectors with independent components is concentrated on the intersection of the axes with the unit sphere, and the spectral measure of vectors with strong dependence in the components tends to be concentrated along the diagonals of the orthants. We illustrate these features for the lagged vectors (X_t, X_{t+1}) of the AR(1) process $X_t = 0.9X_{t-1} + Z_t$ with i.i.d. symmetric noise (Z_t) regularly varying with index 1.8; see the left part of Figure 5.14. In the right part we illustrate the dependence in the tails for pairs of daily log-returns of the S&P500 series.

Example 5.4.6 (A toy model)
If one wants to get an intuitive understanding of the spectral measure, the following example is quite helpful:

$$\mathbf{X} = R\,(\cos\Phi, \sin\Phi), \tag{5.4.13}$$

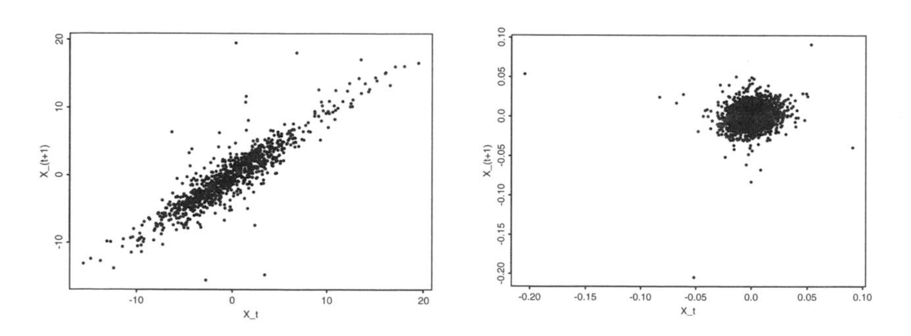

Figure 5.14 Left: *Plot of 1000 lagged vectors* $\mathbf{X}_t = (X_t, X_{t+1})$ *for the AR(1) process* $X_{t+1} = 0.9X_t + Z_t$ *for i.i.d. symmetric regularly varying noise* (Z_t) *with tail index 1.8. The vectors* $\mathbf{X}_t$ *with a large norm* $|\mathbf{X}_t|$ *are typically concentrated along the line* $y = 0.9x$. Right: *Scatterplot of the pairs* (X_t, X_{t+1}) *of the daily log-returns* X_t *of the* S&P500 *series. The extremes in the series do not tend to cluster around the axes. This is an indication of dependence in the tails; for an estimate of the density of the corresponding spectral measure we refer to Figure 5.16.*

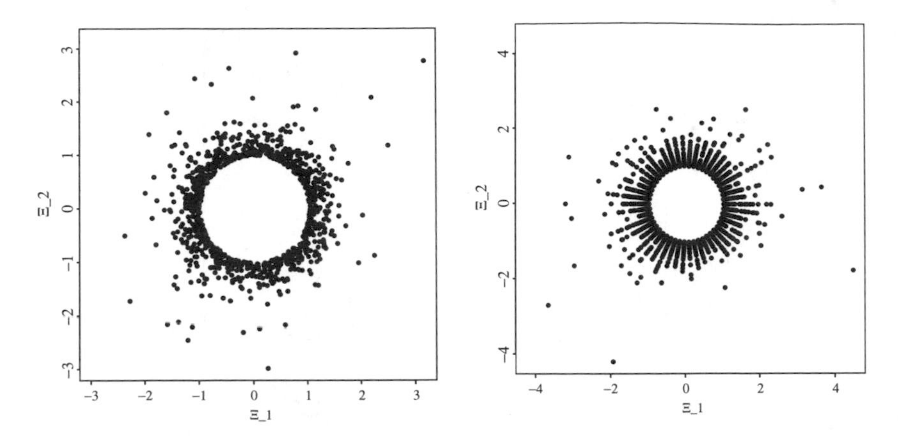

Figure 5.15 *I.I.D. vectors* $\mathbf{X}_i$ *from the model (5.4.13) with tail index* $\alpha = 5$. *Left:* Φ *is uniform on* $(-\pi, \pi]$. *Right:* Φ *has a discrete uniform distribution on the points* $2\pi i/50$.

where the random radius R has distribution $P(R > r) = r^{-\alpha}, r \geq 1$, for some $\alpha > 0$ and is independent of the random angle Φ with distribution on $(-\pi, \pi]$. Choosing the Euclidean norm $|\cdot|$ and exploiting the independence of R and Φ, this vector is immediately seen to be regularly varying with $\mathbf{\Theta} = (\cos\Phi, \sin\Phi)$:

$$\frac{P(|\mathbf{X}| > tx\,, \widetilde{\mathbf{X}} \in S)}{P(|\mathbf{X}| > x)} = \frac{P(R > tx\,, \mathbf{\Theta} \in S)}{P(R > x)} = \frac{P(R > tx)}{P(R > x)} P(\mathbf{\Theta} \in S)$$
$$= t^{-\alpha}\, P(\mathbf{\Theta} \in S),$$

provided $\min(tx, x) \geq 1$. The knowledge of the distribution of Φ allows for some straightforward interpretation of the two-dimensional dependence in the tails; see Figure 5.15 for two examples; cf. Figure 5.16 for estimates of the density of Φ.

Point process convergence to a Poisson process and regular variation

For an i.i.d. sequence of regularly varying $\mathbb{R}^d$-valued vectors $\mathbf{X}_i$ the following elementary relation holds for a Borel set A bounded away from zero with $\mu(\partial A) = 0$:

$$nP(x_n^{-1}\mathbf{X}_1 \in A) = E\left[\sum_{i=1}^{n} I_{\{x_n^{-1}\mathbf{X}_i \in A\}}\right] \to \mu(A). \qquad (5.4.14)$$

By virtue of Poisson's limit theorem we conclude that the binomial sequence $\sum_{i=1}^{n} I_{\{x_n^{-1}\mathbf{X}_i \in A\}}$ converges in distribution to a Poisson random variable with parameter $\mu(A)$. This approach allows for an extension to point processes: let $\varepsilon_{\mathbf{x}}$ be *Dirac measure* at $\mathbf{x}$, i.e., for any set A

$$\varepsilon_{\mathbf{x}}(A) = I_A(\mathbf{x}).$$

It is a well known fact (see Resnick [118, 119]) that regular variation in $\overline{\mathbb{R}}^d \backslash \{\mathbf{0}\}$ in the sense of relation (5.4.10) is actually equivalent to convergence in distribution of the

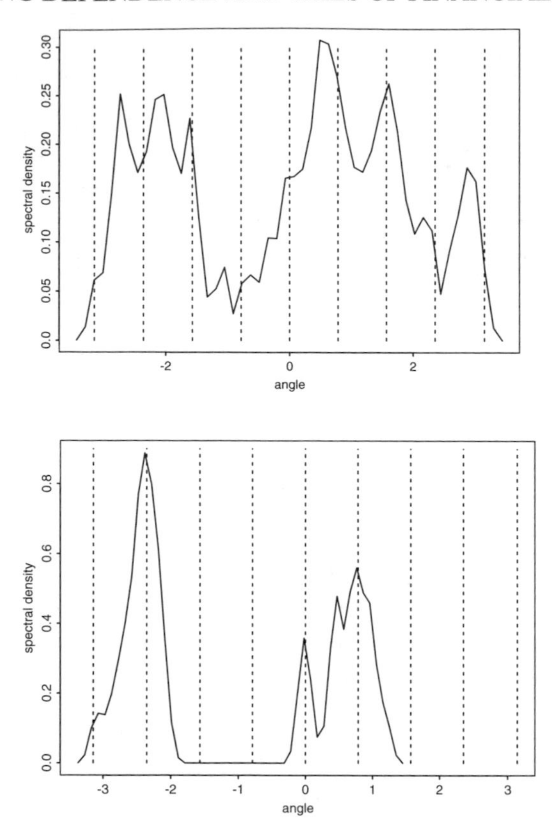

Figure 5.16 Top: *Spectral density estimation for the pairs* (X_t, X_{t+1}) *of the* S&P500 *returns* (X_t). *The vertical lines indicate multiples of* $\pi/4$. Bottom: *Spectral density estimation for the pairs* (X_t, Y_t), *where* (X_t) *is the log-return series of the DAX,* (Y_t) *the corresponding series for the CAC40 for the period from September 21, 1988, until August 24, 1995. See Example 5.4.7 for details.*

point processes

$$N_n = \sum_{i=1}^{n} \varepsilon_{\mathbf{X}_i/x_n} \xrightarrow{d} N, \tag{5.4.15}$$

where N is a Poisson random measure or Poisson process with mean measure μ (PRM(μ)) on $\overline{\mathbb{R}}^d \backslash \{\mathbf{0}\}$, the state space for the point processes N_n and N. Recall that N is PRM(μ) with state space $\overline{\mathbb{R}}^d \backslash \{\mathbf{0}\}$ if $N(A_1), \ldots, N(A_m)$ are independent for disjoint sets A_i and $N(A) \sim \text{Poisson}(\mu(A))$ for every Borel set $A \subset \overline{\mathbb{R}}^d \backslash \{\mathbf{0}\}$ with the interpretation that $N(A) = \infty$ a.s. if $\mu(A) = \infty$. Convergence in distribution as given in (5.4.15) is equivalent to the convergence of the finite-dimensional distributions, i.e.,

$$(N_n(A_1), \ldots, N_n(A_m)) \xrightarrow{d} (N(A_1), \ldots, N(A_m))$$

for any collection of sets $A_1, \ldots, A_m \subset \overline{\mathbb{R}}^d \backslash \{\mathbf{0}\}$ with $\mu(\partial A_i) = 0$. We refer to Kallenberg [87] and Resnick [118, 119] for introductions to the theory of point processes and their weak convergence, and in particular to the latter two references for connections with regular variation.

The point process convergence in (5.4.15) and the particular form of the measure μ for i.i.d. regularly varying $\mathbf{X}_1, \ldots, \mathbf{X}_n$ imply that for any set A bounded away from the origin the number of points $x_n^{-1}\mathbf{X}_i \in A$ is roughly a Poisson($\mu(A)$) random variable. This applies in particular to the sets $A = A(t, S)$ defined in (5.4.9). Thus, regular variation tells us about the likelihood that one of the scaled $x_n^{-1}\mathbf{X}_i$s in an i.i.d. sample falls into the set $A(t, S)$ with distance t from the origin, and S tells us about the direction where the extremal event will happen. The spectral measure $P(\boldsymbol{\Theta} \in \cdot)$ gives a quantitative description of the likelihood of this event.

Finally, regular variation also has an intuitive interpretation in terms of the convergence of componentwise maxima. To explain this, assume for the sake of simplicity that $d = 2$ and $\mathbf{X}_i = (X_i, Y_i)$ has nonnegative components. Write

$$M_n(X) = \max_{i \le n} X_i \quad \text{and} \quad M_n(Y) = \max_{i \le n} Y_i .$$

We also need the sets

$$C_{\mathbf{a}} = \{\mathbf{y} = (y_1, y_2) : y_1 \le a_1, y_2 \le a_2\}^c , \quad \text{where} \quad \mathbf{a} = (a_1, a_2) \in (0, \infty]^2 .$$

Since the $\mathbf{X}_i$s are i.i.d., a Taylor expansion argument yields

$$\begin{aligned} &P\left(x_n^{-1} M_n(X) \le a_1, x_n^{-1} M_n(Y) \le a_2\right) \\ &= \left[P\left(x_n^{-1} X_1 \le a_1, x_n^{-1} Y_1 \le a_2\right)\right]^n \\ &= \left(1 - P\left(x_n^{-1}\mathbf{X}_1 \in C_{\mathbf{a}}\right)\right)^n \sim \exp\left\{ - nP\left(x_n^{-1}\mathbf{X}_1 \in C_{\mathbf{a}}\right)\right\} \\ &\to \exp\{-\mu(C_{\mathbf{a}})\}. \end{aligned}$$

Some standard argument (similar to the weak convergence of probability measures) shows that the convergence of $nP(x_n^{-1}\mathbf{X}_1 \in C_{\mathbf{a}})$ to $\mu(C_{\mathbf{a}})$ for all $\mathbf{a}$ is actually equivalent to regular variation in the sense explained above, and therefore convergence in distribution of the partial maxima of i.i.d. vectors $\mathbf{X}_i$ with nonnegative components is equivalent to the point process convergence (5.4.15); we again refer to Resnick [118, 119] for precise statements.

Some statistical problems

When dealing with real-life data it is important to estimate the measure μ on $\overline{\mathbb{R}}^d \backslash \{\mathbf{0}\}$, the tail index α, and the spectral measure $P(\boldsymbol{\Theta} \in \cdot)$ on the unit sphere $\mathbb{S}^{d-1}$. Although we know in particular cases (such as the GARCH and stochastic volatility models) that the underlying finite-dimensional distributions of the processes are regularly varying, it is most often impossible to describe their characteristics in an accessible form; see Section 5.5.

Estimation problems in one dimension for the tail index have led to explosive growth in the literature, and the multivariate case is by no means easier. To estimate tail characteristics, one must use the rare data points far from the origin. The main difficulty is to choose a sufficiently high threshold to get reliable estimates, in particular when samples are small. An honest discussion can be found so far only in some case studies that are restricted to dimensions 2 or 3; see de Haan and de Ronde [80] for a survey paper and the related thesis of Draisma [54]. A discussion and many references on the estimation of α in the univariate case (also for the dependent case) can be found in Embrechts et al. [58], Chapter 6. Recent progress on the estimation of α for dependent data is reported in Drees [55], Novak [113], and Resnick and Stărică [123]. The latter theory applies to the multivariate case if one considers the norms $|\mathbf{X}_i|$. Nonparametric estimation of μ is the topic of the papers by de Haan and Resnick [78] and Einmahl et al. [57]. Stărică [131] gives an accessible account of multivariate regular variation, considers a particular multivariate GARCH model and applications to the modeling of joint extremal movements of pairs of foreign exchange rates. Multivariate extreme value distributions are treated in the recent book of Kotz and Nadarajah [93] who consider various multivariate parametric models. We refer to the survey paper by Ledford and Tawn [95] for several parametric models and related estimation techniques.

The author believes that multivariate extreme value statistics is not well developed. In particular, for dependent data such as return series, the map contains large white spots. Further investigations are clearly needed. The gap between the theory that was well understood in the 1980s (see, e.g., Resnick [119]), and its statistical applications has not been closed so far. In contrast to the univariate case for which we have by now an almost complete picture and a multitude of real-life applications in insurance, climatology, earthquake modeling, stress testing in engineering, etc., the multivariate case lacks an honest discussion about the limitations and applicability of the methods.

Example 5.4.7 (Spectral density estimation for S&P500, DAX and CAC40)
In this example we assume that $|\cdot|$ is the Euclidean norm. We consider spectral measures $P(\mathbf{\Theta} \in \cdot)$ on $\mathbb{S}^1$. Then we can write $\mathbf{\Theta} = (\cos\Phi, \sin\Phi)$, and if Φ has a density on $(-\pi, \pi]$ we refer to it as the spectral density. In Figure 5.16 we give two examples of spectral density estimation in a two-dimensional setting. In both cases the estimation is based on those 2% of the considered two-dimensional sample for which the Euclidean norm is above the 98% empirical quantile of the lengths of the pairs; we use a standard S+ function for probability density estimation.

In the top graph we consider the pairs (X_t, X_{t+1}) of the S&P500 returns X_t. The density has peaks around $-3\pi/4$, $\pi/4$, $\pi/2$, and π. The first two peaks indicate that an extreme value of X_t is likely to be followed by a value X_{t+1}, roughly of the same absolute size. When this happens, both X_t and X_{t+1} are likely to have the same sign. The peak at $\pi/2$ indicates that a large positive value X_{t+1} is not necessarily preceded by a large value $|X_t|$; it may happen out of the blue. Analogously, the peak at π shows that a large negative value of X_t is not necessarily followed by a large value $|X_{t+1}|$. From the heights of the peaks one can also conclude that joint extremal movements of X_t and X_{t+1} are more likely than independent extremal moves of the components.

In the bottom graph we consider the pairs (X_t, Y_t), where (X_t) is the log-return series of the DAX from September 21, 1988 until August 24, 1995, (Y_t) the corresponding series for the CAC40. The DAX and CAC40 are the major German and French composite stock indices, respectively. The density is concentrated at $-3\pi/4$ and $\pi/4$. This means that extreme movements of (X_t, Y_t) happen in a way such that both components are big at the same time. This strong dependence in the tails is not surprising in light of the strong joint links between the German and French economies,

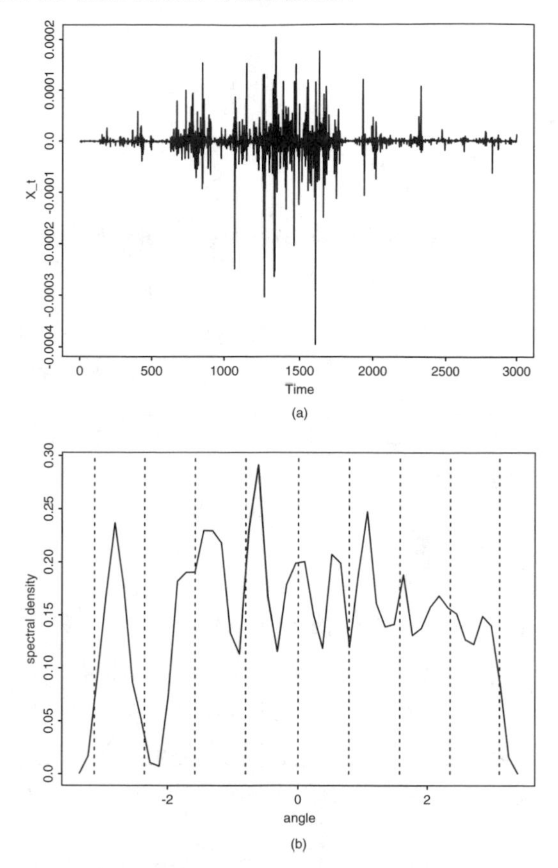

Figure 5.17 *Simulation of 3000 values from a stochastic volatility model* $X_t = \sigma_t Z_t$ (a, c) *and spectral density estimation for the pairs* (X_t, X_{t+1}) (b, d). *The i.i.d. symmetric noise satisfies* $P(Z_t > x) = 0.5x^{-4}$, $x \geq 1$, *and* $\log \sigma_t$ *is a Gaussian linear process. Hence* $P(X_t > x) \sim cx^{-4}$; *see* (5.4.18). (a, b): *The process* $(\log \sigma_t)$ *is fractional ARIMA with difference parameter* $d = 0.49$. *This implies hyperbolic decay of its ACF and in turn very strong dependence in the series. See Brockwell and Davis [29], Section 13.2, for details. The estimated spectral density has various peaks at multiples of* $\pi/4$ *and a deep valley at* $-3\pi/4$. *The latter fact implies that extremes are very unlikely with both* $|X_t|, |X_{t+1}|$ *large and* X_t, X_{t+1} *negative. Any angle between* $-\pi/2$ *and* π *is likely.*

on the one hand, and their dependence on the U.S. economy on the other hand. The large peak at $-3\pi/4$ is a small surprise since it means that, if an extreme pair (X_t, Y_t) occurs, it is quite likely that both X_t and Y_t are negative, whereas it is less likely that both X_t and Y_t are positive. This can be understood as a consequence of the leverage effect mentioned in Section 5.1.3. The small peak at zero indicates that there is a small probability that an extreme movement might be caused only by a positive value of the DAX.[9] In contrast to the S&P500 example, it is quite astonishing

[9] The author has learned from Steve Marron that this peak is statistically insignificant. It appears as an artifact of the S+ procedure used.

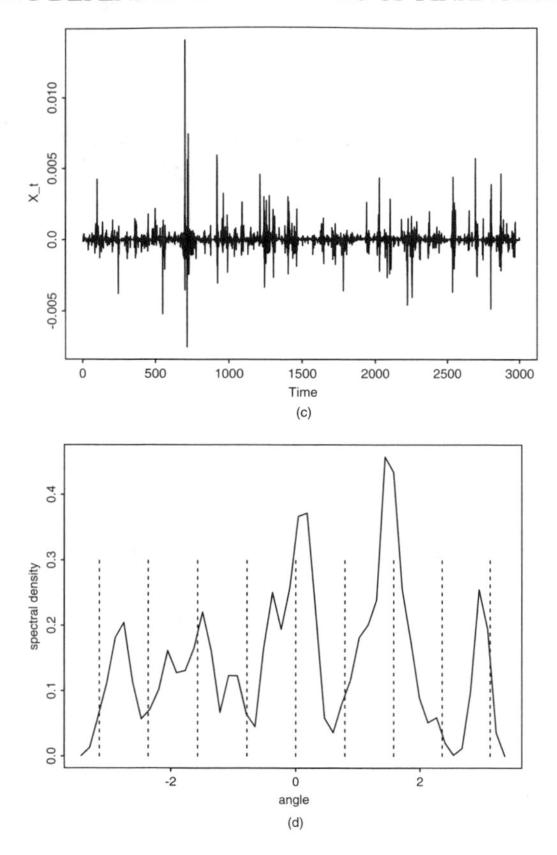

Figure 5.17 *(Continued).* (c, d): *The log-volatility is an* ARMA(1,1) *model* $\log \sigma_t = 0.9 \log \sigma_t + 0.5\eta_{t-1} + \eta_t$ *with i.i.d. standard Gaussian noise. The ACF of this series decays exponentially fast; the series has very short memory. The estimated spectral density has sharp peaks at multiples of* $\pi/2$. *This means that extremes are due to almost independent behavior of* X_t *and* X_{t+1}.

that the DAX-CAC40 density is not spread over the whole interval $(-\pi, \pi]$. This observation implies that for extremal movements it would be enough to study only the DAX.

5.4.3 Operations on regularly varying random vectors

Regular variation is a very flexible tool for describing heavy tails and dependence in the tails. This will be illustrated in the sequel where we will see that regularly varying vectors remain regularly varying after some operations.

Linear combinations

Simple consequences of multivariate regular variation are, for example, that the linear combinations of the components of a regularly varying random vector $\mathbf{X}$ are regularly varying. Indeed, writing

$$A_{\mathbf{x}} = \{\mathbf{y} : (\mathbf{x}, \mathbf{y}) > 1\} \quad \text{for } \mathbf{x} \neq \mathbf{0},$$

where $(\cdot, \cdot)$ denotes the usual inner product, we have

$$\lim_{x\to\infty} \frac{P((\mathbf{x}, \mathbf{X}) > x)}{P(|\mathbf{X}| > x)} = \lim_{x\to\infty} \frac{P(\mathbf{X} \in x A_{\mathbf{x}})}{P(|\mathbf{X}| > x)} = \mu(A_{\mathbf{x}}),$$

hence $(\mathbf{x}, \mathbf{X})$ is univariate regularly varying with the same index as $\mathbf{X}$ provided that $\mu(A_{\mathbf{x}}) > 0$.

Now suppose that the following condition holds:

$$\begin{cases} \text{There exists an } \alpha > 0, \text{ a slowly varying function } L \text{ and a finite-valued function} \\ w \text{ which is not identical zero such that, for all } \mathbf{x} \neq \mathbf{0}, \text{ the limit} \\ \lim_{u\to\infty} \dfrac{P((\mathbf{x}, \mathbf{X}) > u)}{u^{-\alpha} L(u)} = w(\mathbf{x}) \quad \text{exists.} \end{cases} \tag{5.4.16}$$

That is, for all $\mathbf{x}$ with $w(\mathbf{x}) > 0$, the random variable $(\mathbf{x}, \mathbf{X})$ is regularly varying with index α. As noted above, (5.4.16) holds if $\mathbf{X}$ is regularly varying with index α. The converse implication is not obvious, i.e., if $\mathbf{X}$ obeys (5.4.16), then the vector $\mathbf{X}$ is regularly varying with index α and a uniquely determined spectral measure on the unit sphere corresponding to a fixed norm. Here is a partial answer to the problem.

Theorem 5.4.8 (Basrak et al. [7]) *Let* $\mathbf{X}$ *be a random vector in* $\mathbb{R}^d$.

1. *If the random vector* $\mathbf{X}$ *is regularly varying with index* $\alpha > 0$, *then* (5.4.16) *holds with the same* α.
2. *If* $\mathbf{X}$ *satisfies the condition* (5.4.16), *where the* α *is noninteger, then* $\mathbf{X}$ *is regularly varying in the sense of* (5.4.5) *with index* α *and the spectral measure is uniquely determined.*
3. *If* $\mathbf{X}$ *assumes values in* $[0, \infty)^d$ *and satisfies* (5.4.16) *for* $\mathbf{x} \in [0, \infty)^d \backslash \{\mathbf{0}\}$, *where* α *is a noninteger, then* (5.4.5) *holds with index* α *and the spectral measure is uniquely determined.*
4. *If* $\mathbf{X}$ *assumes values in* $[0, \infty)^d$ *and satisfies* (5.4.16), *where* α *is an odd integer, then* (5.4.5) *holds with index* α *and the spectral measure is uniquely determined.*

For integer-valued α and general $\mathbf{X}$ the problem remains open. Meerschaert and Scheffler [102], Example 6.1.35, show that one can find two-dimensional regularly varying vectors $\mathbf{X}_1$ and $\mathbf{X}_2$ with index $\alpha = 1$ and different spectral measures for which (5.4.16) holds with the same function w.

The problem arises from the fact that one has to reconstruct the infinite measure μ in (5.4.10) from its values on the half-spaces $A_{\mathbf{x}}$. Because μ is an infinite measure and the half-spaces do not constitute a π-system, this creates difficulties. Notice that the problem bears some resemblance with various well known results in probability theory where one has to conclude from the behavior of the linear combinations of the components of vectors about the properties of the vector itself. Those include the Cramér–Wold device on convergence in distribution of sequences of vectors and the fact that Gaussianity of all linear combinations implies Gaussianity of the vector. However, it is also well known that α-stability of the linear combinations of a vector with $\alpha \in (0, 1)$ does not necessarily imply α-stability of the vector; see Samorodnitsky and Taqqu [128], Section 2.2.

Example 5.4.9 In Section 5.1.3 we have argued about the interplay between heavy tails and the CLT. It follows from the above theory that, for every fixed n, $X_1 + \cdots + X_n$ is regularly varying with index $\alpha > 0$ if the vector $\mathbf{X} = (X_1, \ldots, X_n)$ is regularly varying with index α. In particular, as $x \to \infty$,

$$\frac{P(X_1 + \cdots + X_n > x)}{P(|\mathbf{X}| > x)} \to \mu(\{\mathbf{x} : x_1 + \cdots + x_n > 1\}).$$

If $\mathbf{X}$ has i.i.d. and nonnegative regularly varying components with positive index then, choosing the max-norm $|\cdot|$, as $x \to \infty$,

$$\frac{P(X_1 + \cdots + X_n > x)}{P(\max(X_1, \ldots, X_n) > x)} \sim \frac{P(X_1 + \cdots + X_n > x)}{nP(X_1 > x)}$$
$$\to \mu(\{\mathbf{x} : x_1 + \cdots + x_n > 1\}) = 1. \qquad (5.4.17)$$

(Compare also with the results in Section 5.4.3.) The assumption of fixed n was crucial. For a general sequence of i.i.d. nonnegative random variables X_i relation (5.4.17) leads to the notion of subexponentiality: X_1 has a subexponential distribution if for $n \geq 2$,

$$\frac{P(X_1 + \cdots + X_n > x)}{P(\max(X_1, \ldots, X_n) > x)} \sim 1;$$

cf. Embrechts et al. [58], Chapter 1. The latter family contains distributions with very heavy tails (such as regularly varying distributions) and moderately heavy tails. A subexponential distribution has finite power moments of all degrees and tails lighter than the exponential distribution. The class of subexponential distributions includes the Weibull distribution with shape parameter less than 1 and the log-normal distribution. The subexponential distributions are considered as prime examples of heavy-tailed distributions. They have gained popularity in various applied probability contexts such as telecommunications, branching, queuing, and insurance mathematics.

The picture in (5.4.17) changes if $x = x_n \to \infty$ and $n \to \infty$. Depending on how fast x_n increases one can get very different answers as to the asymptotic order of the left-hand ratio in (5.4.17). In the i.i.d. case this is well known and is described either by the CLT with normal or infinite variance stable limits ([58], Section 2.2) or by heavy-tailed large deviation behavior ([58], Section 8.6).

Products

Consider two independent nonnegative random variables X and Y. Assume X is regularly varying with index $\alpha > 0$ and $EY^{\alpha+\epsilon} < \infty$ for some $\epsilon > 0$. Then XY is regularly varying with index α, and

$$P(XY > x) \sim EY^{\alpha}\, P(X > x) \quad \text{as } x \to \infty. \qquad (5.4.18)$$

This result was proved by Breiman [28] (in a special case). It is an extremely useful tool for many applied problems where one has to deal with products of random variables. Among others, it has been extensively used in the recent queuing theory literature on telecommunications.

A multivariate version reads as follows:

Theorem 5.4.10 (Resnick and Willekens [124], Basrak et al. [7]) *Assume the $\mathbb{R}^d$-valued random vector $\mathbf{X}$ is regularly varying in the sense of* (5.4.10) *and $\mathbf{A}$ is a random $q \times d$ matrix, independent of $\mathbf{X}$. If $0 < E\|\mathbf{A}\|^{\alpha+\epsilon} < \infty$ for some $\epsilon > 0$, then*

$$\lim_{x\to\infty} \frac{P(\mathbf{AX} \in x\,\cdot)}{P(|\mathbf{X}| > x)} \stackrel{v}{\to} \widetilde{\mu}(\cdot) := E[\mu \circ \mathbf{A}^{-1}(\cdot)]\,,$$

where $\stackrel{v}{\to}$ denotes vague convergence in $\overline{\mathbb{R}}^d \backslash \{\mathbf{0}\}$.

A particular case is the following.

Corollary 5.4.11 *Let $\mathbf{X}$ be regularly varying with index $\alpha > 0$, independent of the random vector $\mathbf{Y} = (Y_1, \ldots, Y_d)'$ which has independent components. Assume that $E|Y_i|^{\alpha+\epsilon} < \infty$ for some $\epsilon > 0$, $i = 1, \ldots, d$. Then the inner product $(\mathbf{Y}, \mathbf{X})$ is regularly varying with index α.*

Products of heavy-tailed, not necessarily regularly varying random variables have been considered in the literature for some time. We refer to Cline and Samorodnitsky [31] and references therein for a recent account on the topic, in particular on products of subexponential random variables.

Example 5.4.12 The investigation of the tails of products of regularly varying random variables can be quite tricky. Assume, for example that the two-dimensional vector $\mathbf{X} = (X_1, X_2)$ is regularly varying with nonnegative components. Then by the definition of regular variation,

$$\begin{aligned} \frac{P(X_1X_2 > t^2x^2)}{P(|\mathbf{X}| > x)} &= \frac{P(x^{-1}\mathbf{X} \in t\,\{\mathbf{y} : y_1y_2 > 1\})}{P(|\mathbf{X}| > x)} \\ &\to \mu(t\,\{\mathbf{y} : y_1y_2 > 1\}) = t^{-\alpha}\,\mu(\{\mathbf{y} : y_1y_2 > 1\}). \end{aligned} \quad (5.4.19)$$

Thus, X_1X_2 is again regularly varying with index $\alpha/2$ *if* $\mu(\{\mathbf{y} : y_1y_2 > 1\}) > 0$. This, however, is not true for many examples of interest. For example, the above Breiman theory tells that X_1X_2 is regularly varying with the same index α if X_1 and X_2 are independent and one of the components in the vector $\mathbf{X}$ has substantially lighter right tail than the other component. This means that the whole mass of the measure μ is concentrated on the axis where the largest component lives. Hence the right-hand side of (5.4.19) is zero.

If both components have approximately the same magnitude of the tail, additional information about X_1 and X_2 is needed to get more precise information about $P(X_1X_2 > x)$. For example, if X_1 and X_2 are i.i.d. with $P(X_i > x) \sim cx^{-\alpha}$ for some $c > 0$, then $P(X_1X_2 > x) \sim c'x^{-\alpha}\log x$, some $c' > 0$; see Rosiński and Woyczyński [126] and Davis and Resnick [46] for more general results on products of i.i.d.

regularly varying random variables. In this case, μ is concentrated on the axes and the right-hand side of (5.4.19) is again zero.

If $X_1 = X_2$, μ is concentrated on the line $x = y$, and hence the right-hand side in (5.4.19) is finite, and so we may conclude what we already know: X_1^2 is regularly varying with index $\alpha/2$. Generally speaking, only in the case of strong tail dependence between X_1 and X_2 one can expect that (5.4.19) gives meaningful information. This is the case for products of the components of any lagged vector (X_0, X_h) of a GARCH process, as follows from the theory developed in Section 5.5.2.

Infinite moving averages

Interesting supplements to the results about the regular variation of linear combinations of the components of a regularly varying vector, given in Section 5.4.3, are results about (in)finite moving average processes, so-called linear processes. It is the best understood class of time series models, including the class of ARMA processes mentioned in Section 5.2.1.

Assume that (Z_t) is an i.i.d. sequence of regularly varying random variables with index $\alpha > 0$ and tail balance condition

$$P(Z_1 > x) = p P(|Z_1| > x)(1 + o(1)) \quad \text{and}$$
$$P(Z_1 \leq -x) = q P(|Z_1| > x)(1 + o(1)),$$

where $p+q = 1, x \to \infty$. Then the vector $(Z_1, \ldots, Z_n)$ is regularly varying as well, where the spectral measure of this distribution is concentrated on the intersection of the unit sphere $\mathbb{S}^{n-1}$ with the axes, see Example 5.4.4. Hence linear combinations of finitely many Z_is are regularly varying, see Section 5.4.3. The same statement remains valid if we consider an infinite moving average of the Z_ts:

$$X_t = \sum_{j=-\infty}^{\infty} \psi_j Z_{t-j}, \tag{5.4.20}$$

where the ψ_js are real coefficients satisfying some condition stronger than square summability, depending on the tail of Z_t. Then the following relation holds (see Embrechts et al. [58], Appendix A3.3; the weakest conditions on (ψ_j) are given in Mikosch and Samorodnitsky [105]):

$$\lim_{x\to\infty} \frac{P(X_1 > x)}{P(|Z_1| > x)} = \sum_{j=-\infty}^{\infty} [p\, (\psi_j)_+^\alpha + q\, (\psi_j)_-^\alpha].$$

This means that X_t is regularly varying. Using the same methods as for fixed t, one can show that the finite-dimensional distributions of (X_t) are regularly varying as well (although no reference seems availabe for general linear processes).

If we consider the noise (Z_t) in (5.4.20) as input into a linear filter with coefficients ψ_j and X_t in (5.4.20) as output after filtering, we see that the property of regular variation remains invariant. This is due to the linearity of the system. We will see later, in Section 5.5, that this invariance is atypical for nonlinear time series, such as GARCH processes.

5.5 The tails

The tails of the finite-dimensional distributions of the stationary process (X_t) determine its extremal behavior. Clearly, there is no unique definition of the tail of a multivariate distribution G; in our context of regularly varying distributions it will be convenient to refer to the collection of probabilities $G(A)$ for A bounded away from zero as the tail of G.

In what follows, we will study conditions under which the finite-dimensional distributions of stochastic volatility and GARCH models are regularly varying. In order to ensure regular variation, we need to require in the stochastic volatility model that the noise is regularly varying. Surprisingly, the GARCH process has regularly varying finite-dimensional distributions even for large classes of light-tailed noise distributions. This is due to a fundamental result of Kesten [88] for SREs.

5.5.1 The stochastic volatility case

We study the stochastic volatility case first because it is uncomplicated. Consider the stochastic volatility process $X_t = \sigma_t Z_t$ with $\log \sigma_t$ being a strictly stationary mean-zero Gaussian linear process as introduced in (5.3.1), independent of the i.i.d. symmetric sequence (Z_t). Clearly, σ_t is log-normally distributed. The log-normal distribution is heavy-tailed in the sense that its right tail is heavier than an exponential tail. The log-normal distribution belongs to the class of subexponential distributions; see Example 5.4.9. For log-returns data the log-normal distribution is too light-tailed and power law behavior of the tails seems more appropriate. This can easily be achieved by assuming that the noise is regularly varying with index α. Then by Breiman's result (5.4.18), as $x \to \infty$,

$$P(X_1 > x) = P(\sigma_1 Z_1 > x) \sim E\sigma_1^\alpha P(Z_1 > x) \quad \text{and}$$
$$P(X_1 \leq -x) \sim E\sigma_1^\alpha P(Z_1 \leq -x).$$

A similar relation for

$$(X_1, \ldots, X_n)' = \mathrm{diag}(\sigma_1, \ldots, \sigma_n)(Z_1, \ldots, Z_n)'$$

can be obtained by the multivariate version of Breiman's result (Theorem 5.4.10) because the i.i.d. vector $(Z_1, \ldots, Z_n)$ is regularly varying and independent of the matrix $\mathrm{diag}(\sigma_1, \ldots, \sigma_n)$.

5.5.2 The GARCH case

For the stochastic volatility model we had to impose a regular variation condition on the noise to get regular variation of its finite-dimensional distributions. For the GARCH process the tail behavior is less obvious. However, it will turn out that the finite-dimensional distributions of a GARCH process are regularly varying under mild conditions on the noise distribution. This is somewhat surprising because, in this case, light-tailed input such as i.i.d. Gaussian noise (Z_t) generates regularly varying output (X_t). This is due to the fact that the squared GARCH process and the squared volatility process can be embedded in a SRE; see Section 5.3.2. Therefore we first consider the more general SRE case.

The tails of the solution to a stochastic recurrence equation

Under general conditions, the stationary solution to the SRE

$$\mathbf{Y}_n = \mathbf{A}_n \mathbf{Y}_{n-1} + \mathbf{B}_n\,, \quad n \in \mathbb{Z}\,, \tag{5.5.1}$$

where $((\mathbf{A}_n, \mathbf{B}_n))$ is an i.i.d. sequence with nonnegative-valued $d \times d$ random matrices $\mathbf{A}_n$ and nonnegative-valued d-dimensional random vectors $\mathbf{B}_n$, satisfies a multivariate regular variation condition. This follows from ingenious work by Kesten [88] on renewal theory for products of i.i.d. random matrices in the general case $d \geq 1$. We give a modification of Kesten's fundamental result, a combination of Theorems 3 and 4 in [88]. In what follows, we use the Euclidean norm $|\cdot|$ in $\mathbb{R}^d$ and $\|\cdot\|$ denotes the corresponding operator norm.

Theorem 5.5.1 (Kesten [88]) *Let $((\mathbf{A}_n, \mathbf{B}_n))$ be an i.i.d. sequence of $d \times d$ matrices $\mathbf{A}_n$ with nonnegative entries and d-dimensional nonnegative-valued random vectors $\mathbf{B}_n \neq \mathbf{0}$ a.s. Assume that the following conditions hold:*

- *For some $\epsilon > 0$, $E\|\mathbf{A}_1\|^\epsilon < 1$.*
- *$\mathbf{A}_1$ has no zero rows a.s.*
- *The set*

$$\{\log \|\mathbf{a}_n \cdots \mathbf{a}_1\| : n \geq 1,\ \mathbf{a}_n \cdots \mathbf{a}_1 > 0 \text{ and } \mathbf{a}_n, \ldots, \mathbf{a}_1 \in \text{the support of } P_{\mathbf{A}_1}\} \tag{5.5.2}$$

  generates a dense group in $\mathbb{R}$ with respect to summation and the Euclidean topology. Here $P_{\mathbf{A}_1}$ denotes the distribution of $\mathbf{A}_1$.
- *There exists a $\kappa_0 > 0$ such that*

$$E\left(\min_{i=1,\ldots,d} \sum_{j=1}^d A_{ij}\right)^{\kappa_0} \geq d^{\kappa_0/2} \tag{5.5.3}$$

  and

$$E\left(\|\mathbf{A}_1\|^{\kappa_0} \log^+ \|\mathbf{A}_1\|\right) < \infty\,. \tag{5.5.4}$$

Then the following statements hold:

1. *There exists a unique solution $\kappa_1 \in (0, \kappa_0]$ to the equation*

$$0 = \lim_{n\to\infty} \frac{1}{n} \log E\|\mathbf{A}_n \cdots \mathbf{A}_1\|^{\kappa_1}\,. \tag{5.5.5}$$

2. *If $E|\mathbf{B}_1|^{\kappa_1} < \infty$, there exists a unique strictly stationary causal solution $(\mathbf{Y}_n)$ to the stochastic recurrence equation* (5.5.1).

3. If $E|\mathbf{B}_1|^{\kappa_1} < \infty$, *then* $\mathbf{Y}_1$ *satisfies the following regular variation condition:*

$$\text{For all } \mathbf{x} \in \mathbb{R}^d \setminus \{\mathbf{0}\}, \quad \lim_{u\to\infty} u^{\kappa_1}\, P((\mathbf{x}, \mathbf{Y}_1) > u) = w(\mathbf{x}) \quad \text{exists} \tag{5.5.6}$$

and is positive for all nonnegative-valued vectors $\mathbf{x} \neq 0$.

In particular, all components of $\mathbf{Y}_1$ *are regularly varying with index* κ_1.

Kesten [88] also treats the case of $\mathbf{A}_n$ and $\mathbf{B}_n$ with general entries; see also Le Page [97] and Klüppelberg and Pergamenchtchikov [91] for some extensions. Goldie [70] gave an elegant alternative proof in the case $d = 1$. It is based on renewal theory and also treats real-valued A_ts and B_ts. In particular, he determined the exact value of the tail constant c_κ in the relation $P(Y_t > x) \sim c_\kappa x^{-\kappa}$. Related work can be found in Diebolt and Guégan [49] and Baxendale and Khasminskii [9]. Various nonlinear models can be embedded in a SRE or can be shown to be close in some sense to the solution of a SRE. Those include bilinear models with light-tailed noise (Basrak et al. [6]) and autoregressive models with ARCH error (Borkovec [18, 19]). These models have regularly varying finite-dimensional distributions as well.

At a first glance, the conditions of Kesten's theorem look mysterious. The case $d = 1$ is more transparent.

Example 5.5.2 (The case $d = 1$)
Assume that $((A_n, B_n))$ is a sequence of i.i.d. $(0, \infty)^2$-valued random vectors. Condition (5.5.2) then simply means that the distribution of the step size of the random walk

$$\log(A_1 \cdots A_n) = \sum_{i=1}^{n} \log A_i, \quad n \geq 1 \tag{5.5.7}$$

is not concentrated on a lattice. This is a typical assumption in renewal theory. Indeed, an inspection of the proof of Goldie [70] (cf. Embrechts et al. [58], Section 8.4, for a special case) shows that renewal theory is applied to this random walk. The assumption $E \log A_1 < 0$ in combination with $EB_1^{\kappa_1} < \infty$ ensures the existence of a stationary version of (Y_t); compare with the discussion in the GARCH(1,1) case before Theorem 5.3.1 where exactly this condition appears. Moreover, the random walk (5.5.7) has negative drift. Now, if $EA_1^\delta < \infty$ for some $\delta > 0$, the conditions $E \log A_1 < 0$ and $EA_1^\epsilon < 1$ are actually equivalent. Therefore part two of the theorem follows.

The function

$$h(\epsilon) = EA_1^\epsilon = E\mathrm{e}^{\epsilon \log A_1} \tag{5.5.8}$$

is convex in ϵ. So far we know that $h(0) = 1$ and $h(\epsilon) < 1$ for small $\epsilon > 0$. Condition (5.5.3) finally ensures that there exists $\kappa_0 > 0$ such that $h(\kappa_0) \geq 1$. By convexity of h we conclude the existence of a unique solution κ_1 to the equation $h(\kappa_1) = 1$. This is the content of part one of the theorem. The moment generating function $h(\epsilon)$ of the random variable $\log A_1$ and the solution κ_1 play a major role in renewal and queuing theory as well as in insurance mathematics. In the latter case,

κ_1 is known as the Lundberg or adjustment coefficient, which appears in the classical Cramér bound of the ruin probability; see [58], Chapter 1. From the Cramér ruin theory it is also well known that an additional moment condition is needed that ensures the finiteness of $h'(\kappa_0)$ (cf. Theorem 1.2.2 in [58]): this is condition (5.5.4).

A comparison of the methods for deriving bounds for the probability of ruin (the tail probability of the supremum of a random walk with negative drift, equivalently, the tail of the stationary distribution in a stable queue) shows that the tools used are quite similar in the one-dimensional case; see [58], Chapter 1 and Section 8.4, or go directly to Goldie's [70] paper. In particular, in both cases an application of the key renewal theorem is key to deriving the desired tail behavior. See also the recent paper of Klüppelberg and Pergamenchtchikov [90] for an extension to more general Markov chains. An alternative method based on large deviation techniques could also be applied; see Nyrhinen [114].

We continue with a simple one-dimensional example. Let (Z_i) be i.i.d. standard normal random variables, α_0 be any positive number. Define $A_t = Z_t^2$, $B_t = \alpha_0 Z_t^2$, and $Y_t = A_t Y_{t-1} + B_t$, i.e.,

$$Y_t = Z_t^2 Y_{t-1} + \alpha_0 Z_t^2, \quad t \in \mathbb{Z}. \tag{5.5.9}$$

It is not difficult to see that the latter SRE describes the squares of a stationary ARCH(1) model with standard normal innovations. Reading through the above comments on one-dimensional SREs it is not difficult to see that Kesten's theorem applies, i.e., $P(Y_1 > x) \sim cx^{-\kappa_1}$ and $EA_1^{\kappa_1} = E[Z_1^2]^{\kappa_1} = 1$. Since the solution κ_1 is unique, $\kappa_1 = 1$ and therefore Y_t has tail index 1. In Figure 5.18 we illustrate this fact.

Choosing $\mathbf{x} = \mathbf{e}_i$ in (5.5.6) as the ith unit vector, it is immediate that the ith component of $\mathbf{Y}_1$ is regularly varying with index κ_1. It is natural to ask whether the vector $\mathbf{Y}_1$ itself is regularly varying. Clearly, (5.5.6) is a special case of (5.4.16) and so an appeal to Theorem 5.4.8 immediately gives the following result.

Corollary 5.5.3 *Under the assumptions of Theorem 5.5.1, the marginal distribution of the unique strictly stationary causal solution* $(\mathbf{Y}_n)$ *of the stochastic recurrence equation* (5.5.1) *is regularly varying in the following sense. If the value* κ_1 *in* (5.5.5) *is not an even integer, then there exists a positive constant* c *and a random vector* $\mathbf{\Theta}$ *with values in the unit sphere* $\mathbb{S}^{d-1}$ *such that*

$$u^{\kappa_1} P(|\mathbf{Y}_1| > t\,u,\ \widetilde{\mathbf{Y}}_1 \in \cdot\,) \stackrel{v}{\rightarrow} c\,t^{-\kappa_1} P(\mathbf{\Theta} \in \cdot\,), \quad \text{as } u \to \infty. \tag{5.5.10}$$

Now iterate the SRE to obtain

$$(\mathbf{Y}_1, \ldots, \mathbf{Y}_m) = (\mathbf{A}_1, \mathbf{A}_2\mathbf{A}_1, \ldots, \mathbf{A}_m \cdots \mathbf{A}_1)\mathbf{Y}_0 + \mathbf{R}_m.$$

The components of $\mathbf{R}_m$ have lighter tails than the components of $\mathbf{Y}_0$ due to the fact that $E\|\mathbf{A}_1\|^{\kappa_1} < \infty$ and $E|\mathbf{B}_1|^{\kappa_1} < \infty$. A standard argument shows that the contribution of $\mathbf{R}_m$ to the tail of $(\mathbf{Y}_1, \ldots, \mathbf{Y}_m)$ is asymptotically negligible. The regular variation of the vector $(\mathbf{Y}_1, \ldots, \mathbf{Y}_m)$ is assured by regular variation of $\mathbf{Y}_0$ and by the multivariate Breiman Proposition 5.4.10.

We conclude:

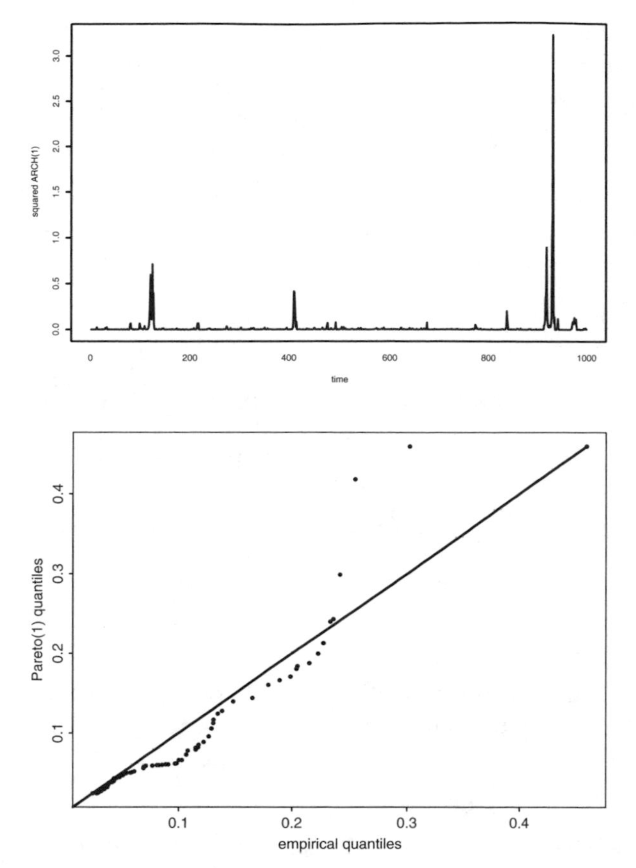

Figure 5.18 Top: *Simulation of 1000 values from the squared* ARCH(1) *SRE* (5.5.9) *with parameter* $\alpha_0 = 0.001$. Bottom: *QQ-plot of the largest 5% values in a simulated sample of size* 2000 *from the same model against the Pareto distribution with tail parameter one. We also indicate the line* $y = x$. *The QQ-plot is in good agreement with the fact that the right tail of* Y_1 *is Pareto-like.*

Corollary 5.5.4 *Under the conditions of Corollary 5.5.3 the finite-dimensional distributions of the stationary solution* $(\mathbf{Y}_t)$ *of* (5.5.1) *are regularly varying with index* κ_1.

The above discussion might have given the impression that the solution to SREs always has regularly varying finite-dimensional distributions. This is not necessarily the case as work by Goldie and Grübel [71] shows: if the distribution of $\|\mathbf{A}_1\|$ is concentrated on a too small compact interval, the tails of $\mathbf{Y}_t$ can be very thin.

The tails of a GARCH process

We have learned in Section 5.3.2 that the vector

$$\mathbf{Y}_t = \left(\sigma_{t+1}^2, \ldots, \sigma_{t-q+2}^2, X_t^2, \ldots, X_{t-p+2}^2\right)'$$

constructed from a GARCH(p, q) process $X_t = \sigma_t Z_t$ can be embedded in the SRE (5.3.5). Therefore an application of Kesten's Theorem 5.5.1 (we refrain from verifying its assumptions) yields the following.

Theorem 5.5.5 (Basrak et al. [8]) *Assume the following conditions:*

1. $\alpha_0 > 0$ *and the Lyapunov exponent* $\gamma < 0$ *of the SRE* (5.5.1) *specified by* (5.3.5).
2. Z_1 *has a positive density on* $\mathbb{R}$ *and either* $E|Z_1|^h < \infty$ *for all* $h < h_0$ *and* $E|Z_1|^{h_0} = \infty$ *for some finite* $h_0 > 0$ *or* $E|Z_1|^h < \infty$ *for all* $h > 0$.
3. *Not all of the parameters* α_i *and* β_i *vanish.*

Then there exist a $\kappa_1 > 0$ *and a finite-valued function* $w(\mathbf{x})$ *such that*

$$\text{for all } \mathbf{x} \in \mathbb{R}^{p+q-1} \setminus \{\mathbf{0}\}, \qquad \lim_{u \to \infty} u^{\kappa_1} P((\mathbf{x}, \mathbf{Y}_1) > u) = w(\mathbf{x}) \quad \text{exists},$$

i.e., $(\mathbf{x}, \mathbf{Y}_1)$ *is regularly varying with index* κ_1. *Moreover, if* $\mathbf{x} \in [0, \infty)^{p+q-1}$ *and* $\mathbf{x} \neq \mathbf{0}$ *then* $w(\mathbf{x}) > 0$.

Furthermore, if κ_1 *is not even, then* $\mathbf{Y}_1$ *is regularly varying with index* κ_1*, i.e., there exist a constant* $c > 0$ *and an* $\mathbb{S}^{p+q-2}$*-valued random vector* $\boldsymbol{\Theta}$ *such that*

$$u^{\kappa_1} P(|\mathbf{Y}_1| > t\,u, \widetilde{\mathbf{Y}}_1 \in \cdot\,) \xrightarrow{v} c\, t^{-\kappa_1} P(\boldsymbol{\Theta} \in \cdot\,), \quad \text{as } u \to \infty.$$

Example 5.5.6 (The GARCH(1, 1) case)
In this case it is not difficult to verify the conditions of Theorem 5.5.5. However, one can directly verify the conditions of Kesten's Theorem 5.5.1. We know from Theorem 5.3.1 that the conditions $\alpha_0 > 0$ and $E \log(\alpha_1 Z_1^2 + \beta_1) < 0$ ensure the existence of a unique strictly stationary solution to the GARCH(1,1) SRE

$$\sigma_t^2 = \alpha_0 + \alpha_1 X_{t-1}^2 + \beta_1 \sigma_{t-1}^2 = \alpha_0 + \left(\alpha_1 Z_{t-1}^2 + \beta_1\right)\sigma_{t-1}^2.$$

Hence $Y_t = \sigma_t^2$ satisfies

$$Y_t = A_t Y_{t-1} + B_t \quad \text{with} \quad A_t = \alpha_1 Z_{t-1}^2 + \beta_1 \quad \text{and} \quad B_t = \alpha_0.$$

Moreover, $E A_1^\epsilon < 1$ for some small ϵ follows from $E \log A_1 < 0$. Condition (5.5.2) is satisfied because Z_1 has a density (actually, it suffices to assume that $\log A_1$ does not have a lattice distribution) and therefore the distribution of the step size of the negative drift random walk $\log(A_n \cdots A_1)$ is nonlattice. Condition (5.5.3) simply turns into $1 \leq E A_1^{\kappa_0}$ for some κ_0. If there exists $h_0 < \infty$ such that $E|Z_1|^{h_0} = \infty$ and $E|Z_1|^h < \infty$ for $h < h_0$, one has $E|Z_1|^h \uparrow \infty$ as $h \uparrow h_0$. Thus one can choose $\kappa_0 < h_0$ sufficiently close to h_0. Alternatively, if $E|Z_1|^h < \infty$ for all $h > 0$, the unbounded support of Z_1 ensures that $E|Z_1|^h \uparrow \infty$ as $h \uparrow \infty$, and therefore one can find κ_0 sufficiently large, which satisfies $1 \leq E A_1^{\kappa_0}$. Condition (5.5.4), i.e., $E A_1^{\kappa_0} \log^+ A_1 < \infty$ is satisfied for the same reason. Hence a unique $\kappa_1 \in (0, \kappa_0]$ exists with the property that

$$E A_1^{\kappa_1} = E\left(\alpha_1 Z_1^2 + \beta_1\right)^{\kappa_1} = 1. \tag{5.5.11}$$

Since B_t is a constant, it immediately follows from Kesten's theorem and Corollary 5.5.4 that the finite-dimensional distributions of the sequence (σ_t^2) are multivariate

Table 5.1 *Estimation Results for κ, Normal Noise*

β_1	0.9	0.8	0.7	0.6	0.5	0.4	0.3	0.2	0.1	0.0
$\widehat{\kappa}$	2.0	12.4	16.4	19.0	21.0	23.0	24.8	26.4	28.2	29.6
sd	0.0003	0.01	0.05	0.08	0.80	0.40	1.00	0.50	0.50	3.00
β_1	0.81	0.82	0.83	0.84	0.85	0.86	0.87	0.88	0.89	
$\widehat{\kappa}$	12.0	11.4	10.8	10.0	9.0	8.0	7.0	5.6	3.8	
sd	0.006	0.008	0.004	0.004	0.002	0.002	0.002	0.001	0.004	

regularly varying with index κ_1. It is not difficult to see that if a nonnegative-valued vector $(Y_1, \dots, Y_d)$ is regularly varying with index $\alpha > 0$, the vector $(Y_1^p, \dots, Y_d^p)$ is regularly varying with index α/p for any $p > 0$. Hence the finite-dimensional distributions of (σ_t) are regularly varying with index $\kappa = 2\kappa_1$. An appeal to Corollary 5.4.11 and some additonal argument give that the finite-dimensional distributions of the sequence of random variables $X_t = \sigma_t Z_t$ are regularly varying with index κ. In the above approach, in contrast to Theorem 5.5.5, no restrictions on the values κ_1 are required.

Example 5.5.7 (A simulation study on the estimation of κ)
In the GARCH(1, 1) case we can in principle calculate the solution κ_1 of the equation (5.5.11) for given parameters α_1 and β_1. This can be done in the naive way by replacing the left-hand expectation in (5.5.11) by a simulated average for various values of κ_1 provided one knows the exact distribution of the noise Z_t. Then one chooses the κ_1-value for which the simulated average is closest to one. A theoretical justification for this approach exists (Csörgő and Teugels [34]). The resulting estimator $\widehat{\kappa}_1$ can be shown to be consistent and asymptotically normal.

In reality we clearly never know the exact distribution of the noise and therefore this approach is doubtful. However, in order to see how this Monte Carlo procedure works we made a small simulation study with fixed $\alpha_1 = 0.1$ and varying β_1. When fitting GARCH(1, 1) models to returns one often obtains small α_1-values, and therefore the chosen value is close to reality.

In Table 5.1 we collected the average $\widehat{\kappa} = 2\widehat{\kappa}_1$-value from 20 repetitions of the Monte Carlo procedure based on 1,000,000 averages of i.i.d. simulated variables $(\alpha_1 Z_t^2 + \beta_1)^{\kappa_1}$. (This sample size was the largest one the author's computer could still handle.) We first choose Z_t standard normal. Despite the large sample size, the variation of the estimator $\widehat{\kappa}$ is very large and therefore the obtained values must be considered with enormous care. Standard deviations (sd) for $\widehat{\kappa}$ of the order 0.5 are not unusual, in particular for larger κ-values. For smaller κ, i.e., larger β_1-values, the estimator performs much better.

In a second experiment we choose the student distribution with four degrees of freedom, scaled to unit variance, for the noise Z_t. The results are reported in Table 5.2.

These results show that there can be a substantial variation in the estimation of κ.[10] Alternatively, one can estimate κ by using classical methods such as Hill's estimator;

[10] An alternative and perhaps more promising approach for calculating κ_1 was suggested by a referee of this chapter. He suggested to use numerical integration for solving the equation $E(\alpha_1 Z_1^2 + \beta_1)^{\kappa_1} = 1$ which is possible for many distributions of Z_1.

Table 5.2 *Estimation Results for κ, Student Noise with 4 Degrees of Freedom*

β_1	0.9	0.8	0.7	0.6	0.5	0.4	0.3	0.2	0.1	0.0
$\widehat{\kappa}$	2.0	4.0	4.4	4.4	4.4	4.4	4.8	4.8	4.6	4.8
sd	1.4×10^{-7}	0.5	0.5	0.5	0.5	1.4	1.5	1.5	2.0	2.0
β_1	0.81	0.82	0.83	0.84	0.85	0.86	0.87	0.88	0.89	
$\widehat{\kappa}$	4.2	4.0	4.0	3.8	3.6	3.4	3.4	3.0	2.6	
sd	0.15	0.13	0.08	0.10	0.08	0.04	0.02	0.005	0.002	

see Drees [55] and Resnick and Stărică [123] for some results on the estimation of the tail index for dependent data. For dependent data, the asymptotic variance of the estimators is typically larger than in the i.i.d. case.

Example 5.5.8 (The IGARCH(1, 1) case)
Recall the definition of an IGARCH model from Example 5.3.7. In this case we have $\alpha_1 + \beta_1 = 1$, and the unique solution to the equation (5.5.11) is $\kappa_1 = 1$. Following the argument above, we conclude that the finite-dimensional distributions of (X_t) are regularly varying with index 2. To be more specific, let $\mathbf{X}_d = (X_1, \ldots, X_d)$. Then there exist a constant $c > 0$ and a random vector with values in $\mathbb{S}^{d-1}$ such that for $t > 0$,

$$x^2 P(|\mathbf{X}_d| > tx\,, \widetilde{\mathbf{X}}_d \in \cdot) \xrightarrow{v} ct^{-2}\, P(\mathbf{\Theta} \in \cdot) \quad \text{as } x \to \infty.$$

In particular, X_t has infinite variance, a property shared by any IGARCH(p, q) process (Engle and Bollerslev [61]). Assume that (X_t) is a stationary IGARCH(p, q) process. Then (σ_t) is stationary as well. Taking expectations in the defining GARCH equation (5.2.7) and observing that $EX_1^2 = E\sigma_1^2$, we obtain

$$\begin{aligned} E\sigma_1^2 &= \alpha_0 + EX_1^2 \sum_{i=1}^{p} \alpha_i + E\sigma_1^2 \sum_{j=1}^{q} \beta_j \\ &= \alpha_0 + E\sigma_1^2 \left[\sum_{i=1}^{p} \alpha_i + \sum_{j=1}^{q} \beta_j \right] = \alpha_0 + E\sigma_1^2. \end{aligned}$$

Since $\alpha_0 > 0$ is necessary for stationarity, the only possible conclusion is that $E\sigma_1^2 = EX_1^2 = \infty$, and similar arguments apply to the case $\sum_{i=1}^{p} \alpha_i + \sum_{j=1}^{q} \beta_j > 1$. It is an open question as to whether any IGARCH(p, q) process is regularly varying with index $\kappa = 2$.

Using Corollary 5.4.11, one can show that the finite-dimensional distributions of a GARCH process are regularly varying.

Corollary 5.5.9 *Assume that the conditions of Theorem 5.5.5 hold. Then there exists a constant $\kappa_1 > 0$ such that the linear combinations of the vectors $(X_1^2, \sigma_1^2, \ldots, X_d^2, \sigma_d^2)$ for any $d \geq 1$ are regularly varying with index κ_1. If κ_1 is not an even integer, then the latter vector is multivariate regularly varying with index κ_1, and the vector $(X_1, \sigma_1, \ldots, X_d, \sigma_d)$ is regularly varying with index $\kappa = 2\kappa_1$. For a* GARCH(1, 1) *process the corollary holds without the restriction on the values of κ_1.*

5.6 Classical measures of dependence: Mixing properties and correlations

In probability theory and statistics different tools have been developed to measure the strength of dependence in a vector or sequence of random objects. Among them, covariances and correlations are the quantities used most frequently by statisticians, time series analysts, and practitioners. These instruments completely describe the dependence structure in a Gaussian world. However, for nonGaussian random variables, vectors and stochastic processes, covariances, and correlations are less meaningful. As we know from a course on elementary probability theory, they describe the degree to which two random variables are linearly dependent. For example, the ACF of a GARCH process (X_t) is zero at all lags; hence, there is no linear dependence between X_t and X_{t+h} for $h \geq 1$. Nevertheless, there is a complicated dependence structure in this process, and therefore the ACF does not help one much.

Among theoretical persons the mixing properties of a stationary sequence are considered as appropriate tools that describe dependence in a very sophisticated way. These conditions are often expressed in terms of σ-fields; we restrict our attention to one particular case.

5.6.1 Strong mixing

Rosenblatt's [125] strong mixing property is one of the most popular mixing conditions in the literature. We focus on the strong mixing condition. The strictly stationary ergodic sequence of random vectors $\mathbf{Y}_t$ is strongly mixing with rate function (ϕ_k), if

$$\sup_{A\in\sigma(\mathbf{Y}_s\,,\,s\leq 0)\,,\,B\in\sigma(\mathbf{Y}_s\,,\,s>k)} |P(A\cap B) - P(A)\,P(B)| \;=:\; \phi_k \to 0 \quad \text{as } k \to \infty.$$

If (ϕ_k) decays to zero at an exponential rate then $(\mathbf{Y}_t)$ is said to be strongly mixing with geometric rate. Then the memory between past and future dies out exponentially fast.

The rate function is closely related to the decay of the autocovariances of a one-dimensional strongly mixing process (Y_t). Indeed, if $E|Y_1|^{2+\delta} < \infty$ for some $\delta > 0$, then

$$|\gamma_Y(h)| = |\mathrm{cov}(Y_0, Y_h)| \leq c\,\phi_h^{\delta/(2+\delta)} \tag{5.6.1}$$

for some constant $c > 0$ and all h. See Ibragimov and Linnik [84], Theorem 17.2.2.

The stochastic volatility case

Since the volatility sequence (σ_t) and the noise (Z_t) are mutually independent, the dependence of the sequence of random variables $X_t = \sigma_t Z_t$ is caused only by the sequence (σ_t). If we assume, as before, that $(\log \sigma_t)$ is a Gaussian ARMA process, one can give some easy conditions in order to describe the dependence in (X_t).

If $(\log \sigma_t)$ is an ARMA process, its autocovariance function decays to zero exponentially fast; see Brockwell and Davis [29], Chapter 3. For a Gaussian stationary sequence this implies that $(\log \sigma_t)$ is strongly mixing with geometric rate (Doukhan [52]). For nonGaussian sequences this statement is in general not true.

Since strong mixing is defined via σ-fields, it transfers to measurable functions of the $\log \sigma_t$s, in particular to the (σ_t) and (σ_t^2) sequences. Since (σ_t) and (Z_t) are independent, we have

$$\begin{aligned}
&|P((\ldots, X_{-1}, X_0) \in A, (X_k, X_{k+1}, \ldots) \in B) \\
&\quad - P((\ldots, X_{-1}, X_0) \in A)\; P((X_k, X_{k+1}, \ldots) \in B)| \\
&\quad = |E[f(\ldots, \sigma_{-1}, \sigma_0) g(\sigma_k, \sigma_{k+1}, \ldots)] - E[f(\ldots, \sigma_{-1}, \sigma_0)]\; E[g(\sigma_k, \sigma_{k+1}, \ldots)]|,
\end{aligned}$$

where

$$f(\ldots, \sigma_{-1}, \sigma_0) = P((\ldots, X_{-1}, X_0) \in A \mid \sigma_s,\ s \le 0),$$

$$g(\sigma_k, \sigma_{k+1}, \ldots) = P((X_k, X_{k+1}, \ldots) \in B \mid \sigma_s,\ s \ge k),$$

and standard results about strong mixing (Doukhan [52]) show that the right-hand side is bounded by $4\phi_k$. Hence the sequences (X_t), $(\log \sigma_t)$, (σ_t), (σ_t^2) basically have the same mixing properties. Again, the stochastic volatility case is easy to handle and does not require sophisticated tools.

The GARCH case

Under general conditions, a stationary GARCH process is strongly mixing with geometric rate. This follows from general results about mixing properties of Markov chains as presented in Meyn and Tweedie [103]. These results can be applied first to the solutions of the SRE (5.3.5) and then to the squared GARCH process (X_t^2). Finally, since we assumed that (Z_t) is i.i.d. symmetric the same arguments as for the stochastic volatility case show that $(|X_t|, \text{sign}(Z_t)) = (|X_t|, \text{sign}(X_t))$ inherits the mixing properties of (X_t^2) and so does $X_t = |X_t| \, \text{sign}(X_t)$.

Theorem 5.6.1 (Boussama [22]) *Consider the SRE* (5.5.1) *with specification* (5.3.5) *through* (5.3.7) *which defines the* GARCH(p, q) *process. Assume that this SRE has a stationary solution, and that Z_t has a positive density in some interval around zero. Then the resulting unique* GARCH(p, q) *process is strictly stationary and strongly mixing with geometric rate.*

Remark 5.6.2 A particular consequence of Theorem 5.6.1 in combination with inequality (5.6.1) is that

$$\gamma_{g(X)}(h) = |\text{cov}(g(X_1), g(X_{1+h}))| \le c\, q^h\,, \quad h \ge 0,$$

for some positive constants c and $q < 1$ provided that $E|g(X_1)|^{2+\delta} < \infty$ for some $\delta > 0$. This means that the ACVFs and ACFs of any measurable function g of a GARCH process, if they are well defined, decay exponentially fast to zero. □

5.6.2 When does the sample ACF describe the ACF?

The mixing properties of a stationary process are precise theoretical tools that describe the dependence in the time series. Notice that, because the definition of strong mixing involves σ-fields, this notion immediately translates to measurable functions of the underlying process, so, for example, to the sequence of products $X_t X_{t+h}$ or to the lagged vectors $(X_t, \ldots, X_{t+h})$.

However, strong mixing is not a property one can detect by statistical tools. Therefore one has to rely on other means. Standard time series analysis deals with the second order structure of time series. This is mainly due to the fact that Gaussian stationary processes are fully described by the mean and the autocorrelations, and time series analysis was, and still is, devoted in its most parts to Gaussian processes. For nonGaussian sequences the correlation structure of a time series has been used for estimation and prediction purposes. Least squares estimation (such as the Yule-Walker estimators for AR-processes) only requires the knowledge of the autocorrelations of the process, and so does prediction of a time series based on the minimization of the mean square prediction error. Thus the autocovariances and autocorrelations have traditionally played an important role.

For nonlinear time series models such as GARCH or stochastic volatility models it is of interest to study the meaning of the autocorrelations and autocovariances. In particular, since one observes empirically that log-returns may have infinite 3rd, 4th, or 5th moments, it is questionable what the sample autocorrelations and sample autocovariances actually mean. Indeed, if the 4th moment of X_t does not exist but $\mathrm{var}(X_t) < \infty$, and if (X_t) is modeled by a GARCH process, say, then the strong law of large numbers gives us for every $h \geq 0$,

$$\gamma_{n,X}(h) = \frac{1}{n}\sum_{t=1}^{n-h}(X_t - \overline{X}_n)(X_{t+h} - \overline{X}_n) \stackrel{\text{a.s.}}{\to} \gamma_X(h) = \mathrm{cov}(X_0, X_h)\,,$$

i.e., the sample autocovariance at lag h consistently estimates the autocovariance at lag h, but since $E[(X_t X_{t+h})^2]$ can be shown to be infinite, the central limit theorem with Gaussian limit is not applicable anymore. Moreover, since it is common practice to look at the sample autocovariances and sample autocorrelations of $|X_t|$ and X_t^2 (as surrogate estimators of the unobserved σ_t and σ_t^2), one would need moments higher than the 4th in order to apply standard central limit theory for strongly mixing sequences. If these moments exist and are finite, standard results (Ibragimov and Linnik [84], Theorem 18.5.3, and Doukhan [52]) ensure the validity of the CLT. We formulate such a result for X_t. Recall the notions of autocorrelation $\rho_X(h)$ at lag h and its sample analogue $\rho_{n,X}(h)$ for a stationary sequence (X_t) from (5.1.1) and (5.1.2).

Theorem 5.6.3 *Assume that one of the following conditions holds:*

(1) (X_t) *is a strongly mixing stochastic volatility process with* $E|X_1 X_{1+h}|^{2+\delta} < \infty$ *for* $h = 0, 1, \ldots, m$, *some* $\delta > 0$ *and a mixing rate function* (ϕ_k) *satisfying*

$$\sum_{k=1}^{\infty} \phi_k^{\delta/(2+\delta)} < \infty. \tag{5.6.1}$$

(2) (X_t) *is a* GARCH(p, q) *process with* $E|X_1|^{4+\delta} < \infty$ *for some* $\delta > 0$ *satisfying the conditions of Theorem 5.6.1.*

Then the central limit theorem holds:

$$\left(n^{1/2}(\gamma_{n,X}(h) - \gamma_X(h))\right)_{h=0,\dots,m} \xrightarrow{d} (V_h)_{h=0,\dots,m},$$

$$\left(n^{1/2}(\rho_{n,X}(h) - \rho_X(h))\right)_{h=1,\dots,m} \xrightarrow{d} \gamma_X^{-1}(0)(V_h - \rho_X(h)\, V_0)_{h=1,\dots,m},$$

where $(V_1, \dots, V_m)$ *is multivariate normal with mean zero and covariance matrix*

$$\left[\sum_{k=-\infty}^{\infty} \operatorname{cov}(X_0 X_i, X_k X_{k+j})\right]_{i,j=1,\dots,m} \qquad \textit{and } V_0 = E\left(X_0^2\right).$$

The proof of these statements follows from the mentioned standard CLT for strongly mixing sequences ([84], Theorem 18.5.3, [52]). The latter result says that the strongly mixing sequence (Y_t) with mixing rate function (ϕ_k) satisfies the CLT with normalization $\sqrt{n}$ if condition (5.6.1) holds and $E|Y_1|^{2+\delta} < \infty$ for some $\delta > 0$. In the case of strong mixing with geometric rate, (5.6.1) is obviously satisfied. In particular, a GARCH(p, q) process satisfying the conditions of Theorem 5.6.1 and a stochastic volatility process with $(\log \sigma_t)$ Gaussian ARMA (see Section 5.6.1) have geometric mixing rates. Therefore (5.6.1) does not appear explicitly in the formulation of (2). For the stochastic volatility process with $(\log \sigma_t)$ Gaussian ARMA, the required moment condition

$$E|X_1 X_{1+h}|^{2+\delta} = E|\sigma_1 \sigma_{1+h}|^{2+\delta} E|Z_1 Z_{1+h}|^{2+\delta} < \infty$$

boils down to $E|Z_1|^{2+\delta} < \infty$ since $\sigma_1 \sigma_{1+h}$ has a lognormal distribution.

For the sequences (X_t) in both parts of Theorem 5.6.3, $\rho_X(h) = \gamma_X(h) = 0$, the covariance matrix of the limiting Gaussian vector is diagonal and, therefore, the limiting vector has independent components. Since the strong mixing property is defined via σ-fields generated by subsequences of (X_t), the mixing rate functions of the sequences $(|X_t|)$ and (X_t^2) are dominated by (ϕ_k) and therefore the same CLT for mixing sequences applies. Thus the results of Theorem 5.6.3 remain valid for the sample ACVF and ACF of the sequences $(|X_t|)$ and (X_t^2): one can everywhere replace X_t by $|X_t|$ (or X_t^2). However, in these cases the limiting Gaussian vectors have dependent components. Moreover, the moment conditions for the sample ACF and ACVF of the squares X_t^2 have to be adjusted correspondingly: $E|X_1 X_{1+h}|^{4+\delta} < \infty$ in the stochastic volatility case and $E|X_1|^{8+\delta} < \infty$ in the GARCH case.

5.7 Stable random vectors

If the condition of finite second moment for the stochastic volatility model/finite fourth moment for the GARCH(p, q) model is not satisfied, classical summation theory for i.i.d. and weakly dependent random variables (see for example Feller [63]) suggests that the limit distributions for the sample autocovariances are infinite variance stable. The theory of infinite variance stable processes is masterfully presented in Samorodnitsky and Taqqu [128].

Recall that the random vector $\mathbf{X}$ and its distribution are said to be α-stable for some $\alpha \in (0, 2]$ if for i.i.d. copies $\mathbf{X}_1, \dots, \mathbf{X}_n$ of $\mathbf{X}$ and any $n \geq 1$ there exists a constant

$\mathbf{a}_n$ such that

$$\mathbf{X}_1 + \cdots + \mathbf{X}_n \stackrel{d}{=} n^{1/\alpha}\mathbf{X} + \mathbf{a}_n \,.$$

Clearly, the case $\alpha = 2$ defines a Gaussian vector. For $\alpha < 2$ the components of $\mathbf{X}$ have infinite variance. Like the Gaussian random vectors, infinite variance stable random vectors have densities which, in general, cannot be written in terms of familiar functions. The best way to summarize the distribution is by means of characteristic functions.

An infinite variance stable vector $\mathbf{X}$ with values in $\mathbb{R}^d$ has characteristic function ([128], Theorem 2.3.1)

$$Ee^{i(\mathbf{x},\mathbf{X})} = \begin{cases} \exp\{-\int_{\mathbb{S}^{d-1}} |(\mathbf{x},\mathbf{y})|^\alpha (1 - i\,\mathrm{sign}((\mathbf{x},\mathbf{y}))\tan(\pi\alpha/2))\Gamma(d\mathbf{y}) + i(\mathbf{x},\mu)\} & \alpha \neq 1, \\ \exp\{-\int_{\mathbb{S}^{d-1}} |(\mathbf{x},\mathbf{y})|(1 + i\,\frac{2}{\pi}\mathrm{sign}((\mathbf{x},\mathbf{y}))\log|(\mathbf{x},\mathbf{y})|)\Gamma(d\mathbf{y}) + i(\mathbf{x},\mu)\} & \alpha = 1. \end{cases}$$

The number $\alpha \in (0, 2)$ (the index of stability), the spectral measure Γ on the unit sphere $\mathbb{S}^{d-1}$ and the location parameter μ uniquely determine the distribution of an infinite variance α-stable random vector $\mathbf{X}$. If $\mathbf{X}$ is α-stable, it is regularly varying with index α and, up to a constant multiple, Γ is the spectral measure appearing in the definition of regular variation.

The stable distributions are the only possible weak limits for (normalized and centered) sums of i.i.d. random vectors, regular variation of the summands with index α being necessary and sufficient for their attraction to an α-stable law; see Rvačeva [127].

An infinite variance α-stable random variable X enjoys a series representation; see Samorodnitsky and Taqqu [128], Chapter 3. In particular,

$$X \stackrel{d}{=} \sum_{k=1}^{\infty} \left(r_k\, \Gamma_k^{-1/\alpha} Y_k - b_k \right), \tag{5.7.1}$$

where (r_k), (Γ_k), (Y_k) are independent, the r_ks are i.i.d. Bernoulli random variables assuming the values ± 1, the Γ_ks are the points of a unit rate homogeneous Poisson process on $(0, \infty)$ and the Y_ks are i.i.d. with $E|Y_1|^\alpha < \infty$. The b_ks are constants which can be chosen as zeros if $\alpha < 1$ or X is symmetric. From this representation it is clear that infinite variance stable variables are functionals of the points of a Poisson process, and therefore it seems natural to use the theory of weak convergence of point processes for studying the asymptotic behavior of the sample autocovariances and sample autocorrelations with infinite variance stable limits. In what follows, we will indicate how one can make this idea work.

5.8 Weak convergence of point processes

5.8.1 A motivating example

In Section 5.4.2 we learned about the equivalence of the regular variation condition on the i.i.d. random vectors $\mathbf{X}_i$, i.e., $nP(x_n^{-1}\mathbf{X}_1 \in \cdot) \stackrel{v}{\to} \mu(\cdot)$ in $\overline{\mathbb{R}}^d \backslash \{\mathbf{0}\}$ with $\mu(t\cdot) = t^{-\alpha}\mu(\cdot)$ for some $\alpha > 0$, all $t > 0$, and the weak convergence of the point processes

$\sum_{i=1}^n \varepsilon_{\mathbf{X}_i/x_n}$ with $P(|\mathbf{X}_1| > x_n) \sim n^{-1}$ to a Poisson random measure PRM(μ). The limiting measure μ in the regular variation condition reappears as the mean measure of the Poisson random measure. The weak convergence of point processes is closely related to the weak convergence of extreme values. One does certainly not exaggerate if one says that modern extreme value theory is nothing but applied point process theory. This claim is supported by the monographs of Leadbetter et al. [94], Resnick [119], Falk et al. [62], and Embrechts et al. [58].

On the other hand, the weak convergence of point processes has also been extensively used in order to derive results of sum-type functionals for the points of the converging point processes, including their sample means, sample autocovariances, and sample autocorrelations. This may sound somewhat paradoxical because point processes count points whereas the mentioned functionals sum points, and it seems that both operations do not have too much in common. However, this method works well when the extremal points in the sample are roughly of the same size as the sum of the points. Early on, Resnick [118] propagated this approach and applied it in joint work with Davis to linear and, more recently, to bilinear processes; see for example [44, 45, 46, 47]. To explain the basic ideas we consider a simple example.

Let (X_i) be i.i.d. positive random variables regularly varying with index $\alpha \in (0, 1)$. It is well known from classical theory for sums of i.i.d. random variables (e.g., Feller [63]) that

$$x_n^{-1}(X_1 + \cdots + X_n) \xrightarrow{d} Y_\alpha \tag{5.8.1}$$

for some positive α-stable random variable Y_α, where (x_n) is a sequence of positive constants satisfying $P(X_1 > x_n) \sim n^{-1}$. We also know (e.g., Resnick [119]; cf. Section 5.4.2) that

$$N_n = \sum_{i=1}^n \varepsilon_{X_i/x_n} \xrightarrow{d} N = \sum_{i=1}^\infty \varepsilon_{\Gamma_i^{-1/\alpha}},$$

where (Γ_i) are the points of a homogeneous Poisson process on $(0, \infty)$, i.e., $\Gamma_i = E_1 + \cdots + E_i$ for an i.i.d. exponential sequence (E_i). It is not difficult to see that $N \sim$ PRM(μ), where $\mu(x, \infty) = cx^{-\alpha}$, $x > 0$, for some $c > 0$. For fixed $\epsilon > 0$ the mapping

$$T_\epsilon(m) = \sum_{i=1}^\infty y_i \, I_{(\epsilon,\infty)}(y_i)$$

acting on the point measures $m = \sum_{i=1}^\infty \varepsilon_{y_i}$ can be shown to be a.s. continuous (in the vague topology underlying the weak convergence of point processes; Resnick [118]). An application of the continuous mapping theorem (e.g., Billingsley [13], Theorem 5.1) yields

$$T_\epsilon(N_n) = x_n^{-1} \sum_{i=1}^n X_i I_{(\epsilon,\infty)}\big(x_n^{-1} X_i\big) \xrightarrow{d} T_\epsilon(N) = \sum_{i=1}^\infty \Gamma_i^{-1/\alpha} I_{(\epsilon,\infty)}\big(\Gamma_i^{-1/\alpha}\big). \tag{5.8.2}$$

Letting $\epsilon \downarrow 0$ the right-hand expression converges a.s. to $\sum_{i=1}^\infty \Gamma_i^{-1/\alpha}$. The series converges since $\Gamma_i/i \to E(\Gamma_1)$ a.s. by the strong law of large numbers. The latter infinite series indeed represents an α-stable random variable Y_α, as indicated in (5.7.1) and proved in Samorodnitsky and Taqqu [128], Chapter 3. Using a Slutsky argument (Theorem 4.2 in Billingsley [13]), it suffices to show that

$$\begin{aligned} x_n^{-1} S_n(\epsilon) &= x_n^{-1} \sum_{i=1}^n X_i - x_n^{-1} \sum_{i=1}^n X_i I_{(\epsilon,\infty)}(x_n^{-1} X_i) \\ &= x_n^{-1} \sum_{i=1}^n X_i I_{(0,\epsilon]}(x_n^{-1} X_i) \end{aligned}$$

makes a negligible contribution to the limit by showing that

$$\lim_{\epsilon \downarrow 0} \limsup_{n \to \infty} E\left[\left(x_n^{-1} S_n(\epsilon)\right)^2\right] = 0.$$

Combining the above arguments, we arrive at

$$x_n^{-1}(X_1 + \cdots + X_n) \stackrel{d}{\to} \sum_{i=1}^\infty \Gamma_i^{-1/\alpha} \stackrel{d}{=} Y_\alpha.$$

The case $\alpha \in [1, 2)$ is more tricky since both sides in (5.8.2) need to be centered if one wants to show that the convergences as $\epsilon \downarrow 0$ combined with $n \to \infty$ work. We refrain from giving details and refer, for example, to Resnick [118].

The classical limit theory for proving (5.8.1) as presented, e.g., in Gnedenko and Kolmogorov [69], Ibragimov and Linnik [84], Feller [63], or Petrov [115] heavily depends on the manipulation of characteristic functions. Classical summation theory and characteristic functions fit nicely because the characteristic function of the sum of independent random variables factorizes into the characteristic functions of the summands. In contrast to the latter method, we used two other powerful tools, the weak convergence of point processes in combination with the continuous mapping theorem. The point process approach has its limitations when Gaussian limits appear, but it is well adjusted to the very heavy tail case when the summands have infinite variance. Or in other words, the point process approach works fine when the extremes in a sample are roughly of the same order as the aggregated sum of the sample. For dependent data, the characteristic function approach to the weak limits of sums becomes more tedious since the characteristic function of the sum does not factorize anymore. The point process approach still works for sums of dependent stationary random variables with infinite variance, as was shown by Davis and Hsing [40] under quite general dependence assumptions. Their approach is also the basis for the results presented in the sequel.

5.8.2 The GARCH case

Recall the definition of a GARCH(p, q) process (X_t) from Section 5.2.3. Since we aim at a weak limit theory for the sample autocovariances and sample autocorrelations of a GARCH process, its absolute values, squares, etc., we need to consider point processes whose points are given by the $(m+1)$-dimensional vectors

$$\mathbf{X}_t(m) = (X_t, \ldots, X_{t+m}), \quad m \geq 0.$$

The motivation for this choice will become more transparent soon. The following result is based on work by Davis and Mikosch [41] for the SRE case; it appears in Mikosch and Stărică [107] for the GARCH(1, 1) case and in Basrak et al. [8] for the general GARCH(p, q) case as the fundamental point process convergence result.

Recall from Corollary 5.5.9 that the finite-dimensional distributions of a GARCH process are regularly varying with index $\kappa > 0$, under general conditions on the noise and the parameters of the process. The vector $\mathbf{X}_t(m)$ and its norm $|\mathbf{X}_t(m)|$ are then regularly varying with index κ. In particular,

$$P(|\mathbf{X}_t(m)| > x) \sim cx^{-\kappa} \quad \text{for some } c > 0.$$

Define a sequence (x_n) of positive numbers such that

$$P(|\mathbf{X}_1(m)| > x_n) \sim n^{-1}.$$

Then, clearly, $x_n \sim c' n^{1/\kappa}$ for some $c' > 0$.

Theorem 5.8.1 *Assume the conditions of Corollary 5.5.9 hold and that $\kappa/2$ is not an even integer. Then the following point process convergence holds:*

$$N_n = \sum_{t=1}^{n} \varepsilon_{\mathbf{X}_t(m)/x_n} \overset{d}{\to} N = \sum_{i=1}^{\infty} \sum_{j=1}^{\infty} \varepsilon_{\Gamma_i^{-1/\kappa} \mathbf{Q}_{ij}}. \tag{5.8.3}$$

Here $\overset{d}{\to}$ denotes convergence in distribution of point processes with state space $\overline{\mathbb{R}}^{m+1} \backslash \{\mathbf{0}\}$, (Γ_i) are the points of a homogeneous Poisson process on $(0, \infty)$, with intensity $\theta > 0$, which is independent of the i.i.d. point processes $\sum_{j=1}^{\infty} \varepsilon_{\mathbf{Q}_{ij}}$, $i \geq 1$. The points $\mathbf{Q}_{ij}$ satisfy $\sup_j |\mathbf{Q}_{ij}| = 1$. Their distribution is described in [41].

The intensity θ is nothing but the extremal index of the sequence $(|\mathbf{X}_t(m)|)$ mentioned in Section 5.1.2; see [41]. To make the latter theorem more accessible it pays to go back to the original papers by Davis and Hsing [40] and Davis and Mikosch [41]. There two assumptions were needed (which are hidden in the formulation of Corollary 5.5.9 and Theorem 5.6.1):

- The finite-dimensional distributions of (X_t) are regularly varying with positive index $\kappa > 0$.
- (X_t) satisfies a mild mixing condition that is satisfied for strong mixing.

The structure of the limiting point process is closely related to the regular variation condition on $\mathbf{X}_1(m)$: for $t > 0$,

$$u^{\kappa} P(|\mathbf{X}_1(m)| > tu, \quad \mathbf{X}_1(m)/|\mathbf{X}_1(m)| \in \cdot) \overset{v}{\to} ct^{-\kappa} P(\mathbf{\Theta} \in \cdot), \quad \text{as } u \to \infty. \tag{5.8.4}$$

This means that, in an asymptotic sense, the radial and the spherical parts of the vector $\mathbf{X}_1(m)$ behave independently provided $|\mathbf{X}_1(m)|$ is sufficiently far away from the origin. (This fact is nicely illustrated in Example 5.4.6.) The radial part in the regular variation condition (5.8.4) translates into the point process of the radii $\Gamma_i^{-1/\kappa}$, which is independent of the points $\mathbf{Q}_{ij}$. The latter describe the spherical distribution of the limiting points in space. The distribution of the $\mathbf{Q}_{ij}$ is determined by the spectral measure $P(\boldsymbol{\Theta} \in \cdot)$ but it is not very transparent: in Davis and Mikosch [41] it is given in implicit form via Laplace functionals.

According to the strong law of large numbers the points $\Gamma_i^{-1/\kappa}$ converge to zero as $i \to \infty$. Hence there is an infinite number of balls with random radii $\Gamma_i^{-1/\kappa}$ in any neighborhood of the origin. In these balls we can find the random points $\Gamma_i^{-1/\kappa}\mathbf{Q}_{ij}$. See Figure 5.19 for an illustration.

For the investigation of the sample ACF it will be important to study the lagged vectors

$$x_n^{-2}\mathbf{Y}_t(m) = x_n^{-2}\big(X_t^2, X_t X_{t+1}, \ldots, X_t X_{t+m}\big). \tag{5.8.5}$$

A continuous mapping argument and the point process convergence of Theorem 5.8.1 show that the processes with points consisting of the lagged vectors of these products also converge weakly to a point process whose components are generated in exactly the same way as the products in (5.8.5):

$$\mathbf{Z}_{ij} = \Gamma_i^{-2/\kappa}\big(\big[Q_{ij}^{(0)}\big]^2, Q_{ij}^{(0)}Q_{ij}^{(1)}, \ldots, Q_{ij}^{(0)}Q_{ij}^{(m)}\big).$$

Corollary 5.8.2 *Under the assumptions of Theorem 5.8.1,*

$$\sum_{t=1}^{n} \varepsilon_{\mathbf{Y}_t(m)/x_n^2} \xrightarrow{d} \sum_{i=1}^{\infty}\sum_{j=1}^{\infty} \varepsilon_{\mathbf{Z}_{ij}}\,. \tag{5.8.6}$$

Since sample autocovariances and sample autocorrelations are built from the products $X_t X_{t+h}$, Corollary 5.8.2 turns out to be the crucial tool for proving weak convergence of these statistical tools; see Section 5.9.1 for details.

5.8.3 The stochastic volatility case

We consider the stochastic volatility model $X_t = \sigma_t Z_t$ with i.i.d. symmetric noise (Z_t) satisfying

$$P(Z_t > x) \sim 0.5\, d x^{-\alpha} \quad \text{for some } \alpha, d > 0. \tag{5.8.7}$$

As before we assume that $(\log \sigma_t)$ is a linear process with i.i.d. standard normal noise (η_j):

$$\log \sigma_t = \sum_{j=0}^{\infty} \psi_j \eta_{t-j}, \quad t \in \mathbb{Z}.$$

The only condition needed is square summability of (ψ_j) ensuring the a.s. existence of the infinite series above.

It was mentioned in Example 5.4.12 that the (symmetric) random variable $Z_1 Z_2$ is regularly varying:

$$P(Z_1 Z_2 > x) \sim c x^{-\alpha} \log x \quad \text{for some } c > 0. \tag{5.8.8}$$

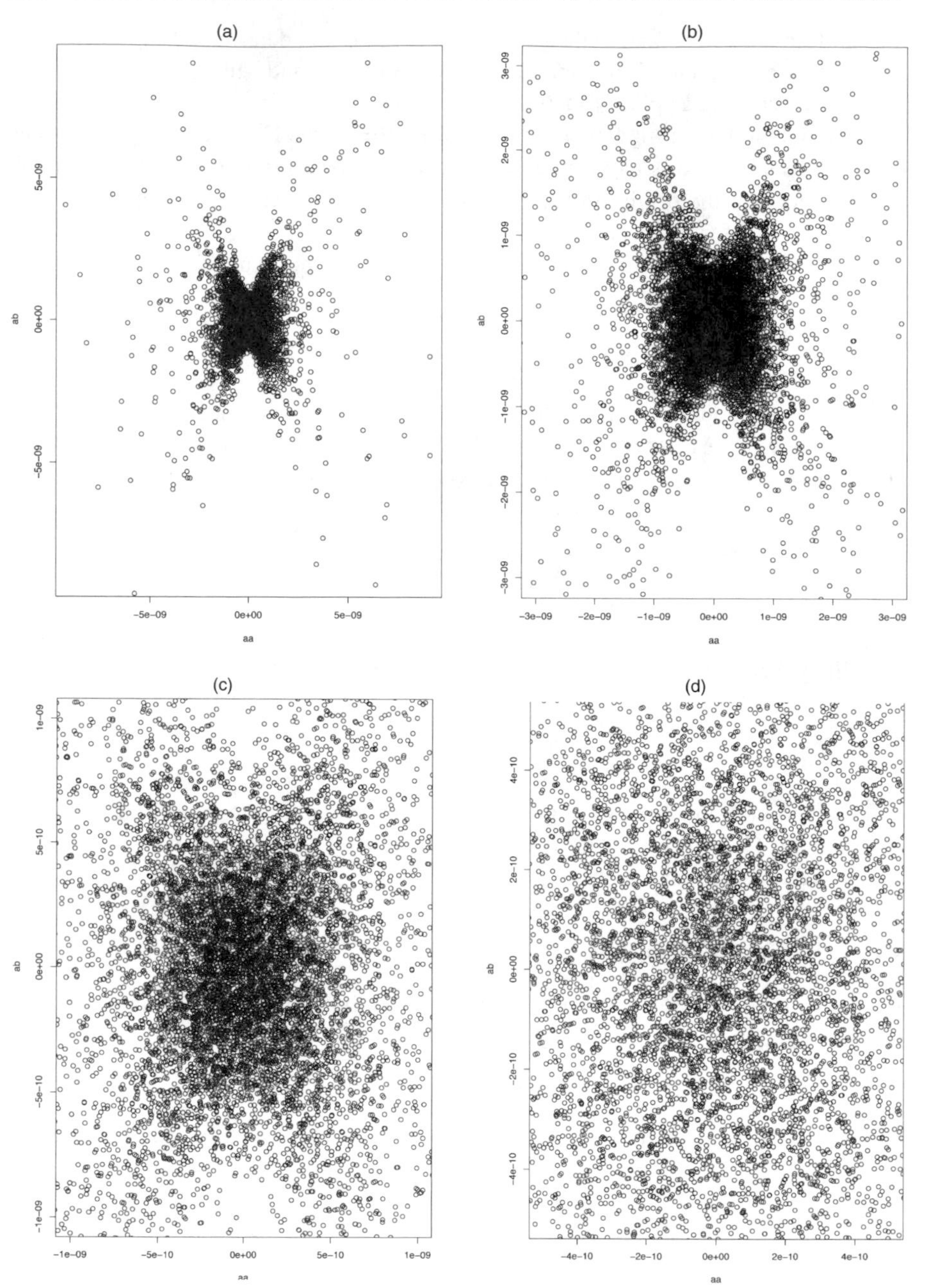

Figure 5.19 *An attempt to visualize the limiting point process in Theorem 5.8.1 on different scales for the* ARCH(1) *process* $X_t = (0.001 + X_{t-1}^2)^{0.5} Z_t$ *with i.i.d. standard normal noise* (Z_t). *Then* $P(X_t > x) \sim c\,x^{-2}$; *see Example 5.5.2. We plot the scaled pairs* $(X_t, X_{t+1})/\sqrt{n}$ *for* $n = 10,000$ (top left). *In the remaining graphs we zoom in this figure to smaller scales around the origin. The shape of the limiting points on larger scale* (top graphs) *reminds one of the form of a butterfly.*

An application of Breiman's formula (5.4.18) implies that the random variables $X_1 X_h$ are regularly varying with index α:

$$\begin{cases} P(X_1X_h > x) &= P(Z_1Z_2\sigma_1\sigma_h > x) \sim E[\sigma_1\sigma_h]^\alpha \; P(Z_1Z_2 > x), \\ P(X_1X_h \le -x) &= P(Z_1Z_2\sigma_1\sigma_h \le -x) \sim E[\sigma_1\sigma_h]^\alpha \; P(Z_1Z_2 \le -x). \end{cases} \tag{5.8.9}$$

We introduce two normalizing sequences (a_n) and (b_n):

$$P(|Z_1| > a_n) \sim n^{-1} \quad \text{and} \quad P(|Z_1Z_2| > b_n) \sim n^{-1}.$$

By (5.8.7) and (5.8.8) it is clear that there exist positive constants c' and c'' such that $a_n \sim c' n^{1/\alpha}$ and $b_n \sim c''(n \log n)^{1/\alpha}$.

As for the GARCH model the point processes constructed from the points

$$\mathbf{Y}_{n,t}(m) = \left(a_n^{-1}X_t, b_n^{-1}X_tX_{t+1}, \ldots, b_n^{-1}X_tX_{t+m}\right)$$

can be shown to have weak limits. This is the content of the next result, which will become the basis for studying the asymptotic behavior of the sample ACF for the stochastic volatility process in Section 5.9.2.

Theorem 5.8.3 (Davis and Mikosch [42]) *Under the above assumptions on the noise (Z_t) and the volatility sequence (σ_t),*

$$N_n = \sum_{t=1}^{n} \varepsilon_{\mathbf{Y}_{n,t}(m)} \xrightarrow{d} N = \sum_{i=0}^{m} \sum_{k=1}^{\infty} \varepsilon_{r_{k,i}\Gamma_{k,i}^{-1/\alpha}\mathbf{e}_i}, \tag{5.8.10}$$

where $\xrightarrow{d}$ denotes convergence in distribution of point processes with state space $\overline{\mathbb{R}}^{m+1}\backslash\{\mathbf{0}\}$, and $\mathbf{e}_i$ is the ith unit vector in $\mathbb{R}^{m+1}$. The sequences of points $(\Gamma_{k,i})$, $i = 0, 1, \ldots, m$ constitute independent homogeneous Poisson processes on $(0, \infty)$, usually with different intensities, depending on the dependence structure of (σ_t). The random variables $r_{k,i}$ are i.i.d. symmetric Bernoulli random variables with values ± 1, independent of $(\Gamma_{k,i})$.

Notice the crucial difference between the latter result and its analog in the GARCH case (Corollary 5.8.2): the limiting points of the products of the stochastic volatility process are concentrated along the axes (see Figure 5.20), whereas the limiting points of the products of the GARCH process are spread everywhere in the state space. The superposition of independent PRMs (interpreted as PRM with state space $\overline{\mathbb{R}}^{m+1}\backslash\{\mathbf{0}\}$) is a PRM whose mean measure μ is just the sum of the individual mean measures. Thus (N_n) has the same weak limit N as the processes $\widetilde{N}_n = \sum_{i=1}^{n} \varepsilon_{\mathbf{X}_i/x_n}$ with i.i.d. $\mathbb{R}^{m+1}$-valued points $\mathbf{X}_i$ with independent components and a regularly varying distribution satisfying $nP(x_n^{-1}\mathbf{X}_1 \in \cdot) \xrightarrow{v} \mu(\cdot)$. Thus, although the dependence between the σ_ts can be arbitrarily strong in terms of autocorrelations or mixing coefficients, the limiting PRM in (5.8.10) is very much the same as if the components of the vectors $\mathbf{Y}_{n,t}(m)$, i.e., the products X_tX_{t+i}, were independent.

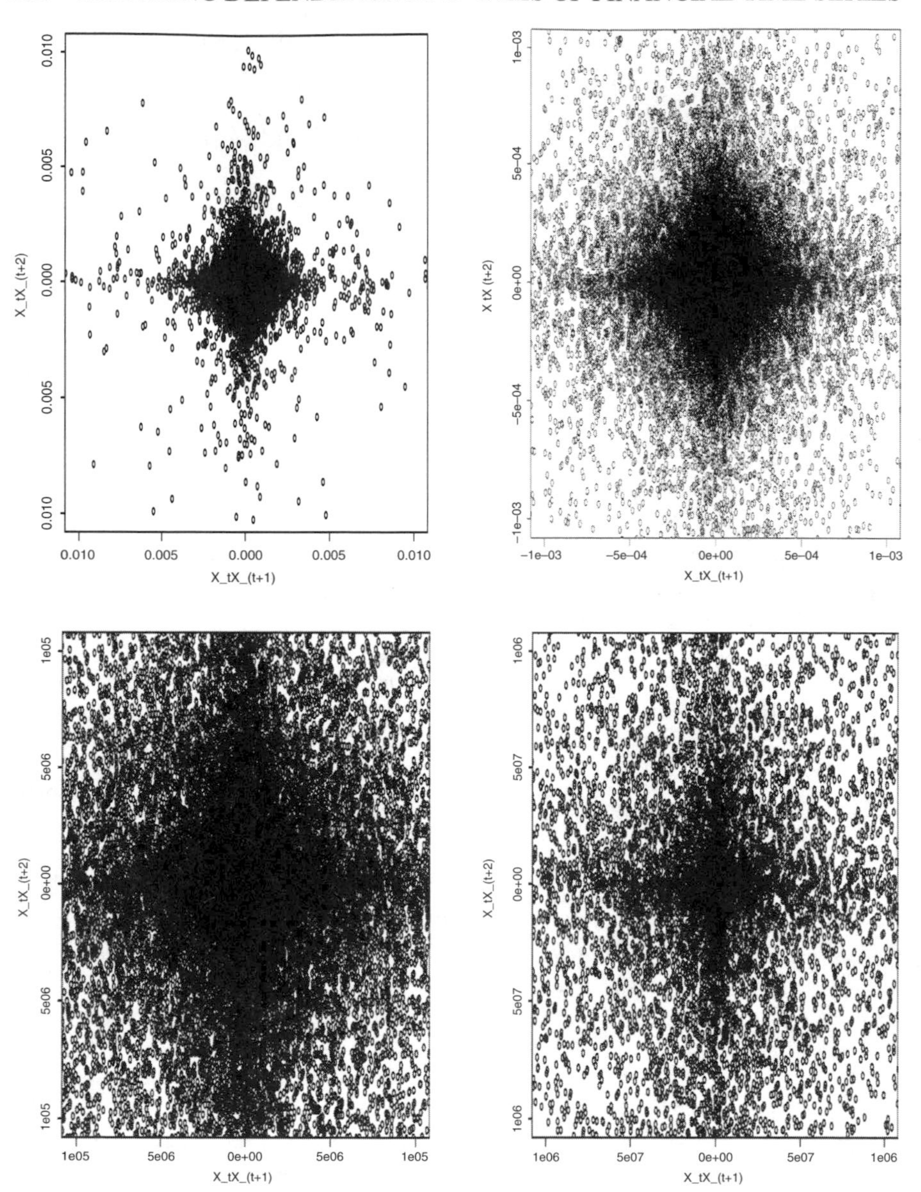

Figure 5.20 *An attempt to visualize the limiting point process in Theorem 5.8.3 on different scales for the stochastic volatility model* $X_t = e^{Y_t} Z_t$, $Y_t = 0.9Y_{t-1} + \eta_t$, *for independent noise sequences* (η_t) *and* (Z_t). *The sequence* (η_t) *is i.i.d. standard Gaussian, the sequence* (Z_t) *is i.i.d. symmetric regularly varying with tail index 2, resulting in regular variation of the products* $X_t X_{t+h}$ *with index 2. We plot the pairs of scaled products* $(X_t X_{t+1}, X_t X_{t+2})/\sqrt{n \log n}$ *for* $n = 100,000$ (top left) *and zoom in to smaller scale in the remaining graphs. The shape of the cloud of limiting points is in agreement with the theoretical result: the limiting points are concentrated along the axes.*

5.9 Asymptotic behavior of the sample autocorrelations

In this section we are interested in the situation when standard central limit theory for the ACVF and ACF as described in Theorem 5.6.3 breaks down, i.e., when the marginal distribution of the time series is very heavy-tailed. Then other techniques are needed to study their asymptotic behavior. It will turn out that point process techniques are, then, the right tools to be used.

For convenience, in order to illustrate the method, we define the modified sample autocovariance $\widetilde{\gamma}_{n,X}(h)$ and the modified sample autocorrelation $\widetilde{\rho}_{n,X}(h)$ at lag h of (X_n) without the usual centering by the sample mean $\overline{X}_n$:

$$\widetilde{\gamma}_{n,X}(h) = \frac{1}{n}\sum_{t=1}^{n} X_t X_{t+h} \quad \text{and} \quad \widetilde{\rho}_{n,X}(h) = \frac{\widetilde{\gamma}_{n,X}(h)}{\widetilde{\gamma}_{n,X}(0)}\,.$$

Since we are only interested in asymptotic results, the change of the summation limit from $t = n - h$ to $t = n$ is inessential.

The sample autocovariances and sample autocorrelations of a stationary sequence (X_t) are functionals of sum-type of products $X_t X_{t+h}$. It is therefore an attractive idea to use the point process convergence results of Section 5.8 in order to derive suitable weak limits of the (normalized) sums $\sum_{t=1}^{n} X_t X_{t+h}$ by summing up the points in the limiting point processes for the $(X_t X_{t+h})$ sequence. This approach is similar to the one for the sample mean, as explained in Section 5.8.1. This continuous mapping argument can really be shown to work in various situations. One of the first successful applications of this method can be found in Resnick [118] who showed that point process convergence can be used to derive infinite variance stable limits for sums of i.i.d. random variables. Using similar point process techniques, Davis and Resnick [44, 45, 46] studied the weak limit behavior of sums and sample autocorrelations of linear processes $X_t = \sum_j \psi_j Z_{t-j}$ with i.i.d. centered regularly varying noise (Z_t) with index $\alpha < 2$. They showed the surprising result that the sample autocorrelation of an infinite variance linear process at lag h estimates the quantity $\sum_j \psi_j \psi_{j+h} / \sum_j \psi_j^2$ which, in the case of a finite variance linear process (X_t), represents the autocorrelation $\mathrm{cov}(X_0, X_h)$. Thus, although the ACF at lag h is not defined anymore — because of infinite variance marginals — its sample analogue still estimates a quantity that can be interpreted as population autocorrelation. For a long time these results have been considered as a justification for the belief that, in principle, there is no problem with the sample ACF for heavy-tailed time series.

However, once one leaves the class of linear processes the sample autocorrelations can behave in a very unpredictable way. For example, Davis and Resnick [47], Resnick [120], Basrak et al. [6], and Resnick and van den Berg [122] study various pitfalls of sample autocorrelation behavior for bilinear and other nonlinear time series models with very heavy tails. In particular, if such a process does not have sufficiently high moments the sample autocorrelations can have nondegenerate weak limits. Thus the sample ACF would not estimate anything reasonable in this case. A similar statement can be shown to be true for GARCH processes, whereas the sample ACF of stochastic volatility models behaves in a way that is very similar to the linear process case mentioned above. Davis and Mikosch [43] give a comparative study on sample ACF behavior for linear, GARCH, and stochastic volatility processes with heavy tails.

5.9.1 The GARCH case

The basic point process idea for dealing with the sample autocorrelations of heavy-tailed processes has already been explained above. Now recall the basic point process result of (5.8.6) for the products $X_t X_{t+h}$ of a GARCH process (X_t):

$$\sum_{t=1}^{n} \varepsilon_{x_n^{-2}(X_t^2, X_t X_{t+1}, \ldots, X_t X_{t+m})} \stackrel{d}{\rightarrow} \sum_{i=1}^{\infty} \sum_{j=1}^{\infty} \varepsilon_{\Gamma_i^{-2/\kappa}\left(\left[Q_{ij}^{(0)}\right]^2, Q_{ij}^{(0)} Q_{ij}^{(1)}, \ldots, Q_{ij}^{(0)} Q_{ij}^{(m)}\right)}.$$

For a fixed h we sum up the points in the point processes on both sides of this limiting relation and hope that we can achieve a convergence result of type

$$[n x_n^{-2}]\, \widetilde{\gamma}_{n,X}(h) = x_n^{-2} \sum_{t=1}^{n} X_t X_{t+h} \stackrel{d}{\rightarrow} \sum_{i=1}^{\infty} \Gamma_i^{-2/\kappa} \sum_{j=1}^{\infty} Q_{ij}^{(0)} Q_{ij}^{(h)}.$$

Compare the right-hand side with the series representation (5.7.1) of a $\kappa/2$-stable random variable. Indeed, calculating, for example, the Laplace transform of the right-hand infinite series, the latter can be shown to represent a $\kappa/2$-stable random variable; see Davis and Hsing [40]. However, this method only works for values $\kappa < 2$. If $\kappa > 2$, $E X_0 X_h$ exists and is finite, i.e., the autocorrelation at lag h is well defined. Then one has to center $\widetilde{\gamma}_{n,X}(h)$ with $E X_0 X_h$, i.e., one has to consider $n x_n^{-2}(\widetilde{\gamma}_{n,X}(h) - E X_0 X_h)$ in order to achieve a weak convergence result.

The results are summarized in the following theorem, which was proved in Davis and Mikosch [41] for the SRE and ARCH(1) cases, for the GARCH(1,1) case in Mikosch and Stărică [107] and can be found for the general GARCH(p, q) case in Basrak et al. [8]. Recall that the normalizing constants x_n are of the asymptotic order $n^{1/\kappa}$, which explains the order of the normalization in the following result.

Theorem 5.9.1 *Assume that* (X_t) *is a* GARCH(p, q) *process with regularly varying finite-dimensional distributions of index* $\kappa > 0$ *satisfying the conditions*

(1) If $\kappa \in (0, 2)$, *then*

$$\left(n^{1-2/\kappa} \gamma_{n,X}(h)\right)_{h=0,\ldots,m} \stackrel{d}{\rightarrow} (V_h)_{h=0,\ldots,m},$$

$$\left(\rho_{n,X}(h)\right)_{h=1,\ldots,m} \stackrel{d}{\rightarrow} (V_h / V_0)_{h=1,\ldots,m},$$

where the vector $(V_0, \ldots, V_m)$ *is jointly* $\kappa/2$*-stable in* $\mathbb{R}^{m+1}$.

(2) If $\kappa \in (2, 4)$ *then*

$$\left(n^{1-2/\kappa}(\gamma_{n,X}(h) - \gamma_X(h))\right)_{h=0,\ldots,m} \stackrel{d}{\rightarrow} (V_h)_{h=0,\ldots,m},$$

$$\left(n^{1-2/\kappa}(\rho_{n,X}(h) - \rho_X(h)\right)_{h=1,\ldots,m} \stackrel{d}{\rightarrow} \gamma_X^{-1}(0)(V_h - \rho_X(h)\, V_0)_{h=1,\ldots,m},$$

where $(V_0, \ldots, V_m)$ *is jointly* $\kappa/2$*-stable in* $\mathbb{R}^{m+1}$.

The limiting random vectors can be expressed in terms of the Γ_i and $\mathbf{Q}_{ij}$ used for the formulation of Theorem 5.8.1. For more details, see [41], Theorem 3.5. Notice that $\gamma_X(h) = \rho_X(h) = 0$, $h \geq 1$, for the GARCH(p, q) model.

The conclusions of Theorem 5.9.1 remain valid for other functions of (X_t) including linear combinations of powers of the components, in particular for $(|X_t|)$ and (X_t^2). In the latter case the moment conditions have to be modified: $\kappa \in (0, 4)$ in part (1) and $\kappa \in (4, 8)$ in part (2). Notice that $\gamma_{|X|}(h) \neq 0$ and $\gamma_{X^2}(h) \neq 0$ (when they are well defined) and therefore centering in the limit relations of part (2) is needed; in part (1) centering of the sample ACVFs and ACVs $\gamma_{n,|X|}, \gamma_{n,X^2}, \rho_{n,|X|}, \rho_{n,X^2}$ is not needed. These statements follow from the fact that $(|X_t|)$ and (X_t^2) inherit the strong mixing property with the same (geometric) rate function (ϕ_k) as well as joint regular variation of the finite-dimensional distributions from the (X_t) process, and point process convergence follows from the continuous mapping theorem.

The case $\kappa > 4$ is covered by the CLT in part (2) of Theorem 5.6.3 with rate $\sqrt{n}$. It is worthwhile to compare the rates in both theorems. If $\kappa \in (2, 4)$ the convergence of the sample ACF to the ACF can be very slow, the slower the closer κ to 2. This is surprising if one recalls the corresponding result for linear processes (Davis and Resnick [46]), where the rate of convergence of the sample ACF is the faster the smaller the index of regular variation of X_t. Moreover, for linear processes the sample ACF never converges to a nondegenerate weak limit, but this is the case for the GARCH process when $\kappa \in (0, 2)$.

The results of Theorem 5.9.1 are not of great practical value, but more of a philosophical nature. They tell us that, if we were to model log-returns by a GARCH process with infinite fourth moment — a situation not untypical for many real-life data — we could not use the standard central limit theory to construct asymptotic confidence bands for the sample ACF. In particular, the use of the sample ACF for squared returns would suffer from the lack of moments. A consequence would be that rates of convergence in these results are extremely slow. Therefore, and because the limiting distributions are very heavy-tailed, asymptotic confidence bands would be much wider than the classical $1/\sqrt{n}$-bands. In particular, they could be wider than the estimated autocorrelations. We do not know much about the limit distributions, in particular we do not know all parameters of the stable distributions involved, and we know little about the spectral measure of the limiting stable vector. These facts make the asymptotic theory for the sample ACF not very pleasant because it is dependent on Monte-Carlo simulations: one would have to derive asymptotic confidence bands for the ACF depending on the choice of GARCH parameters. We refer to Figure 5.21 and Figure 5.22 in order to get some impression of the asymptotic confidence bands one has to expect in this case and for comparison with the stochastic volatility model.

5.9.2 The stochastic volatility case

We consider the same stochastic volatility model (X_t) as in Section 5.8.3. The continuous mapping approach to the converging point processes also works in this situation.

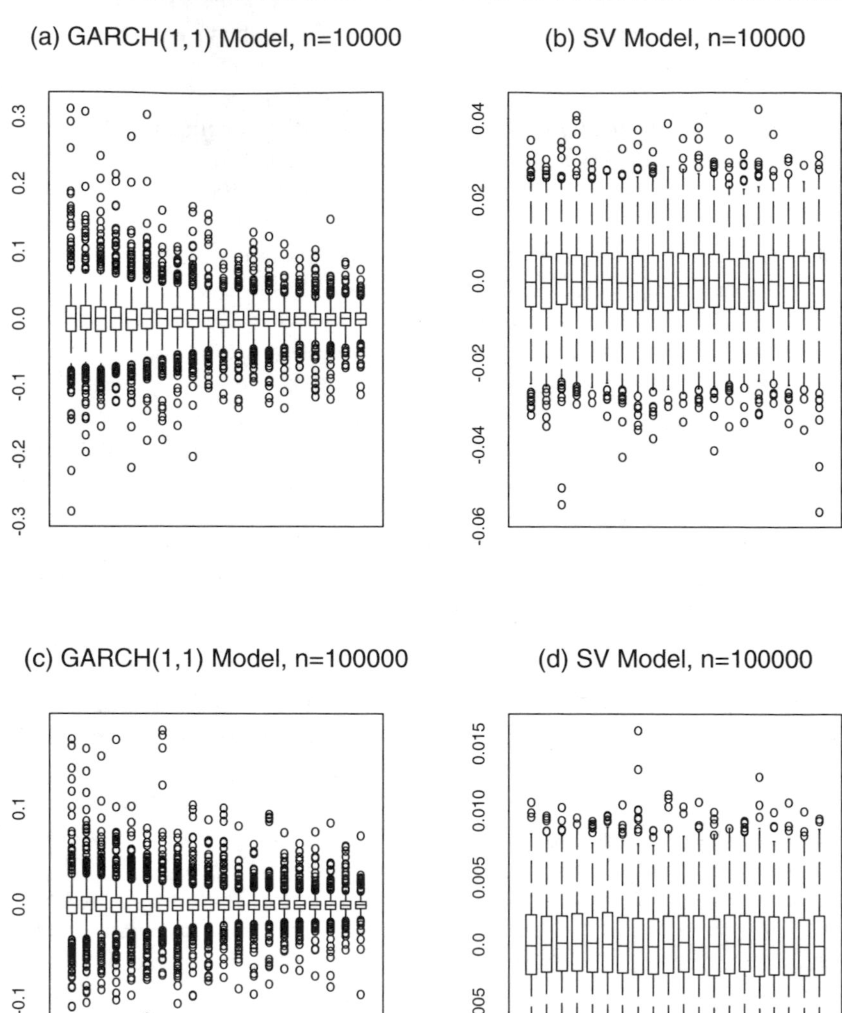

Figure 5.21 *Boxplot comparison of the asymptotic distributional behavior of the sample ACF for a GARCH* (left) *and a stochastic volatility model* (right). *Boxplots are generated for the first* 20 *lags of the sample ACF. At each lag, the boxplot describes the distribution of the sample autocorrelation from* 1000 *independent* GARCH(1, 1) *samples. The parameters and noise distributions are chosen in such a way that both time series have tail index* 3, *which is not untypical for return series. Despite the large (and highly unrealistic) sample sizes* $n = 10{,}000$ *and* $n = 100{,}000$ *the distribution of the sample ACF of the GARCH model is unusually wide and hardly changes when the sample size is increased, although the sample ACF is consistent in this case. For the stochastic volatility model the sample ACF is* $\sqrt{n}$*-consistent.*

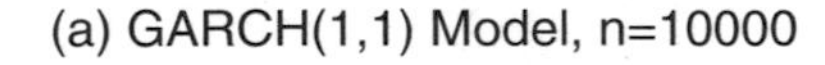

(b) SV Model, n=10000

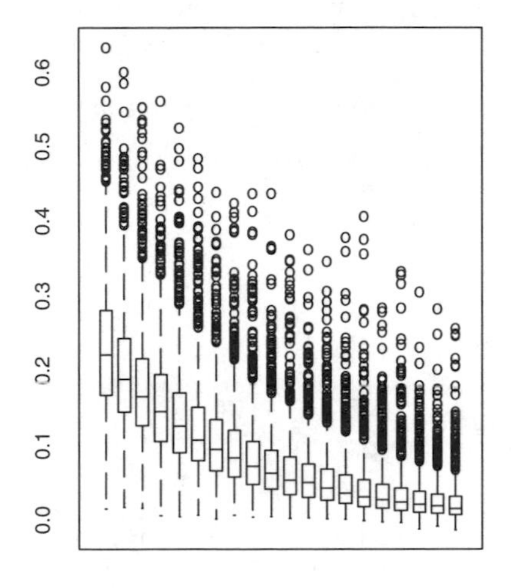

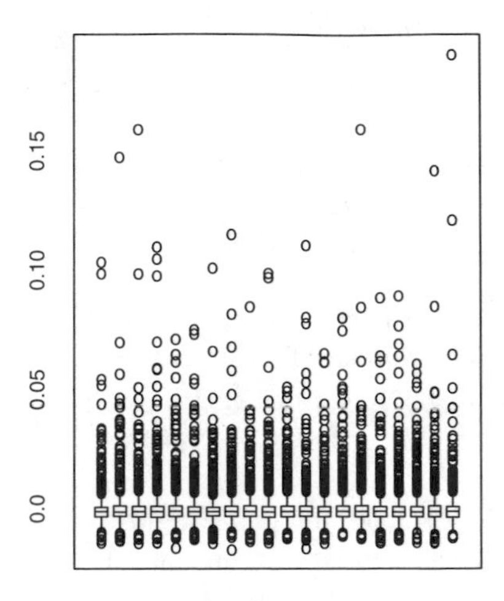

(c) GARCH(1,1) Model, n=100000

0.0
0.1
0.2
0.3
0.4
0.5
0.6

(d) SV Model, n=100000

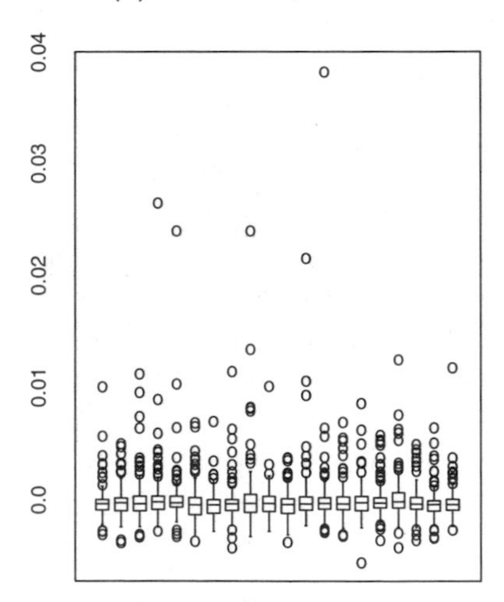

Figure 5.22 *Boxplot comparison of the asymptotic distributional behavior of the sample ACF for the squares of the GARCH and the stochastic volatility models from Figure 5.21. Since the tail index of the model is* 3 *in both cases, it is* 1.5 *for the squared time series. The sample autocorrelations of the squared GARCH process have nondegenerate weak limits in this case. On the other hand, the rate of convergence to zero of the sample ACF for the squared stochastic volatility model is faster than the classical* $\sqrt{n}$*-rates.*

Recall from Theorem 5.8.3 the limit relation

$$N_n = \sum_{t=1}^{n} \varepsilon_{(a_n^{-1} X_t, b_n^{-1} X_t X_{t+1}, \ldots, b_n^{-1} X_t X_{t+m})} \xrightarrow{d} N = \sum_{i=0}^{m} \sum_{k=1}^{\infty} \varepsilon_{r_{k,i} \Gamma_{k,i}^{-1/\alpha} \mathbf{e}_i},$$

where $a_n \sim c n^{1/\alpha}$ and $b_n \sim c'(n \log n)^{1/\alpha}$. For $h = 0$ a continuous mapping argument suggests that

$$[n a_n^{-2}]\, \widetilde{\gamma}_{n,X}(0) = a_n^{-2} \sum_{t=1}^{n} X_t^2 \xrightarrow{d} \sum_{k=1}^{\infty} \Gamma_{k,0}^{-2/\alpha}.$$

The right-hand series represents a positive $\alpha/2$-stable random variable for $\alpha < 2$; see (5.7.1). Similarly, for fixed $h \geq 1$, it seems plausible that

$$[n b_n^{-1}]\, \widetilde{\gamma}_{n,X}(h) = b_n^{-1} \sum_{t=1}^{n} X_t X_{t+h} \xrightarrow{d} \sum_{k=1}^{\infty} r_{k,h} \Gamma_{k,h}^{-1/\alpha}.$$

The α-stable limit on the right-hand side exists for $\alpha < 2$.

These heuristic arguments can be made to work even under weaker assumptions on the distribution of the noise Z_t than required here, see [42].

Theorem 5.9.2 (Davis and Mikosch [42]) *Assume (X_t) satisfies the conditions of Theorem 5.8.3. If $\alpha \in (0, 2)$, then*

$$n \left(n^{-2/\alpha} \gamma_{n,X}(0), (n \log n)^{-1/\alpha} \gamma_{n,X}(1), \ldots, (n \log n)^{-1/\alpha} \gamma_{n,X}(m) \right) \xrightarrow{d} (V_h)_{h=0,\ldots,m},$$

where $V_0, \ldots, V_m$ are independent random variables, V_0 is $\alpha/2$-stable and V_h, $h = 1, \ldots, m$, is α-stable. In addition,

$$((n/\log n)^{1/\alpha} \rho_{n,X}(h))_{h=1,\ldots,m} \xrightarrow{d} (V_h / V_0)_{h=1,\ldots,m}\,. \tag{5.9.1}$$

Notice that in the stochastic volatility case the normalization for the sample autocovariance $\gamma_{n,X}(0)$ has to be stronger than for the other sample autocovariances $\gamma_{n,X}(h)$, $h > 0$. This implies that the sample ACF $\rho_{n,X}(h)$ at lag h converges to zero at a rate that is faster the smaller α; the normalization $a_n^2/b_n \sim \widetilde{c}\,(n/\log n)^{1/\alpha}$ in (5.9.1) compares favorably to the classical $\sqrt{n}$-consistency results for the sample ACF. Not only the result, but also the proof of it, are very close to the linear process case; see Davis and Resnick [46]. However, one always estimates zero in the case of a stochastic volatility process in contrast to a linear process that is distinct from an i.i.d. sequence.

The discussion at the end of Davis and Mikosch [42] also indicates that the sample ACFs of the absolute and squared returns of a stochastic volatility model behave in a very similar way. This means that the sample ACFs of the absolute and squared returns quickly converge to zero. This can be seen in Figure 5.21 and Figure 5.22, where we compare the large sample behavior of the sample ACF for GARCH and stochastic volatility models.

5.10 Do GARCH and stochastic volatility models explain the stylized facts of log-returns?

After this long theoretical excursion one may want to look back at the starting point of our discussion in Section 5.1. There we discussed some of the typical properties of log-returns. Now it is worthwhile to ask whether the above theory can explain any of the stylized facts.

5.10.1 A case study for GARCH

Mikosch and Stărică [107] discuss the extremal properties of GARCH processes, their sample ACF behavior and the use of GARCH modeling for tick-by-tick return data. We include a part of their discussion here. We focus on a series of 70,000 30-minute foreign exchange (FX) rate data of Japanese Yen (JPY) against U.S. Dollar (USD) between 1992 and 1996, kindly provided by Olsen and Associates (Zürich). Since tick-by-tick data are measured at unevenly spaced instants of time, the time scale has been transformed into so-called Olsen's θ-time. The basic idea of this transformation is to avoid seasonalities and to stabilize the flow of return data over time. In a mathematical sense, the aim of this time transformation is to construct a stationary time series. We refer to the recent book by Dacorogna et al. [35] on high density financial data; Shiryaev [130] gives a mathematical explanation for θ-time.

Using quasi-maximum likelihood estimation (see Section 5.2.3), a GARCH(1, 1) process with the following parameters was fitted to the FX data:

$$\alpha_0 = 10^{-7}, \quad \alpha_1 = 0.11, \quad \beta_1 = 0.88. \tag{5.10.1}$$

The estimated value $\varphi_1 = \alpha_1 + \beta_1 = 0.99$ suggests the use of an IGARCH process (Example 5.3.7). Values φ_1 close to one have been frequently observed for longer return series; see, for example, Engle and Bollerslev [61], Baillie and Bollerslev [2], or Guillaume et al. [75].

The residuals

Given the estimated parameters $\widehat{\alpha}_0$, $\widehat{\alpha}_1$, and $\widehat{\beta}_1$ of a GARCH(1, 1) process one can calculate the residuals

$$\widehat{Z}_t = X_t/\widehat{\sigma}_t, \quad \text{where} \quad \widehat{\sigma}_t^2 = \widehat{\alpha}_0 + \widehat{\alpha}_1 \widehat{\sigma}_{t-1}^2 + \widehat{\beta}_1 X_{t-1}^2.$$

Clearly, one has to choose initial values $\widehat{\sigma}_0$ and X_0. Empirical evidence from GARCH(1, 1) simulations shows that the values $\widehat{\sigma}_t$ for large t are quite insensitive to the initial value.

The bottom graph in Figure 5.23 presents the residuals of the FX log-returns with parameters (5.10.1) and their density. The latter shows that the distribution is roughly symmetric and very much peaked around zero (leptokurtic). A comparison of the graphs in Figure 5.23 shows that the residuals are much less volatile than the original FX series. The residuals are nicely fitted by a student distribution with (roughly) four degrees of freedom, both in its center and in the tails; see Figure 5.24. The tails of a student random variable S with four degrees of freedom are Pareto-like: $P(S > x) \sim c\,x^{-4}$. In the literature, student and generalized Pareto distributions were fitted to the residuals; see for example Baillie and Bollerslev [2] or Frey and McNeil [64].

The tails

The discussion in Example 5.5.6 suggests to calculate the tail index κ corresponding to a GARCH(1, 1) process as the solution to the equation $E(\alpha_1 Z_1^2 + \beta_1)^{\kappa/2} = 1$.

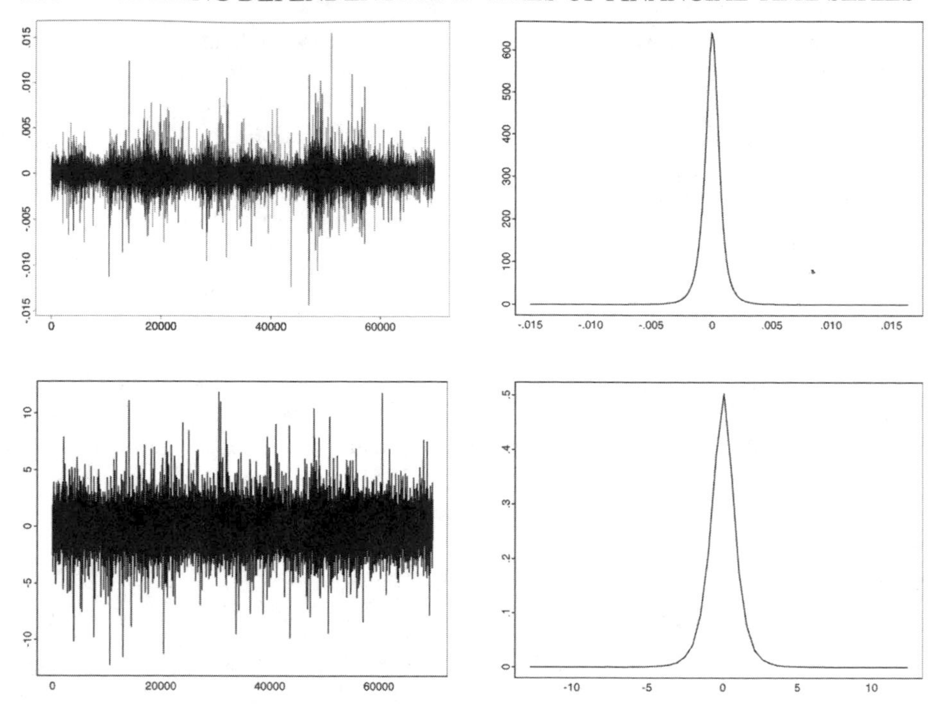

Figure 5.23 Top: *All* 70,000 *JPY-USD FX log-returns* (left) *and their density* (right). Bottom: *The residuals of the JPY-USD FX log-returns* (left) *and their density* (left). *The scale difference on the x-axis when compared with the FX density is due to the standardization* $\text{var}(Z_1) = 1$.

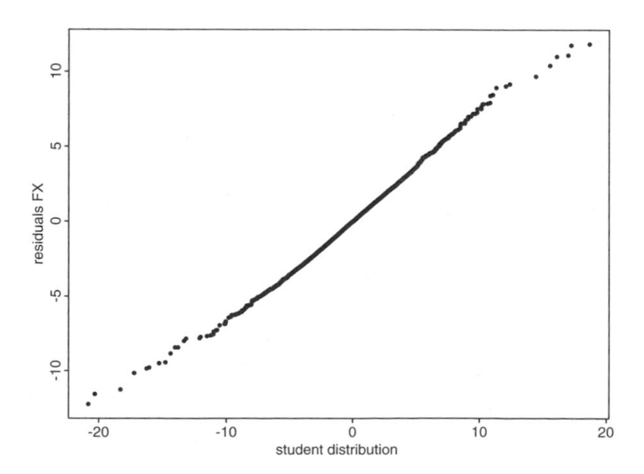

Figure 5.24 *QQ-plot of the* GARCH(1, 1) *residuals against the quantiles of a student distribution with* 4 *degrees of freedom.*

Proceeding in a naive way, we replace the parameters α_1 and β_1 by their estimators and, because we do not know the distribution of the noise (Z_t), we choose the empirical distribution of the residuals $(\widehat{Z}_t)$ as the distribution of the Z_ts. Then we solve the equation

$$\frac{1}{n}\sum_{i=1}^{n}(\alpha_1 \widehat{Z}_i^2 + \beta_1)^{\kappa/2} = 1 \tag{5.10.2}$$

numerically, resulting in an estimate of $\widehat{\kappa} = 2.2$. A partial theoretical basis for this approach, showing asymptotic normality for $\widehat{\kappa}$, was given by Pitts et al. [116] in the more general context of estimating the adjustment coefficient for ruin bounds in insurance mathematics. However, there are two crucial differences between our approach and theirs: they do not assume that parameters of the underlying distribution have to be estimated and they replace the residuals $\widehat{Z}_i$ by i.i.d. Z_is. The latter deviations have been correctly criticized by Berkes et al. [12] who give a precise asymptotic theory for the estimation of κ through the solution of (5.10.2) based on the residuals $\widehat{Z}_i$.

Alternatively, assuming for Z_1 a student distribution with $d = 3.25, 3.5, 3.75, 4$ degress of freedom and variance 1, we calculated the solution $\widehat{\kappa}$ to (5.10.2) from 1 million i.i.d. simulated Z_ts, resulting in estimates of $\widehat{\kappa} = 2.3, 2.4, 2.5$, and 2.5, respectively. Again, these values have to be taken with caution since the variance of $\widehat{\kappa}$ can be quite large, as documented in Table 5.2.

From the theory in Section 5.5.2 we can conclude that the GARCH(1, 1) process with parameters (5.10.1) estimated from the FX log-returns has distributional tails

$$P(\sigma_1 > x) \sim c_0 x^{-2.2} \quad \text{and} \quad P(X_1 > x) \sim E[Z_1|^{2.2}\, P(\sigma_1 > x),$$

for some $c_0 > 0$. This implies that $\text{var}(X_1) < \infty$, but $EX_1^4 = \infty$.

The sample ACF

Figure 5.25 shows the sample ACFs of the FX log-returns, their absolute values, and squares at the first 300 lags. Pointwise confidence bands were derived from 1000 independent simulations of the sample ACFs for a GARCH(1, 1) process with parameters (5.10.1). The corresponding i.i.d. noise was generated from the empirical distribution of the residuals of the FX log-returns. The interpretation of the sample ACFs in Figure 5.25 very much depends on how heavy the tails of the X_ts are. If we accept the hypothesis of a Pareto-like tail for the finite-dimensional distributions of the X_ts, we can apply the asymptotic theory of Theorem 5.9.1 about the sample ACF behavior. In particular, we conclude that the sample ACFs of the X_ts and $|X_t|$s consistently estimate their deterministic counterparts, but the normalized sample ACFs converge to infinite variance 1.1-stable limits at the rate $n^{-1+\kappa/2} = n^{0.1}$. Notice that $n^{0.1} = 70{,}000^{0.1} = 3.05$. Thus, despite the large sample size, one can hardly speak of asymptotic confidence bands for $\rho_{n,X}(h)$ and $\rho_{n,|X|}(h)$. This observation is supported by the width of the bands in Figure 5.25. The slow rate of convergence of the sample ACFs in combination with the heavy tails of the limit distributions raises serious questions about the meaning and quality of these estimators. This remark applies

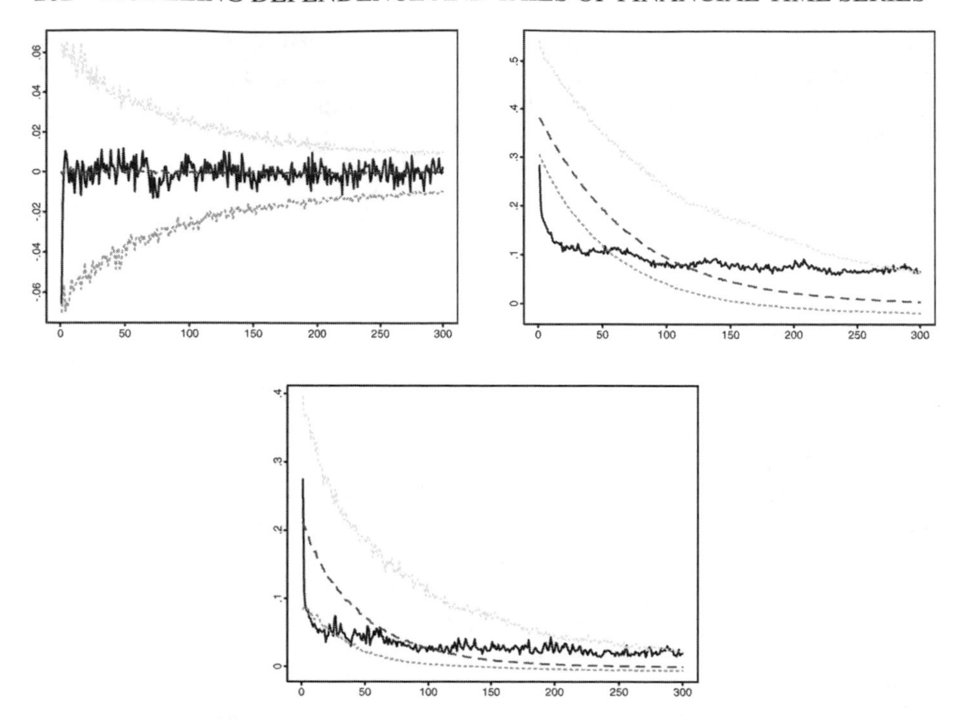

Figure 5.25 *The sample ACFs for the FX log-returns* (top, left)*, their absolute values* (top, right) *and squares* (bottom). *Each graph contains pointwise (simulation based) confidence bands for the distribution of the sample ACFs for the sequences* (X_t), $(|X_t|)$, *and* (X_t^2), *respectively, where* (X_t) *is a* GARCH(1, 1) *process with parameters* (5.10.1). *In each graph, the solid lines and the dotted line indicate the* 2.5%- *and* 97.5%-*quantiles, and the median, respectively, of the distribution of the sample ACF at a fixed lag. For the returns, the sample ACF is inside the* 95% *confidence; it is negative at the first few lags with very small absolute values, and it is very close to the median curve which is indistinguishable from zero. In the top right graph, the sample ACF of the absolute returns at the first* 50 *lags stays below the* 2.5%-*quantile prescribed by the fitted* GARCH(1, 1) *model, then it stays almost constant up to lag* 300 *and remains in the* 95%-*confidence band. A similar picture can be observed for the squared returns in the bottom graph: after a fast decay from inside the confidence bands at the first lags, the sample ACF is below the* 2.5%-*quantile of the distribution prescribed by the GARCH, and then it is almost constant and stays inside the confidence band.*

even more to the sample ACFs of the squares. In this case, $\rho_{n,X^2}(h)$ would converge in distribution to a nondegenerate limit, i.e., it does not estimate anything.

The sample ACFs at the first 50 lags, say, of the absolute FX log-returns do not fall within the 95% confidence bands suggested by the fitted GARCH(1, 1) process. This means that, even when accounting for the statistical uncertainty due to the estimation procedure, the GARCH(1, 1) model does not describe the second order dependence structure of the absolute FX log-returns sufficiently accurately. Even if we take into account that the sample ACF of the squares is not meaningful, the sample ACF of the absolute returns, despite its big statistical uncertainty, should decay to zero roughly at

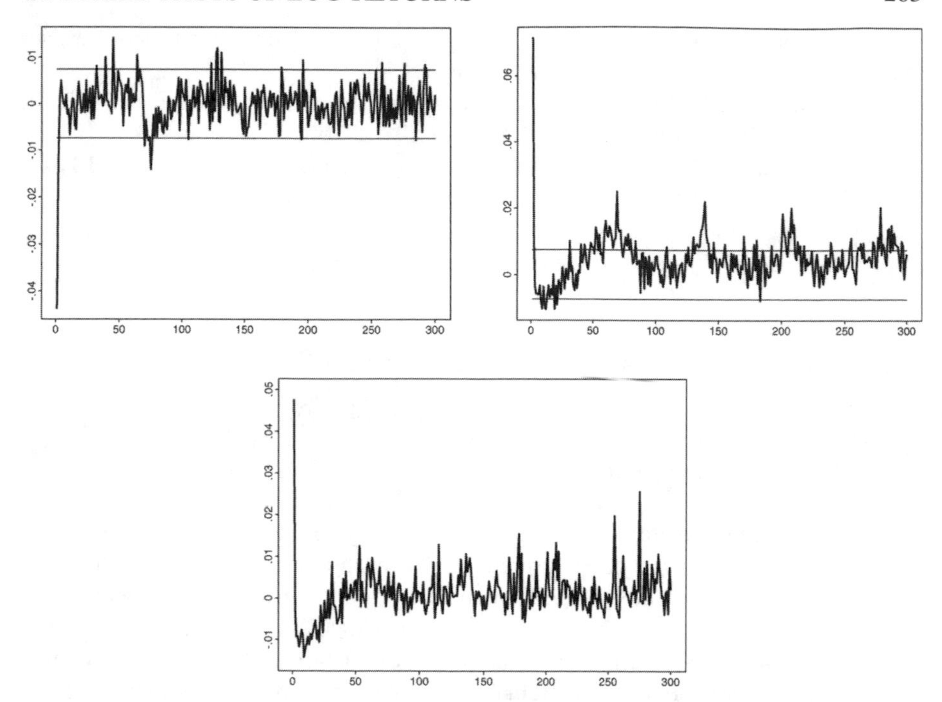

Figure 5.26 *The sample ACFs of the residuals of the FX log-returns* (top, left)*, their absolute values* (top, right) *and squares* (bottom, left)*. The straight lines in the two upper graphs indicate the* $\pm 1.96/\sqrt{n}$ *asymptotic confidence bands for an i.i.d. sequence with finite second moment. In the lower graph we refrain from giving* $(1/\sqrt{n})$*-confidence bands because* Z_1 *possibly has an infinite 4th moment. Compare with the sample ACFs of the FX log–returns in Figure 5.25, in particular observe the differences in scale.*

an exponential rate. Exponential decay of the ACF of the $|X_t|$s follows from the strong mixing property with geometric rate of the GARCH(1, 1) process; see Remark 5.6.2. For reasons of comparison it would be desirable to get analytic expressions for the theoretical ACF of the $|X_t|$s but this seems impossible even for Gaussian noise.

On the other hand, the sample ACFs for the residuals, their absolute values and squares (Figure 5.26) are quite convincing in the sense that they behave very much like the sample ACF of a finite variance i.i.d. sequence or of a moving average process with very small parameters. This compliance with the theoretical requirements of the model is a remarkable feature of the GARCH(1, 1) process and contributed greatly to its success.

The extremal index

In this section we try to judge the quality of the fit of the GARCH(1, 1) model with parameters (5.10.1) as regards some basic extremal characteristics of the FX log-return series. For this reason we estimate the extremal index as a simple quantity characterizing the clustering behavior of high level exceedances of a time series; see Section 5.1.2 for a brief introduction.

As a consequence of the point process convergence outlined in Section 5.8.2, the normalized partial maxima $M_n(X)$, $M_n(\sigma)$ and $M_n(|X|)$ of the sequences (X_t), (σ_t) and $(|X_t|)$, respectively, have a Frechét limit distribution with parameter κ, see Mikosch and Stărică [107] for details. In particular, choosing the normalizing constants $x_n(A)$ for any stationary sequence (A_t) in such a way that $P(A_t > x_n(A)) \sim n^{-1}$ and writing $\Phi_\kappa(x) = \exp\{-x^{-\kappa}\}$, $x > 0$, for the standard Fréchet distribution, one obtains for every $y > 0$,

$$P(M_n(X) \le x_n(X)\, y) \to [\Phi_\kappa(y)]^{\theta_X},$$

$$P(M_n(\sigma) \le x_n(\sigma)\, y) \to [\Phi_\kappa(y)]^{\theta_\sigma},$$

$$P(M_n(|X|) \le x_n(|X|)\, y) \to [\Phi_\kappa(y)]^{\theta_{|X|}}.$$

Here θ_A denotes the extremal index of any stationary sequence (A_t). We mentioned in Section 5.1.2 that $1/\theta_A$ has interpretation as mean cluster size of high threshold exceedances in the sequence (A_t). Although it is possible to give analytical expressions for the extremal indices of the X-, $|X|$-, etc., sequences in the GARCH(1, 1) case, these expressions are too difficult for numerical evaluations. Therefore, one depends on statistical estimators of the extremal index. We compared the estimators for the simulated fitted GARCH(1, 1) process with those for the FX log-returns, using the so-called blocks method explained in Embrechts et al. [58], Section 8.1. It depends on the number of the upper order statistics in the sample that has to be small compared to the sample size.

Figure 5.27 displays the results for estimating the extremal indices θ_X of the JPY-USD FX log-returns, θ_Z of their residuals and θ_σ of the volatility sequence, based on the GARCH(1, 1) fit with parameters (5.10.1). The 95% confidence bands and the median were obtained from 400 independent GARCH(1, 1) simulations with parameters (5.10.1) and the empirical distribution of the FX residuals for the distribution of the Z_ts.

The estimates for θ_X and θ_σ of the FX log-returns lie above the 97.5% quantile, indicating that they do not agree with the fitted GARCH(1, 1) model. Alternatively, the expected cluster size is smaller than for the fitted GARCH(1, 1) model, i.e., there is less dependence in the tails for the returns and volatilities than for the prescribed GARCH(1, 1) model. Nevertheless, the estimated values $\widehat{\theta}_\sigma \approx 0.1$ are small and show that the sequence (σ_t) contains large clusters of high threshold exceedances with expected cluster size ≈ 10. The clustering effect for the extremes of the (X_t) sequence is less pronounced, but still gives values $\widehat{\theta}_X \approx 0.6$.

The corresponding estimates for θ_Z lie within the 95% bands for an i.i.d. sequence. This implies that the residuals very much behave like an i.i.d. sequence with extremal index 1. As for the calculation of κ from (5.10.2), one has to be cautious with interpretations because there is no theory for the estimation of θ_Z based on the residuals of a GARCH(1, 1) model.

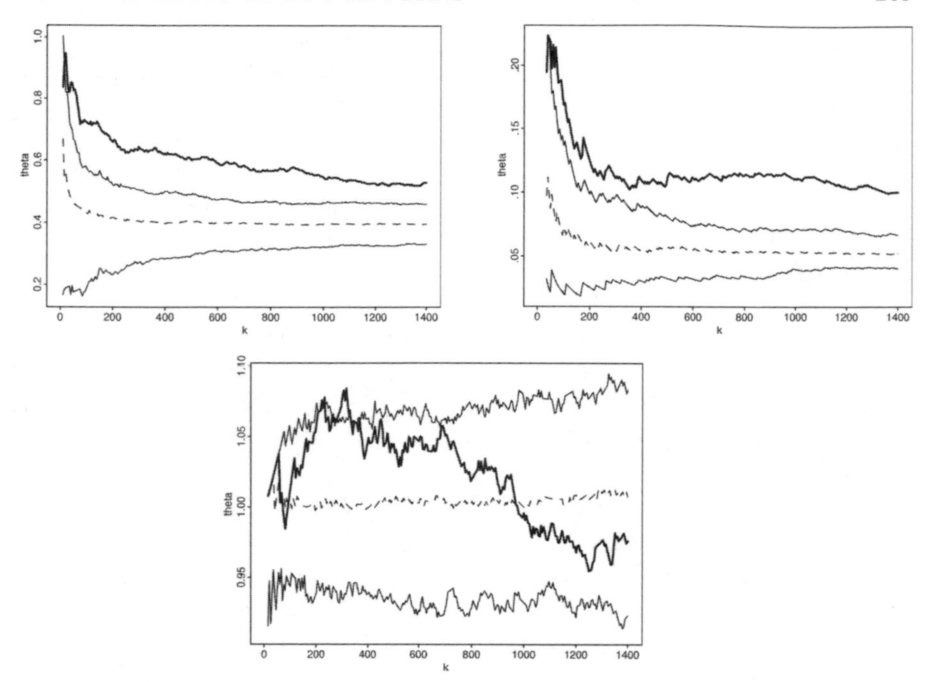

Figure 5.27 *Estimates of the extremal indices* θ_X (top left), θ_σ (top right), *and* θ_Z (bottom) *as a function of the k upper order statistics.* (Top): *The three lower curves correspond to the* 97.5%, 50% *and* 2.5% *quantiles of the estimators for the* GARCH(1, 1) *model with parameters* (5.10.1). *The upper curve presents the* θ *estimates for the FX log-returns. They do not fall within the* 95% *confidence bands.* (Bottom): *In the lower graph the estimates for* θ_Z *fall into the* 95% *confidence bands.*

Concluding remarks

- The GARCH(1, 1) fit to various log-return series is rather convincing, even for large series, if one takes the structure of the residuals as a basis for the judgment of the quality of the fit. The fitted residuals behave very much like an i.i.d. sample, in particular their sample ACF and the extremal index have the properties of an i.i.d. sequence. This shows that the GARCH model mimics some of the essential dependence structure of the returns X_t. In order to judge the goodness of fit of GARCH models, it would be desirable to have reliable goodness of fit tests for the residuals. So far such results are available only in the ARCH(p) case (Horváth et al. [83]). The latter paper contains the asymptotic theory for Kolmogorov-Smirnov type goodness of fit tests based on the residuals. Unfortunately, the proposed result is of limited practical relevance since the limiting process is not distribution free, i.e., it depends on the distribution of the noise (Z_t) and the ARCH parameters. Nevertheless, this result shows that the naive approach to deal with the residuals as if they were i.i.d. can lead to substantial deviations from the classical limit results.

- On the other hand, the GARCH(1, 1) model does not accurately capture either the extremal behavior or the correlation structure as described by the sample ACF of return data.
- Given we accept a GARCH process as a reasonable model for returns for a certain period, the sample autocorrelations are extremely unreliable tools for model selection and validation. In various practical situations one has to deal with models that do not have sufficiently high moments. Then the standard theory for the sample ACF (Gaussian limit distribution and $\sqrt{n}$-rates of convergence) breaks down and the sample ACFs of returns, absolute and squared returns are either poor estimators of their theoretical counterparts (slow convergence rates) or meaningless (nondegenerate limit distributions).
- A general conclusion may be: avoid using the sample ACF to draw conclusions about the dependence structure of log-return data.
- Even when accounting for the mentioned larger than usual statistical uncertainty, the GARCH(1, 1) model (and more generally any GARCH model) cannot explain the effect of almost constant sample ACFs for the absolute returns. An alternative explanation will be given in Section 5.11.1: nonstationarity of the data.

5.10.2 The stochastic volatility model

Neither the GARCH nor the stochastic volatility model have a theoretical grounding in economic or financial theory. They are ad hoc models for explaining the changing stochastic volatility of log-returns by a simple model. By the efforts of its inventors the GARCH model has become more popular than the stochastic volatility model although, as the above discussion might have shown, a GARCH process is an extremely difficult theoretical object. Much of its popularity is due the fact that it is tailor-made for the purposes of parameter estimation.

As we have seen above, the stochastic volatility model can explain regularly varying tails and dependence in a flexible way. In particular, the stochastic volatility can be chosen in such a way that the ACF of its absolute and squared values converges arbitrarily slowly to zero. The GARCH model fails to explain this stylized fact. Attempts were made to incorporate long range dependence in the GARCH model by introducing additional parameters, resulting in models such as FIGARCH or LM-ARCH, but they fell short in achieving this goal; see the discussions in Giraitis, Kokoszka, and Leipus [66] and Mikosch and Stărică [109]; Giraitis et al. [68] propose an alternative mathematical approach to modeling long memory by allowing for negative values in the volatility sequence (σ_t).

It is not clear which of the two models would win in a fair competition; perhaps both would lose. In particular, there do not exist rigorous statistical means to distinguish between them. However, there is a striking difference between the sample ACF behaviors for the absolute and squared processes that might be useful for discriminating between the two models. In the GARCH case, the sample ACF has the slower convergence rates the fewer moments exist and, eventually, if moments above a certain threshold do not exist, the sample ACF has a nondegenerate distributional limit. This is in contrast to the stochastic volatility model, where the sample ACF converges faster to zero the fewer moments (below two) exist. In particular, if the data came from a stochastic volatility model and we chose $r > 0$ sufficiently large, the sample ACF of $|X_t|^r$ should approach zero quickly. This, however, cannot be observed in general. Although there are some time series such as the S&P500 returns for which the sample ACF dies out quickly, at all lags for $r = 3$ or 4; for other return series (CAC40, DAX, and FX rates) one does not observe this effect. This may rule out the stochastic volatility model or refers to another property: nonstationarity of the data. This is the topic of the next section.

5.11 Nonstationarity in log-return series

There are two particular anomalies of log-returns that have puzzled the author for some time. One of them is the long range dependence (LRD) one observes in absolute returns and their squares. The other one is the IGARCH effect (the sum of the estimated parameters of a GARCH model add up to a number close to one). LRD and IGARCH effects are present in medium and large return samples.

We start with a discussion of the notion of LRD in Section 5.11.1. It has become quite popular not only among statisticians and time series analysts, but also among practitioners who use the tools of time series analysis in their daily work. LRD is one explanation for the slow decay of the sample ACF we see in return data. Another explanation is nonstationarity of the data. In Section 5.11.1 we indicate that standard tools of time series analysis (sample ACF, periodogram) can behave very much in the same way both for long range dependent stationary sequences and for nonstationary sequences. In Section 5.11.2 we indicate that the IGARCH effect might be due to nonstationarity as well.

5.11.1 The LRD effect

What is LRD, and how can one detect it?

Over the last few years the phenomenon of long range dependence or long memory has attracted some attention. There exists a long and steadily increasing list of publications of theoretical and empirical nature that aim at the probabilistic modeling or the statistical detection of LRD in areas as diverse as hydrology, meteorology, climatology, telecommunications and finance; see Doukhan et al. [53] for a recent theoretical discussion of LRD. Roughly speaking, LRD means that there is dependence in a time series over an unusually long period of time. There exist various definitions of LRD for a (second-order) stationary process (X_t); see Beran [10]. One way to define it is via the decay rate of the ACF: (X_t) is long range dependent if

$$\sum_{h=0}^{\infty} |\rho_X(h)| = \infty\,, \tag{5.11.1}$$

short range dependent otherwise. For example, ARMA processes with finite variance have exponentially decaying ACF (Brockwell and Davis [29], Chapter 3), hence they are short range dependent. Fractional ARIMA processes with difference parameter $d \in (0, 0.5)$ exhibit LRD ([29], Section 13.2). An alternative definition of LRD is possible via the spectral density f_X of a stationary time series. If the latter exists, it satisfies the relation

$$\rho_X(h) = [\mathrm{var}(X_0)]^{-1} \int_{(-\pi,\pi]} e^{ih\lambda} f_X(\lambda) d\lambda, \quad h \in \mathbb{Z},$$

see [29], Section 4.3. Under some subtle conditions LRD in the sense of the ACF, i.e., (5.11.1), translates into the fact that the spectral density has a singularity at zero, and vice versa. The presence of a singularity of $f_X(\lambda)$ at zero can be taken as another definition of LRD in (X_t). In particular, if one assumes that for some constant $c_\rho > 0$

$$\rho_X(h) \sim c_\rho h^{2(H-1)} \quad \text{for large } h \text{ and some } H \in (0.5, 1), \tag{5.11.2}$$

(in this case (5.11.1) holds) then under some further subtle conditions for some constant $c_f > 0$

$$f_X(\lambda) \sim c_f \lambda^{-2H+1} \quad \text{for frequencies } \lambda > 0 \text{ near zero}, \tag{5.11.3}$$

and vice versa. Exact conditions for the equivalence of (5.11.2) and (5.11.3) can be found in Zygmund [136]. The constant $H \in (0.5, 1)$ is called the Hurst coefficient, and LRD is sometimes referred to as the Hurst phenomenon. The closer H to 1 the slower the rate of $\rho_X(h)$ to zero as $h \to \infty$, i.e., the longer the range of dependence in the time series. Alternatively, the closer H to one the more pronounced is the singularity of $f_X(\lambda)$ at zero, i.e., the stronger the LRD in terms of the spectral density.

A naive way to detect LRD by statistical means is to replace the ACF $\rho_X(h)$ by the sample ACF $\rho_{n,X}(h)$ and the spectral density by its empirical estimator, the periodogram:

$$I_{n,X}(\lambda) = \frac{1}{n} \left| \sum_{t=1}^{n} \mathrm{e}^{-i\lambda t} X_t \right|^2, \quad \lambda \in [0, \pi].$$

It is common practice to evaluate $I_{n,X}$ at the Fourier frequencies

$$\lambda_j = 2\pi j/n \in (0, \pi).$$

If (X_t) is stationary and ergodic and $\mathrm{var}(X_0) < \infty$, the sample ACF estimates the ACF consistently. Although the periodogram does not consistently estimate the spectral density (it has a nondegenerate limit distribution), under general conditions it is asymptotically unbiased: $EI_{n,X}(\lambda) \to f_X(\lambda)$ for every fixed $\lambda \in (0, \pi)$. Using smoothing techniques, $f_X(\lambda)$ can be consistently approximated; see [29], Section 10.4, or Priestley [117], Chapter 7. Asymptotic normality for the sample ACF and the smoothed periodogram can be derived under certain moment and mixing conditions.

However, the assumptions of stationarity, ergodicity, and mixing cannot be verified satisfactorily on real-life data: one has to believe them. In Figure 5.28 we show an example of a stationary time series for which the stationarity assumption is not plausible.

An explanation of the LRD effect: structural breaks

Now recall the stylized ACF-fact for log-return series (X_t) from Section 5.1.2, which we supplement with the corresponding stylized fact in terms of the periodogram:

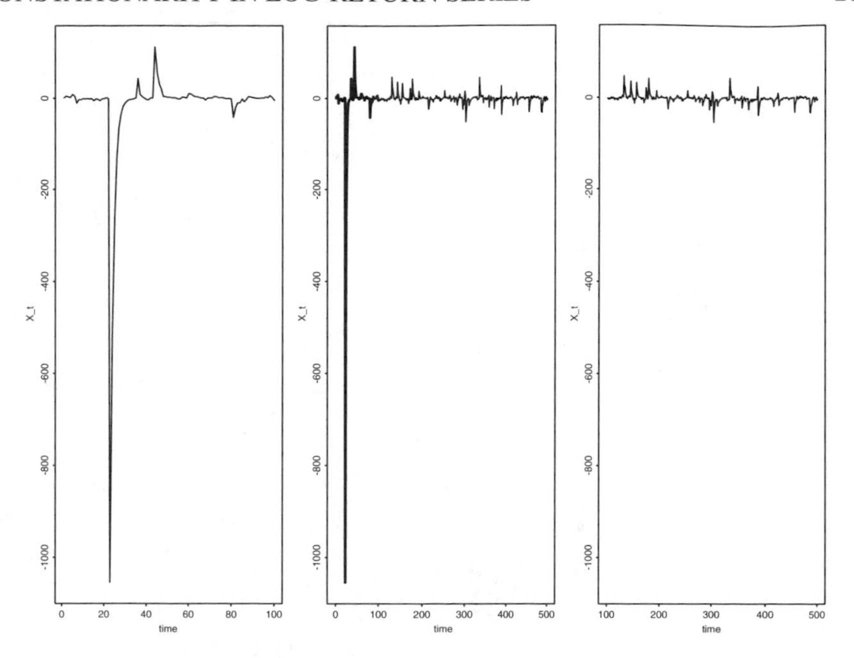

Figure 5.28 *We see a realization of* 500 *values of the stationary* AR(1) *process* $X_t = 0.5X_{t-1} + Z_t$ *with i.i.d. Cauchy noise* (Z_t) (middle). *The left* (right) *graph visualizes the first* 100 (*last* 400) *values. If one looks at the whole series the first piece gives one the impression that the series is not stationary. When the time series in the right graph is analyzed the assumption of stationarity is plausible, while, when in possession of the whole series, a structural break seems more likely.*

- The sample ACF $\rho_{n,X}$ of the data is negligible at almost all lags with a possible exception at the first lag. The smoothed periodogram is flat at all frequencies.
- The sample ACFs $\rho_{n,|X|}$ and ρ_{n,X^2} of the absolute and squared returns are all positive, decay relatively fast at the first lags, and tend to stabilize around a positive value for larger lags. The periodograms $I_{n,|X|}$ and I_{n,X^2} blow up at zero.

Whereas the first property indicates that (X_t) constitutes white noise, the second property hints at LRD in the $|X|$- and X^2-sequences. We refer to the latter phenomenon as the LRD effect of log-return data. We illustrate the LRD effect for the S&P500 series for the sample ACF in Figure 5.4 and for the periodogram in Figure 5.29.

The LRD effect is typical for longer time series: only for them it makes sense to look at the sample ACF at larger lags. Indeed, by construction, $\rho_{n,X}(h)$ is reliable (consistent estimator) only for lags h small compared to n, say $h \leq n/4$ as recommended by Box and Jenkins [23], p. 33. Thus, only for samples of several thousand points one is allowed to look at the sample ACF at lags 250, 300, 350, etc., and a similar remark applies to the periodogram for small frequencies. Only for such time series one can make statistical inference on the LRD phenemenon. Therefore, the overwhelming

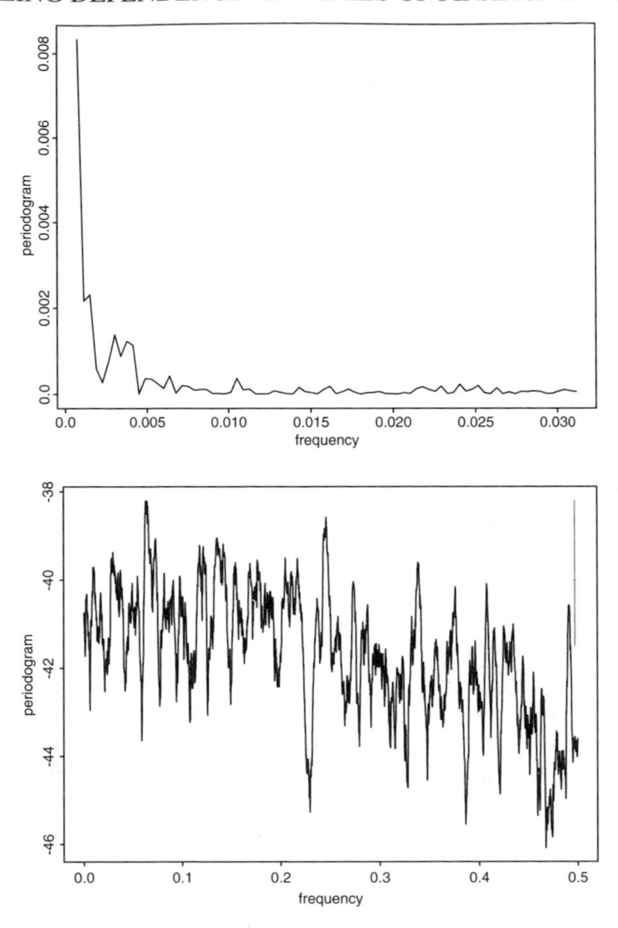

Figure 5.29 *The log-periodogram of the* S&P500 *log-returns* (bottom)*, and the periodogram of the absolute log-returns close to zero* (top).

proportion of time series used in the econometrics literature to document the LRD property covers decades of economic activity. The latter are inevitably marked by turbulent years of crises and, possibly, structural changes. To give some examples: Ding et al. [51] discuss the slow decay of the sample ACF of powers of absolute daily returns in the S&P 1928 to 1990; Ding and Granger [50] consider daily returns of the Japanese Nikkei index 1970 to 1992, the FX rate DEM/USD 1971 to 1992, and Chevron stock 1962 to 1991; Bollerslev and Mikkelsen [17] use the S&P500 daily returns 1953 to 1990 to fit their FIGARCH model while Breidt et al. [27] fit a stochastic volatility model with LRD in the volatility sequence to the daily returns of the value-weighted market index 1962 to 1987.

We have tried to argue that the statistical detection of LRD requires long time series. On the other hand, the critical reader might have some serious doubts about such a statistical analysis: in decades of economic activity, structural breaks are likely, and, therefore, the stationarity assumption is questionable. Under the circumstances of changing structure in the time series, statistical tools can easily fool one. To illustrate this point we show how the LRD effect builds up in a nonstationary time series $(X_t)_{t=1,\dots,2000}$ containing one structural break. The first and second 1000 points of the series are generated from two different GARCH(1, 1) models with corresponding parameters:

$$\alpha_0 = 0.17 \times 10^{-5}, \quad \alpha_1 = 0.1, \quad \beta_1 = 0.5\,, \tag{5.11.4}$$

$$\alpha_0 = 0.17 \times 10^{-5}, \quad \alpha_1 = 0.1, \quad \beta_1 = 0.8. \tag{5.11.5}$$

The noise sequences of the two models are independent and standard normal. The structural break in this series is obvious by a simple eyeball inspection; see the right-hand top graph in Figure 5.30. In the left column one can find the sample ACF of the absolute values $|X_t|$ and the corresponding periodogram for the first 1000 points: the sample ACF decays to zero exponentially fast and the periodogram is a smooth function for all frequencies including zero. For the second piece of the data one gets a similar picture; we omit it. This is in agreement with the theoretical properties of a stationary GARCH(1, 1) model; see the remarks in Section 5.10.1. The picture changes dramatically for the concatenated time series: both the sample ACF and the periodogram exhibit the LRD effect; see the right column of Figure 5.30. The parameters β_1 in the two pieces only differ by 0.3. By choosing a larger difference of the parameters or by concatenating pieces from more than two models, the LRD effect becomes even more striking.

The LRD effect of the sample ACF of the absolute values develops even though we know there is no LRD in the data: the time series is nonstationary. The explanation lies in the change in mean of the underlying series. This can be seen by some simple analysis as outlined in Mikosch and Stărică [106, 108]. Assume we have a sample $X_1, \dots, X_n$ that consists of two pieces from different stationary GARCH(1, 1) models with finite variance, say $X_1^{(1)}, \dots, X_{[np]}^{(1)}$ and $X_{[np]+1}^{(2)}, \dots, X_n^{(2)}$ for some $p \in (0, 1)$. Then the ergodicity of the two different models implies that the sample ACVF of $(|X_t|)$ at a fixed lag h converges as $n \to \infty$ to the quantity

$$\begin{aligned} \gamma_{n,|X|}(h) \stackrel{P}{\to}\; & p\, \gamma_{|X^{(1)}|}(h) + (1-p)\, \gamma_{|X^{(2)}|}(h) \\ & + p\,(1-p)\,(E|X^{(1)}| - E|X^{(2)}|)^2, \end{aligned} \tag{5.11.6}$$

where $\gamma_{|X^{(1)}|}(h)$ and $\gamma_{|X^{(2)}|}(h)$ are the theoretical ACVFs corresponding to the first and second model, and analogous results are evident for the sample ACF. Moreover, if the two subsamples are uncorrelated and $\lambda_{j_n} = 2\pi j_n/n \to 0$ sufficiently fast, then the periodogram $I_{n,|X|}(\lambda_{j_n})$ satisfies, as $n \to \infty$,

$$\begin{aligned} E I_{n,|X|}(\lambda_{j_n}) \sim\; & p\; 2\pi f_{|X^{(1)}|}(\lambda_{j_n}) + (1-p)\; 2\pi f_{|X^{(2)}|}(\lambda_{j_n}) + \\ & \frac{2}{n\lambda_{j_n}^2}(E|X^{(1)}| - E|X^{(2)}|)^2(1 - \cos(2\pi j_n p)), \end{aligned} \tag{5.11.7}$$

where $f_{|X^{(1)}|}(\lambda)$ and $f_{|X^{(2)}|}(\lambda)$ are the spectral densities corresponding to the first and second model.

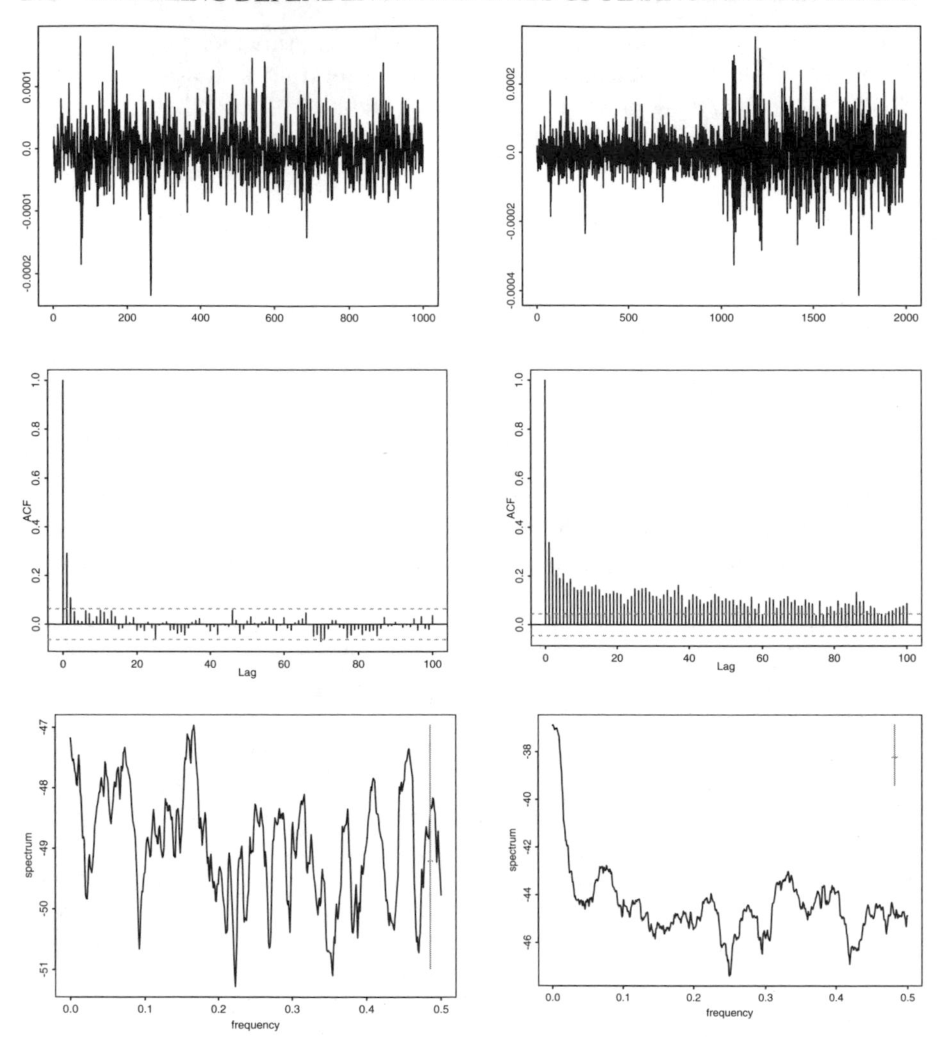

Figure 5.30 Left: *1000 simulated values of the* GARCH(1, 1) *model* (5.11.4) (top), *the sample ACF* (middle) *and the log-periodogram* (bottom) *of the absolute values* $|X_t|$. *Both tools indicate short range dependence.* Right: *Concatenated time series from two different* GARCH(1, 1) *models with parameters* (5.11.4) *and* (5.11.5) (top), *the sample ACF* (middle) *and the log-periodogram* (bottom) *for the absolute values. The vertical lines at the right end of the graphs for the estimated spectrum indicate the width of* 95% *asymptotic pointwise confidence bands.*

These simple arguments explain the LRD effect in simulated data with one break at a deterministic instant of time as illustrated in Figure 5.30, and similar calculations apply to the case of finitely many different models in a time series. Indeed, since the GARCH(1, 1) is strongly mixing with geometric rate, the sample ACF of its absolute values and squares decays to zero exponentially fast (see Section 5.10.1); and its spectral density is everywhere continuous. Thus (5.11.6) explains

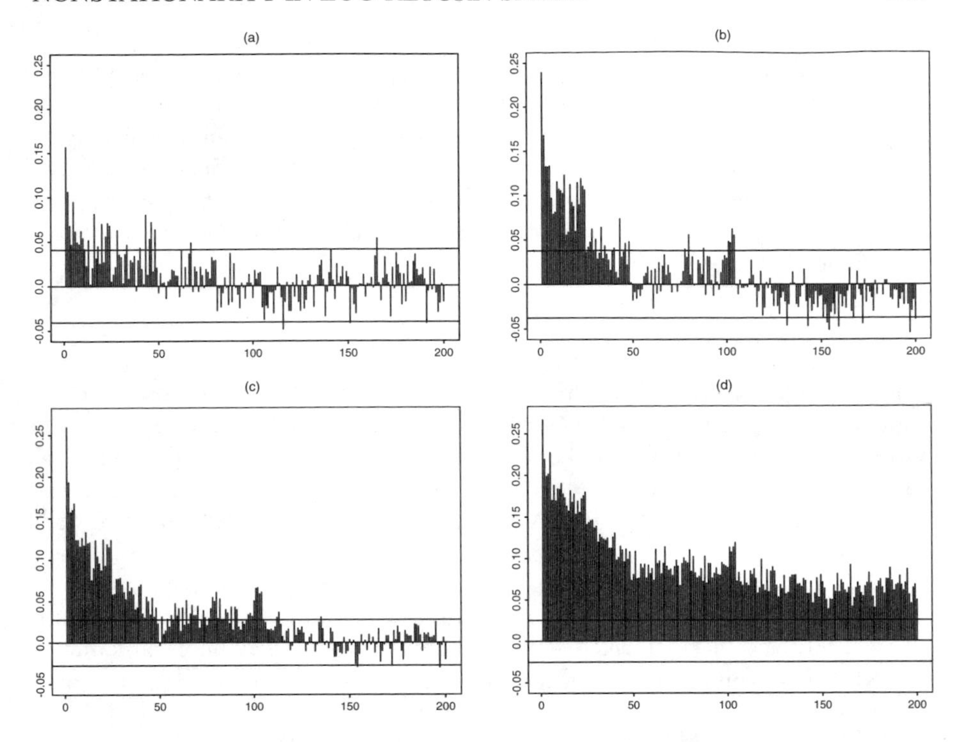

Figure 5.31 *The sample ACF for the absolute log-returns of the first* 9 *and* 11 (a, b), *20 and 24 years* (c, d) *of the S&P500 data.*

that the sample ACF decays quickly at the first lags and then approaches the positive value $p(1-p)(E|X^{(1)}| - E|X^{(2)}|)^2$. This is exactly what we observe for log-returns. Relation (5.11.7) explains that we observe a blow-up of the periodogram at the origin because the spectral densities $f_{|X^{(i)}|}(\lambda)$ are continuous at zero and the third term in (5.11.7) can be quite large for small frequencies, due to the factor $(n\lambda_{j_n}^2)^{-1}(E|X^{(1)}| - E|X^{(2)}|)$ for $n\lambda_{j_n}^2 \to 0$.

In real-life data we would not expect to have abrupt breaks of the structure, with a few exceptions, but we can easily imagine a long time series such as the S&P500 being modeled through the concatenation of smaller pieces from different models. This would give some theoretical ground for the LRD effect in long log-return series. We illustrate the build-up of the LRD effect in the S&P500 series by concatenation of several years of data in Figure 5.31.

When dealing with long financial time series, structural breaks often have well known reasons. Mikosch and Stărică [106] considered tests for detecting breaks in the spectral structure of squared returns in the S&P500 and in FX rate series. It turns out that various of the detected change points can be identified as concrete economic events such as the beginning of recession periods in the U.S. Change point analyses for ARCH-type models have been suggested by various other authors; see for example Kokoszka and Leipus [92]. A reasonable alternative is a locally stationary model as advocated by Dahlhaus and coworkers where parameters change when time evolves. We refer to Dahlhaus [36, 37], Dahlhaus and Neumann [38], and Dahlhaus and Polonik [39] for a recent survey paper. Recently, Granger and Stărică [73] and Stărică (personal communication) have argued and provided statistical evidence that absolute returns can be reasonably modeled in a minimalistic way by $|X_t| = e^{\mu(t)} Z_t$ where $\mu(t)$ is a deterministic (step or smooth) function changing over time and Z_t is a regularly varying i.i.d. sequence. The latter model reduces the problem of detecting dependence to estimation of $\mu(t)$ and the problem of heavy tails to estimation of the distribution of Z_t.

5.11.2 The IGARCH effect

The LRD effect is one anomaly that can easily be explained by nonstationarity of the time series. Throughout this chapter we have witnessed another striking property of GARCH modeling: the IGARCH effect; see Example 5.3.7 and Example 5.5.8 and Figure 5.12 in the S&P500 case. From the previous discussion we know that long return series are usually fitted by GARCH models whose estimated parameters add up to a value close to 1. From Example 5.5.8 we know that an IGARCH(1, 1) model has regularly varying finite-dimensional distributions with index 2, hence infinite variance. This is in agreement with numerical calculations; see Section 5.10.1. It is immediately clear that LRD and IGARCH exclude each other since the usual definition of LRD is based on the second-order characteristics such as the ACF which requires finite variance.

On the other hand, if one estimates the tail parameter κ of log-return series from the data (see Section 5.1.1 for a short introduction) one typically obtains values between 3 and 5; see Figure 5.3 for the S&P500 case and Chapter 6 in Embrechts et al. [58] for more statistical evidence. This indicates that the GARCH model is not flexible enough to explain the dependence and tails present in real-life data. One possible reason for the IGARCH effect is nonstationarity of the data. The empirical evidence for the IGARCH effect comes from considering long log-return series; see Bollerslev et al. [16] and the references therein. For example, Bollerslev and Mikkelsen [17] fit a GARCH(1, 1) process to the S&P500 daily returns 1953 to 1990 with estimated $\hat{\varphi}_1 = \widehat{\alpha}_1 + \widehat{\beta}_1 = 0.995$. Breidt et al. [27] obtain $\hat{\varphi}_1 = 0.999$ for the daily returns of the value-weighted market index 1962 to 1987.

While studies of daily asset log-returns have frequently found IGARCH behavior, studies with higher-frequency data over shorter time spans have often uncovered weaker persistence. For example, Baillie and Bollerslev [3] report on the estimation of a GARCH(1, 1) model for hourly log-returns on FX rates of GBP, DEM, CHF, and JPY with respect to USD January 1986 to July 1986 the values $\hat{\varphi}_1^{GBP} = 0.606$, $\hat{\varphi}_1^{DEM} = 0.568$, $\hat{\varphi}_1^{CHF} = 0.341$, $\hat{\varphi}_1^{JPY} = 0.717$.

The IGARCH effect can also be seen in a time series consisting of two GARCH pieces with parameters given in (5.11.4) and (5.11.5); see Figure 5.32 for statistical evidence. The parameters were estimated by Gaussian quasi-MLE; see Section 5.2.3. We fit a GARCH(1, 1) model to the increasing samples $X_1, \dots, X_{150+k*50}$, $k = 1, \dots, 37$. For sample sizes less than 1000, the sum of the theoretical parameters is $\varphi_1 = \alpha_1 + \beta_1 = 0.6$. Notice that the values $\widehat{\varphi}_1$ are not even close to 0.6 for

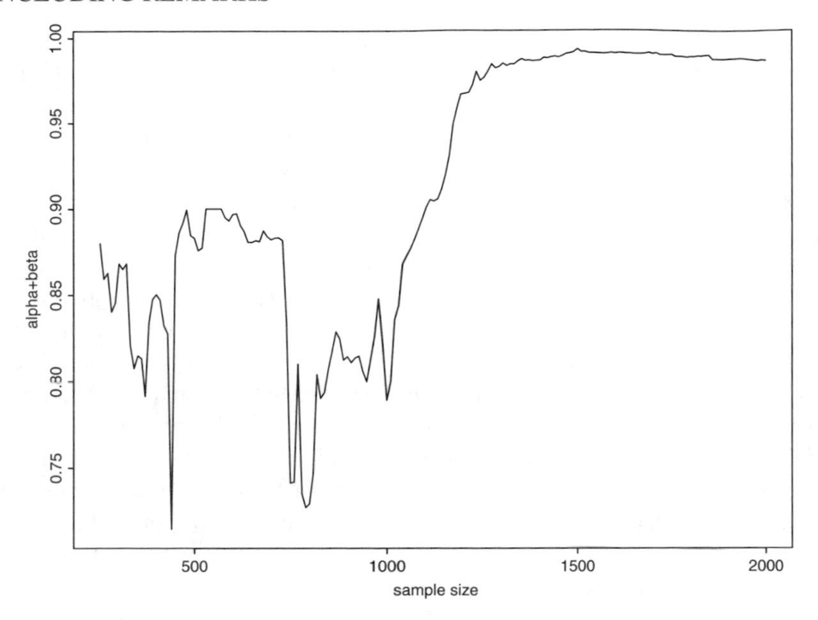

Figure 5.32 *The estimated values* $\varphi_1 = \alpha_1 + \beta_1$ *for a* GARCH(1, 1) *model fitted to samples increasing by* 50 *values from the simulated series at the right top of Figure 5.30. The labels on the* x*-axis indicate the size of the sample used in the estimation procedure.*

$n = 1000$ although the estimator is $\sqrt{n}$-consistent. This behavior is not untypical; it is not an artifact of a badly chosen sample path, but due to the properties of the quasi-MLE procedure. The graph in Figure 5.32 clearly shows how the switch of regimes (at $t = 1000$) makes the sum increase to 1. This is in agreement with Whittle estimation theory for the GARCH(1, 1) parameters in nonstationary sequences; see Mikosch and Stărică [106]. However, the Whittle estimator is poor for tail parameters between 4 and 8 and even nonconsistent for tail parameters less than 4 (Mikosch and Straumann [110]). It would be desirable to get theoretical results for the IGARCH effect in nonstationary time series for the Gaussian quasi-MLE procedure, which is much less sensitive with respect to the heaviness of the tails of the marginal distribution.

5.12 Concluding remarks

Hopefully, the above discussion has convinced the reader that modeling of log-return data is not an easy matter. If one browses through the literature on financial data one can find a large multitude of methods and proposals. It seems that *any* more recent technique of modeling or statistical analysis has been used to describe the data, including deterministic (chaotic) time series, dynamical systems, neural networks, genetic algorithms, infinite variance stable processes, stochastic differential equations, Malliavin calculus, self-similar processes, long memory processes, Lévy processes, fractional Brownian motion, . . . As a matter of fact, financial time series such as the daily log-returns of the S&P500 have attracted the attention of various researchers who have applied their, sometimes very distinct, methods to them and obtained different answers. Certainly, i.i.d. Gaussian noise as prescribed for log-returns by the Black-Scholes model and long memory processes with Pareto tails are worlds apart.

The described state of the art allows one to draw several conclusions. If very different models have been taken for one and the same data set — the S&P500 is such a time series — it seems that we do not know too much about the structure of the data. The usual approach in time series analysis and mathematical finance is to assume a physical model for the evolution of the price process as if the latter followed a law similar to Newton's gravitation or Maxwell's electro-magnetic fields. Mathematical models have certainly always been influenced by physics, and they work quite well in that field, as we know. By now, there is no sufficient economic explanation for a GARCH or stochastic volatility model or even geometric Brownian motion. These models have been taken for mathematical or statistical convenience, and one must not be surprised if they do not always do what they are supposed to. The existing models are black box models; they do not take additional economic or political information into account. It may be doubted whether one can build a time series model which *would* use such information.

Despite its drawbacks, the GARCH and related models (some people have counted more than 150 modifications) have the advantage of being *simple.* The fact that the residuals of GARCH processes fitted to real-life data are "almost independent" in many cases shows that these models can be used for forecasting the conditional log-return distribution over a short time horizon. They are certainly not good models over a long period of time; as we have seen, parameters change and one has to reestimate.

The question as to whether even a GARCH(1, 1) model might be too sophisticated cannot be answered at the moment. Granger and Stărică [73] looked at the absolute returns of the S&P index from 1928 to 2000 and fitted a simple model of the form

$$\log|X_t| = \mu(t) + \sigma(t)\,\varepsilon_t, \tag{5.12.1}$$

where (ε_t) is an i.i.d. sequence or an ARMA process, $\mu(t)$ and $\sigma(t)$ are deterministic functions. They show by some goodness of fit tests that $\mu(t)$ stays constant over long periods of time, several years in general, and $\sigma(t)$ does not vary very much at all. If this model were more appropriate than any other it would mean that there is not much structure in absolute log-returns at all. Granger and Stărică [73] compare the short term prediction power of (5.12.1) with a classical long memory FARIMA model. It turns out that the model (5.12.1) with updated $\mu(t)$ and constant σ does a better job than fractional ARIMA models. This is perhaps not surprising because a long memory model has to take into account the whole evolution of the time series from its beginning to its end whereas the updated model (5.12.1) would react to more recent developments.

Over the last years the need for a better modeling of financial data has become quite obvious. The Value at Risk (VaR) calculation that is required by the financial regulatory authorities has called for an efficient way of calculating 1% or 5% quantiles in the profit and loss (P&L) distribution of a bank. Typically, the latter corresponds to a linear combination of hundreds or thousands of different, perhaps correlated, prices/returns of assets. There are two common ways to handle the problem: historical simulation (i.e., Monte-Carlo simulation of the distribution of the portfolio value based on historical data, often 1 year) or assuming Gaussianity of the data. In the latter case one only needs the correlations to determine the distribution of a linear combination of jointly Gaussian random variables. The historical simulation corresponds to the empirical distribution of the data, and therefore the tails can never be worse than those suggested by the data. This means that an extrapolation outside the data is impossible, an event worse than happened in the past is excluded. On the other hand, Gaussianity of the data is much too optimistic as regards the tails of the distributions, and one uses inadequate quantities (correlations) to describe dependence in a highly nonGaussian environment.

Multivariate regular variation is an elegant alternative to correlations in order to describe dependence in the extremes of multivariate data. However, it involves the estimation of the index of regular variation and of the spectral distribution. Conclusions about the tail of a linear combination of a regularly varying vector of returns, in particular about the very low quantiles, are then not straightforward.

The demand for a more realistic modeling of financial data is also implied by the fact that in many cases one has a huge amount of data (so-called high density or high frequency data), not just daily or weekly returns, but hundreds of transactions every day. Olsen & Associates in Zürich organized two big conferences on the issue. It seems that traditional time series models or stochastic differential equations are inappropriate in this case but it is not clear what is suitable. Moreover, it has been discussed whether more data mean more information or simply more noise, but a clear answer has not been found although, from a statistical point of view, one might hope that more data might improve the accuracy of estimation. The practical experience of the Olsen group has been summarized in the recent book by Dacorogna et al. [35].

A major challenge is the multivariate case. Although various models for multivariate log-returns (including GARCH and stochastic volatility models) exist, their mathematical/statistical tractability is to a large extent an open question. The curse of dimensionality also concerns multivariate extremes. Although a theory exists in this case and statistical case studies have been done in two or three dimensions, it is questionable how the theory can be applied when one deals with portfolios of several hundred or thousand assets. Unfortunately, there is no Gaussian distribution in multivariate extreme value theory.

The previous discussion on the theoretical properties of the GARCH and stochastic volatility models and their fit to real-life data might have given the right impression:

Return data are not easily fitted by any model. They change structure at any time.

Return data contain trends, change variance, and dependence structure. Therefore it is impossible, for example, to tell whether the LRD effect is due to nonstationarity, LRD stationarity or just an artifact of too small confidence bands, ignoring the heavy tails of the distributions. Although extreme value theory, extreme value statistics, and time series analysis are potentially useful tools for financial data, one has to be careful with the conclusions to be drawn. To make extreme value statistics work one needs large samples. This in turn makes the assumption of structural breaks or changes in the data more likely and, therefore, the statistical analyses less reliable. One of the very difficult tasks of modeling financial data consists of finding out when such a break happens and how long a period of relative stability in the data persists. In the opinion of the author, if effects like IGARCH or LRD show up in return data one should not trust a unique model, but break the data into pieces and try to find out what is the underlying statistical/economic reason for the changing structure.

There are more questions than answers, which will keep the academic community and practitioners busy for the next few years.

Acknowledgment

I take pleasure in thanking Richard A. Davis for reading this long chapter during the Workshop on Stable Processes in Oberwolfach in October 2001 and for his critical suggestions, which led to an improvement on presentation and content. Richard and I started our collaboration on the probabilistic and statistical issues of the extremes, the dependence in the tails and the tail behavior of financial time series models when I was visiting Colorado State University (Fort Collins) in 1997. Two further collaborators, Bojan Basrak and Catalin Stărică, have further contributed to the better understanding of the theory on GARCH processes. I am grateful to my Ph.D. student Daniel Straumann for very carefully reading this chapter. Sid Resnick did an excellent refereeing job. His critical remarks forced me to think about the presentation and formulation at many places. I am very much indebted to Aad van der Vaart who read the chapter very carefully and gave me very detailed advice in order to improve upon correctness and clarity of the presentation.

Parts of this chapter were written when I visited the Newton Institute in Cambridge during the thematic period on uncertainty and risk in July 2001. Financial support by the Newton Institute, the European Research Mobility Programme DYNSTOCH, and a Danish Research Council (SNF) grant is gratefully acknowledged. A first version of this chapter was used as lecture notes for the summer school for French speaking Swiss PhD students in Crans Montana (Switzerland) in September 2000 and a workshop on time series at the Mathematical Center of Luminy (France) in April 2001. The final version of this chapter was the basis for lectures at the Semstat Meeting in Gothenburg in December 2001.

Bibliography

[1] BABILLOT, M., BOUGEROL, P. AND ELIE, L. (1997) The random difference equation $X_n = A_n X_{n-1} + B_n$ in the critical case. *Ann. Probab.* 25, 478–493.

[2] BAILLIE, R.T. AND BOLLERSLEV, T. (1989) The message in daily exchange rates: a conditional-variance tale. *J. Business and Economic Statistics* 7, 297–304.

[3] BAILLIE, R.T. AND BOLLERSLEV, T. (1990) Intra-day and inter-market volatility in foreign exchange rates. *Review of Economic Studies* 58, 565–585.

[4] BAILLIE, R.T., BOLLERSLEV, T. AND MIKKELSEN, H.O. (1996) Fractionally integrated generalized autoregressive conditional heteroskedasticity. *J. Econometrics* 74, 3–30.

[5] BARNDORFF-NIELSEN, O.E. AND SHEPHARD, N. (2001) Modelling by Lévy processes for financial econometrics. In: Barndorf-Nielsen, O.E., Mikosch, T. and Resnick, S.I. (Eds.) *Lévy Processes: Theory and Applications,* pp. 283–318. Birkhäuser, Boston.

[6] BASRAK, B., DAVIS, R.A. AND MIKOSCH, T. (1999) The sample ACF of a simple bilinear process. *Stoch. Proc. Appl.* 83, 1–14.

[7] BASRAK, B., DAVIS, R.A. AND MIKOSCH, T. (2002) A characterization of multivariate regular variation. *Ann. Appl. Probab.* 12, 908–920.

[8] BASRAK, B., DAVIS, R.A. AND MIKOSCH. T. (2002) Regular variation of GARCH processes. *Stoch. Proc. Appl.* 99, 95–116.

[9] BAXENDALE, P.H. AND KHASMINSKII, R.Z. (1998) Stability index for products of random transformations. *Adv. Appl. Probab.* 30, 968–988.

[10] BERAN, J. (1994) *Statistics for Long–Memory Processes.* Monographs on Statistics and Applied Probability, No. 61. Chapman and Hall, New York.

[11] BERKES, I., HORVÁTH, L. AND KOKOSZKA, P. (2003) GARCH processes: structure and estimation. *Bernoulli*, 9, 201–228.

[12] BERKES, I., HORVÁTH, L. AND KOKOSZKA, P. (2003) Estimation of the maximal moment exponent of a GARCH(1,1) sequence. *Econometric Theory*, 19, 565–586.

[13] BILLINGSLEY, P. (1968) *Convergence of Probability Measures.* Wiley, New York.

[14] BINGHAM, N.H., GOLDIE, C.M. AND TEUGELS, J.L. (1987) *Regular Variation.* Cambridge University Press, Cambridge.

[15] BOLLERSLEV, T. (1986) Generalized autoregressive conditional heteroskedasticity. *J. Econometrics* 31, 307–327.

[16] BOLLERSLEV, T., CHOU, R.Y. AND KRONER, K.F. (1992) ARCH modeling in finance: a review of the theory and empirical evidence. *J. Econometrics* 52, 5–59.

[17] BOLLERSLEV, T. AND MIKKELSEN, H.O. (1996) Fractionally integrated generalized autoregressive conditional heteroskedasticity. *J. Econometrics* 74, 3–30.

[18] BORKOVEC, M. (1999) *Large Fluctuations in Financial Models.* PhD Thesis. TU Munich.

[19] BORKOVEC, M. (2000) Extremal behavior of the autoregressive process with ARCH(1) errors. *Stoch. Proc. Appl.* 85, 189–207.

[20] BOUGEROL, P. AND PICARD, N. (1992) Strict stationarity of generalized autoregressive processes. *Ann. Probab.* 20, 1714–1730.

[21] BOUGEROL, P. AND PICARD, N. (1992) Stationarity of GARCH processes and of some nonnegative time series. *J. Econometrics* 52, 115–127.

[22] BOUSSAMA, F. (1998) *Ergodicité, mélange et estimation dans le modelès GARCH.* PhD Thesis, Université 7 Paris.

[23] BOX, G.E.P. AND JENKINS, G.M. (1976) *Time Series Analysis: Forecasting and Control.* Holden-Day, San Francisco.

[24] BRANDT, A. (1986) The stochastic equation $Y_{n+1} = A_n Y_n + B_n$ with stationary coefficients. *Adv. Appl. Probab.* 18, 211–220.

[25] BRANDT, A., FRANKEN, P. AND LISEK, B. (1990) *Stationary Stochastic Models.* Wiley, Chichester.

[26] BREIDT, F.J. AND CARRIQUIRY, A.L. (1996) Improved quasi-maximum likelihood estimation for stochastic volatility models. In: *Modelling and Prediction. Proceedings Hsinchu, 1994,* pp. 228–247. Springer, New York.

[27] BREIDT, F.J., CRATO, N. AND DE LIMA, P. (1998) The detection and estimation of long memory in stochastic volatility. *J. Econometrics* 83, 325–348.

[28] BREIMAN, L. (1965) On some limit theorems similar to the arc-sin law. *Theory Probab. Appl.* 10, 323–331.

[29] BROCKWELL, P.J. AND DAVIS, R.A. (1991) *Time Series: Theory and Methods,* 2nd edition. Springer, New York.

[30] BROCKWELL, P.J. AND DAVIS, R.A. (1996) *Introduction to Time Series and Forecasting.* Springer, New York.

[31] CLINE, D.B.H. AND SAMORODNITSKY, G. (1994) Subexponentiality of the product of independent random variables. *Stoch. Proc. Appl.* 49, 75–98.

[32] COMTE, F. AND LIEBERMAN, O. (2003) Asymptotic theory for multivariate GARCH processes. *J. Multivar. Anal.* 84, 61–84.

[33] CROUHY, M., GALAI, D. AND MARK, R. (2001) *Risk Management.* McGraw-Hill, New York.

[34] CSÖRGŐ, S. AND TEUGELS, J. (1990) Empirical Laplace transform and approximation of compound distributions. *J. Appl. Probab.* 27, 88–101.

[35] DACOROGNA, M.M., GENCAY, R., MÜLLER, U.A., OLSEN, R.B. AND PICTET, O.V. (2001) *An Introduction to High-Frequency Finance.* Academic Press, New York.

[36] DAHLHAUS, R. (1997) Fitting time series models to nonstationary processes. *Ann. Statist.* 25, 1–37.

[37] DAHLHAUS, R. (2000) A likelihood approximation for locally stationary processes. *Ann. Statist.* 28, 1762–1794.

[38] DAHLHAUS, R. AND NEUMANN, M. (2001) Locally adaptive fitting of semiparametric models to nonstationary time series. *Stoch. Proc. Appl.* 91, 277–308.

[39] DAHLHAUS, R. AND POLONIK, W. (2002) Empirical spectral processes and nonparametric maximum likelihood estimation for time series. In: Dehling, H.G., Mikosch, T. and Sørensen, M. (2002) *Empirical Process Techniques for Dependent Data,* pp. 275–298. Birkhäuser, Boston.

[40] DAVIS, R.A. AND HSING, T. (1995) Point process and partial sum convergence for weakly dependent random variables with infinite variance. *Ann. Probab.* 23, 879–917.

[41] DAVIS, R.A. AND MIKOSCH, T. (1998) The sample autocorrelations of heavy-tailed processes with applications to ARCH. *Ann. Statist.* 26, 2049–2080.

[42] DAVIS, R.A. AND MIKOSCH, T. (2001) Point process convergence of stochastic volatility processes with application to sample autocorrelations. *J. Appl. Probab.* Special Volume 38A, A Festschrift for David Vere-Jones, 93–104.

[43] DAVIS, R.A. AND MIKOSCH, T. (2001) The sample autocorrelations of financial time series models. In: Fitzgerald, W.J., Smith, R.L., Walden, A.T. and Young, P.C. (Eds.) *Nonlinear and Nonstationary Signal Processing,* pp. 247–274. Cambridge University Press, Cambridge (U.K.).

[44] DAVIS, R.A. AND RESNICK, S.I. (1985) Limit theory for moving averages of random variables with regularly varying tail probabilities. *Ann. Probab.* 13, 179–195.

[45] DAVIS, R.A. AND RESNICK, S.I. (1985) More limit theory for the sample correlation function of moving averages. *Stoch. Proc. Appl.* 20, 257–279.

[46] DAVIS, R.A. AND RESNICK, S.I. (1986) Limit theory for the sample covariance and correlation functions of moving averages. *Ann. Statist.* 14, 533–558.

[47] DAVIS, R.A. AND RESNICK, S.I. (1996) Limit theory for bilinear processes with heavy tailed noise. *Ann. Appl. Probab.* 6, 1191–1210.

[48] DIACONIS, P. AND FREEDMAN, D. (1999) Iterated random functions. *SIAM Review* 41, 45–76.

[49] DIEBOLT, J. AND GUÉGAN, D. (1993) Tail behaviour of the stationary density of general nonlinear autoregressive processes of order 1. *J. Appl. Probab.* 30, 315–329.

[50] DING, Z. AND GRANGER, C.W.J. (1996) Modeling volatility persistence of speculative returns: A new approach. *J. Econometrics* 73, 185–215.

[51] DING, Z., GRANGER, C.W.J. AND ENGLE, R. (1993) A long memory property of stock market returns and a new model. *J. Empirical Finance* 1, 83–106.

[52] DOUKHAN, P. (1994) *Mixing. Properties and Examples.* Lecture Notes in Statistics 85. Springer Verlag, New York.

[53] DOUKHAN, P., OPPENHEIM, G. AND TAQQU, M.S. (Eds.) (2003) *Long Range Dependence*. Birkhäuser, Boston.

[54] DRAISMA, G. (2000) *Parametric and Semi-Parametric Methods in Extreme Value Theory.* PhD Thesis, Tinbergen Institute, Erasmus University Rotterdam.

[55] DREES, H. (2000) Weighted approximations of tail processes for β-mixing random variables. *Ann. Appl. Probab.* 10, 1274–1301.

[56] DURBIN, J. AND KOOPMAN, S.J. (1997) Monte Carlo maximum likelihood estimation for non-Gaussian state space models. *Biometrika* 84, 669–684.

[57] EINMAHL, J.H.J., HAAN, L. DE AND PITERBARG, V.I. (2001) Nonparametric estimation of the spectral measure of an extreme value distribution. *Ann. Statist.* 29, 1401–1423.

[58] EMBRECHTS, P., KLÜPPELBERG, C. AND MIKOSCH, T. (1997) *Modelling Extremal Events for Insurance and Finance.* Springer, Berlin.

[59] ENGLE, R.F. (1982) Autoregressive conditional heteroscedastic models with estimates of the variance of United Kingdom inflation. *Econometrica* 50, 987–1007.

[60] ENGLE, R.F. (Ed.) (1995) *ARCH Selected Readings.* Oxford University Press, Oxford (UK).

[61] ENGLE, R.F. AND BOLLERSLEV, T. (1986) Modelling the persistence of conditional variances. With comments and a reply by the authors. *Econometric Rev.* 5, 1–87.

[62] FALK, M., HÜSLER, J. AND REISS, R.-D. (1994) *Laws of Small Numbers: Extremes and Rare Events.* Birkhäuser, Basel.

[63] FELLER, W. (1971) *An Introduction to Probability Theory and Its Applications.* Vol. II. Second edition. Wiley, New York.

[64] FREY, R. AND MCNEIL, A.J. (2000) Estimation of tail-related risk measures for heteroscedastic financial time series: an extreme value approach. *J. Empir. Finance* 7, 271–300.

[65] GHYSELS, E., HARVEY, A.C. AND RENAULT, E. (1996) Stochastic volatility. In: *Handbook of Statistics* 14, pp. 119–191. North-Holland, Amsterdam.

[66] GIRAITIS, L., KOKOSZKA, P. AND LEIPUS, R. (2000) Stationary ARCH models: dependence structure and central limit theorem. *Econometric Theory* 16, 3–22.

[67] GIRAITIS, L. AND ROBINSON, P.M. (2001) Whittle estimation of ARCH models. *Econometric Theory* 17, 608–631.

[68] GIRAITIS, L., ROBINSON, P.M. AND SURGAILIS, D. (2000) A model for long memory conditional heteroscedasticity. *Ann. Appl. Probab.* 10, 1002–1024.

[69] GNEDENKO, B.V. AND KOLMOGOROV, A.N. (1954) *Limit Distributions for Sums of Independent Random Variables.* Addison-Wesley, Cambridge, Mass.

[70] GOLDIE, C.M. (1991) Implicit renewal theory and tails of solutions of random equations. *Ann. Appl. Probab.* 1, 126–166.

[71] GOLDIE, C.M. AND GRÜBEL, R. (1996) Perpetuities with thin tails. *Adv. Appl. Probab.* 28, 463–480.

[72] GOLDSHEID, I.YA. (1991) Lyapunov exponents and asymptotic behaviour of the product of random matrices. In: *Lecture Notes in Math.* 1486, pp. 23–37. Springer, Berlin.

[73] GRANGER, C.W.J. AND STĂRICĂ, C. (2000) Non-stationarities in stock returns. Technical Report. Department of Mathematics, Chalmers University Gothenburg.

[74] GREENWOOD, P. AND RESNICK, S.I. (1979) A bivariate stable characterisation and domains of attraction. *J. Multivar. Anal.* 9, 206–221.

[75] GUILLAUME, D.M., PICTET O.V. AND DACOROGNA, M.M. (1995) On the intra-daily performance of GARCH processes. Olsen & Associates (Zürich). Internal Document DMG.1994-07-31.

[76] HAAN, L. DE AND RESNICK, S.I. (1977) Limit theory for multivariate sample extremes. *Z. Wahrscheinlichkeitstheorie verw. Geb.* 40, 317–337.

[77] HAAN, L. DE AND RESNICK, S.I. (1979) Derivatives of regularly varying functions in $\mathbb{R}^d$ and domains of attraction of stable processes. *Stoch. Proc. Appl.* 8, 349–355.

[78] HAAN, L. DE AND RESNICK, S.I. (1993) Estimating the limit distribution of multivariate extremes. *Commun. Statistics: Stochastic Models* 9, 275–309.

[79] HAAN, L. DE, RESNICK, S.I., ROOTZÉN, H. AND VRIES, C. DE (1989) Extremal behaviour of solutions to a stochastic difference equation with applications to ARCH processes. *Stoch. Proc. Appl.* 32, 213–224.

[80] HAAN, L. DE AND RONDE, J. DE (1998) Sea and wind: multivariate extremes at work. *Extremes* 1, 7–45.

[81] HARVEY, A.C., RUIZ, E. AND SHEPHARD, N. (1994) Multivariate stochastic variance models. *Rev. Econom. Stud.* 61, 247–264.

[82] HEYDE, C.C. (1997) *Quasi-Likelihood and Its Application: A General Approach to Optimal Parameter Estimation.* Springer Series in Statistics. Springer, New York.

[83] HORVÁTH, L., KOKOSZKA, P. AND TEYSSIÉRE, G. (2001) Empirical process of squared residuals of an ARCH sequence. *Ann. Statist.* 29, 445–469.

[84] IBRAGIMOV, I.A. AND LINNIK, YU.V. (1971) *Independent and Stationary Sequences of Random Variables.* Wolters–Noordhoff, Groningen.

[85] JACQUIER, E., POLSON., N.G. AND ROSSI, P.E. (1994) Bayesian analysis of stochastic volatility models (with discussion). *J. Busin. Econom. Statist.* 12, 371–417.

[86] JOHNSON, N.L. AND KOTZ, S. (1972) *Distributions in Statistics: Continuous Multivariate Distributions.* Wiley, New York.

[87] KALLENBERG, O. (1983) *Random Measures,* 3rd edition. Akademie–Verlag, Berlin.

[88] KESTEN, H. (1973) Random difference equations and renewal theory for products of random matrices. *Acta Math.* 131, 207–248.

[89] KINGMAN, J.F.C. (1973) Subadditive ergodic theory. *Ann. Probab.* 1, 883–909.

[90] KLÜPPELBERG, C. AND PERGAMENCHTCHIKOV, S. (2001) Renewal theory for functionals of a Markov chain with compact state space. *Ann. Probab.* To appear.

[91] KLÜPPELBERG, C. AND PERGAMENCHTCHIKOV, S. (2002) The tail of the stationary distribution of a random AR(q) model. *Ann. Appl. Probab.* To appear.

[92] KOKOSZKA, P. AND LEIPUS, R. (2000) Change-point estimation in ARCH models. *Bernoulli* 6, 513–539.

[93] KOTZ, S. AND NADARAJAH, S. (2000) *Extreme Value Distributions. Theory and Applications.* Imperial College Press, London.

[94] LEADBETTER, M.R., LINDGREN, G. AND ROOTZÉN, H. (1983) *Extremes and Related Properties of Random Sequences and Processes.* Springer, Berlin.

[95] LEDFORD, A.W. AND TAWN, J.A. (1997) Modelling dependence within tail regions. *J. Roy. Statist. Soc. B* 59, 475–499.

[96] LEE, S.W. AND HANSEN, B.E. (1994) Asymptotic theory for the GARCH(1,1) quasi-maximum likelihood estimator. *Econometric Theory* 10, 29–52.

[97] LE PAGE, E. (1983). Théorèmes de renouvellement pour les produits de matrices aléatoires. Équations aux différences aléatoires. Séminaires de probabilités Rennes 1983. *Publ. Sém. Math.* Univ. Rennes I, 116 pp.

[98] LUMSDAINE, R.L. (1996) Consistency and asymptotic normality of the quasi-maximum likelihood estimator in IGARCH(1,1) and covariance stationary GARCH(1,1) models. *Econometric Theory* 64, 575–596.

[99] MANDELBROT, B. (1963) The variation of certain speculative prices. *J. Busin. Univ. Chicago* 36, 394–419.

[100] MANDELBROT, B. (1997) *Fractals and Scaling in Finance.* Discontinuity, concentration, risk. Selecta Volume E. Selected works by Benoit B. Mandelbrot. Springer, New York.

[101] MANDELBROT, B. (2001) Scaling in financial prices I. Tails and dependence. *Quantitative Finance* 1, 113–123.

[102] MEERSCHAERT, M.M. AND SCHEFFLER, P. (2001) *Limit Distributions for Sums of Independent Random Variables: Heavy Tails in Theory and Practice.* Wiley, New York.

[103] MEYN, S.P. AND TWEEDIE, R.L. (1993) *Markov Chains and Stochastic Stability.* Springer, London.

[104] MIKOSCH, T., GADRICH, T., KLÜPPELBERG, C. AND ADLER, R. (1995) Parameter estimation for ARMA models with infinite variance innovations. *Ann. Statist.* 23, 305–326.

[105] MIKOSCH, T. AND SAMORODNITSKY, G. (2000) The supremum of a negative drift random walk with dependent heavy-tailed steps. *Ann. Appl. Probab.* 10, 1025–1064.

[106] MIKOSCH, T. AND STĂRICĂ, C. (1999) Change of structure in financial time series, long-range dependence and the GARCH model. Technical Report. Available under www.math.ku.dk/~mikosch.

[107] MIKOSCH, T. AND STĂRICĂ, C. (2000) Limit theory for the sample autocorrelations and extremes of a GARCH(1,1) process. *Ann. Statist.* 28, 1427–1451. An extended version is available under www.math.ku.dk/~mikosch.

[108] MIKOSCH, T. AND STĂRICĂ, C. (2000) Is it really long memory we see in financial returns? In: Embrechts, P. (Ed.) *Extremes and Integrated Risk Management*, pp. 149–168. Risk Books, London.

[109] MIKOSCH, T. AND STĂRICĂ, C. (2003) Long range dependence effects and ARCH modeling. In: Doukhan, P., Oppenheim, G. and Taqqu, M.S. (Eds.) *Long Range Dependence*. Birkhäuser, Boston. pp. 439–460.

[110] MIKOSCH, T. AND STRAUMANN, D. (2002) Whittle estimation in a heavy-tailed GARCH(1,1) model. *Stoch. Proc. Appl.* 187–222.

[111] MITTNIK, S. AND RACHEV, S. (2000) *Stable Paretian Models in Finance.* Wiley, New York.

[112] NELSON, D.B. (1990) Stationarity and persistence in the GARCH(1, 1) model. *Econometric Theory* 6, 318–334.

[113] NOVAK, S. (2000) Estimating Value-at-Risk from dependent data. Technical Report.

[114] NYRHINEN, H. (2001) Finite and infinite time ruin probabilities in a stochastic environment. *Stoch. Proc. Appl.* 92, 265–286.

[115] PETROV, V.V. (1975) *Sums of Independent Random Variables.* Springer, Berlin.

[116] PITTS, S.M., GRÜBEL, R. AND EMBRECHTS, P. (1996) Confidence bounds for the adjustment coefficient. *Adv. Appl. Probab.* 28, 802–827.

[117] PRIESTLEY, M.B. (1981) *Spectral Analysis and Time Series.* Academic Press, London.

[118] RESNICK, S.I. (1986) Point processes, regular variation, and weak convergence. *Adv. Appl. Prob.* 18, 66–138.

[119] RESNICK, S.I. (1987) *Extreme Values, Regular Variation, and Point Processes.* Springer, New York.

[120] RESNICK, S.I. (1997) Heavy tail modeling and teletraffic data. With discussion and a rejoinder by the author. *Ann. Statist.* 25, 1805–1869.

[121] RESNICK, S.I. (2004) Modeling data networks. In: *Extreme Values in Finance, Telecommunications, and the Environment*, Finkenstädt, B. and Rootzén, H., Eds., CRC/Chapman & Hall, Boca Raton, FL.

[122] RESNICK, S.I. AND BERG, E. VAN DEN (2000) A test for nonlinearity of time series with infinite variance. *Extremes* 3, 145–172.

[123] RESNICK, S.I. AND STĂRICĂ, C. (1998) Tail index estimation for dependent data. *Ann. Appl. Probab.* 8, 1156–1183.

[124] RESNICK, S.I. AND WILLEKENS, E. (1991) Moving averages with random coefficients and random coefficient autoregressive models. *Commun. Statistics: Stochastic Models* 7, 511–525.

[125] ROSENBLATT, M. (1956) A central limit theorem and a strong mixing condition. *Proc. Nat. Acad. Sci. U.S.A.* 42, 43–47.

[126] ROSIŃSKI, J. AND WOYCZYŃSKI, W.A. (1987) Multilinear forms in Pareto-like random variables and product random measures. *Colloq. Math.* 51, 303–313.

[127] RVAČEVA, E.L. (1962) On domains of attraction of multi-dimensional distributions. *Select. Transl. Math. Statist. and Probability.* American Mathematical Society, Providence, RI 2, 183–205.

[128] SAMORODNITSKY, G. AND TAQQU, M.S. (1994) *Stable Non–Gaussian Random Processes. Stochastic Models with Infinite Variance.* Chapman and Hall, London.

[129] SHEPHARD, N. (1996) Statistical aspects of ARCH and stochastic volatility. In: Cox, D.R., Hinkley, D.V. and Barndorff-Nielsen, O.E. (Eds.) *Likelihood, Time Series with Econometric and Other Applications.* Chapman and Hall, London.

[130] SHIRYAEV, A.N. (1999) *Essentials of Stochastic Finance — Facts, Models, Theory.* World Scientific, Singapore.

[131] STĂRICĂ, C. (2000) Multivariate extremes for models with constant conditional correlations. *J. Emp. Finance* 6, 513–553.

[132] TAYLOR, S.J. (1986) *Modelling Financial Time Series.* Wiley, Chichester.

[133] TONG, Y.L. (1990) *The Multivariate Normal Distribution.* Springer, New York.

[134] VERVAAT, W. (1979) On a stochastic difference equation and a representation of nonnegative infinitely divisible random variables. *Adv. Appl. Probab.* 11, 750–783.

[135] WEISS, A.A. (1986) Asymptotic theory for ARCH models: estimation and testing. *Econometric Theory* 2, 107–131.

[136] ZYGMUND, A. (1968) *Trigonometric Series, I, II.* Second edition. Cambridge University Press, London.

CHAPTER 6

Modeling Data Networks

Sidney Resnick
Cornell University

Contents

1-58488-411-8/04/$0.00+$.50

Data networks offer a fascinating, if somewhat potentially frustrating, setting for many applied probability and extreme value techniques to be applied. We survey some of the basic models and statistical techniques for fitting the models. We point out some of the shortcomings in the models and in the statistical techniques. The required range of techniques is broad. The ability to contribute in internet time is questionable.

6.1 Introduction

The story begins around 1993 with the publication of what is now known as the Bellcore study [54, 100, 164]. Traditional queueing models had thrived on assumptions of exponentially bounded tails, Poisson inputs and lots of independence. Collected network data studied at what was then Bellcore (now Telcordia) exhibited properties that were inconsistent with traditional queueing models. These anomalies were also found in world wide web downloads in the Boston University study [28, 29, 30, 31, 32, 33, 36]. The unusual properties found in the data traces included:

- Self-similarity (ss) and long-range dependence (LRD) of various transmission rates:
  - packet counts per unit time
  - www bits/time
- Heavy tails of quantities such as
  - file sizes
  - transmission rates
  - transmission durations
  - CPU job completion times
  - call lengths

The Bellcore study in early 1990s resulted in a paradigm shift worthy of a sociological study to understand the frenzy to jump on and off various bandwagons, but after some resistance to the presence of long range dependence, there was widespread acceptance of the statement that packet counts per unit time exhibit self similarity and long range dependence. Research goals then shifted from detection of the phenomena to greater understanding of the causes. The challenges were:

- Explain the origins and effects of long-range dependence and self-similarity.
- Understand some connections between self-similarity, long range dependence, and heavy tails. Use these connections to find an explanation for the perceived long range dependence in traffic measurements.
- Begin to understand the effect of network protocols and architecture on traffic. The simplest models, such as the featured infinite source Poisson model, pretend protocols, and controls are absent. This is an ambitious goal.
- Say something useful for the purposes of capacity planning.

6.1.1 The infinite node Poisson model

Attempts to explain long range dependence and self-similarity in traffic rates centered around the paradigm: heavy tailed file sizes cause LRD in network traffic. Specific models must be used to explain this and the two most effective and simple models were:

Superposition of on/off processes [74, 75, 88, 109, 110, 120, 153, 157, 164]. This is described as follows: imagine a source/destination pair. The source sends at unit rate for a random length of time to the destination and then is silent or inactive for a random period. Then the source sends again and when finished is silent and so on. So the transmission schedule of the source follows an alternating renewal or on/off structure. Now imagine the traffic generated by many source/destination pairs being superimposed and this yields the overall traffic.

The infinite source Poisson model, sometimes called the M/G/∞ input model [72, 76, 89, 90, 109, 121, 137, 142]. Imagine infinitely many potential users connected to a single server that processes work at constant rate r. At a Poisson time point, some user begins transmitting work to the server at constant rate which, without loss of generality, we take to be rate one. The length of the transmission is random with heavy tailed distribution. The length of the transmission may be considered to be the size of the file needing transmission.

Both models have their adherents, and the two models are asymptotically equivalent in a manner noone (to date) has made fully transparent. We will focus on the infinite source Poisson model.

Some good news about the model:

- It is somewhat flexible and certainly simple.
- Since each node transmits at unit rate, the overall transmission rate at time t is simply the number of active users $N(t)$ at t. From classical M/G/∞ queueing theory, we know $N(t)$ is a Poisson random variable with mean $\lambda\mu_{\text{on}}$ where λ is the rate parameter of the Poisson process, and μ_{on} is the mean file size or mean transmission length.
- The length of each transmission is random and heavy tailed.
- The model offers a very simple explanation of long range dependence being caused by heavy tailed file sizes.
- The model predicts traffic aggregated over users and accumulated over time $[0, T]$ is approximated by either a Gaussian process (fractional Brownian motion or FBM) or a heavy tailed stable Lévy motion [109]. Thus the two approximations are very different in character but at least both are self-similar.

Some less good news about the model:

- The model does not fit collected data traces all that well.
  - The constant transmission rate assumption is clearly wrong. Each of us knows from personal experience that downloads and uploads do not proceed at a constant rate.

  - Not all times of transmissions are Poisson. Identifying Poisson time points in the data can be problematic. Some are machine triggered and these will certainly not be Poisson. While network engineers rightly believe in the invariant that behavior associated with humans acting independently can be modeled as a Poisson process, it is highly unlikely that, for example, subsidiary downloads triggered by going to the CNN website (imagine the calls to DoubleClicks ads) would follow a Poisson pattern.

- There is no hope that this simple model can successfully match fine time scale behavior observed below, say, 100 milliseconds. Below this time scale threshold, observational studies speculate that traffic exhibits multifractal characteristics.
- The model does not take into account admission and congestion controls such as TCP. How can one incorporate a complex object such as a control mechanism into an informative probability model?

6.1.2 Broad issues (BIs) to consider for data network modeling

Before considering further the infinite node Poisson model, consider the following issues related to data network modeling and data analysis.

BI 1. The problem of research time scales: how do the disciplines of applied probability, statistics, and applied mathematics make contributions to data network analysis and planning in internet time? The rapid pace of development makes it difficult for mathematical research to find a niche. If you propose a research project lasting perhaps 2 to 3 years to an internet engineer, their eyes will glaze over. The world will be a different place by the time the project is completed and your project results may or may not have applicability. How many of us remember using Mosaic as our browser? The year 1994 may be fresh in some of our minds, but is the dark ages of the internet. Several of the data sets analyzed in [72] are quite old (of the order of 7 years and the clock is still ticking) and this raises the question of their relevance. Are these measurements an accurate reflection of current traffic? Maybe not. If not, are the methods collected and developed to analyze the old data sets at least useful for analyzing current measurements? Hopefully.

Is the most meaningful contribution we can make that we simply help cause and justify paradigm shifts with explanations that may lag behind empirical, experimental, and heuristic developments?

Consider the following sobering thought. When the monolithic AT&T spun off Lucent and Lucent wound up with Bell Laboratories, the newly constituted AT&T Labs-Research in Florham Park did not stock the building with researchers and engineers from the applied probability community. (My informal count in 1999 of people whose primary affiliation was applied probability was two, though of course there are many broad and talented people who defy easy classification.) Why did the expertise of the applied probability community not appeal to the organizers of the Florham Park facility?

BI 2. Insider vs. outsider: do you:

(a) Analyze data that is already available, say on the web (e.g., the ITA web site at http://ita.ee.lbl.gov/html/traces.html) or that you have bootlegged by hook or by crook? This is often what academics (including the author) have done. However, such data may be old (dangerous when the internet time clock is ticking), badly suited for the purposes of the study, or just plain dirty. Note, for example, the UCB data analyzed in [72] is not very suitable for testing the Poisson assumption, or

(b) Design a network experiment to get the data you want. This typically requires cooperative net administrators and some hardware and software expertise more typically found in the computer science and electrical engineering communities. It may require you to go beyond your local area network to something of the scale of World Net, UUNet, etc.

BI 3. How can you discern the influence of network architectures and protocols? How do you model the complexities of something like TCP [92] realistically? Can you get a reasonably accurate model that encourages analytic analysis, or if mathematical analysis is too difficult, is the model accurate enough that simulation and experimentation yield fruitful information? Some beginning but not entirely satisfactory attempts to grapple with the dynamics of TCP are contained in [6, 17, 67, 116, 117, 118, 152].

BI 4. A broad variety of techniques is useful and a wide background is required. Teamwork may be the way to go. Here is a partial list of skills and expertise that your team's toolbox should have:

(a) Applied probability and statistics:

- Stochastic processes: FBM, Lévy stable motion, and fractional motions
- Poisson point process theory
- weak convergence
- heavy tailed analysis
- long-range dependence
- self-similarity and multifractality
- extreme value theory and associated statistical techniques
- estimation methods such as maximum likelihood, exploratory tools using graphical techniques
- time series analysis
- queueing theory

(b) Applied mathematics:

- wavelets (seem to be the right tool for examining phenomena on different time scales)
- numerical methods
- design and implementation of simulation tool
- computing

BI 5. Black box vs. structural modeling: the philosophy behind black box modeling is to provide a broad class of models with enough parameters to fit a variety of data sets. The emphasis is on finding something that fits according to some criterion. For example, the classical Box-Jenkins ARMA approach to fitting time series models is to difference the data until it looks stationary and then fit an ARMA(p, q) model of the form

$$X_n = \sum_{i=1}^{p} \phi_i X_{n-i} + \sum_{i=0}^{q} \theta_i Z_{n-i},$$

where one specifies p, q, ϕ, and θ to yield a model matching the L_2 sample moments of the data. Traditionally, $\{Z_n\}$ is white noise.
One may try to adapt this class of models to a heavy tailed context, in which case it is natural to suppose $\{Z_n\}$ is i.i.d. and heavy tailed. However, while such highly structured methods may have some appeal in economics and finance, in other contexts the approach may have drawbacks.
Consider the following problems with the approach.

(a) The dependence structure in heavy tailed models is complex, and there is no reason why a highly structured linear model will model the dependence correctly. Generally, you do not get good fits for dependent heavy tailed data with ARMA modeling and one could go out on a limb and assert that the only heavy tailed data that ARMA fits are simulated data from the ARMA model.

(b) Even if the black box approach of fitting something like an ARMA worked acceptably, there would be a tendency to say "so what," since this does not provide fundamental insights into system dynamics. A method that finds a pattern in a bunch of numbers but ignores physics, structure, and system dynamics is not so revealing.

(c) In a rapidly changing environment, spending excessive time just fitting the data to a black box model may not be so useful, since the next data set generated from similar mechanisms may not be fit to the same model. One should not overemphasize this point, but keep in mind the goal is to understand system dynamics and not just to successfully fit a model to the data.

These objections have caused the internet community to emphasize structural modeling, which incorporates idealized features of the network. The typical internet trace collects packet headers that contain rather detailed information, and this can hopefully be utilized to good effect.

6.2 How do heavy tails cause long range dependence?

Understanding the connection between heavy tails and long range dependence requires a context. For the simplest explanations one can choose either the superposition of on/off processes or the infinite node Poisson model, and our preference is for the latter.

6.2.1 The infinite node Poisson model

The simplest model that explains the paradigm that heavy tails induce long range dependence is the infinite source Poisson model, sometimes called the M/G/∞ input model.

In this model, there is potentially an infinite number of sources capable of sending work to the server. Imagine that transmission sources turn on and initiate sessions or connections at Poisson time points $\{\Gamma_k\}$ with rate λ. The lengths of sessions $\{L_n\}$ are i.i.d. nonnegative random variables with common distribution F_{on} and during a session, work is transmitted to the server at constant rate. As a normalization, we assume the transmission rate is 1. Assume

$$1 - F_{\text{on}}(t) := \bar{F}_{\text{on}}(t) = t^{-\alpha} L(t), \quad t \to \infty; \tag{6.2.1}$$

for some slowly varying function L, that is,

$$\lim_{t\to\infty} \frac{\bar{F}_{\text{on}}(tx)}{\bar{F}_{\text{on}}(t)} = x^{-\alpha}, \quad x > 0.$$

In practice, empirical estimates of α usually range between 1 and 2 [101, 164]. However, studies of file sizes sometimes report measurements of $\alpha < 1$ [7, 142]. The assumption of a fixed unit transmission rate is one of the more unrealistic aspects of the model, which accords neither with anyone's personal experience nor with measurement studies. Later, in Section 6.8, we mention how to modify the model in the interests of greater realism so that either (i) transmission rates are random and possibly dependent on the size of the file to be transmitted; or (ii) it is assumed that cumulative input from a source follows a random multifractal process. However, for the present, in the interests of simplicity and for tractability, the fixed transmission rate will be assumed.

Note that in the case of $0 < \alpha < 1$, which we will term the very heavy tailed case, both the mean and the variance of F_{on} are infinite. In case $1 < \alpha < 2$, termed merely the heavy tailed case, the variance of F_{on} is infinite but

$$\mu_{\text{on}} = E(L_1) = \int_0^\infty \bar{F}_{\text{on}}(t)dt < \infty.$$

The processes of primary interest for describing this system are the following:

$$\begin{aligned} N(t) &= \text{number of sessions in progress at } t \\ &= \text{number of busy servers in the M/G/}\infty\text{ model} \\ &= \sum_{k=1}^{\infty} 1_{[\Gamma_k \le t < \Gamma_k + L_k]} \end{aligned} \tag{6.2.2}$$

and

$$A(t) = \int_0^t N(s)ds = \text{cumulative input in } [0, t], \tag{6.2.3}$$

$$r = \text{release rate or the rate at which the server works off the offered load.}$$

Note that expressing $A(t)$ as an integral gives $N(t)$ the interpretation of instantaneous input rate at time t. So realizations of $N(t)$ correspond to data traces of packet counts per unit time.

Stability requires us to assume that the long term input rate should be less than the output rate, so we require

$$\lambda\mu_{\text{on}} < r.$$

This means the content or buffer level process $\{X(t), t \geq 0\}$, which satisfies

$$dX(t) = N(t)dt - r1_{[X(t)>0]}dt,$$

is regenerative with finite mean regeneration times and achieves a stationary distribution.

6.2.2 *Connection between heavy tails and long range dependence*

The common explanation for long range dependence in the total transmission rate by the system is that high variability causes long range dependence, where we understand that high variability means heavy tails. The long range dependence resulting from the heavy tailed distribution F_{on} can be easily seen for the infinite node Poisson model.

Assume that $1 < \alpha < 2$. To make our argument transparent, we consider the following background. For each t, $N(t)$ is a Poisson random variable. Why? When $1 < \alpha < 2$, $N(\cdot)$ has a stationary version on $\mathbb{R}$, the whole real line. Assume

$$\sum_k \epsilon_{\Gamma_k} = \text{PRM}(\lambda dt)$$

is a homogeneous Poisson random measure on $\mathbb{R}$, with rate λ. Then

$$M := \sum_k \epsilon_{(\Gamma_k, L_k)} = \text{PRM}(\lambda dt \times F_{\text{on}}) \tag{6.2.4}$$

is a two-dimensional Poisson random measure on $\mathbb{R} \times [0, \infty)$ (e.g., [139]) with mean measure $\lambda dt \times F_{\text{on}}(dx)$, and

$$\begin{aligned} N(t) &= \sum_k 1_{[\Gamma_k \leq t < \Gamma_k + L_k]} \\ &= M(\{(s, l) : s \leq t < s + l\} = M(B) \end{aligned}$$

is Poisson because it is the two-dimensional Poisson process M evaluated on the region B, see Figure 6.1. Note B is the region in the (s, l)-plane to the left of the vertical line through $(t, 0)$ and above the -45 degree line through $(t, 0)$. The mean

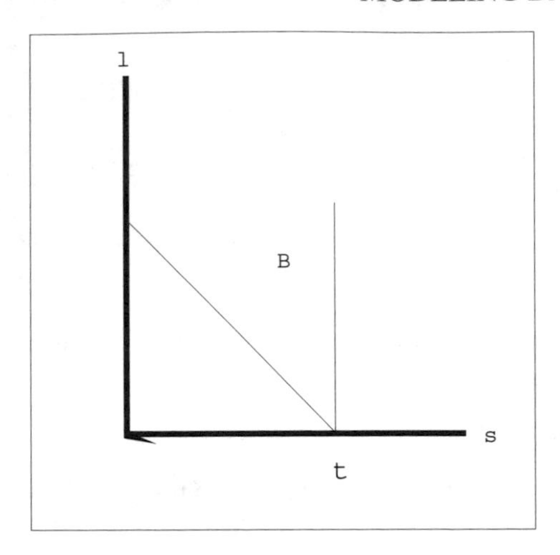

Figure 6.1 *The region B.*

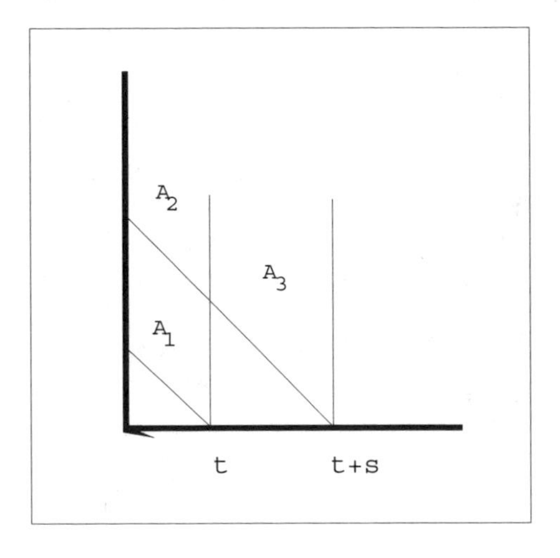

Figure 6.2 *The regions A_1, A_2, and A_3.*

of $M(B)$ is

$$E(M(\{(s,l) : s \leq t < s+l\})) = \int\int_{\{(s,l):s\leq t<s+l\}} \lambda ds \bar{F}_{\text{on}}(dl)$$
$$= \int_{s=-\infty}^{t} \bar{F}_{\text{on}}(t-s)\lambda ds = \lambda\mu_{\text{on}}. \quad (6.2.5)$$

Understanding the relation between $\{N(t)\}$ and the random measure M allows us to compute easily the covariance function. Refer to Figure 6.2. Recall that $N(t)$ corresponds to points to the left of the vertical through $(t, 0)$ and above the -45 degree line through $(t, 0)$, with a similar interpretation for $N(t+s)$. The process

$\{N(t), t \in \mathbb{R}\}$ is stationary with covariance function

$$\mathrm{Cov}(N(t), N(t+s)) = \mathrm{Cov}(M(A_1) + M(A_2), M(A_2) + M(A_3)),$$

and because $M(A_1)$ and $M(A_3)$ are independent, the previous expression reduces to

$$= \mathrm{Cov}(M(A_2), M(A_2)) = \mathrm{Var}(M(A_2)).$$

For a Poisson random variable, the mean and the variance are equal, and therefore the above equals

$$= E(M(A_2)) = \int_{u=-\infty}^{t} \lambda du \; \bar{F}_{\mathrm{on}}(t+s-u)$$
$$= \lambda \int_s^\infty \bar{F}_{\mathrm{on}}(v)dv \sim \frac{\lambda}{\alpha - 1} s^{-(\alpha-1)} L(s),$$

where we used Karamata's theorem for regularly varying functions [14, 42, 139]. To summarize, we find that

$$\begin{aligned} \mathrm{Cov}(N(t), N(t+s)) &= \lambda \int_s^\infty \bar{F}_{\mathrm{on}}(v)dv \\ &\sim \frac{\lambda}{\alpha-1} s^{-(\alpha-1)} L(s) \\ &= \frac{\lambda}{\alpha-1} s \bar{F}_{\mathrm{on}}(s), \quad s \to \infty. \end{aligned} \tag{6.2.6}$$

The slow decay of the covariance as a function of the lag s characterizes long range dependence.

6.3 Further implications of the simple model: is network traffic stable Lévy motion or fractional Brownian motion?

Our simple Poisson based model offers a compelling explanation of how heavy tailed file sizes induce long range dependence in the traffic rates. To decide if our model is an accurate enough reflection of reality, however, we need to see how well data measurements fit the model. So we require a partial catalogue of features of the model, in order to see if such features are found in data measurements. In this section, based on [109], we analyze what the model predicts about the cumulative traffic process.

6.3.1 Background

There is strong belief in the self-similar nature of aggregate traffic rates, at least at time scales above a certain threshold. Empirical [7, 164] and theoretical [74, 75, 76, 157] evidence supports the heavy tailed explanation of the self-similarity. However, various authors diverge in their conclusions about the marginal distributions of cumulative traffic. There exists theoretical interest in the conclusion that marginal distributions should be heavy tailed [93, 137] and spirited interest and evidence for Gaussian marginal distributions [101]. The latter result coincides with empirical observations that traffic looks Gaussian on links that are heavily loaded.

The infinite source Poisson model with heavy tailed file sizes allows cumulative traffic at large time scales to look either heavy tailed or Gaussian, depending on whether the rate at which transmissions are initiated (crudely referred to as the connection rate) is moderate or quite large.

The process describing offered traffic is $A(t)$, the cumulative input in $[0, t]$ by all sources. Recall from (6.2.2) and (6.2.3) that the model assumes unit rate transmissions and $A(t)$ is the integral of $N(s)$ over $[0, t]$. For large T, we think of $(A(Tt), t \geq 0)$ as the process on large time scales. The results show that if the connection rate $\lambda(\cdot)$ is allowed to depend on T in such a way that it has a growth rate in T that is moderate (in a manner to be made precise), then $A(T\cdot)$ looks like an α-stable Lévy motion, while if the connection rate grows faster than a critical value, $A(T\cdot)$ looks like a fractional Brownian motion. We make these statements precise by adopting a heavy traffic outlook and think of a family of models indexed by T, where the T-th model has connection rate $\lambda(T)$ and file size distribution F_{on}. Depending on growth rates, the T-th model is approximated by either Lévy stable motion or FBM.

Let $(\Gamma_k, -\infty < k < \infty)$ be the points of the rate λ homogeneous Poisson process on $\mathbb{R}$, labeled so that $\Gamma_0 < 0 < \Gamma_1$ and hence $\{-\Gamma_0, \Gamma_1, (\Gamma_{k+1} - \Gamma_k, k \neq 0)\}$ are i.i.d. exponentially distributed random variables with parameter λ. The random measure that counts the points is denoted by $\sum_{k=-\infty}^{\infty} \epsilon_{\Gamma_k}$ and is a Poisson random measure with mean measure $\lambda\mathbb{L}$, where $\mathbb{L}$ stands for Lebesgue measure. The communication system has an infinite number of nodes or sources, and at time Γ_k a connection is made and some node begins a transmission at constant rate to the server. As a normalization, this constant rate is taken to be unity. The lengths of transmissions are random variables L_k. Assume $L_{\text{on}}, L_1, L_2, \ldots$ are i.i.d. and independent of (Γ_k), and

$$P(L_{\text{on}} > x) = \bar{F}_{\text{on}}(x) = x^{-\alpha} L(x), \quad x > 0, \quad 1 < \alpha < 2, \tag{6.3.1}$$

where L is a slowly varying function. Since $\alpha \in (1, 2)$, the variance of L_{on} is infinite and its mean μ_{on} is finite. We will need the quantile function

$$b(t) = (1/\bar{F}_{\text{on}})^{\leftarrow}(t) =: \inf\left\{x : \frac{1}{1 - F_{\text{on}}(x)} \geq t\right\}, \quad t > 0, \tag{6.3.2}$$

which is regularly varying with index $1/\alpha$. Recall the two-dimensional Poisson random measure M defined by (6.2.4), which is a counting function on $\mathbb{R} \times [0, \infty]$ corresponding to the points $\{(\Gamma_k, L_k)\}$ and has mean measure $\lambda\mathbb{L} \times F_{\text{on}}$; cf. [139].

To remind us we consider the T-th model, we sometimes subscript quantities by T. So for example, the number of active sources at t, or the overall transmission rate at t is denoted by either $N(t)$ or $N_T(t)$. We will consider a family of Poisson processes indexed by the scaling parameter $T > 0$ such that the intensity $\lambda = \lambda(T)$ goes to infinity as $T \to \infty$. The intensity $\lambda = \lambda(T)$ will be referred to as the connection rate for the T-th model.

Recall that heavy tailed transmission times L_k induce long range dependence in N; the precise expression of this is (6.2.6). High variability in transmission times causes long range dependence in the rate at which work is offered to the system.

6.3.2 The critical input rate

Recall that $\lambda = \lambda(T)$ is the parameter governing the connection rate in the T-th model, and suppose $\lambda = \lambda(T)$ is a nondecreasing function of T. We phrase our condition first in terms of the quantile function b defined in (6.3.2). The asymptotic behavior of $A_T(\cdot)$ depends on whether

$$\text{Slow Growth Condition 1:} \quad \lim_{T\to\infty} \frac{b(\lambda T)}{T} = 0,$$

or

$$\text{Fast Growth Condition 2:} \quad \lim_{T\to\infty} \frac{b(\lambda T)}{T} = \infty$$

holds. Notice that $b(\cdot)$ is regularly varying with index $1/\alpha$.

There is an alternative, more intuitive, way to express the conditions.

Lemma 1 *Assume F_{on} satisfies (6.3.1). Consider the stationary version of the input rate $N_T(\cdot)$.*

1. The slow growth condition 1 is equivalent to either of the two conditions

$$\lim_{T\to\infty} \lambda T\ \bar F_{\text{on}}(T) = 0 \quad \textit{or} \quad \lim_{T\to\infty} \textit{Cov}\,(N_T(0), N_T(T)) = 0. \tag{6.3.3}$$

2. The fast growth condition 2 is equivalent to either of the two conditions

$$\lim_{T\to\infty} \lambda T \bar F_{\text{on}}(T) = \infty \quad \textit{or} \quad \lim_{T\to\infty} \textit{Cov}\,(N_T(0), N_T(T)) = \infty. \tag{6.3.4}$$

If we think of the scaled process $N_T(t) = N(Tt)$, then the covariance appearing in (6.3.3) and (6.3.4) is the lag 1 covariance of $N_T(\cdot)$. So, as we proceed through our family of models indexed by T, under slow growth, the lag 1 covariance is diminishing at large scales, and under fast growth the lag 1 covariance is getting very strong.

Proof. In the case of Condition 1, there exists a function $0 < \epsilon(T) \to 0$ such that $T\epsilon(T) \to \infty$ and $b(\lambda T) = T\epsilon(T)$. Thus

$$\lambda T \sim 1/\bar F_{\text{on}}(T\epsilon(T)). \tag{6.3.5}$$

Therefore, Condition 1 implies

$$\lambda T\ \bar F_{\text{on}}(T) \sim \bar F_{\text{on}}(T)/\bar F_{\text{on}}(T\epsilon(T)) \to 0. \tag{6.3.6}$$

Conversely, if $\delta(T) := \lambda T \bar F_{\text{on}}(T) \to 0$, then using $b^{\leftarrow}(T) \sim 1/\bar F_{\text{on}}(T)$, we get

$$\frac{b(\lambda T)}{T} \sim \frac{b(\delta(T) b^{\leftarrow}(T))}{b(b^{\leftarrow}(T))} \to 0,$$

and so Condition 1 and (6.3.6) are equivalent. Similarly, Condition 2 is the same as

$$\lambda T\ \bar F_{\text{on}}(T) \to \infty. \tag{6.3.7}$$

To get the equivalence in terms of the covariances, use (6.2.6). □

We will need the following facts proven in [72]. If Condition 1 holds, then

$$\lim_{T\to\infty} \frac{\lambda T^2 \bar F_{\text{on}}(T)}{b(\lambda T)} = 0, \tag{6.3.8}$$

and if Condition 2 holds, this limit is infinite.

6.3.3 α-stable approximations for the infinite source Poisson model under slow growth

We now assume Condition 1 holds and show why at large time scales, A is approximately an α-stable Lévy motion; that is, a process with stationary independent increments with α-stable marginals [64, 151]. The following is the result under the slow growth condition.

Theorem 1 *If Condition* 1 *holds, then the process* $(A(Tt), t \geq 0)$ *describing the total accumulated input in* $[0, Tt]$, $t \geq 0$, *satisfies the limit relation*

$$X^{(T)}(\cdot) := \frac{A(T\cdot) - T\lambda\mu_{\text{on}}(\cdot)}{b(\lambda T)} \stackrel{fidi}{\rightarrow} X_\alpha(\cdot), \tag{6.3.9}$$

where $X_\alpha(\cdot)$ *is an* α*-stable Lévy motion. Here* $\stackrel{fidi}{\rightarrow}$ *denotes convergence of the finite dimensional distributions.*

Remark. The mode of converge cannot be extended to J_1 convergence in the Skorokhod space $D[0, \infty)$. This follows, for example, from [93] who show that a sequence of processes with a.s. continuous sample paths cannot converge in distribution in $(D[0, \infty), J_1)$ to a process with a.s. discontinuous sample paths. A thorough discussion of this phonomena is in [162]; see also [137].

Here is an outline of some elements of the proof.

6.3.3.1 The basic decomposition

We start by giving a useful decomposition of the random variable $A(T)$ corresponding to a decomposition of $(-\infty, T] \times [0, \infty)$:

$$\begin{aligned} R_1 &:= \{(s, y) : 0 < s \leq T,\ 0 < y,\ s + y \leq T\}, \\ R_2 &:= \{(s, y) : 0 < s \leq T,\ T < s + y\}, \\ R_3 &:= \{(s, y) : s \leq 0,\ 0 < s + y \leq T\}, \\ R_4 &:= \{(s, y) : s \leq 0,\ T < s + y\}, \end{aligned} \tag{6.3.10}$$

(see Figure 6.3). Rewrite $A(T)$, using (6.2.3), as

$$\begin{aligned} A(T) &= \sum_k L_k 1_{[(\Gamma_k, L_k) \in R_1]} + \sum_k (T - \Gamma_k) 1_{[(\Gamma_k, L_k) \in R_2]} \\ &\quad + \sum_k (L_k + \Gamma_k) 1_{[(\Gamma_k, L_k) \in R_3]} + \sum_k T 1_{[(\Gamma_k, L_k) \in R_4]} \\ &=: A_1 + A_2 + A_3 + A_4 \,. \end{aligned} \tag{6.3.11}$$

Recall the definition of the PRM M from (6.2.4) with mean measure $\lambda\mathbb{L} \times F_{\text{on}}$. Note that A_i is a function of the points of M in region R_i, and because the R_is are disjoint, $A_i, i = 1, \ldots, 4$, are mutually independent. Calculations as in (6.2.5) and use of

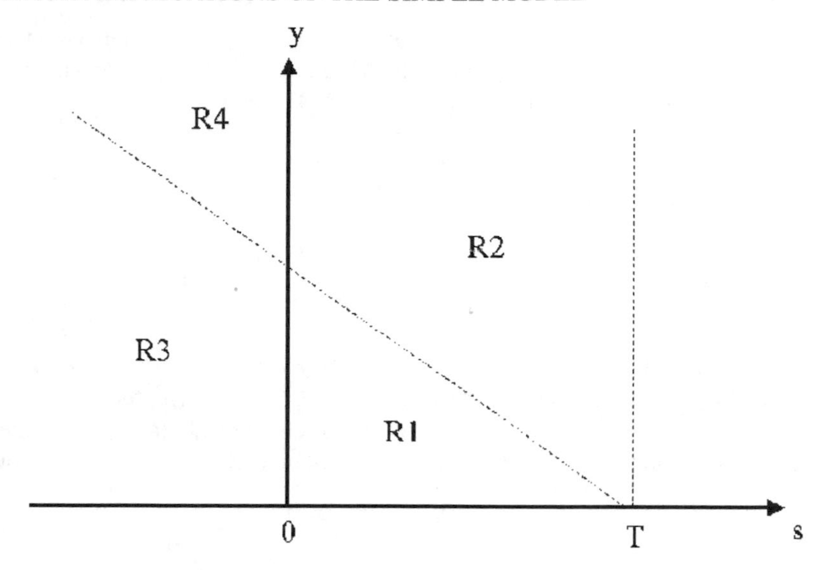

Figure 6.3 *The regions R_1, R_2, R_3, and R_4.*

Karamata's theorem give that as $T \to \infty$

$$\begin{aligned}
\lambda m_1 &:= EM(R_1) = \lambda \int_0^T F_{\text{on}}(T-s)\,ds \sim \lambda T, \\
\lambda m_2 &:= EM(R_2) = \lambda \int_0^T \bar{F}_{\text{on}}(T-s)\,ds \sim \lambda\, \mu_{\text{on}}, \qquad (6.3.12) \\
\lambda m_3 &:= EM(R_3) = \lambda \int_{s=-\infty}^0 \int_{y=-s}^{-s+T} F_{\text{on}}(dy)\,ds \sim \lambda \mu_{\text{on}} \\
\lambda m_4 &:= EM(R_4) = \lambda \int_{s=-\infty}^0 \int_{y=-s+T}^{\infty} F_{\text{on}}(dy)\,ds = \lambda \int_T^\infty \bar{F}_{\text{on}}(u)du \\
&\sim \lambda T \bar{F}_{\text{on}}(T)/(\alpha-1) \to 0.
\end{aligned}$$

So the mean measure $EM(\cdot)$ restricted to R_i is finite for $i = 1, \dots, 4$, which implies that the points of $M\big|_{R_i}$ can be represented as a Poisson number of i.i.d. random vectors [140, page 341]:

$$M|_{R_i} \stackrel{d}{=} \sum_{k=1}^{P_i} \epsilon_{(t_{k,i},\, j_{k,i})}, \quad i = 1, \dots, 4,$$

where P_i is a Poisson random variable with mean λm_i, which is independent of the i.i.d. pairs $(t_{k,i}, j_{k,i})$, $k \geq 1$, with common distribution

$$\left.\frac{\lambda \mathbb{L}(ds) F_{\text{on}}(dy)}{\lambda m_i}\right|_{R_i} = \left.\frac{\mathbb{L}(ds) F_{\text{on}}(dy)}{m_i}\right|_{R_i}, \qquad (6.3.13)$$

for $i = 1, \ldots, 4$. Notice that the distributions of $((t_{k,i}, j_{k,i}))$ are independent of λ, which only enters into the specification of the mean of P_i, $i = 1, \ldots, 4$. This means that for fixed T, we can represent A_i as the sum of a Poisson number of i.i.d. random variables,

$$\begin{aligned} A_1 &\stackrel{d}{=} \textstyle\sum_{k=1}^{P_1} j_{k,1}\,, & A_2 &\stackrel{d}{=} \textstyle\sum_{k=1}^{P_2}(T - t_{k,2}), \\ A_3 &\stackrel{d}{=} \textstyle\sum_{k=1}^{P_3}(j_{k,3} + t_{k,3}), & A_4 &\stackrel{d}{=} \textstyle\sum_{k=1}^{P_4} T. \end{aligned} \tag{6.3.14}$$

6.3.3.2 Moments of the summands

We need information about the moments of the summands in (6.3.14). All the variables are bounded by T, so all moments exist, and we derive the asymptotic form of the moments as $T \to \infty$. For notational ease, we often suppress the dependence on T in the notation, and let (t_i, j_i) be random variables with the same distribution as $(t_{k,i}, j_{k,i})$, for $i = 1, \ldots, 4$.

From (6.3.12) and (6.3.13), observe that for $l \geq 1$,

$$\begin{aligned} Ej_1^l &= \int_0^T \int_{y=0}^{T-s} y^l \frac{\mathbb{L}(ds)F_{\text{on}}(dy)}{m_1} \sim \frac{1}{T}\int_{s=0}^{T}\int_{y=0}^{T-s} y^l\ F_{\text{on}}(dy)\,ds \\ &= \frac{1}{T}\int_{s=0}^{T}\left(\int_{y=0}^{s} y^l F_{\text{on}}(dy)\right)ds. \end{aligned} \tag{6.3.15}$$

For $l = 1$, since $\int_0^s yF_{\text{on}}(dy) \to \mu_{\text{on}}$, we have

$$Ej_1 \to \mu_{\text{on}}\,. \tag{6.3.16}$$

For $l > \alpha$, we have, using first a change of variables and then Karamata's theorem, that

$$\begin{aligned} \frac{Ej_1^l}{T^l \bar{F}_{\text{on}}(T)} &\sim \int_0^1\int_0^s y^l\ \frac{F_{\text{on}}(T dy)}{\bar{F}_{\text{on}}(T)}\ ds \sim \int_0^1\int_0^s y^l\ \alpha y^{-1-\alpha}\ dy\ ds \\ &= \frac{\alpha}{(l-\alpha)(l-\alpha+1)}. \end{aligned} \tag{6.3.17}$$

We also note that (6.3.16) and (6.3.17) imply

$$\frac{\text{Var}(j_1)}{T^2\bar{F}_{\text{on}}(T)} \sim \int_0^1\int_0^s y^2\ \alpha y^{-1-\alpha}\ dy\ ds\ = \frac{\alpha}{(2-\alpha)(3-\alpha)} =: \sigma_1^2 \tag{6.3.18}$$

and

$$\limsup_{T\to\infty} \frac{E|j_1 - Ej_1|^3}{T^3\bar{F}_{\text{on}}(T)} \leq \limsup_{T\to\infty} \frac{4\big(Ej_1^3 + (Ej_1)^3\big)}{T^3\bar{F}_{\text{on}}(T)} = \text{const}\,. \tag{6.3.19}$$

Similar calculations for $T - t_2$ give that for $l \geq 1$,

$$\begin{aligned} E(T - t_2)^l &= \int_{s=0}^{T} \int_{y=T-s}^{\infty} (T - s)^l \frac{\mathbb{L}(ds)F_{\text{on}}(dy)}{m_2} \\ &\sim \frac{1}{\mu_{\text{on}}} \int_{s=0}^{T} \int_{y=T-s}^{\infty} (T - s)^l \; F_{\text{on}}(dy)\, ds \\ &= \frac{1}{\mu_{\text{on}}} \int_0^T u^l \bar{F}_{\text{on}}(u) du, \end{aligned}$$

and therefore, for $l \geq 1$, as $T \to \infty$, from Karamata's theorem

$$\begin{aligned} \frac{E(T - t_2)^l}{T^{l+1} \bar{F}_{\text{on}}(T)} &\sim \frac{1}{\mu_{\text{on}}} \int_0^1 x^l \frac{\bar{F}_{\text{on}}(Tx)}{\bar{F}_{\text{on}}(T)} dx \\ &\sim \frac{1}{\mu_{\text{on}}} \int_0^1 x^{l-\alpha} dx = \frac{1}{\mu_{\text{on}}(l - \alpha + 1)}. \end{aligned} \tag{6.3.20}$$

This implies that

$$\frac{\text{Var}(T - t_2)}{T^3 \bar{F}_{\text{on}}(T)} \sim \frac{1}{\mu_{\text{on}}(3 - \alpha)} =: \sigma_2^2 \tag{6.3.21}$$

and

$$\limsup_{T \to \infty} \frac{E|T - t_2 - E(T - t_2)|^3}{T^4 \bar{F}_{\text{on}}(T)} \leq \text{const}\,. \tag{6.3.22}$$

Finally,

$$\begin{aligned} E(j_3 + t_3)^l &= \int_{s=-\infty}^{0} \int_{y=-s}^{T-s} (y + s)^l \frac{\mathbb{L}(ds)F_{\text{on}}(dy)}{m_3} \\ &\sim \frac{1}{\mu_{\text{on}}} \int_{-\infty}^{0} \int_{y=-s}^{T-s} (y + s)^l \; F_{\text{on}}(dy)\, ds\,, \end{aligned}$$

and thus

$$\begin{aligned} \frac{E(j_3 + t_3)^l}{T^{l+1} \bar{F}_{\text{on}}(T)} &\sim \frac{1}{\mu_{\text{on}}} \int_{s=-\infty}^{0} \int_{y=-s}^{1-s} (y + s)^l \; \frac{F_{\text{on}}(T dy)}{\bar{F}_{\text{on}}(T)} \, ds \\ &\sim \frac{1}{\mu_{\text{on}}} \int_{s=-\infty}^{0} \int_{y=-s}^{1-s} (y + s)^l \; \alpha y^{-1-\alpha} \; dy \; ds. \end{aligned} \tag{6.3.23}$$

It follows that

$$\frac{\text{Var}(j_3 + t_3)}{T^3 \bar{F}_{\text{on}}(T)} \sim \frac{1}{\mu_{\text{on}}} \int_{s=-\infty}^{0} \int_{y=-s}^{1-s} (y + s)^2 \; \alpha y^{-1-\alpha} \; dy \; ds =: \sigma_3^2$$

and

$$\limsup_{T \to \infty} \frac{E|j_3 - t_3 - E(j_3 - t_3)|^3}{T^4 \bar{F}_{\text{on}}(T)} \leq \text{const}\,.$$

To compute σ_3^2, observe that

$$\sigma_3^2 := \frac{1}{\mu_{\rm on}} \int_{s=0}^{\infty} \int_{y=s}^{1+s} (y-s)^2 \alpha y^{-\alpha-1} dy ds$$

$$= \frac{1}{\mu_{\rm on}} \int_{y=0}^{1} \alpha y^{-\alpha-1} \Big[\int_{s=0}^{y} (y-s)^2 ds \Big] dy$$

$$+ \frac{1}{\mu_{\rm on}} \int_{y=1}^{\infty} \alpha y^{-\alpha-1} \Big[\int_{s=y-1}^{y} (y-s)^2 ds \Big] dy$$

$$= \frac{1}{\mu_{\rm on}} \int_{y=0}^{1} \alpha y^{-\alpha-1} \Big[\frac{y^3}{3} \Big] dy + \frac{1}{\mu_{\rm on}} \int_{y=1}^{\infty} \alpha y^{-\alpha-1} \Big[\frac{1}{3} \Big] dy$$

$$= \frac{1}{\mu_{\rm on}} \Big[\frac{\alpha}{3(3-\alpha)} + \frac{1}{3} \Big] = \frac{1}{\mu_{\rm on}(3-\alpha)}. \tag{6.3.24}$$

This may not be remarkably thrilling, but is quite useful.

6.3.3.3 α-Stable limits: one-dimensional convergence

We show under Condition 1 that $A(T)$ is asymptotically an α-stable random variable by showing that $A_1(T) = A_1$ is asymptotically stable and $A_i(T) = A_i$, $i = 2, 3, 4$ are asymptotically negligible.

It is relatively easy to see that

$$A_i / b(\lambda T) \stackrel{P}{\to} 0, \quad i = 2, 3, 4. \tag{6.3.25}$$

We restrict ourselves to the case $i = 2$; a similar argument works for $i = 3, 4$. By (6.3.20), (6.3.8), and Condition 1,

$$EA_2 = EP_2\, E(T - t_2) = [\lambda m_2]\, E(T - t_2) \sim (\text{const})\, \lambda T^2 \bar{F}_{\rm on}(T) = o(b(\lambda T)).$$

Thus it remains to consider A_1. Recall the representation of A_1 given in (6.3.14). We start with the following decomposition:

$$A_1 - \lambda \mu_{\rm on} T = \sum_{k=1}^{P_1} (j_{k,1} - Ej_1) + Ej_1\, [P_1 - EP_1] + [EA_1 - \lambda \mu_{\rm on} T]$$

$$= A_{11} + A_{12} + A_{13}.$$

By (6.3.16), $Ej_1 \sim \mu_{\rm on}$. Since P_1 is Poisson with mean $\lambda m_1 \to \infty$, it satisfies the central limit theorem, i.e.,

$$[\lambda m_1]^{-1/2}\, [P_1 - \lambda m_1] \stackrel{d}{\to} N(0, 1). \tag{6.3.26}$$

We conclude that

$$A_{12} = O_P([\lambda T]^{1/2}) = o_P(b(\lambda T)), \tag{6.3.27}$$

since

$$\lim_{T\to\infty} \frac{\sqrt{\lambda T}}{b(\lambda T)} = \lim_{s\to\infty} \frac{s^{1/2}}{b(s)} = \lim_{s\to\infty} s^{1/2-1/\alpha}/L(s)$$

and $1 < \alpha < 2$ implies $\frac{1}{2} < \frac{1}{\alpha} < 1$.

By (6.3.14) and (6.3.26), A_{11} is a sum of approximately $\lambda m_1 \sim \lambda T$ i.i.d. summands. Under Condition 1, $b(\lambda T)/T \to 0$, so that for any $x > 0$ fixed, we eventually have $T - b(\lambda T)x > 0$. Therefore, from (6.3.13)

$$\begin{aligned}
\lambda T P(j_1 > b(\lambda T)\,x) &= \lambda T \int\!\!\int_{\substack{0\le s\le T\\ 0\le s+y\le T\\ y>b(\lambda T)x}} \frac{ds\,F_{\text{on}}(dy)}{m_1}\\
&= \lambda T \int_{s=0}^{T-b(\lambda T)x} ds \int_{y=b(\lambda T)x}^{T-s} \frac{F_{\text{on}}(dy)}{m_1}\\
&= \lambda T\Big[\frac{1}{m_1}\bar{F}_{\text{on}}(b(\lambda T)x)(T-b(\lambda T)x)\\
&\quad -\frac{1}{m_1}\int_0^{T-b(\lambda T)x}\bar{F}_{\text{on}}(T-s)ds\Big]\\
&\sim \left(1-\frac{b(\lambda T)x}{T}\right)\lambda T\ \bar{F}_{\text{on}}(b(\lambda T)\,x)\\
&\quad -\frac{b(\lambda T)}{T}\int_x^{T/b(\lambda T)}\lambda T\ \bar{F}_{\text{on}}(b(\lambda T)\,s)\,ds\\
&\sim x^{-\alpha}.
\end{aligned}$$

Following an argument using point processes [107, 108, 127, 128, 137, 138, 139] we get for $t \ge 0$,

$$Y^{(T)}(\cdot) := \big(b(\lambda T)\big)^{-1}\sum_{k=1}^{[\lambda T\cdot]}(j_{k,1}-Ej_1) \Rightarrow X_\alpha(\cdot) \quad \text{in } \mathbb{D}[0,\infty), \tag{6.3.28}$$

where the limit is a totally skewed α-stable Lévy random motion ($p = 1,\ q = 0$). In fact, by independence, we may couple (6.3.26) and (6.3.28) to get joint convergence

$$\left(Y^{(T)}(\cdot), \frac{P_1}{\lambda T}\right) \Rightarrow (X_\alpha(\cdot), 1) \quad \text{in } \mathbb{D}[0,\infty)\times\mathbb{R}.$$

Using composition and the continuous mapping theorem, one obtains

$$\begin{aligned}
\big(b(\lambda T)\big)^{-1}A_{11} &= Y^{(T)}(P_1/(\lambda T))\\
&= \big(b(\lambda T)\big)^{-1}\sum_{i=1}^{P_1}(j_{k,1}-Ej_1) \Rightarrow X_\alpha(1).
\end{aligned} \tag{6.3.29}$$

It remains to consider A_{13}. By (6.3.15) and Karamata's theorem,

$$\begin{aligned}
A_{13} &= E(A_1) - \lambda\mu_{\text{on}}T = Ej_1\ EP_1 - \lambda T\,\mu_{\text{on}}\\
&= \lambda\int_0^T\left[\int_0^s y\ F_{\text{on}}(dy) - \mu_{\text{on}}\right]ds = -\lambda\int_0^T\int_s^\infty y\ F_{\text{on}}(dy)\,ds\\
&\sim -(\text{const})\ \lambda T^2\bar{F}_{\text{on}}(T) = o(b(\lambda T)).
\end{aligned} \tag{6.3.30}$$

The last limit relation follows from (6.3.8). Combining the limit relations (6.3.25), (6.3.27), (6.3.29), and (6.3.30) we conclude that $A(T)$ has the desired α-stable limit.

6.3.3.4 α-Stable limits: finite dimensional convergence

We restrict ourselves to a sketch of the convergence of the two-dimensional distributions; the general case is analogous. Suppose $t_1 < t_2$. The same arguments as for the one-dimensional convergence show that it suffices to consider the joint convergence of $[b(\lambda T)]^{-1}(A_1(Tt_i) - \lambda T t_i \, \mu_{\text{on}})$, $i = 1, 2$. We can write

$$\begin{aligned} A_1(Tt_2) &= A_1(Tt_1) + \sum_{Tt_1 < \Gamma_k \le Tt_2} L_k 1_{[\Gamma_k + L_k \le Tt_2]} + \sum_{\Gamma_k \le Tt_1} L_k 1_{[Tt_1 < \Gamma_k + L_k \le Tt_2]} \\ &=: A_1(Tt_1) + A_{21}(T(t_2 - t_1)) + A_{22} \,. \end{aligned}$$

Observe that $A_1(Tt_1)$ and $A_{21}(T(t_2 - t_1))$ are independent and that $A_{21}(T(t_2 - t_1)) \stackrel{d}{=} A_1(T(t_2 - t_1))$. Hence the proof of the two-dimensional distributions follows from the one-dimensional convergence if one can show that $[b(\lambda T)]^{-1} A_{22} \stackrel{P}{\to} 0$. This follows by arguments similar to earlier ones as one shows $EA_{22} = o(b(\lambda T))$ using (6.3.8).

6.3.4 FBM approximations for the infinite source Poisson model under fast growth

We now study why fast connection rates associated with strong correlations of $N_T(\cdot)$ imply fractional Brownian motion limits.

Recall that a mean-zero Gaussian process $(B_H(t), t \ge 0)$ with a.s. continuous sample paths is called fractional Brownian motion if it has covariance structure

$$\operatorname{Cov}(B_H(t), B_H(s)) = \frac{\sigma_H^2}{2}(|t|^{2H} + |s|^{2H} - |t - s|^{2H}) \quad \text{for some } \sigma_H > 0, \, H \in (0, 1).$$

The case $H = 1/2$ corresponds to Brownian motion and, if $H \in (1/2, 1)$, the autocovariance function of the increment process $(B_H(t) - B_H(t-1))_{t=1,2,\dots}$, the so-called fractional Gaussian noise, exhibits long range dependence. The following is the result under the fast growth condition.

Theorem 2 *If Condition 2 holds, then the process $(A(Tt), t \ge 0)$ describing the total accumulated input in $[0, Tt]$, $t \ge 0$, satisfies the limit relation*

$$\frac{A(T\cdot) - \lambda \mu_{\text{on}} T(\cdot)}{[\lambda T^3 \bar{F}_{\text{on}}(T) \sigma^2]^{1/2}} \Rightarrow B_H(\cdot).$$

Here $\Rightarrow$ denotes weak convergence in $(\mathbb{D}[0, \infty), J_1)$, B_H is standard fractional Brownian motion, $H = (3 - \alpha)/2$ and σ^2 is given by

$$\begin{aligned} \sigma^2 &= \frac{\alpha}{(2-\alpha)(3-\alpha)} + \frac{2}{\mu_{\text{on}}(3-\alpha)} + \frac{1}{\alpha - 1} \\ &= \frac{1}{3-\alpha}\left[\frac{\alpha}{2-\alpha} + \frac{2}{\mu_{\text{on}}}\right] + \frac{1}{\alpha - 1}. \end{aligned} \tag{6.3.31}$$

Remark. Notice that $H = (3 - \alpha)/2 \in (0.5, 1)$. Hence the corresponding fractional Gaussian noise sequence of B_H exhibits long range dependence. This is in contrast to Theorem 1 where the limiting process, α-stable Lévy motion, has independent increments.

As for Theorem 1, the decomposition of Section 6.3.3.1 will be the key for deriving the Gaussian limit. We omit details and refer to [109]. Unlike the slow growth case, each A_i, $i = 1, \ldots, 4$ contributes to the Gaussian limits. Fast growth means there are more contributions in $[0, T]$ making it difficult for a single contribution to dominate as is typical of heavy tailed limits. The moment conditions of Subsection 6.3.3.2 allow one to prove asymptotic normality using the Lyapunov condition [64, page 286] and [141, page 319].

6.3.5 *Covariance calculations for the infinite source Poisson model*

It is interesting to note that the second order structure of

$$A_T(T\cdot) = \int_0^{T\cdot} N_T(s)ds$$

converges to that of FBM. This happens even under the slow growth condition when $A_T(T\cdot)$ is approximated by Lévy stable motion. This emphasizes how dangerous it is to judge a process [book] by its second order properties [cover].

For $0 \le s < t$ write (suppressing the subscript T)

$$\begin{aligned}
E(A(Tt) - A(Ts))^2 &= E\int_{Ts}^{Tt} N(x)dx \int_{Ts}^{Tt} N(y)dy \\
&= T^2 \int_{x=s}^{t}\int_{y=s}^{t} E(N(Tx)N(Ty))dxdy \\
&= 2T^2 \int\int_{s\le x<y\le t} E(N(Tx)N(Ty))dxdy \\
&= 2T^2 \int_{x=s}^{t}\int_{y=x}^{t} (\mathrm{Cov}(N(Tx), N(Ty)) + \lambda^2\mu_{\mathrm{on}}^2)dxdy \\
&= 2T^2\Big\{\int_{x=s}^{t}\int_{y=x}^{t}\Big(\lambda\int_{w=T(y-x)}^{\infty} \bar{F}_{\mathrm{on}}(w)dw\Big)dxdy \\
&\quad + \int_{x=s}^{t}\int_{y=x}^{t} \lambda^2\mu_{\mathrm{on}}^2 dxdy\Big\} \\
&= I_{s,t} + II_{s,t},
\end{aligned}$$

where we used (6.2.6). Integrate $II_{s,t}$ to get

$$II_{s,t} = 2T^2\lambda^2\mu_{\mathrm{on}}^2\frac{(t-s)^2}{2} = T^2\lambda^2\mu_{\mathrm{on}}^2(t-s)^2.$$

For $I_{s,t}$ we have

$$\begin{aligned}
I_{s,t} &= 2T^3\int_{x=s}^{t}\int_{y=x}^{t}\Big(\lambda\int_{y-x}^{\infty}\bar{F}_{\mathrm{on}}(Tw)dw\Big)dxdy \\
&= 2T^3\lambda\int_{x=s}^{t}\int_{y=0}^{t-x}\Big(\int_{w=y}^{\infty}\bar{F}_{\mathrm{on}}(Tw)dw\Big)dxdy,
\end{aligned}$$

and as $T \to \infty$, this is asymptotic to

$$\sim 2T^{3-\alpha}\lambda L(T)\int_{x=s}^{t}\left(\int_{y=0}^{t-x}\frac{y^{-(\alpha-1)}}{\alpha-1}dy\right)dx$$

$$= 2T^{3-\alpha}L(T)\lambda\int_{x=0}^{t-3}\frac{x^{2-\alpha}}{(2-\alpha)(\alpha-1)}dx = 2T^{3-\alpha}\frac{L(T)\lambda(t-s)^{3-\alpha}}{(2-\alpha)(3-\alpha)(\alpha-1)}.$$

We conclude

$$\lim_{T\to\infty}\frac{I_{s,t}}{\lambda T^3\bar{F}_{\text{on}}(T)} = \frac{2}{(3-\alpha)(2-\alpha)(\alpha-1)}(t-s)^{3-\alpha} = K(\alpha)(t-s)^{3-\alpha} \quad (6.3.32)$$

and

$$II_{s,t} = T^2\lambda^2\mu_{\text{on}}^2(t-s)^2. \quad (6.3.33)$$

Now observe

$$\begin{aligned}\text{Cov}\big(A(Ts), A(Tt)\big) &= E\big(A(Ts)A(Tt)\big) - EA(Ts)A(Tt)\\ &= E\big(A(Ts)A(Tt)\big) - Ts\lambda\mu_{\text{on}}Tt\lambda\mu_{\text{on}}\end{aligned}$$

(since $E\int_0^t N(u)du = \int_0^t EN(u)du = \int_0^t \lambda\mu_{\text{on}}du = t\lambda\mu_m$)

$$= \frac{1}{2}(EA(Ts)^2 + EA(Tt)^2 - E(A(Tt) - A(Ts))^2) - T^2\lambda^2\mu_{\text{on}}^2 st$$

$$= \frac{1}{2}(I_{0,s} + I_{0,t} - I_{s,t} + II_{0,s} + II_{0,t} - II_{s,t}) - T^2\lambda^2\mu_{\text{on}}^2 st$$

$$= \frac{1}{2}(I_{0,s} + I_{0,t} - I_{s,t}) + \left\{\frac{1}{2}T^2\lambda^2\mu_{\text{on}}^2[s^2 + t^2 - (t-s)^2] - T^2\lambda^2\mu_{\text{on}}^2 st\right\}$$

$$= \frac{1}{2}(I_{0,s} + I_{0,t} - I_{s,t}) + 0,$$

and so

$$\lim_{T\to\infty}\frac{\text{Cov}(A(Ts), A(Tt))}{\lambda T^3\bar{F}_{\text{on}}(T)} = K(\alpha)[s^{3-\alpha} + t^{3-\alpha} - (t-s)^{3-\alpha}].$$

If we set $3-\alpha = 2H$ or $\frac{3-\alpha}{2} = H$ then since $1 < \alpha < 2$, we get $\frac{1}{2} < H < 1$.

The conclusion: the second order structure of

$$\frac{A_T(T\cdot)}{\sqrt{\lambda T^3\bar{F}_{\text{on}}(T)}}$$

converges to that of FBM with $H = \frac{3-\alpha}{2}$, but we do not have weak convergence under the slow growth condition.

6.4 Does the model fit the data? Checking for Poisson, independence, and stationarity. Formal and informal statistical techniques

Now we have a model with some fairly well understood characteristics. Does it do a satisfactory job of explaining reality? Here are several statistical techniques that help decide.

6.4.1 How do you identify Poisson time points and validate the choice statistically?

Not all data is particularly well suited to identifying Poisson time points. Points of a homogeneous Poisson process have the following characteristics:

- Interpoint distances are i.i.d.,
- Interpoint distances are exponentially distributed.

The Q&D methods of checking these characteristics that are most common are:

- Check that the sample autocorrelation function (ACF) of interpoint distances is approximately zero for a reasonable number of lags.
- Check the exponential distribution postulate by a QQ-plot, which plots theoretical quantiles of the exponential distribution against sample quantiles of the empirical distribution function, to check if the plot is approximately linear.

The expectation is that:

- The behavior of lots of humans acting independently is often well modelled by a Poisson process, but
- Initiation times of machine triggered downloads or transmissions cannot be modelled as a Poisson process.

Example 1 The UCB data; HTTP sessions via modem. The UC Berkeley data is an 18-day trace collected in November 1996. It contains the home IP HTTP traffic processed by UC Berkeley during this period. It is available at http://ita.ee.lbl.gov/html/contrib/UCB.home-IP-HTTP.html. In [72], we analyzed a 3-hour peak portion of the data, with mean traffic rate of 341 kbit/s. The data content consists of: initiation time of a file transfer, file size, transfer times of a request, and IP address of client. Due to the nonstationarity (more on this later) and the diurnal cycle, we restricted the analysis to several hours of peak traffic on a weekday, i.e., the period 5 to 8 p.m. on Thursday, November 7. This part of the trace consists of about 80,000 requests. It is essential to select the period for analysis carefully.

The Figure 6.4 displays both the ACF and QQ-plots for times between requests. From the plots, the data does not look convincingly independent (too many spikes protruding from the magic window in the ACF plot) nor exponentially distributed (the plot deviates alarmingly from a straight line). Further analysis of the data would be necessary to identify a subset of points as Poisson times. Probably a Poisson cluster process would better fit the data.

6.4.2 Checking heavy tailed data for independence

6.4.2.1 The sample autocorrelation function

Although there are many pitfalls associated with using the sample autocorrelation function for nonGaussian data [25, 37, 41, 61, 126, 130, 131, 135], it is still the most common method for checking for independence. The sample ACF of the stationary

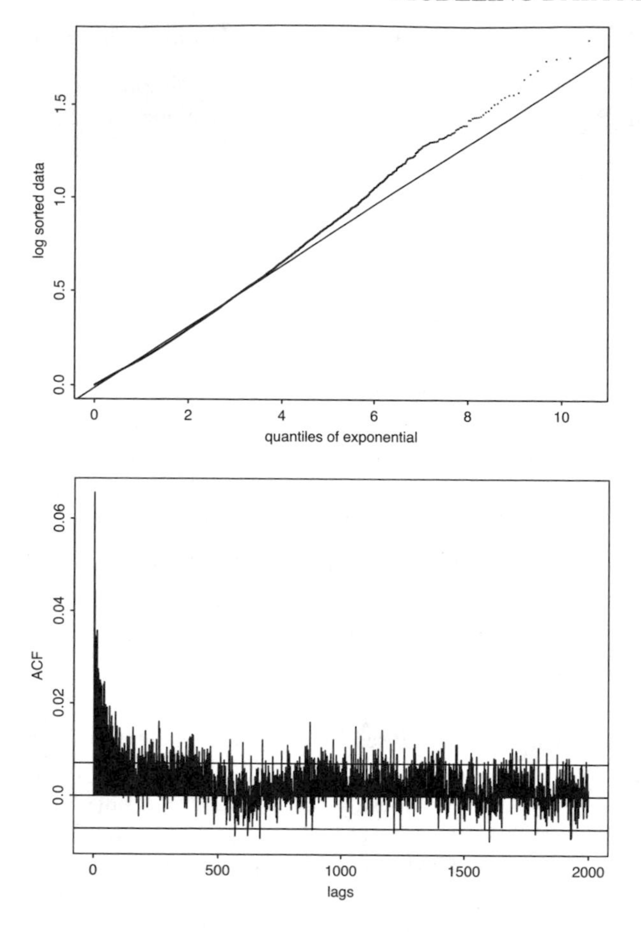

Figure 6.4 *UCB interarrival times of requests: (top) QQ-plot against exponential distribution, (bottom) autocorrelation function of interarrival times.*

sequence $X_1, X_2, \ldots$ is defined as

$$\hat{\rho}(h) = \frac{\sum_{i=1}^{n-h}(X_i - \bar{X})(X_{i+h} - \bar{X})}{\sum_{i=1}^{n}(X_i - \bar{X})^2}, \quad h = 1, 2, \ldots,$$

and the method is to plot $\hat{\rho}(h)$, for various lags h and check if the values are all close to zero. Typically we plot at lags $h = 1, \ldots, 25$ and this should be adequate to assess evidence against independence. Of course, an essential point is to give meaning to the phrase close to zero. Consider the following two cases.

The variances are finite: then standard L_2 theory applies and Bartlett's formula from classical time series analysis [18] provides asymptotic normality for $\hat{\rho}(h)$ and under the null hypothesis of independence, one constructs for each $\hat{\rho}(h)$ a 95% confidence interval. This leads to the magic window that many mature statistical packages such as Splus automatically plot. If the ACF spikes at various lags

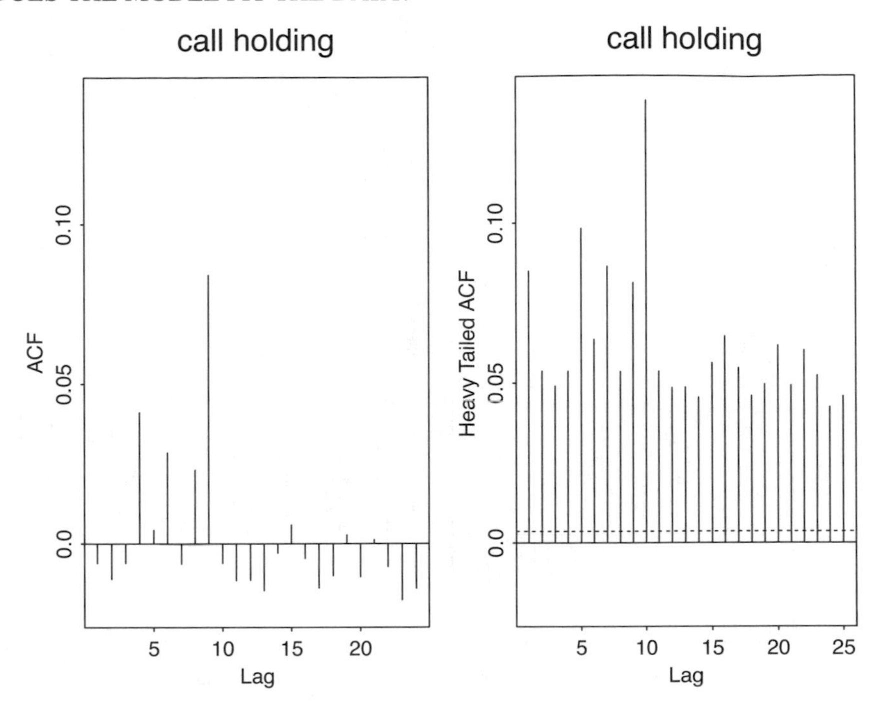

Figure 6.5 *Classical ACF (left) and heavy tailed modification (right) for call holding time data.*

protrude from the magic window less than 5% of the lags, we shrug comfortably and think so far, so good. No evidence against independence has turned up using this technique.

The variances are infinite: if the data is heavy tailed with infinite variance, the mathematical correlations do not exist. However, under the null hypothesis of independence, formulas of Davis and Resnick [18, 38, 39, 40] provide for asymptotic distributions of $\hat{\rho}(h)$ given by the distribution of the ratio of stable random variables. The distribution cannot be calculated explicitly, but percentiles of the distribution can be easily simulated and incorporated into a routine. Then, a magic window is plotted and the procedure outlined in (a) can be carried out with this new magic window.

Example 2 Consider 4045 telephone call holding times indexed according to the time of initiation of the call. The range of the call holding data is (2288,11714735). Figure 6.5 shows the classical sample ACF on the left for the call holding data side by side with the heavy tailed modification on the right which does not center the data by $\bar{X}$. The right graph of the heavy tailed sample ACF has a dotted line drawn at height $h = 0.0035$ and the interval $[0, h]$ is a 95% confidence window analogous to the one given by Bartlett's formula in classical time series. The confidence window is drawn based on the assumption that the data is independent and has Pareto tails. According

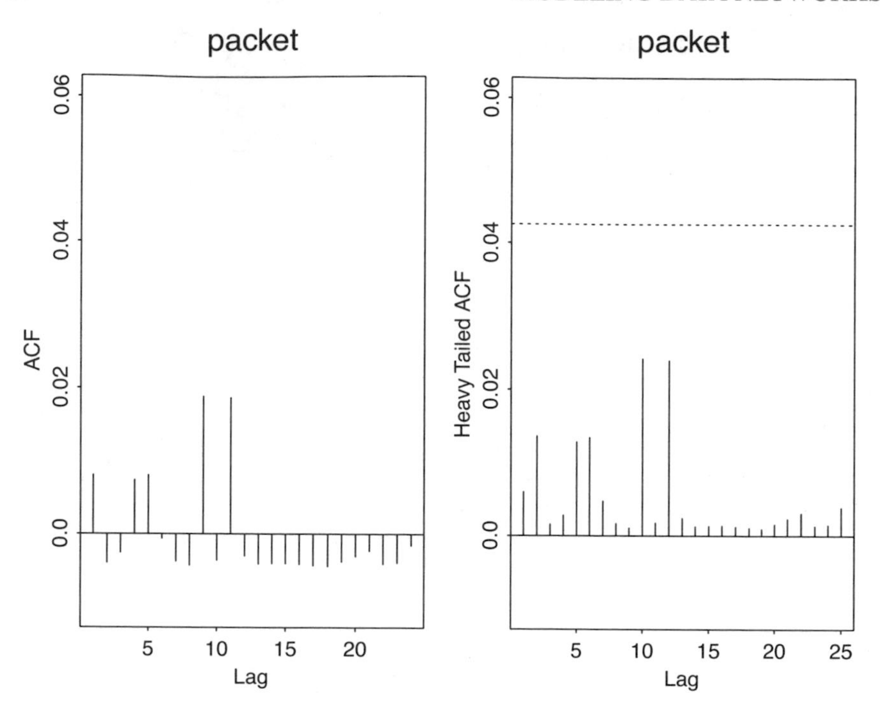

Figure 6.6 *Classical ACF (left) and heavy tailed modification (right) for packet interarrival data.*

to [40, Theorem 3.3], h is given by

$$h = l\alpha^{1/\alpha} \frac{n^{-1/\alpha}}{\log n}$$

where l is the quantile satisfying

$$P[U/V \leq l] = .95$$

for $U,\ V$ independent positive stable random variables with indices α and $\alpha/2$. In the case of the call holding data, we used the estimated value of $\alpha = .97$. The quantile l was estimated by simulation. The position of h relative to the heights of the sample heavy tailed ACF values casts serious doubts on the assumption of independence. Figure 6.6 exhibits comparable graphs for the packet interarrival data. The right heavy tailed graph does not offer evidence against the hypothesis of independence.

6.4.2.2 Two Q&D methods

Here are two additional techniques that are often helpful.

(i) Transform. If data is heavy tailed, take a function of the data (say the log) to get a lighter tail. The advantage is that this puts you in a possibly more familiar domain and you can test for independence using your favorite method. You could, for example, use the classical ACF method with Bartlett's formula. The disadvantage is that you

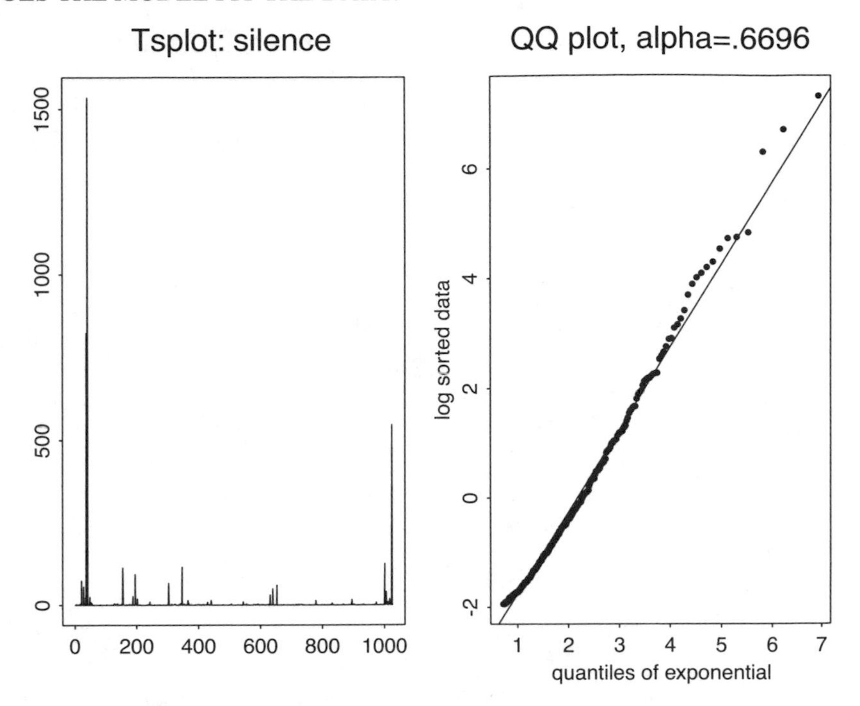

Figure 6.7 *Silence: tsplot (left) and qq-plot (right).*

may obscure the importance of large values, which is exactly what makes the data interesting in the first place.

(ii) The subset method. See [61, 126, 136]. This is a special case of cross-validation. Split the data into, say, two subsets. Plot the ACF of each half separately. If the data comes from an i.i.d. model, the two plots should look exactly the same even for moderate sample sizes of, for instance, 200. A discussion of how to make this more formal is in [136].

Example 3 (Silence) This data consists of 1026 times between transmissions of packets at a terminal. It is clearly not independent. Figure 6.7 gives the time series plot and a QQ-plot (more later) estimating α.

Suppose, against good advice, that we try to fit a black box time series model to the data. Good available techniques [61, 125] for fitting linear models suggest an autoregression of order 9

$$X_n = \sum_{i=1}^{9} \phi_i X_{n-i} + Z_n, \quad n = 1, 2, \dots,$$

and where $\{Z_n\}$ are i.i.d.. There are many good ways to estimate the coefficients. We used the linear programming method [58, 59, 60, 61]. A goodness of fit test for the AR(9) model is to fit the coefficients and then estimate the residuals. If the model is correct, the residuals are close to i.i.d. However, that is not the case here. Plotting the

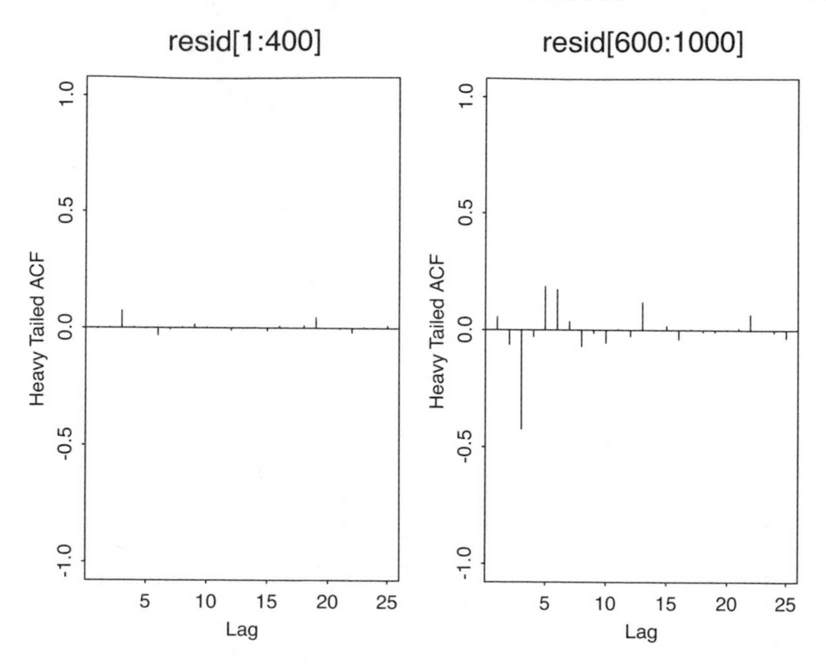

Figure 6.8 *Sample acf for two disjoint subsets of silence of length 400.*

sample correlation functions of two disjoint blocks of length 400 gives very different plots, showing a serious problem with the fit.

6.4.3 Stationarity

Network traffic data is often not stationary over the course of, say, 24 hours. There are diurnal cycles and periods when intensity is high and periods when it is low. This is to be expected and discerning this problem does not require any particular genius.

Example 4 (Ericsson) The Ericsson data trace consists of time stamps of starts and completions of the TCP connections that correspond to HTTP file transfers to and from a corporate Ericsson WWW server which holds home pages and information primarily directed to about 2000 company users at Ericsson facilities around the world. The recording started Thursday, October 15, 1998, at 15:20, and ended Friday, October 16, at 15:49. The information extracted from the data gives the times of connection starts, connection durations, number of bytes transferred (from server to user as well as the opposite direction), and client identification for each connection. The data set is quite nonstationary, and hence a more stationary subset needs to be selected.

Figure 6.9 displays traffic rates resulting from HTTP requests to the Erisson server over approximately a 24-hour period. The lack of stationarity is clear.

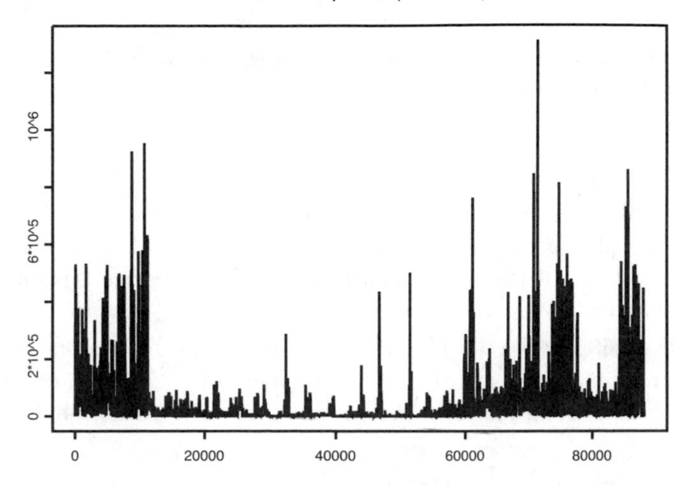

Figure 6.9 *Ericsson traffic rates over 24 hours; number of bytes/second.*

How does one cope with the lack of stationarity? To date, the most common method is to think that since there is so much data, we might as well take a block of the copious data which looks stationary. A common rule of thumb is to restrict data to time spans of at most 4 hours.

Obviously, a more sophisticated approach would be to account for nonstationary behavior in the model.

6.5 Does the model fit the data? How to detect heavy tails

There are a variety of techniques — none foolproof — that are available for deciding when a heavy tailed model is appropriate:

- Hill plots (a modified maximum likelihood estimation method) and refinements (smoothing, plotting on different scales).
- QQ-plots of the empirical quantiles of the log-transformed data against corresponding exponential quantiles.
- Mean residual life plots from exteme value theory, whose popularity is puzzling since they do not work when $\alpha < 1$.
- Extreme value theory techniques such as
  - The Dekkers, Einmahl, de Haan moment estimator [51] of $1/\alpha$.
  - Peaks over threshold modeling leading to fitting a generalized Pareto distribution by the method of maximum likelihood to the exceedances in the data. There is excellent Splus software called EVIS (extreme values in statistics) available (free) from Alexander McNeil's web site (http://www.math.ethz.ch/~mcneil/software.html) as well as the appealing package XTREMES by Reiss and

Thomas [124]. A drawback to this, and many other similar methods, is sensitivity to the choice of threshold.

There is not universal agreement on terminology, but we will say that a random variable X has a heavy (right) tail if

$$P[X > x] \sim x^{-\alpha} L(x), \quad x \to \infty, \tag{6.5.1}$$

where $L(x)$ is slowly varying. The existing statistical techniques would be hard pressed to discern with discrimination the form of $L(x)$ and for many people in the engineering community the phrase *heavy tails* means Pareto or Pareto beyond some point or asymptotically Pareto; that is, $L(x)$ is constant. We will try to resist this simplification.

Recall the following cases:

(i) Very heavy tails: $0 < \alpha < 1$. In this case, assuming the random variable X is positive, both the mean and the second moment of X are infinite. This case is somewhat rare, but file size distributions have been fitted with α in this range. See [7, 142].

(ii) Heavy tails with infinite second moment: $1 < \alpha < 2$. This is a frequently observed case in which the mean is finite but the second moment is infinite. On/off cycle distributions have been fitted with this case [165], and it allows a stationary renewal process to be defined since finite means prevail.

(iii) Heavy tails with finite variance: $\alpha > 2$. This case is typical of financial data.

We now consider the methods outlined above in more detail. We suppose $\{X_n, n \geq 1\}$ is a stationary sequence and that

$$P[X_1 > x] = x^{-\alpha} L(x), \quad x \to \infty$$

where L is slowly varying and $\alpha > 0$. Consistency of estimates of α can usually be proven under just stationarity, but asymptotic normality usually requires the i.i.d. assumption or mixing conditions.

6.5.1 *The Hill estimator and Hill plot*

Let

$$X_{(1)} > X_{(2)} > \cdots > X_{(n)}$$

be the order statistics of the sample $X_1, \ldots, X_n$. We pick $k < n$ and define the Hill estimator [78] of $1/\alpha$ based on $k + 1$ upper order statistics to be

$$H_{k,n} = \frac{1}{k} \sum_{i=1}^{k} \log \frac{X_{(i)}}{X_{(k+1)}}.$$

The number of upper order statistics used in the estimation is $k + 1$. The Hill plot is the plot of

$$\left((k, H_{k,n}^{-1}), 1 \leq k < n\right).$$

If the process is i.i.d., or a linear MA(∞) or is an ARCH process from economics or more generally satisfies mixing conditions, then the Hill estimator is consistent for $1/\alpha$ in the sense that

$$H_{k,n} \xrightarrow{P} \alpha^{-1}$$

as $n \to \infty$, $k/n \to 0$. The Hill plot should have a stable regime sitting at height roughly α. There is a voluminous literature. See [48, 79, 106, 132, 133, 134, 146, 147]. In the i.i.d. case, under a second order regular variation condition, $H_{k,n}$ is asymptotically normal with asymptotic mean $1/\alpha$ and asymptotic variance $1/\alpha^2$. See [34, 35, 44, 47, 49, 50, 66, 122].

6.5.1.1 The Hill estimator in practice

In practice, the Hill estimator is used as follows: we make the *Hill plot*, of

$$\left\{\left(k, H_{k,n}^{-1}\right), 1 \leq k \leq n\right\}$$

and hope the graph looks stable so you can pick out a value of α.

Sometimes this works beautifully and sometimes the plots are not very revealing. Consider Figure 6.10, which shows two cases where the procedure is heart-warming. The top row are time series plots. The top left plot is 4045 simulated observations from a Pareto distribution with $\alpha = 1$ and the top right plot is 4045 telephone call holding times indexed according to the time of initiation of the call. Both plots are scaled by division by 1000. The range of the Pareto data is (1.0001, 10206.477) and the range of the call holding data is (2288, 11714735). The bottom two plots are Hill plots $\{(k, H_{k,n}^{-1}), 1 \leq k \leq 4045\}$, the bottom left plot being for the Pareto sample and the bottom right plot for the call holding times. After settling down, both Hill plots

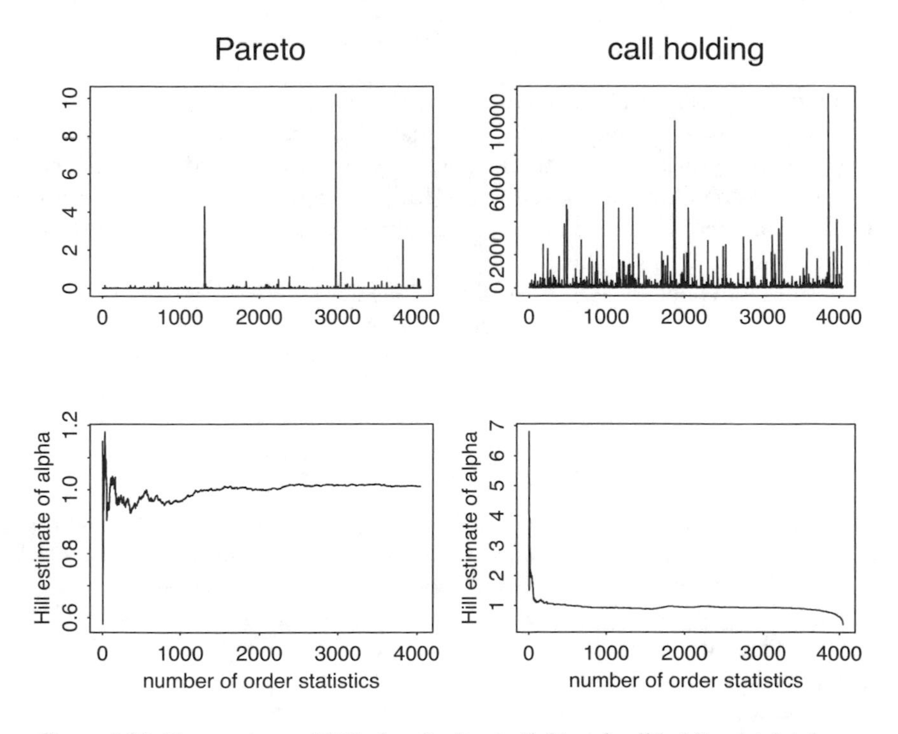

Figure 6.10 *Time series and Hill plots for Pareto (left) and call holding (right) data.*

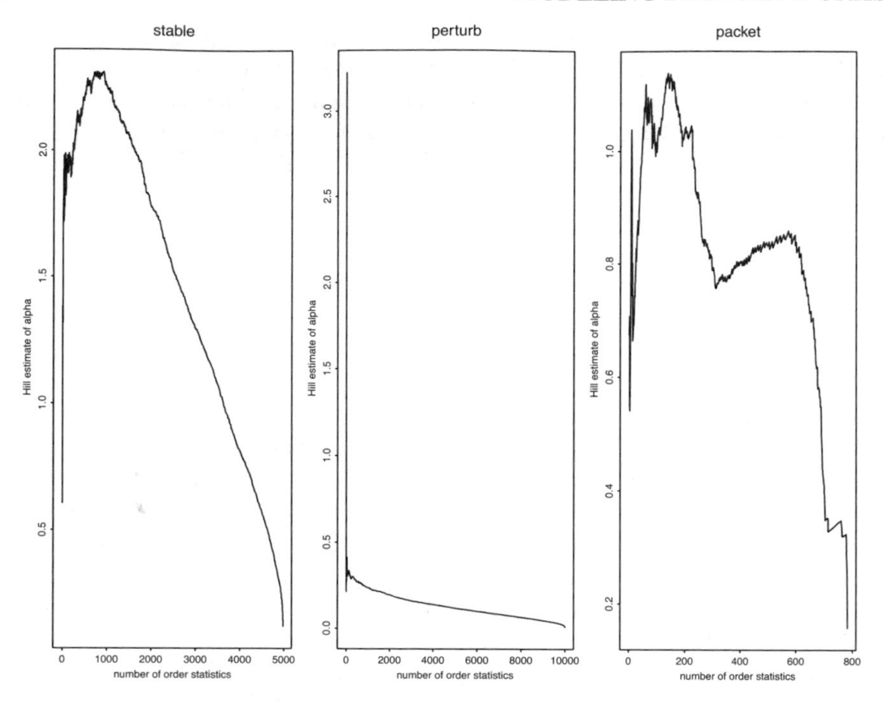

Figure 6.11 *A Hill horror plot.*

are gratifyingly stable and are in a tight neighborhood. The Hill plot for the Pareto seems to nail $\alpha = 1$ correctly and the estimate in the call holding example seems to be between .9 and 1. (So in this case, not only does the variance not exist but the mean appears to be infinite as well.) The Hill plots could be modified to include a confidence interval based on the asymptotic normality of the Hill estimator. McNeil's Hill plot function does just this.

The Hill plot is not always so revealing. Consider Figure 6.11, one of many Hill horror plots. The left plot is for a simulation of size 10,000 from a symmetric α–stable distribution with $\alpha = 1.7$. One would have to be paranormal to discern the correct answer of 1.7 from the plot. The middle plot is for a sample of size 10,000, called *perturb*, from the distribution tail

$$1 - F(x) \sim x^{-1}(\log x)^{10}, \quad x \to \infty,$$

so that $\alpha = 1$. The plot exhibits extreme bias and comes nowhere close to indicating the correct answer of 1. The problem, of course, is that the Hill estimator is designed for the Pareto distribution and thus does not know how to interpret information correctly from the factor $(\log x)^{10}$ and merely readjusts its estimate of α based on this factor rather than identifying the logarithmic perturbation. The third plot is 783 real data called packet representing interarrival times of packets to a server in a network. The problem here is that the graph is volatile, and it is not easy to decide what the estimate should be. The sample size may just be too small.

A summary of difficulties when using the Hill estimator include:

1. How do you get a point estimate from a graph? What value of k do you use?
2. The graph may exhibit considerable volatility and/or the true answer may be hidden in the graph.
3. The Hill estimate has optimality properties only when the underlying distribution is close to Pareto. If the distribution is far from Pareto, there may be outrageous errors, even for sample sizes like 1,000,000.
4. The Hill estimator is not location invariant. A shift in location does not affect the tail index but may throw the Hill estimator into a tizzy.

The lack of location invariance means the Hill estimator can be surprisingly sensitive to changes in location. Figure 6.12 illustrates this. The top plots are time series plots of 5000 i.i.d. Pareto observations where the true $\alpha = 1$. The two right plots on top have the Pareto observations shifted by 1 and then 2. The bottom two plots are the corresponding Hill plots. Shifting by larger and larger amounts soon produces a completely useless plot.

For point 1, several previous studies advocate choosing k to minimize the asymptotic mean squared error of Hill's estimator [73, 122]. In certain cases, the asymptotic form of this optimal k can be expressed, but such a form requires one to know the distribution rather explicitly, and it is not clear how much value one gets from an asymptotic formula. There are adaptive methods and bootstrap techniques that try to overcome these problems; it remains to be seen if they will enter the research community's toolbox.

For point 2, there are simple smoothing techniques that always help to overcome the volatility of the plot and plotting on a different scale frequently overcomes the difficulty associated with the stable example. These techniques are discussed in the next paragraph.

6.5.1.2 Variant 1: The smooHill plot

The Hill plot often exhibits extreme volatility, which makes finding a stable regime in the plot more guesswork than science, and to counteract this Resnick and Stărică [134] developed a smoothing technique yielding the smooHill plot: pick an integer u (usually 2 or 3) and define

$$smoo\ \hat{\alpha}_{k,n,u} = \frac{1}{\frac{1}{(u-1)k}\sum_{j=k+1}^{uk} H_{j,n}}.$$

In the i.i.d. case when a second order regular variation condition holds, the asymptotic variance of $smooH_{k,n}$ is less than that of the Hill estimator, namely:

$$\frac{1}{\alpha^2}\frac{2}{u}\left(1 - \frac{\log u}{u}\right).$$

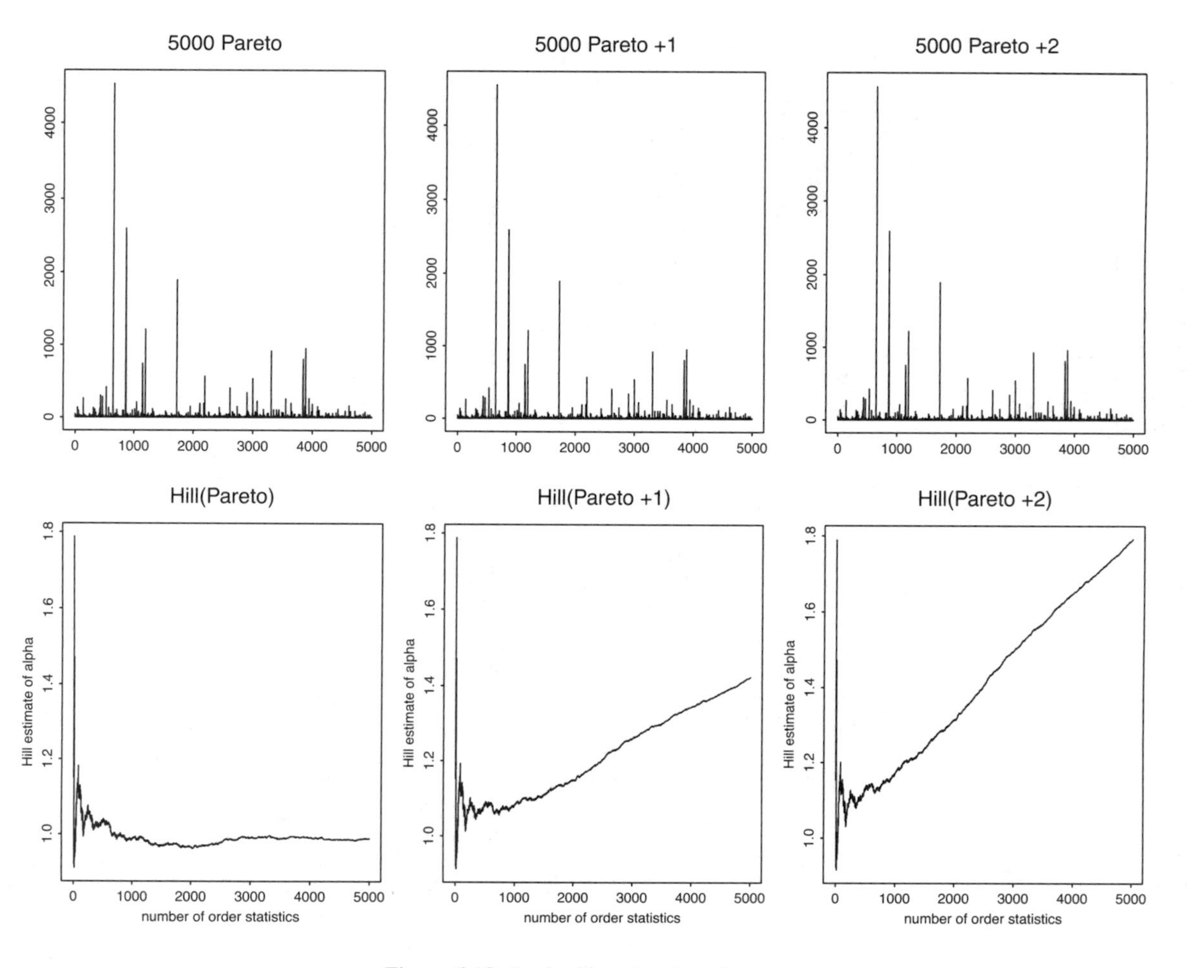

Figure 6.12 *Lack of location invariance.*

6.5.1.3 Variant 2: Alt plotting; changing the scale

As an alternative to the Hill plot, it is sometimes useful to display the information provided by the Hill or smooHill estimation as

$$\{(\theta, H^{-1}_{\lceil n^{\theta}\rceil,n}), 0 \leq \theta \leq 1\},$$

and similarly for the smooHill plot, where we write $\lceil y \rceil$ for the smallest integer greater or equal to $y \geq 0$. We call such plots the alternative Hill plot abbreviated AltHill and the alternative smoothed Hill plot abbreviated AltsmooHill. The alternative display is sometimes revealing since the initial order statistics get shown more clearly and cover a bigger portion of the displayed space. Unless the distribution is Pareto, the AltHill plot spends more of the display space in a small neighborhood of α than the conventional Hill plot [52].

Figure 6.13 compares several Hill plots for 5000 observations from a stable distribution with $\alpha = 1.7$. Plotting on the usual scale is not revealing and the alt plot is more informative.

6.5.2 Dynamic and static QQ-plots

Suppose $\{X_1, \ldots, X_n\}$ are i.i.d. with distribution F. As we did for the Hill plots, pick k upper order statistics

$$X_{(1)} > X_{(2)} > \cdots X_{(k)},$$

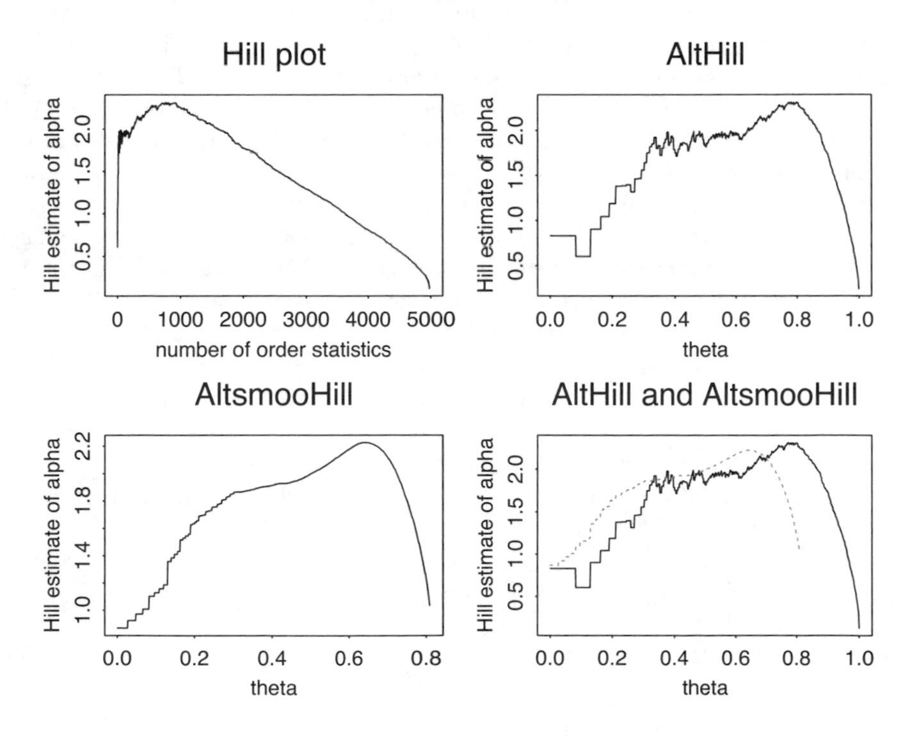

Figure 6.13 *Stable,* $\alpha = 1.7$.

neglect the rest and consider these as the order statistics of exceedances above $X_{(k+1)}$. The distribution of these exceedances should be roughly Pareto if F is heavy tailed. Take the log transform to make this sample of exceedances approximately exponentially distributed and plot empirical quantiles of this sample of size k against the theoretical quantiles of the exponential density. This plot

$$\left\{\left(-\log(1-\frac{j}{k+1}\right), \log X_{(k-j+1)}, 1 \leq j \leq k\right\} \tag{6.5.2}$$

should yield approximately a straight line with slope=$1/\alpha$ if F satisfies (6.5.1). The slope of the least squares line through the points is an estimator called the QQ-estimator [94]. Computing the slope, we find that the QQ-estimator is given by

$$\widehat{\alpha^{-1}}_{k,n} = \frac{\frac{1}{k}\sum_{i=1}^{k}\left(-\log\left(\frac{i}{k+1}\right)\right)\log\left(\frac{X_{(i)}}{X_{(k+1)}}\right) - \frac{1}{k}\sum_{i=1}^{k}\left(-\log\left(\frac{i}{k+1}\right)\right)H_{k,n}}{\frac{1}{k}\sum_{i=1}^{k}\left(-\log\left(\frac{i}{k+1}\right)\right)^2 - \left(\frac{1}{k}\sum_{i=1}^{k}\left(-\log\left(\frac{i}{k+1}\right)\right)\right)^2}. \tag{6.5.3}$$

There are two different plots one can make based on the QQ-estimator. There is the dynamic QQ-plot obtained from plotting $\{(k, 1/\widehat{\alpha^{-1}}_{k,n}, 1 \leq k \leq n\}$, which is similar to the Hill plot. Another plot, the static QQ-plot, is obtained by choosing and fixing k, plotting the points in (6.5.2) and putting the least squares line through the points while computing the slope as the estimate of α^{-1}.

The QQ-estimator is consistent if $k \to \infty$ and $k/n \to 0$. Under an additional second order condition on $F(x)$ and further restriction on $k(n)$, it is asymptotically normal with asymptotic mean $1/\alpha$ and asymptotic variance $2/\alpha^2$. [10, 94]. This is larger than the asymptotic variance of the Hill estimator but bias and volatility of the plot seem to be more of an issue than asymptotic variance. The volatility of the QQ-plot always seems to be less than that of the Hill estimator. As with the Hill estimator, sensitivity to choice of k is an important issue. It is rare that another technique gives a clearer answer to the question, is the data heavy tailed?

Figure 6.14 compares the Hill plot with the dynamic QQ-plot for the call holding data and Figure 6.15 does the same thing for the packet interarrival data. Figure 6.16 gives two static QQ-plots for the call holding data, one using $k = 3500$ and the other using $k = 1500$, yielding estimates of α of 0.95 and 0.98, respectively. For this data set, the estimators are unusually insensitive to the choice of k; the Hill plots and the dynamic QQ-plots are quite stable and the static QQ-plots do not change much as k varies. Figure 6.17 gives two static QQ-plots for the packet interarrival data. The data set is only 783 in length and now there is some sensitivity to k.

The QQ-plot is often a valuable technique to gauge if a heavy tailed model is appropriate. One can plot

$$\left\{\left(-\log\left(1-\frac{j}{n+1}\right), \log X_{(n-j+1)}\right), 1 \leq j \leq n\right\}$$

using all the data. If the plot from some point on looks linear, this is evidence that a heavy tailed model is applicable. In many examples, there are a number of points at the end of the plot that do not seem to follow the line. The number of such points is of the order of 10.

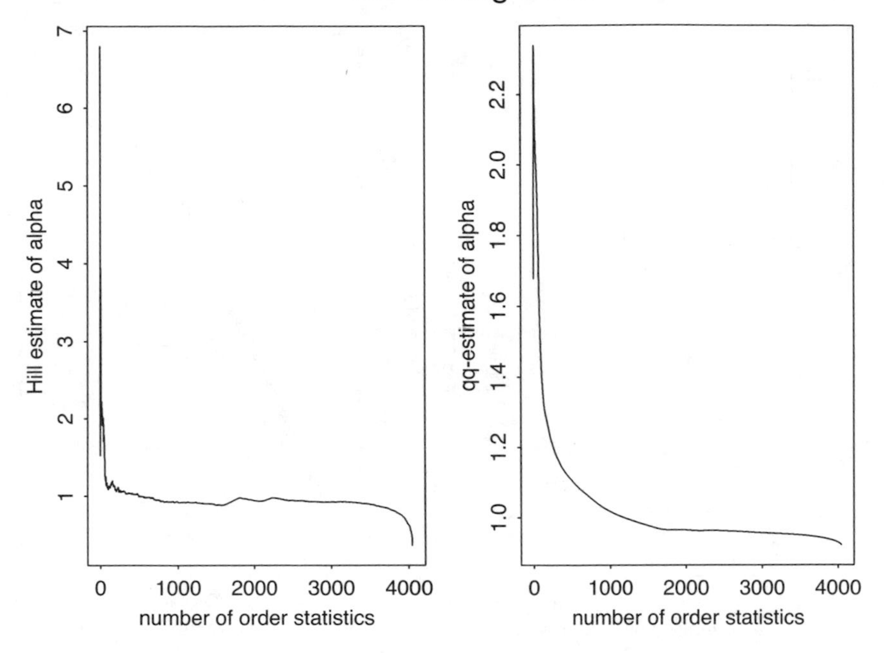

Figure 6.14 *Hill and QQ-plots for call holding times.*

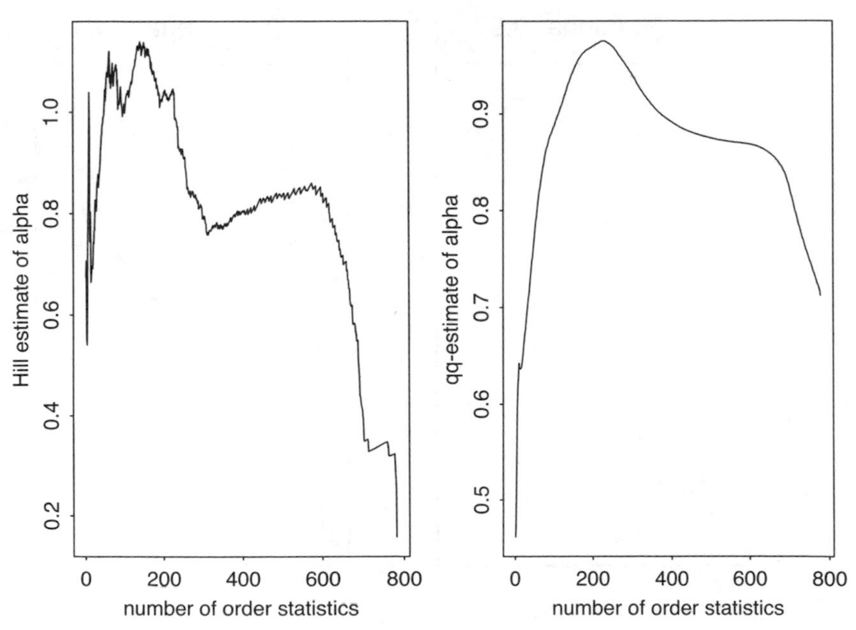

Figure 6.15 *Hill and QQ-plots for packet interarrivals.*

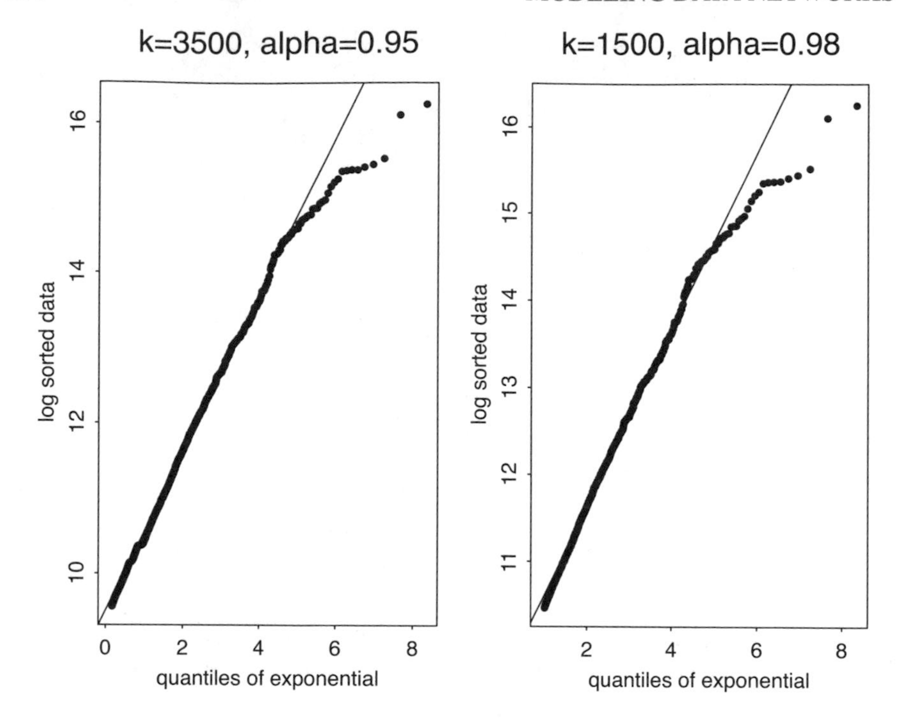

Figure 6.16 *Static QQ-plots for call holding times.*

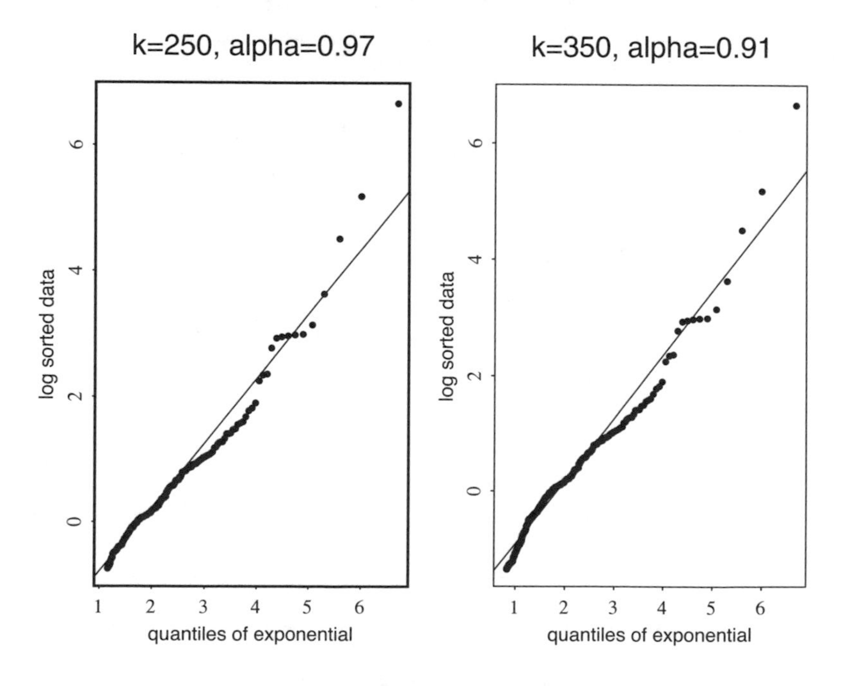

Figure 6.17 *Static QQ-plots for packet interarrivals.*

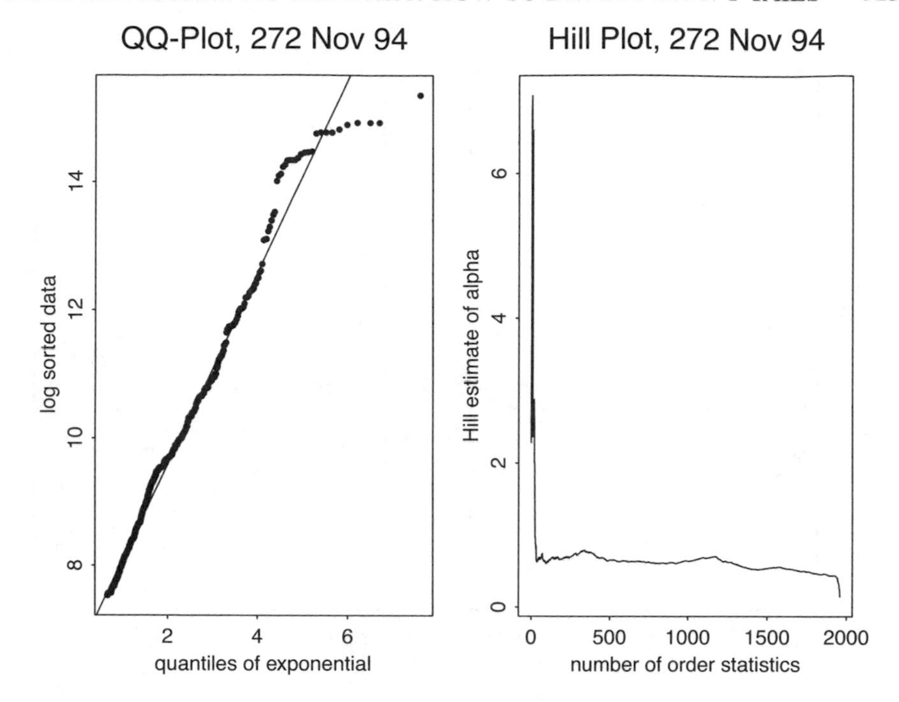

Figure 6.18 *Hill and static QQ-plots for BU file size data.*

An interesting example of heavy tailed data is the Boston University study of web use from 1994 [30, 31, 32, 33, 36] which, among other things, kept track of sizes of files downloaded in a WWW session. For the month of November, 1994, in the lab in Room 272, Figure 6.18 shows both a classical Hill plot and a static QQ-plot. The mean of this data appears infinite since the estimate of $\alpha \approx 0.66$.

One example where the Hill plot requires careful inspection but the QQ-plot is immediately informative is the following data set called *zjz*, which represents file sizes requested from a server at the University of North Carolina. The QQ-plots of Figure 6.20 are informative, allowing an estimate of $\alpha = 1.42$ for $k = 18{,}000$. The size of the data set is 131,943. The Hill plots in Figure 6.19 are initially less clear because there is no clear region where the plot is stable. After the QQ-estimate is obtained, the Hill plot serves to confirm the QQ-estimate.

6.5.3 The Dekkers, Einmahl, De Haan moment estimator

Ignoring location and scale, we know that the extreme value distributions [24, 42, 98, 139] can be parameterized as a one parameter family

$$G_\gamma(x) = \exp\{-(1+\gamma x)^{-\gamma^{-1}}\}, \quad \gamma \in \mathbb{R},\ 1+\gamma x > 0. \tag{6.5.4}$$

The Dekkers, Einmahl, De Haan moment estimator $\hat{\gamma}$ [49, 50, 51] is designed to estimate γ for a random sample in the domain of attraction of G_γ. When $\gamma > 0$, this

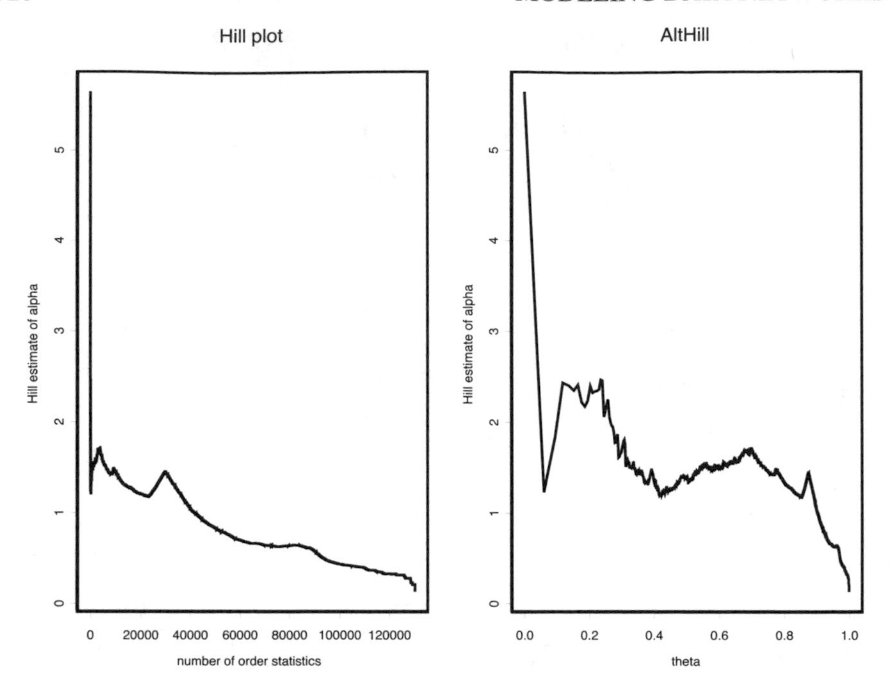

Figure 6.19 *Hill and altHill plots for the zjz file size data.*

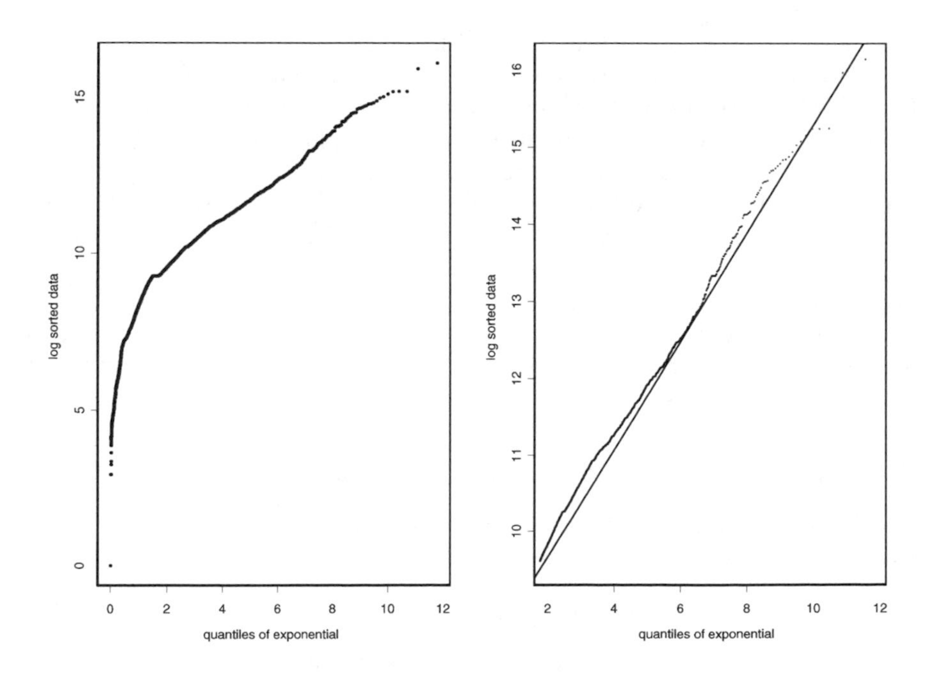

Figure 6.20 *QQ-plots for zjz.*

is the same thing as estimating $\gamma = 1/\alpha$. Recall that the exponential, normal, gamma densities, and many others are in the $D(G_0)$, the domain of attraction of the Gumbel distribution. This provides another method of deciding when a distribution is heavy tailed or not. If $\hat{\gamma}$ is negative or very close to zero, there is considerable doubt that heavy tailed analysis should be applied.

The moment estimator is defined as follows: as usual, let $X_{(1)} \geq X_{(2)} \geq \cdots \geq X_{(n)}$ be the order statistics from a random sample of size n. Define for $r = 1, 2$

$$H_{k,n}^{(r)} = \frac{1}{k} \sum_{i=1}^{k} \left(\log \frac{X_{(i)}}{X_{(k+1)}} \right)^r,$$

so that $H_{k,n}^{(1)}$ is the Hill estimator. Define

$$\hat{\gamma}_n = H_{k,n}^{(1)} + 1 - \frac{1/2}{1 - (H_{k,n}^{(1)})^2 / H_{k,n}^{(2)}}. \tag{6.5.5}$$

Then assuming $F \in D(G_\gamma)$, we have consistency

$$\hat{\gamma}_n \xrightarrow{P} \gamma,$$

as $n \to \infty$ and $k/n \to 0$. Furthermore, under an additional condition and a further restriction on k,

$$\sqrt{k}(\hat{\gamma} - \gamma) \Rightarrow N,$$

where N is a normal random variable with zero mean and variance

$$\sigma(\gamma) = \begin{cases} 1 + \gamma^2, & \text{if } \gamma \geq 0, \\ (1-\gamma)^2(1-2\gamma)\left(4 - 8\frac{1-2\gamma}{1-3\gamma} + \frac{(5-11\gamma)(1-2\gamma)}{(1-3\gamma)(1-4\gamma)}\right), & \text{if } \gamma < 0. \end{cases}$$

The asymptotic variance of the moment estimator exceeds that of the Hill estimator when $\gamma > 0$, so from the point of view of asymptotic variance, there is no reason to prefer it. However, the moment estimator discerns a light tail more effectively than the Hill estimator, and thus it is often useful to apply the moment estimator to see if $\gamma \leq 0$, which would rule out heavy tail analysis.

Figure 6.21 compares the effectiveness of the moment estimator for discerning a light tail ($\gamma = 0$) with that of the Hill estimator $\alpha = \infty$. The Hill estimator does not seem very reliable for this purpose. Both estimators are applied to the same random sample of size 1000 taken from an exponential distribution with unit mean.

Figure 6.22 shows the moment estimator applied to the call holding data on the left and the packet interarrival data on the right. Keep in mind when comparing these graphs with previous graphs that $\gamma = 1/\alpha$.

6.5.4 Peaks over threshold method

A standard technique from extreme value theory is the peaks over threshold method [56, 124]. Assuming that the underlying distribution F is in a domain of attraction of an extreme value distribution G_γ given by (6.5.4), the excesses over a high threshold are approximately distributed as a generalized Pareto distribution. In most cases,

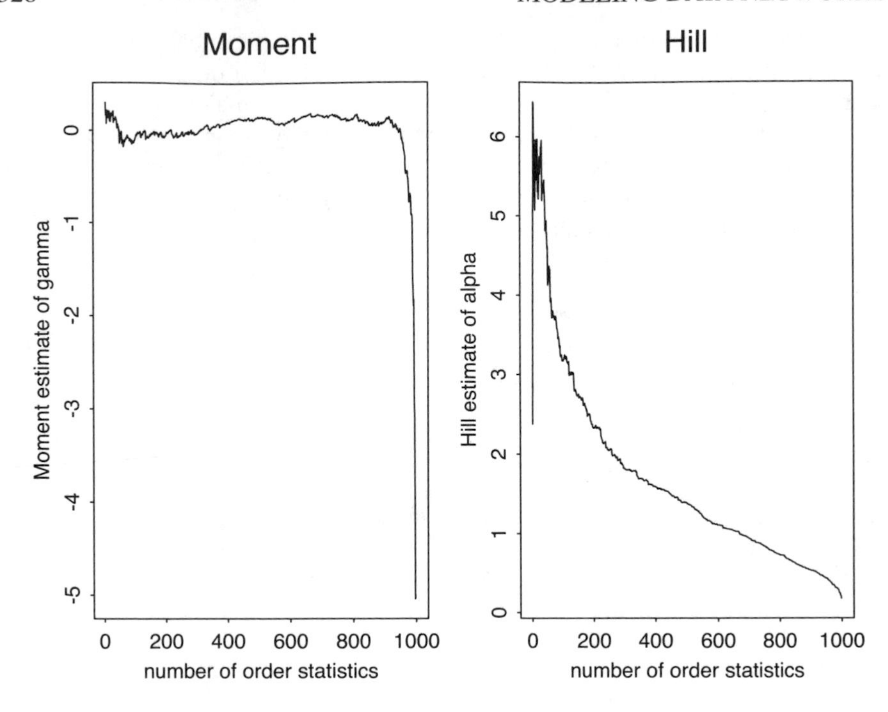

Figure 6.21 *Unit exponential data.*

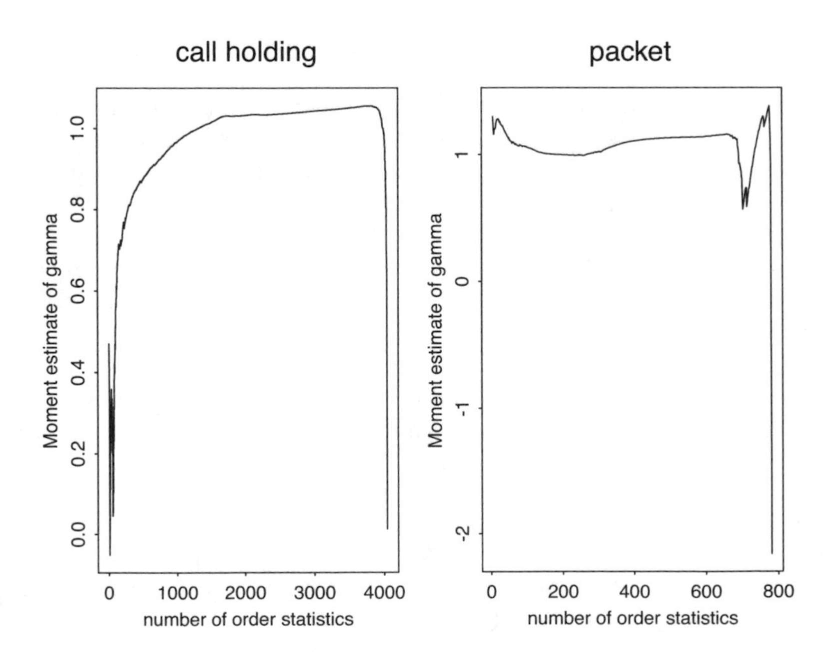

Figure 6.22 *Moment estimator applied to call holding and packet data.*

Table 6.1 *Point estimates of the extreme value parameter for cumulative traffic ± standard deviation.*

Data set	$\gamma(= 1/\alpha)$
simM/G/∞	$-.13 \pm .03$
BUburst 10s	$-.36 \pm .13$
BUburst 1s	$.17 \pm .03$
UCB 10s	$.05 \pm .18$
UCB syn 10s	$-.60 \pm .14$
Munich lo TX	$.09 \pm .04$
Munich lo RX	$-.03 \pm .05$
Munich hi .1s	$.17 \pm .12$
Munich hi .01s	$.10 \pm .03$
Ericsson	$-.47 \pm .09$
Eri syn 1s	$-.31 \pm .12$

the parameters of the generalized Pareto can be estimated by maximum likelihood estimation. This certainly works in the heavy tailed case where F satisfies (6.5.1).

Example 5 The theory presented in the previous Section 6.3 concluded that if the model is correct, then the distribution of cumulative traffic aggregated over users and time is approximable either by FBM or by a stable Lévy motion. In [72] several data sets were examined to see if there were examples where cumulative input looked heavy tailed. The cumulative traffic was calculated over disjoint time blocks to get a sample, and then the POT method was used to estimate the extreme value index γ. In order to conclude the stable Lévy motion approximation is possibly correct, the POT estimation would need to find a positive value for γ that was greater than 1/2 because this would correspond to $0 < \alpha < 2$. Such a value was not found, as the summary from Table 6.1 below, taken from [72], demonstrates.

6.6 Does the model fit the data? Long range dependence, self-similarity, Hurst phenomenon

6.6.1 Long range dependence

A stationary L_2 sequence $\{\xi_n, n \geq 1\}$ possesses long range dependence (LRD) if

$$\mathrm{Cov}(\xi_n, \xi_{n+h}) \sim h^{-\beta} L(h), \quad h \to \infty \tag{6.6.1}$$

for $0 < \beta < 1$ and $L(\cdot)$ slowly varying [11]. Set $\gamma(h) = \mathrm{Cov}(\xi_n, \xi_{n+h})$ and $\rho(h) = \gamma(h)/\gamma(0)$ for the covariance and correlation functions of the stationary process $\{\xi_n\}$. There is no universal agreement about terminology, and sometimes LRD is taken to mean that covariances are not summable: $\sum_h |\gamma(h)| = \infty$, whereas short range dependence means $\sum_h |\gamma(h)| < \infty$.

LRD, like the property of heavy tails, has acquired a mystical, almost religious, significance and generated controversy. Researchers argue over whether it exists, whether it matters if it exists or not, or whether analysts have been fooled into mistaking some other phenomena like shifting levels, undetected trend [95], or nonstationarity for long range dependence. Discussions about this have been going on since (at least) the mid 1970s in hydrology [13, 15, 16, 19, 148, 149, 150], in finance [111, 112] and in data network modeling [21, 22, 53, 65, 77, 99, 119, 154]. Think of it as one more modeling decision that needs to be made. Since LRD is an asymptotic property, models that possess LRD presumably have different asymptotic properties from those models where LRD is absent, although even this is sometimes disputed.

The definitions of short and LRD are designed so that the differences in properties show up clearest under block averaging [27]. Define the block average operator

$$A_m : \mathbb{R}^\infty \mapsto \mathbb{R}^\infty$$

by

$$A_m(x_1, x_2, \ldots) = \left(\frac{1}{m} \sum_{i=(k-1)m+1}^{km} x_i, k \geq 1 \right),$$

so that A_m maps a sequence into the averages over successive blocks of length m. Suppose

$$\{\xi_n^{(m)}\} = A_m(\{\xi_n\})$$

is the block averaged $\{\xi_n\}$ sequence (which is still L_2 weakly stationary), and

$$\gamma^{(m)}(h) = \operatorname{Cov}\left(\xi_n^{(m)}, \xi_{n+h}^{(m)}\right)$$

is the covariance function, and $\rho^{(m)}(h)$ is the corresponding correlation function. Under short range dependence

$$\rho^{(m)}(h) \to 0, \quad |h| \geq 1, \ m \to \infty,$$

and the second order structure of $\{\sqrt{m}\xi_n^{(m)}, n \geq 1\}$ converges, as $m \to \infty$, to that of white noise. However, under LRD defined in (6.6.1),

$$\rho^{(m)}(h) \to \tilde{\rho}(h) = \frac{1}{2}[(h+1)^{2-\beta} - 2h^{2-\beta} + (h-1)^{2-\beta}], \quad m \to \infty, \tag{6.6.2}$$

which is the correlation function of fractional Gaussian noise (FGN). The correlation function of FGN exhibits slow decay characteristic of long range dependence since

$$\tilde{\rho}(h) \sim H(2H-1)h^{2H-2} = \left(1 - \frac{\beta}{2}\right)(1-\beta)h^{-\beta}, \quad h \to \infty, \tag{6.6.3}$$

where $H = 1 - \frac{\beta}{2}$ is called the Hurst parameter (but be wary of this terminology when we get to the R/S statistic).

6.6.2 Self-similarity in discrete time; connections with long range dependence

There is a tendency to muddle LRD and the related notion of self-similarity, so some care in distinguishing between the concepts is worthwhile.

Let $\boldsymbol{X} = (X_1, X_2, \ldots)$ be weakly stationary. Call $\boldsymbol{X}$ distributionally self-similar if there exists $H > 0$ such that for every $m = 1, 2, \ldots$

$$m^{1-H} A_m \boldsymbol{X} \stackrel{d}{=} \boldsymbol{X} \tag{6.6.4}$$

as random elements of $\mathbb{R}^\infty$. This is equivalent to

$$m A_m \boldsymbol{X} \stackrel{d}{=} m^H \boldsymbol{X}$$

where $m A_m \boldsymbol{X}$ is the block aggregated process.

Call $\boldsymbol{X}$ second order self-similar if the second order structure of $m^{1-H} A_m \boldsymbol{X}$ and $\boldsymbol{X}$ are the same for every m. This means

$$(i)\ \rho((A_m \boldsymbol{X})_k, (A_m \boldsymbol{X})_{k+h}) = \rho(X_k, X_{k+h}),$$

$$(ii) \qquad \mathrm{Var}((m^{1-H} A_m \boldsymbol{X})_1) = \mathrm{Var}(X_1).$$

It turns out that (ii) implies (i) and provided $1/2 < H < 1$, a second order self-similar process has the same correlation structure as FGN, and any Gaussian distributionally self-similar process with $1/2 < H < 1$ is FGN.

Asymptotic self-similarity in discrete time: call the weakly stationary process $\boldsymbol{X} = (X_1, X_2, \ldots)$ asymptotically second order self-similar if as $m \to \infty$

$$\rho^{(m)}(h) = \rho((A_m \boldsymbol{X})_k, (A_m \boldsymbol{X})_{k+h}) \to \frac{1}{2}[(h+1)^{2H} - 2h^{2H} - (h-1)^{2H}], \quad h = 0, \pm 1, \ldots. \tag{6.6.5}$$

Call H the self-similarity parameter. Such processes have a correlation structure such that after aggregation over blocks of length m, the resulting process has only approximately the correlation structure of FGN. Any stationary process with long range dependence as defined in (6.6.1) is asymptotically second order self-similar. This includes fractionally differenced ARIMA models.

6.6.3 Self-similarity in continuous time

A continuous time stochastic process $(X(t)), t \geq 0)$ is self-similar with self-similarity parameter $H > 0$ (abbreviated H-ss) if for every $a > 0$

$$X(a\cdot) \stackrel{d}{=} a^H X(\cdot).$$

This means that the finite dimensional distributions of the process $(X(at), t \geq 0)$ with its dilated time are the same as those of the rescaled process $a^H X(\cdot)$ where the state space is stretched.

Brownian motion is H-ss with $H = \frac{1}{2}$, which immediately gives an example of a nonstationary process which is H-ss. In fact a strictly stationary process cannot be H-ss unless it is identically zero. However, there is the following connection between stationarity and self-similarity [96, 97]:

(a) If $\{X(t), t > 0\}$ is H-ss, then $\{Y(t), t \in \mathbb{R}\}$ defined by

$$Y(t) = e^{-tH} X(e^t)$$

is strictly stationary.

(b) If $\{Y(t), t \in \mathbb{R}\}$ is strictly stationary, then $\{X(t), t > 0\}$ defined by

$$X(t) = t^H Y(\log t),$$

is H-ss.

The transformation applied to Brownian motion to get a stationary process yields the Ornstein-Uhlenbeck process.

Stationary increments: a process $\{Y(t), t \geq 0\}$ has stationary increments if for any $s > 0$

$$\{Y(s+t) - Y(s), t \geq 0\} \stackrel{d}{=} \{Y(t) - Y(0), t \geq 0\}.$$

A process Y that is both H-ss and has stationary increments will be abbreviated H-sssi.

Suppose $\{Y(t), t \geq 0\}$ is H-sssi and suppose $E(Y(t)^2) = t^{2H} E(Y(1)^2) < \infty$ and $EY(1) = 0$.
Then

(a) If $H \in (0, 1]$ ([151])

$$\mathrm{Cov}\big(Y(t_1), Y(t_2)\big) = \frac{EY(1)^2}{2}\big(t_1^{2H} + t_2^{2H} - |t_1 - t_2|^{2H}\big). \tag{6.6.6}$$

(b) Define for $j \geq 0$

$$X_j = Y(j+1) - Y(j).$$

Then $\{X_j, j \geq 0\}$ is weakly stationary and has the correlation structure of FGN.

Extremal processes [139] are good examples of H-ss processes that do not have stationary increments. The range of H can be $(0, \infty)$. There are many extremal processes that are not L_2 or even L_1. Self-similarity is a distributional property, not a moment property.

6.6.3.1 Fractional Brownian motion

For $0 < H \leq 1$, a Gaussian process with correlation function

$$\frac{\sigma^2}{2}(|t_1|^{2H} + |t_2|^{2H} - |t_1 - t_2|^{2H}) \tag{6.6.7}$$

for $\sigma > 0$ will be called fractional Brownian motion (FBM) and denoted $(B_H(t), t \geq 0)$. This correlation function corresponds to the covariance in (6.6.6).

This helps explain the temptation to model traffic with FBM. Increments of FBM give FGN, a stationary sequence with LRD provided $H \in (0.5, 1)$.

6.6.3.2 Asymptotic self-similarity in continuous time

This is a notion developed by Lamperti [96, 97]. See also [55, 114, 115, 160, 161] for extensions.

Suppose $\{X(t), t \geq 0\}$ is a stochastic process such that $X(0) = 0$ and for each $t > 0$

$$X(t) \text{ is a nondegenerate random variable.} \tag{6.6.8}$$

Assume $\{\xi_n, n \geq 0\}$ is a sequence of random variables and $b(t), t \geq 0$, is a measurable, nonnegative function such that as $s \to \infty$, $b(s) \to \infty$ and

$$\frac{\xi_{[st]}}{b(s)} \Rightarrow X(t) \tag{6.6.9}$$

in the sense of convergence of finite dimensional distributions. Then for some $H > 0$, $b(\cdot)$ is regularly varying with index H and $X(\cdot)$ is H-ss. Furthermore $X(\cdot)$ is H-sssi provided for $\{\xi_n, n \geq 0\}$ is stationary. This result, known as Lamperti's theorem, shows that limits of partial sum processes of stationary random variables will be self-similar.

Here is a summary and some additional facts: suppose $\{Y(t), t \geq 0\}$ is H-sssi: (a) If $EY(1)^2 < \infty$, then $0 < H \leq 1$. (b) If $H = 1$, then $Y(t) = tY(1)$ almost surely. (c) If $X_j = Y(j+1) - Y(j)$, for $j \geq 1$, then $\{X_j\}$ has the covariances of fgn and hence has long range dependence if $\frac{1}{2} < H < 1$.

This leads to methods of constructing long range dependent processes.

(i) Fractionally differenced ARMA processes [18].

(ii) Increments of H-sssi processes with $\frac{1}{2} < H < 1$. Processes with the H-sssi property can be constructed by various means:

   (a) Stochastic integration [151],

   (b) Lamperti's theorem.

6.6.4 *Should we be happy?*

There are obvious problems with all this when you think it over:

(i) LRD is defined by covariances for weakly stationary variables.

(ii) What happens if the variables have infinite variance? What if the process is nonGaussian so that covariances are relatively uninformative?

(iii) There exist examples of a LRD sequence $\{X_n\}$ and a function $h : \mathbb{R} \mapsto \mathbb{R}$ such that $\{h(X_n)\}$ does not have long range dependence and vice versa. This is disturbing if you think of LRD as an intrinsic property describing dependence. The definition in terms of product moments falls short of capturing this.

6.6.5 *Statistical techniques and exploratory methods*

We mention some (not all) traditional techniques to identify when a model with LRD is appropriate as well as some recent wavelet methods. See [1, 3, 5, 12, 158, 159].

6.6.5.1 *The sample ACF*

The most common, ubiquitous Q&D method, assuming you are convinced the data comes from a stationary process, is to graph the sample ACF. The plot should not decline rapidly. Classical time series data that one encounters in ARMA (Box-Jenkins)

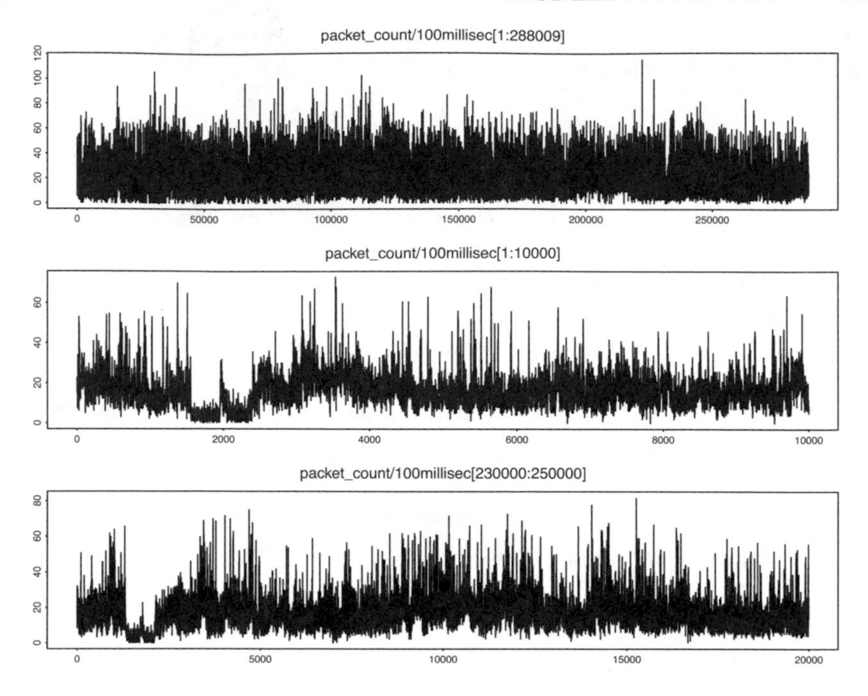

Figure 6.23 *Time series plots for Company X data. Top: full data set. Middle: first 10,000 data. Bottom: last 20,000 observations.*

modeling exercises has a sample ACF that is essentially zero after a few lags, and ACF plots of financial or teletraffic data are often in stark contrast.

Example 6 (Company X) This trace is packet counts per 100 milliseconds $= 1/10$ second for Financial Company X's wide area network link including U.S.–U.K. traffic. It consists of 288,009 observations corresponding to 8 hours of collection from 9 a.m. to 5 p.m. Figure 6.23 shows time series plots. The top plot of the whole data set does not raise any alarms about lack of stationarity, but this is partly due to the muddy plotting resulting from the abundance of data. The middle plot shows the first 10,000 observations and the bottom plot displays the last 20,000. With reduced data size, the last two plots raise some question whether stationarity is appropriate, but this has not been pursued.

Figure 6.24 shows the ACF plot for 2000 lags. There is little hurry for the plot to approach zero. (Do not try to model this with ARMA.)

Example 4 (continued). Recall that the Ericsson data did not look very stationary. In an attempt to counteract the nonstationarity, a subset of the last 28,122 observations was considered. The time series and ACF plots are given in Figure 6.25. The ACF plot is clearly not settling down as it would with an ARMA model, but is also not declining like a power law.

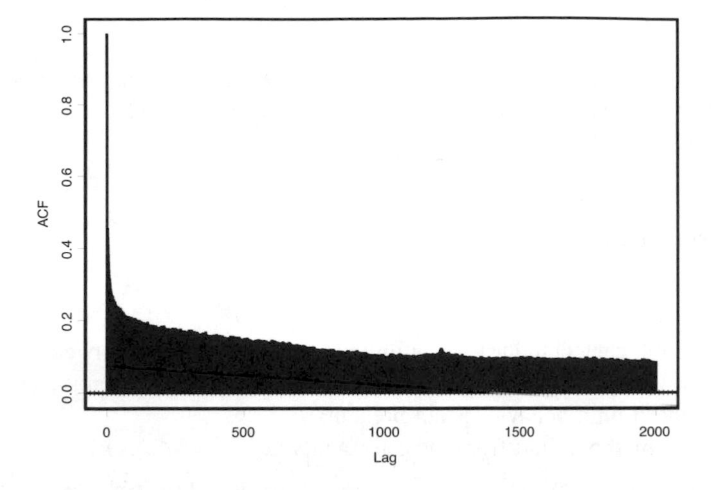

Figure 6.24 *Sample autocorrelation plot for Company X data for 2000 lags.*

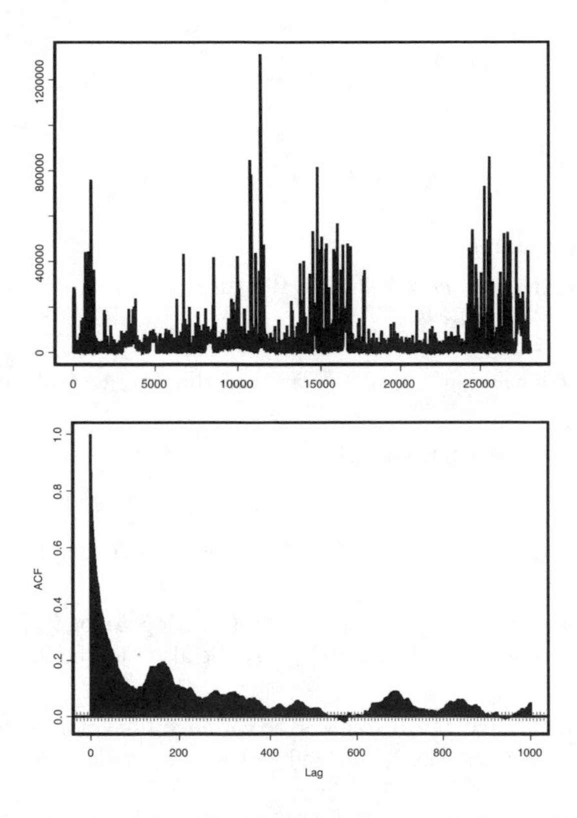

Figure 6.25 *Time series plot (left) of the 28,122 last Company X data and ACF plot (right) of these data.*

6.6.5.2 The variance-time plot

Suppose $\boldsymbol{X} = \{X_n\}$ is a weakly stationary sequence with long range dependence whose ACF satisfies (6.6.1). If we block average over a block of length m, the variance v_m of $(X_1 + \cdots + X_m)/m$ satisfies

$$v_m \sim (\text{const})\, m^{-\beta} L(m). \tag{6.6.10}$$

Note for short range dependence

$$v_m \sim (\text{const})\, \frac{1}{m}. \tag{6.6.11}$$

We seek a graphical technique that capitalizes on the differences between (6.6.10) and (6.6.11) and that will identify $\beta \in (0, 1)$, or at least make sure that the usual square root of n central limit behavior is not present.

Here is the method. Start from measurements $\{x_j, j = 1, \ldots, n\}$.

(i) Compute the block averages $x_j^{(m)}$, $j = 1, \ldots, [n/m]$ over blocks of length m.

(ii) Use $(x_j^{(m)}, j = 1, \ldots, [n/m])$ to estimate v_m by computing the sample variance

$$\hat{v}_m = \frac{1}{[n/m]} \sum_{j=1}^{[n/m]} \left(x_j^{(m)} - \bar{x}^{(m)}\right)^2,$$

where

$$\bar{x}^{(m)} = \sum_{j=1}^{[n/m]} x_j^{(m)}/[n/m]$$

is the sample average.

(iii) Plot $\{(\log m, \log \hat{v}_m), m = 1, 2, \ldots\}$. Because

$$\log v_m \sim (\text{const}) - \beta \log m,$$

the plot of $\{(\log m, \log \hat{v}_m), m \geq$ some threshold value $m_0\}$ should be roughly linear.

(iv) Put the least squares line through

$$\{(\log m, \log \hat{v}_m), m \geq m_0\}.$$

The threshold m_0 can be determined by inspection frequently (or at least hopefully). The slope of this line is an estimate of $-\beta$.

(v) If short-range behavior is present, the plot of $((\log m, \log \hat{v}_m), m \geq m_0)$ would be linear with slope approximately -1, so we also plot the line of that slope for comparison.

Example 6 (continued). Figure 6.26 gives the variance-time plot for the initial 20,000 measurements of the Company X data and yields an estimate of $\hat{\beta} = 0.196$.

6.6.5.3 The Hurst phenomenon and the R/S statistic

What attributes does one expect from data that is generated by a long range dependent process? Because of slowly decreasing covariances, such data will exhibit long periods

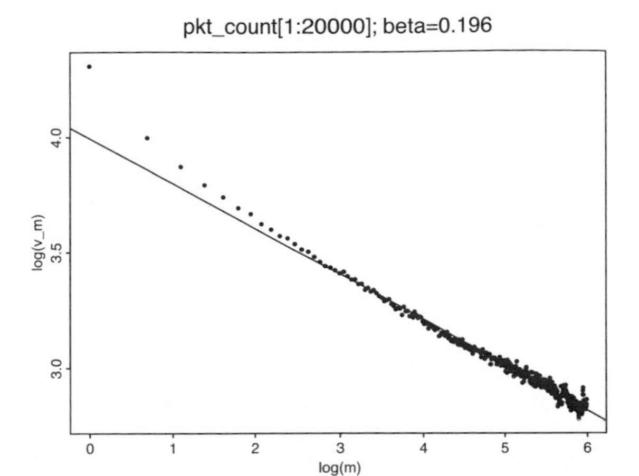

Figure 6.26 *Variance-time plot of the initial 20,000 measurements in the Company X data.*

where the level is high relative to the mean and long periods where it is low. (This may, of course, make LRD difficult to distinguish statistically from a shifting level model.)

The origin of the R/S statistic is in hydrology [80, 81] and comes from the following thought experiment (which to the unitiated may seem a bit like artificial mumbo jumbo): suppose inputs to a reservoir are $X_1, \ldots, X_T$ and think of X_i as the input at the ith time interval. The level at time zero is x, and we are allowed to withdraw water at constant rate $r > 0$. In time periods $1, \ldots, T$ we want the reservoir never to be empty, never to overflow, and we want to allow maximum possible usage of the water, which we interpret as meaning that the final water level at T should be x. Let R be the size of the reservoir. Find R, x, r.

Let $S_i = \sum_{j=1}^{i} X_j$ be the cumulative input in the first i periods and set $S_0 = 0$. The following are the constraints.

(a) The initial amount equals the final amount in the reservoir:

$$x = x + S_T - Tr.$$

This immediately implies $r = \frac{S_T}{T} = \bar{X}_T$.

(b) The never dry condition:

$$x + S_i - ir \geq 0, \quad i = 0, \ldots, T$$

or

$$x + \wedge_{i=0}^{T}(S_i - ir) \geq 0.$$

So we can set

$$x = - \wedge_{i=0}^{T} \left(S_i - \frac{i}{T} S_T \right).$$

(c) The condition that capacity R is never exceeded:

$$x + S_i - ir \leq R, \quad i = 0, \ldots, T$$

or

$$x + \vee_{i=0}^{T}(S_i - ir) \leq R$$

or

$$\vee_{i=0}^{T}\left(S_i - \frac{i}{T}S_T\right) - \wedge_{i=0}^{T}\left(S_i - \frac{i}{T}S_T\right) \leq R.$$

We set

$$R = R_T = \vee_{i=0}^{T}\left(S_i - \frac{i}{T}S_T\right) - \wedge_{i=0}^{T}\left(S_i - \frac{i}{T}S_T\right),$$

which is called the adjusted range.

To make a statistic that is scale invariant, we divide by

$$\mathcal{S} = \mathcal{S}_T = \sqrt{\frac{1}{T}\sum_{i=1}^{T}(X_i - \bar{X}_T)^2}$$

and then $R/\mathcal{S}$ is scale and location invariant. $R/\mathcal{S}$ is called the rescaled adjusted range.

Hurst [80, 81] found empirically that for Nile river flows, as a function of T, $R/\mathcal{S}$ grew roughly like $(const)T^H$ where $H = 0.75$. However in 1951, W. Feller [63] proved that if $\{X_j\}$ are i.i.d. with finite variance, then

$$\frac{R_T/\mathcal{S}_T}{\sqrt{T}} \Rightarrow L, \quad T \to \infty$$

where L is a nondegenerate random variable, and thus, if Hurst's data had come from an i.i.d. model, the growth rate of $R/\mathcal{S}$ should be $T^{1/2}$. This discrepancy is known as the Hurst phenomenon.

The modern way to understand Feller's result is by means of the theory of weak convergence.

Theorem 3 *Suppose $\{X_n\}$ are random variables such that for some constants $\mu_n, b_n > 0$, there exists a process $X(\cdot)$ in $D[0, 1]$ such that*

$$X_n(t) := \frac{\sum_{i=1}^{[nt]}(X_i - \mu_n)}{b_n} \Rightarrow X(t)$$

in $D[0, 1]$. Define

$$W_j = \sum_{i=1}^{j} X_i - \frac{j}{n}\sum_{i=1}^{n} X_i = \sum_{i=1}^{j}(X_i - \bar{X}_n)$$

and

$$R_n = \bigvee_{j=0}^{n} W_j - \bigwedge_{j=0}^{n} W_j.$$

Assuming $\vee_{t=0}^{1}$ *and* $\wedge_{t=0}^{1}$ *operators are almost surely continuous, we have*

$$\frac{R_n}{b_n} \Rightarrow \bigvee_{0 \le t \le 1} (X(t) - tX(1)) - \bigwedge_{0 \le t \le 1} (X(t) - tX(1)).$$

Additionally, if for $\beta_n > 0$

$$\left(X_n(\cdot), \frac{\mathcal{S}_n}{\beta_n} \right) \Rightarrow (X(\cdot), \mathcal{S}),$$

then

$$\frac{R_n/\mathcal{S}_n}{b_n/\beta_n} \Rightarrow \mathcal{S}^{-1} \left(\bigvee_{0 \le t \le 1} (X(t) - tX(1)) - \bigwedge_{0 \le t \le 1} (X(t) - tX(1)) \right).$$

For example, suppose in $D[0, 1]$

$$X_n(t) \Rightarrow B(t),$$

a Brownian motion. Then $B^{(0)}(t) = B(t) - tB(1)$ is a Brownian bridge. When short range dependence is present we frequently have $b_n = \sqrt{n}$ and $\mathcal{S}_n^2 \to \sigma^2$, some constant; this would certainly be true in the i.i.d., finite variance case. Then we would have

$$\frac{R_n/\mathcal{S}_n}{\sigma\sqrt{n}} \Longrightarrow \bigvee_{t=0}^{1} B^{(0)}(t) - \bigwedge_{t=0}^{1} B^{(0)}(t).$$

If we let $R_n^{(0)} = \frac{R_n/\mathcal{S}_n}{\sigma\sqrt{n}}$, so that $R_n/\mathcal{S}_n = \sigma R_n^{(0)} \sqrt{n}$, we would have

$$\log R_n/\mathcal{S}_n = \frac{1}{2} \log n + \log \sigma R_n^{(0)}.$$

Plotting $\{(\log n, \log R_n/\mathcal{S}_n), n \ge$ some threshold$\}$ should approximately give a line of slope $1/2$, and plotting

$$\left(\frac{\log R_n/\mathcal{S}_n}{\log n}, n \ge 1 \right)$$

should give a graph that settles in a neighborhood of $1/2$.

However, this is frequently not the case.

6.6.5.4 Trying to explain the Hurst phenomenon

Suppose the following conditions hold:

(1) The $R/\mathcal{S}$ statistic has a limit distribution:

$$\frac{R_n/\mathcal{S}_n}{b_n} \Rightarrow \chi, \tag{6.6.12}$$

where χ is a nondegenerate random variable, and

(2) b_n is a scaling function satisfying

$$b_n \sim n^J L(n) \in RV_J; \tag{6.6.13}$$

that is, $t^J L(t)$ is regularly varying with index J. The regular variation index J is called the Hurst exponent (as opposed to the Hurst parameter discussed just after (6.6.3). One hopes J always equals the self-similarity parameter H, since the R/S procedure estimates J, but, alas, this is not always the case.

The $R/\mathcal{S}$ estimation procedure: to estimate J we follow the outline in the Brownian limit example above. Define

$$\chi_n = \frac{R_n/\mathcal{S}_n}{b_n},$$

and then $\chi_n \Rightarrow \chi$ and

$$\log \chi_n = \log(R_n/\mathcal{S}_n) - J \log n - \log L(n) \Rightarrow \log \chi,$$

so

$$\frac{\log(R_n/\mathcal{S}_n)}{\log n} - J = \frac{\log \chi_n + \log L(n)}{\log n} = o_p(1).$$

The strategy, then, is to plot

$$\left\{ \left(m, \frac{\log(R_m/\mathcal{S}_m)}{\log m} \right), 1 \le m \le \text{length(dataset)} \right\}$$

and this should asymptote at J.

To see how this works in practice, Figure 6.27 displays two $R/\mathcal{S}$ plots for simulated data. On the left is the $R/\mathcal{S}$ plot for 5000 N(0,1) i.i.d. variables and on the right is the $R/\mathcal{S}$ plot for 5000 i.i.d. stable variables with $\alpha = .2$ and skewness 1. Note both plots hug the height 1/2, which is the correct self-similarity parameter for normal but not the self-similarity parameter for stable, which would be $1/.2 = 5$.

As a bonus, here is the $R/\mathcal{S}$ plot in Figure 6.28 for the last 5000 measurements in the Ericsson trace. The plot is definitely not indicating 1/2.

So, from these examples, the $R/\mathcal{S}$ statistic may detect something, but it may not be the self-similarity parameter. In particular, with the example of the i.i.d. stable random variables, $R/\mathcal{S}$ definitely indicated 1/2 and not anything related to the self-similarity parameter. This requires some further explanation.

Notation: If $\{X(t), 0 \le t \le 1\}$ is a stochastic process such that $X(0) = 0$, we write

$$X^{(0)}(t) = X(t) - tX(1), \quad 0 \le t \le 1$$

for the bridge version. If B is a Brownian motion, $B^{(0)}$ is Brownian bridge, and if X is a stable Lévy motion, then $X^{(0)}$ is a stable bridge. We also write

$$X^{\vee}(t) = \bigvee_{0 \le s \le t} X(s), \quad X^{\wedge}(t) = \bigwedge_{0 \le s \le t} X(s).$$

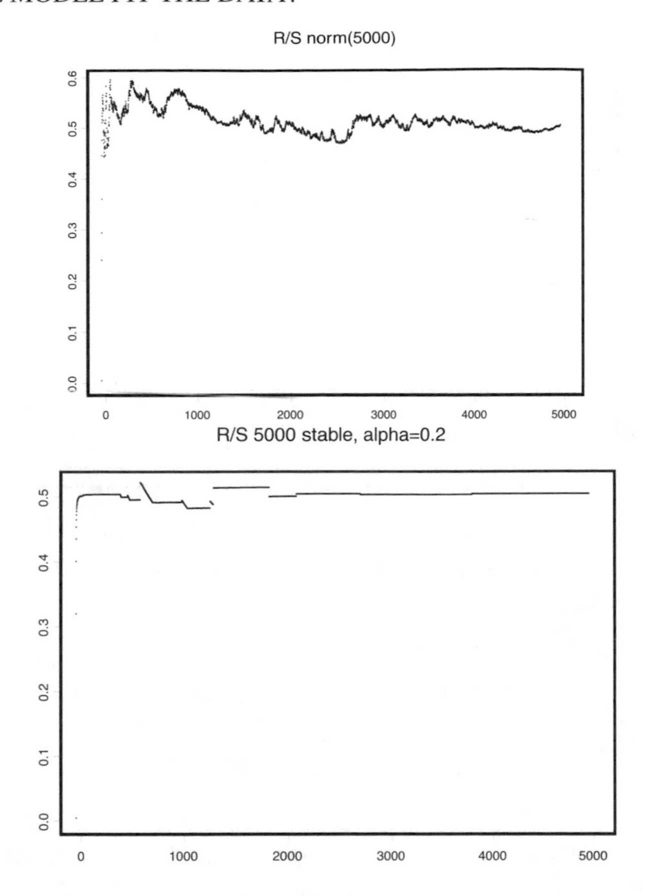

Figure 6.27 *$R/\mathcal{S}$ plots $\{(m, (\log m)^{-1} \log(R_m/\mathcal{S}_m)), 1 \leq m \leq n\}$ for i.i.d. normal (left) and i.i.d. stable (right) with $\alpha = 0.2$ and skewness 1.*

For instance, we have

$$\left(B^{(0)}\right)^{\vee}(1) - \left(B^{(0)}\right)^{\wedge}(1) = \bigvee_{0 \leq t \leq 1} (B(t) - tB(1)) - \bigwedge_{0 \leq t \leq 1} (B(t) - tB(1)).$$

Consider the following cases:

(a) If $\{X_n\}$ are i.i.d. with finite variance, since $R/\mathcal{S}$ is location and scale invariant, we might as well suppose $EX_n = 0$, $EX_n^2 = 1$. Then $\mathcal{S}_n \xrightarrow{P} 1$ and

$$\frac{R_n/\mathcal{S}_n}{\sqrt{n}} \Rightarrow \left(B^{(0)}\right)^{\vee}(1) - \left(B^{(0)}\right)^{\wedge}(1).$$

This certainly does not explain the Hurst phenomenon.

(b) Moran [113] speculated long range dependence could perhaps be explained by heavy tails and/or infinite variance but, alas, this is not true [104]. If $\{X_n\}$ are

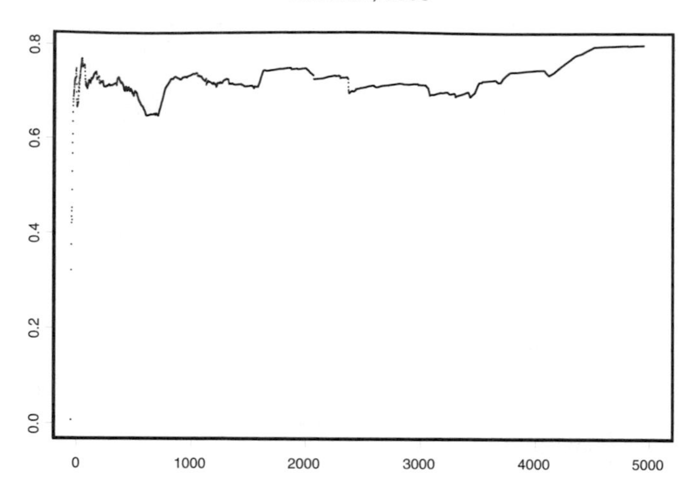

Figure 6.28 *R/S plot* $\{(m, (\log m)^{-1} \log(R_m/\mathcal{S}_m)),\ 1 \le m \le n\}$ *for last 5000 measurements in the Ericsson trace.*

i.i.d. and in the domain of attraction of an α-stable distribution, then

$$\frac{R_n/\mathcal{S}_n}{\sqrt{n}} \Rightarrow \chi, \tag{6.6.14}$$

where χ is constructed from the stable bridge. This too does not explain the Hurst phenomenon; note the disturbing fact that $J = \frac{1}{2}$ but the self-similarity parameter $H = \frac{1}{\alpha}$.

(c) Suppose $\{Y(t), t \ge 0\}$ is H-sssi with nice paths, say in $D[0, \infty)$. Set $X_k = Y(k) - Y(k-1), k \ge 1$ and $S_n = \sum_{i=1}^{n} X_i$. Note $S_n = Y(n)$ since $Y(0) = 0$ from self-similarity. Then using self-similarity, it is easy to check

$$\begin{aligned} \frac{R_n}{n^H} &\overset{d}{=} \bigvee_{k=0}^{n} \left(Y\left(\frac{k}{n}\right) - \frac{k}{n} Y(1) \right) - \bigwedge_{k=0}^{n} \left(Y\left(\frac{k}{n}\right) - \frac{k}{n} Y(1) \right) \\ &\to \left(Y^{(0)}\right)^{\vee}(1) - \left(Y^{(0)}\right)^{\wedge}(1) = R^*. \end{aligned}$$

If also $\{X_i\}$ is ergodic with $EX_1^2 < \infty$, so that

$$\mathcal{S}_n^2 \to \mathrm{Var}(X_1),$$

then

$$\frac{R_n/\mathcal{S}_n}{n^H} \Rightarrow R^*/\sqrt{\mathrm{Var}(X_1)}.$$

The conclusion is that if (a) $Y(\cdot)$ is H-sssi with regular paths, (b) $(Y(k)-Y(k-1), h \ge 1)$ is ergodic with $EY(1)^2 < \infty$, then $J = H$, and the Hurst exponent equals the self-similarity parameter. Asymptotic versions of this result could be developed.

Recent discussions of ways to fool the R/S statistic are in [69, 70]. See also [13].

6.6.5.5 *Wavelet methods*

The recently developed wavelet estimation methodology works under a variety of assumptions. If the process under investigation is stationary with finite variance and long range dependence, the method yields an estimate of the long range dependence parameter β in (6.6.1). When applied to the increments of a self-similar process, the wavelet method yields an estimate of H.

The wavelet method has gained adherents because it provides an appealing compromise between low computational cost and good statistical performance. It does not require an exact parametric model and because it is based on identification of scaling in a log-log diagram, it is possible to judge the range of scales on which the model fits. It is also robust to smooth nonstationarities depending on the filtering capabilities of the wavelet chosen.

Important ideas are in [1, 2, 3, 158, 159]; here is the briefest outline. Let ψ be a reference or mother wavelet; that is, a smooth function, well-localized in both position and frequency which satisfies the admissibility and moment conditions $\int t^n \psi(t)dt = 0,\ n = 0, 1 \dots N$. The phrase well-localized means the function has compact support or at least is rapidly decaying. The moment conditions allow the wavelet transform to filter means and smooth trends and to make the procedure robust against certain departures from stationarity. Define the location-scale family ψ_{ba} by

$$\psi_{ba}(t) := \frac{1}{\sqrt{a}}\psi\left(\frac{t-b}{a}\right), \qquad b \in \mathbb{R},\ a > 0.$$

Define the wavelet transform $W_\psi X$ of a process X by

$$\begin{aligned} W_\psi X(b, a) &:= \int \psi_{ba}(t)X(t)dt \\ &= \int \sqrt{a}\psi(s)X(as+b)ds, \quad (b, a) \in \mathbb{R} \times \mathbb{R}^+. \end{aligned} \tag{6.6.15}$$

If the process $X(\cdot)$ is stationary, or has stationary increments, then the process $W_\psi X(\cdot, a)$ is again stationary from (6.6.15). If a process Y is H-sssi with finite second moments, then

$$E|W_\psi Y(b, a)|^2 = a^{2H+1}E|W_\psi Y(b/a, 1)|^2.$$

Define $X(t) = Y(t+\Delta) - Y(t)$ for $\Delta > 0$. By a change of variables, a Taylor expansion and self-similarity,

$$W_\psi X(b, a) \sim -a^{H-1/2}\Delta \int \psi'(s)Y(s)ds, \quad a \to \infty, \tag{6.6.16}$$

and if second moments exist,

$$\mathrm{Cov}(X(0), X(a)) = E|W_\psi X(b, a)|^2 \sim Ka^{2H-1}, \quad a \to \infty. \tag{6.6.17}$$

For comparison, note that if $X = B_H$, FBM, then for t fixed and $a \to \infty$ we have for a constant $\sigma > 0$

$$\mathrm{Cov}(X(t), X(t+a)) = \sigma((t+a)^{2H} + t^{2H} - a^{2H}) \sim 2\sigma H t a^{2H-1}.$$

Now suppose only that $X(\cdot)$ is a LRD stationary process satisfying the analogue of (6.6.1), then

$$\mathrm{Cov}(X(0), X(a)) = E|W_\psi X(b,a)|^2 \sim K a^{-\beta}, \quad a \to \infty, \tag{6.6.18}$$

which agrees with (6.6.17) for FGN. The wavelet estimator is a regression method based on (6.6.17) or (6.6.18), which makes use of the fact that the wavelet transform is less correlated than the process.

In practice, wavelet coefficients are computed on a dyadic grid, so define

$$d(j,k) = W_\psi X(2^j k, 2^j), \quad j, k \in \mathbb{Z}.$$

By (6.6.18),

$$\mathrm{Cov}(X(0), X(2^j)) = E|d(j,k)|^2 \sim K 2^{-j\beta}, \quad j \to \infty, \tag{6.6.19}$$

which suggests averaging coefficients at fixed scale,

$$\mathrm{Cov}(X(0), X(2^j)) = \frac{1}{n_j} \sum_k |d_{j,k}|^2,$$

where n_j is the number of available coefficients at scale 2^j. The parameter β in (6.6.18), or equivalently H, can then be estimated from a linear regression in the log-log diagram of $\{(j, \frac{1}{n_j} \sum_k |\hat{d}_{j,k}|^2)\}$, where $\hat{d}_{j,k}$ is the estimated wavelet coefficient obtained from the data.

There are refinements to reduce bias induced by the log-transform and to replace an ordinary least squares by weighted least squares [3, 159].

Example 7 (Munich Trace) The Munich low data set contains measurements of cell rates for both the sending (TX) and receiving (RX) directions of an ATM link. The data was collected on Wednesday, November 12, 1997 (TX) and Wednesday, December 17, 1997 (RX) with a temporal resolution of 2 seconds, i.e., the total number of cells that passed the ATM link every 2 seconds was recorded. Recorded traffic was pure IP (mainly HTTP, FTP, and NNTP) data traffic, without any audio/video components. A shorter sample covering the period between 10 a.m. until 1 p.m. was selected for analysis. The shorter period was chosen to obtain a roughly stationary data set.

The estimated cumulative traffic looks neither Gaussian nor stable so the approximations in Theorems 1 and 2 do not seem in evidence. There is a clear and strong LRD, which persists over a wide range of scales as seen by autocorrelation functions and wavelet regression plots. In both traces, the estimates of H are close to 0.95. See Figure 6.29.

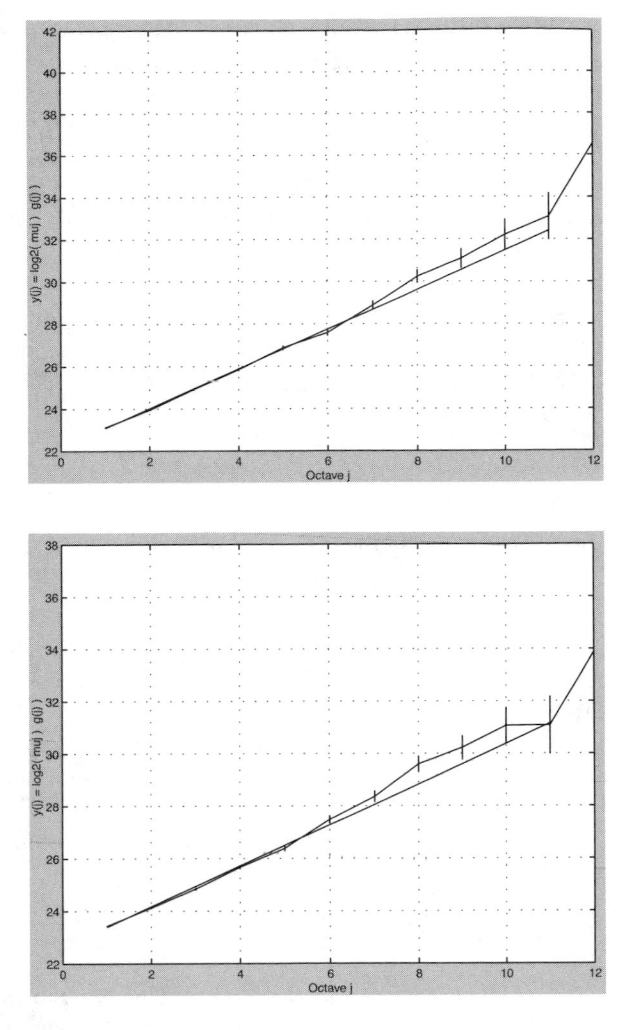

Figure 6.29 *Wavelet regression plots for the Munich traces: receiving (RX) on left and transmitting (TX) on right.*

6.7 Does the model fit the data? Small time scales: Hölder exponents and multifractality

It is now rather widely accepted (but not without some controversy) that network traces examined at time scales above a certain resolution appear statistically self-similar. The limiting processes discussed in Theorems 1 and 2 are self-similar. Lamperti's theorem [96] says to expect self-similarity whenever a process is obtained as a distributional limit by dilating time linearly and scaling space. (See the discussion following (6.6.8) and (6.6.9).)

The observed self-similarity seems to hold above a resolution of about 100 milliseconds and not below. At smaller thresholds, certain researchers fervently believe there is evidence for chaotic behavior or extreme intermittency, which may be typical of multifractal phenomena. [62, 68, 119, 143, 144, 145, 163]. Current statistical goals are to verify if multifractality is really present and develop methods for dealing with it statistically. The probabilistic goal is to develop model-based explanations for why the multifractality should be present [23, 105, 107, 129]. It is at the small time scales that one must seek the influence of protocols and network architecture.

There seem to be two literatures, which have not intersected much. One grows out of the Gaussian process tradition and defines Hölder exponents in a mean square sense [83, 84, 85, 86]. The other is path based. A nice summary is [143]. We give a brief summary of both approaches.

6.7.1 *Second order definition*

The semivariogram V of a second order process $\{X(t)\}$ is defined by

$$V(t, \tau) = \frac{1}{2} E(X(t+\tau) - X(t))^2,$$

and contains the same information as the covariance function. If $X(\cdot)$ is H-ss, then for $\theta > 0$, we have $\theta^{-2H} V(\theta t, \theta \tau) = V(t, \tau)$. If $X(\cdot)$ also has stationary increments, V does not depend on its first variable, $V(t, \tau) = V(\tau)$, and if it is also self-similar, then $V(\tau) = V(1)\tau^{2H} := c\tau^{2H}$. If the process is also a centered Gaussian process, this means that it is an FBM.

A stationary process $X(\cdot)$ has local Hölder index $H_o(t)$ at t (in the mean square sense) if the semivariogram satisfies

$$V(t, \tau) = c\tau^{2H_o(t)} + o\big(\tau^{2H_o(t)}\big), \quad \text{as } \tau \to 0, \tag{6.7.1}$$

for each t. For fBm, $H_o(\cdot)$ is constant and $H = H_o(t)$. This is significant because if empirical estimates of $H_o(t)$ for various values of t either reject the constancy of $H_o(\cdot)$ or indicate, for instance, that $H_o(0)$ is not equal to the estimate of H, then we have evidence against approximation to the cumulative traffic by FBM. For a Gaussian process, the Hölder index process gives precise information on sample paths [4, 82], on the rate of convergence of nonparametric estimates of the covariance function [86] and on the asymptotic behavior of wavelet coefficients of X [83]. A final preliminary comment is that this approach is limited to L_2 processes and has no direct relevance to heavy tailed processes.

One method to estimate $H_o(t)$ can be based on simple empirical quadratic variation [26, 71, 85, 91]. Assuming the underlying process $X(\cdot)$ has stationary increments allows one to focus on $t = 0$. At scale $1/n$, define

$$V_n = \frac{1}{2} \sum_{k=1}^{n} (X(k/n) - X((k-1)/n))^2 .$$

Table 6.2 *Comparison of estimated values for H and $H_o(0)$ for various data sets.*

Data set	$\hat{H}$	$\hat{H}_o(0)$
simM/G/∞	$.90 \pm .01$	.88
BUburst 10s	$.89 \pm .02$	.73
BUburst 1s	$.81 \pm .01$	.87
UCB 10s	$.58 \pm .03$	.65
UCB syn 10s	$.95 \pm .07$	1.36
Munich lo TX	$.89 \pm .01$	.86
Munich lo RX	$.97 \pm .01$	.85
Munich hi .1s	$1.02 \pm .03$	.66
Munich hi .01s	$1.03 \pm .04$	.56
Ericsson	$.88 \pm .02$	1.21
Eri syn 1s	$1.48 \pm .02$	1.51

Suitably normalized, this converges with probability 1,

$$\lim_{n\to\infty} n^{2H_o(0)-1} V_n = c,$$

where c appears in (6.7.1).

The quadratic variation estimator is not scale invariant and may converge slowly, so there are proposed refinements that weight the squared terms and then take ratios and that also filter certain polynomial trends. In [72], the quadratic variation based estimator of $H_o(0)$ discussed in [85, page 437] was used to study various traces. Empirical evidence suggested that $H_o(t)$ was not constant and in no case was it found that $H = H_o(0)$. The estimator of $H_o(0)$ is designed for Gaussian processes and more experience in traffic contexts is required to assess its reliability.

Table 6.2 from [72] compares estimated values of H and $H_o(0)$ for various traces. The agreement is not good.

6.7.2 *Pathwise definition*

Since self-similarity is a distributional property, it seems like mixing apples and oranges to consider small time scale properties in the pathwise sense. In the literature, two types of Hölder exponents have been considered, namely, one based on exponential growth rate and one based on polynomial approximation. We consider the first.

The Hölder exponent based on exponential growth rate of the function $x(\cdot)$ at t is defined as

$$h_x(t) := \liminf_{\epsilon\downarrow 0} \frac{\log \sup_{u:|u-t|\le\epsilon} |x(u)-x(t)|}{\log \epsilon}. \tag{6.7.2}$$

For nondecreasing functions, the definition of $h_x(t)$ in (6.7.2) simplifies to

$$\begin{aligned} h_x(t) &= \liminf_{\epsilon\downarrow 0} \frac{\log[(x(t+\epsilon)-x(t)) \wedge (x(t)-x(t-\epsilon))]}{\log \epsilon} \\ &= \liminf_{\epsilon\downarrow 0} \frac{\log(x(t+\epsilon)-x(t-\epsilon))}{\log \epsilon}. \end{aligned} \tag{6.7.3}$$

The Hölder exponent of the sum of two functions satisfies the following inequality. For two functions x and y, we have

$$h_{x+y}(t) \geq h_x(t) \wedge h_y(t). \tag{6.7.4}$$

Furthermore, equality holds if $h_x(t) \neq h_y(t)$ or if x and y are monotone. We are, of course, interested in Hölder exponents of sums of functions because of the need to superimpose traffic streams.

How do we decide which values of the Hölder exponent are common? One way to do this is with the multifractal spectrum. The multifractal spectrum of the Hölder exponent of the function x for the Hölder exponent based on exponential growth rate is

$$d_x(a) = \dim(\{t > 0 : h_x(t) = a\}), \quad a \in [0, \infty),$$

where for a set Λ, $\dim(\Lambda)$ is the Hausdorff dimension (cf. Chapter 2 of [57]) of Λ.

An increasing Lévy process $X(\cdot)$ has the interesting property that, restricted to any interval, it has a nonrandom multifractal spectrum for the Hölder exponent based on exponential growth rate [87]. In fact the spectrum is linear and is thus nontrivial. Empirically, spectra that have been estimated from traffic, have never been observed to be linear. Neither do observed spectra correspond to the monofractal structure of FBM.

Note that if we require the increasing Lévy process $X(\cdot)$ to also be an α-stable Lévy motion, then $X(\cdot)$ is $1/\alpha$-ss (a distributional property) as well as a random multifractal with a nonrandom multifractal spectrum.

6.8 A model for large and small time scales

The good news about the infinite source Poisson model is that it provides a simple, compelling reason why long range dependence exists in the traffic data. The not so good news is:

1. The model does not fit the data all that well.
   (a) The constant transmission rate assumption is clearly wrong.
   (b) Not all times of transmission initiation can be modeled as Poisson. Presumably humans acting independently can have the times of their acts modeled as a Poisson process but machine triggered events, in say web browsing, will never be Poisson.
   (c) There is no hope the simple model can successfully match fine time scale behavior observed below, say, 100 milliseconds, which is speculated to be multifractal.
2. The model predicts that cumulative traffic can be approximated by either FBM or stable Lévy motion. In practice, traffic is not observed to be heavy tailed, and while some report traffic at heavily loaded links is Gaussian, there is the disturbing evidence that Hölder exponents are not constant and not equal to the self-similarity parameters, as would be the case if the FBM approximation were valid.

3. The traffic data is often not stationary. There are always time of day effects. This has, so far, been ignored.

This is quite a laundry list of defects needing repair. Let us make a start and at least try to address 1.(c).

6.8.1 *A more general model appropriate for the study of small and large time scales*

Here are the ingredients of a more general model.

1. Denote the time when the k-th transmission begins by Γ_k. The sequence $\{\Gamma_k\}$ is strictly increasing to ∞.
2. The size of the file required to be transmitted at Γ_k is J_k, and we assume $J_k > 0$.
3. There is a transmission schedule, denoted by $A_k(\cdot)$, where $A_k(t)$ denotes the amount of data transmitted in time t after the kth transmission has begun. It is a nondecreasing càdlàg function starting at zero and increasing to ∞, which vanishes on the negative real axis.

With these assumptions, the accumulated traffic in $[0, T]$ aggregated over the different users is

$$A(t) = \sum_{k=1}^{\infty} A_k(t - \Gamma_k) \wedge J_k. \tag{6.8.1}$$

Because the assumption of unit transmission rate is highly arbitrary and restrictive, there have been efforts to remove this deficiency. A nonconstant but deterministic transmission schedule has been considered in [93, 137] and with a constant input rate it was shown that a Lévy stable motion approximation could be fashioned. A transmission schedule that was linear with randomly chosen slope (chosen not necessarily independent of the file size) is considered in [108]. Related results about renewal reward processes are given in [102, 103, 123, 156]. However, none of these efforts captures the small time scale properties. Our model allows a random, time-dependent transmission schedule $A_k(\cdot)$ and random file size J_k chosen at the beginning of each transmission. The length of transmission L_k is obtained as a function of these two random quantities and is readily defined as

$$L_k = A_k^{\leftarrow}(J_k) := \inf\{s : A_k(s) \geq J_k\}.$$

This model suggests that the fine time scale behavior of network traffic results from individual transmission schedules exhibiting multifractality. This results in multifractal behavior for the cumulative traffic process at the microscopic level, and still gives a stable Lévy motion as the macroscopic approximation. This may not be good news if you believe this approximation has never been seen in practice, but at least this setup allows multifractal behavior on fine time scales and self-similar scaling at large time scales. The suggestion that the multifractal behavior of the cumulative traffic process results from similar behavior of the individual input processes, presumably due to extreme intermittency and blocking caused by congestion at various nodes, needs to be verified by empirical study of individual, user level input processes.

The following summarizes findings in [107].

6.8.1.1 Small time scale behavior

The small time scale behavior of the cumulative traffic process $A(\cdot)$ requires the following further minimal assumptions on the transmission schedule $\{A_k\}$:

4. Suppose $\{A_k\}$ are identically distributed with stationary increments.
5. The multifractal spectrum of $A_k(\cdot)$ is not degenerate at a single point. This ensures that we consider processes with paths that show real multifractal behavior.
6. The multifractal spectrum of $A_k(\cdot)$ restricted to any (nonrandom) interval is nonrandom.

If A_k is, for example, an increasing Lévy process, then, restricted to any interval, it has a nonrandom multifractal spectrum for the Hölder exponent based on exponential growth rate [87].

If the assumptions (1) to (6) hold, then with probability 1,

$$d_A = d_{A_1}. \tag{6.8.2}$$

So the aggregate traffic process $A(\cdot)$ inherits the multifractal structure of the individual transmission schedules.

6.8.1.2 Large time scale behavior

Small time scale behavior was a path by path analysis. Large time scale analysis requires distributional assumptions.

7. Suppose $\{\Gamma_k\}$ forms a homogeneous Poisson process with intensity parameter λ.
8. Assume $\{(A_k, J_k) : k \geq 1\}$ are i.i.d. and independent of $\{\Gamma_k\}$.
9. Assumptions about the joint distribution of $(A_1(\cdot), J_1)$: Suppose
   (a) there exists a regularly varying function $\sigma(\cdot)$ with index $H > 0$;
   (b) there exists a proper random process χ (that is, $\chi(t)$ is nondegenerate for each t) with stationary increments, taking values in $\mathbb{D}[0, \infty)$.

   Then suppose that for each fixed $\epsilon > 0$,

$$\frac{1}{\bar{F}_J(\sigma(T))} P\left[\frac{J_1}{\sigma(T)} > \epsilon, \frac{A_1(T\cdot)}{\sigma(T)} \in \cdot\right] \stackrel{w}{\rightarrow} \epsilon^{-\alpha_J} P[\chi \in \cdot]$$

   on $\mathbb{D}[0, \infty)$ where "$\stackrel{w}{\rightarrow}$" denotes weak convergence and we assume $1 < \alpha_J < 2$.
10. Technical: for all $\gamma > 0$, assume

$$\lim_{\epsilon \downarrow 0} \limsup_{T \to \infty} \frac{1}{\bar{F}_J(\sigma(T))} P\left[\frac{J_1}{\sigma(T)} \leq \epsilon, \frac{L_1}{T} > \gamma\right] = 0.$$

11. Technical: assume

$$E[\chi(1)^{-\alpha_J}] < \infty.$$

These assumptions lead to certain consequences and there are several simplifying conditions that make the assumptions easier to check. Here is extra information that should make the framework clearer.

(a) The limit χ is H-self-similar. To see this, one must elaborate the methods used to prove Lamperti's theorem [96, 55, 161, 160, 115, 114].

(b) The limit χ being càdlàg, self-similar, and nondegenerate implies that $H > 0$.

(c) The limit χ being nondecreasing implies $H \geq 1$.

(d) If $H = 1$, then $\chi(t) = t\chi(1)$ is linear and no interesting structure exists so we exclude this possibility.

(f) The distribution tail of J_1 is regularly varying with index $-\alpha_J$:

$$\bar{F}_J(x) = 1 - F_J(x) = x^{-\alpha_J} L(x), \quad x > 0.$$

(g) Cumulative traffic on large time scales looks like stable Lévy motion. We describe this in more detail next.

The precise statement for (g) is as follows: suppose (1) to (3) and (7) to (11) hold with $1 < \alpha_J < 2$ and define

$$Y_T(t) = \frac{A(Tt) - \lambda T t E(J_1)}{b_J(T)},$$

where

$$b_J(T) = \left(\frac{1}{1 - F_J}\right)^{\leftarrow}(T)$$

is the usual quantile function. Then we have

$$Y_T \stackrel{fidi}{\to} Z_{\alpha_J}, \tag{6.8.3}$$

where Z_{α_J} is a mean 0, skewness 1, α_J-stable Lévy motion.

Here are some simpler sufficient conditions for (6.8.3) to hold: (i) J_1 and A_1 are independent, (ii) J_1 has a tail of index α_J, (iii) A_1 is itself a proper H-ss process, and (iv) $E[A_1(1)^{-\rho}] < \infty$ for some $\rho > \alpha_J$.

Note that if $A_1(\cdot)$ is a $\frac{1}{H}$-stable Lévy motion, then (iii) and (iv) are automatically satisfied.

6.8.2 A model more amenable to statistics

Section 6.8 gives a model that successfully exhibits both small scale multifractality and large scale self-similarity. However, it is not easy to do statistics with this model. Here we will see what can be achieved by dropping the assumption of fixed unit transmission rate and allowing the rate to be random. We will examine whether the rate is independent of the transmission duration. Some of this material is adapted from [108].

It is difficult to conclude from evidence in measured data that the rate and the length of the transmission are always independent, but in certain cases we may reasonably assume that the rate and the length of the transmission are at least asymptotically independent. As an example, we consider the BUburst dataset considered by [72]. This is data processed from the original 1995 Boston University data described in the report by [36] and catalogued at the internet traffic archive (ITA) web site www.acm.org/sigcomm/ITA/.

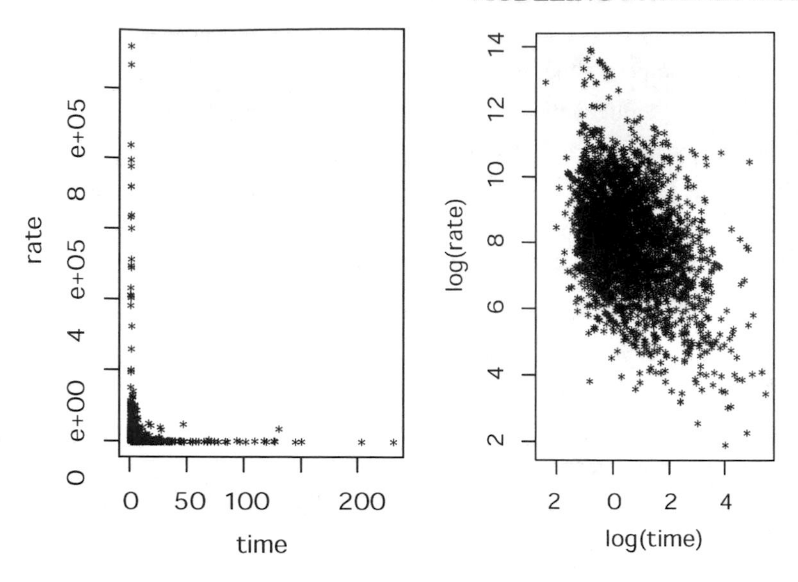

Figure 6.30 *Plot of the length of transmission against the rate of the transmission of the BUburst data in (left) natural scale and in (right) log-log scale.*

A plot of the transmission length against the transmission rate (Figure 6.30) shows that most of the data pairs hug the axes, which suggests the variables are at least asymptotically independent. However, if we plot the data in the log scale on both the axes, then a weak linear dependence is observable and the correlation coefficient between the two variables after log transform is approximately -0.379, which argues against an independence assumption. We consider the log transform to make the variables have finite second moment, so that correlation coefficient becomes meaningful.

The Hill estimates obtained for the transmission length, the transmission rate, and the size of the transmitted file are 1.407, 1.138, and 1.157, respectively. The corresponding Hill plots are given in Figure 6.31.

For each of the variables, the first column gives the Hill plots, the second column gives the AltHill plots (see Section 6.5.1) and the third plot, named the Stărică plot, is an exploratory device suggested in [155, Section 7] to help decide on the number of upper order statistics to be used. (It was designed for higher dimensional problems, but is basically a one-dimensional technique.) It uses the fact that for a random variable X with Pareto tail of parameter α, we have

$$\lim_{T\to\infty} TP\left[\frac{X}{T^{\frac{1}{\alpha}}} > r\right] = r^{-\alpha}.$$

For every k, we estimate the left-hand side by

$$\hat{\nu}_{n,k}((r,\infty]) = \frac{1}{k}\sum_{i=1}^{n} \mathbf{I}\left[\frac{X_i}{(n/k)^{1/\hat{\alpha}_{n,k}}} > r\right].$$

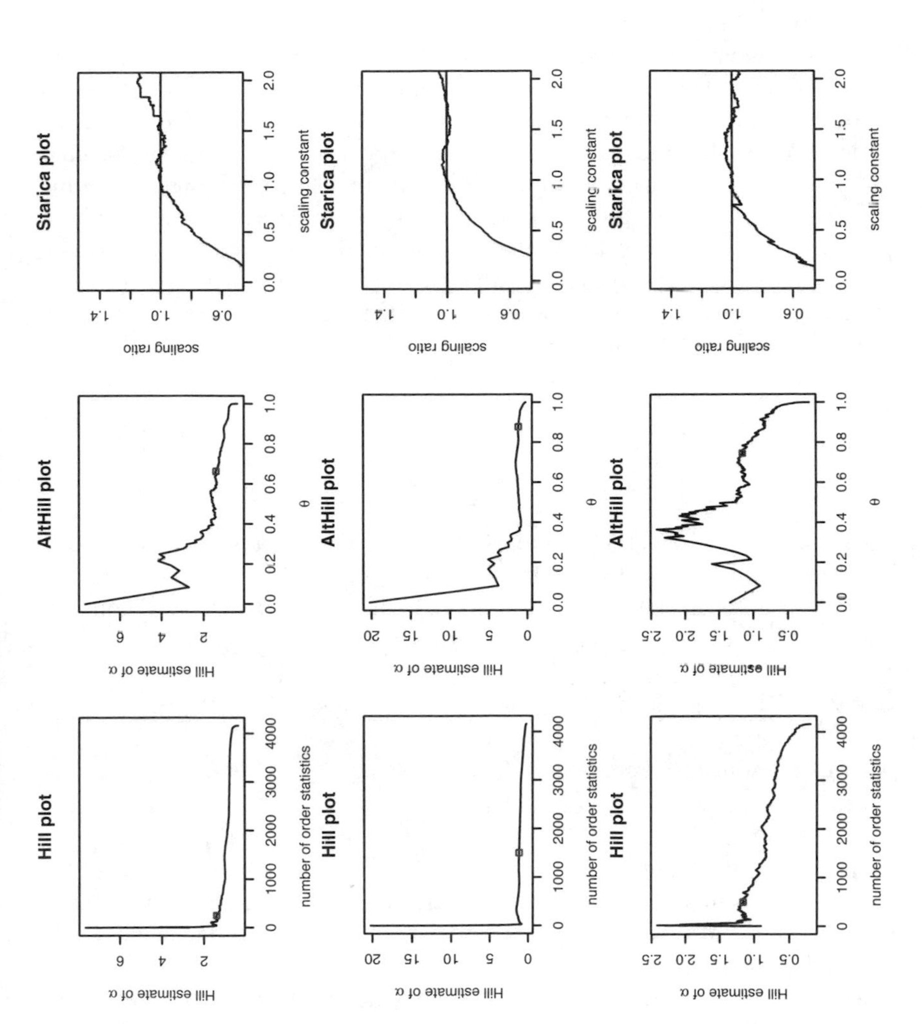

Figure 6.31 *Hill plots of transmission length (top), transmission rate (middle), and transmitted file size (bottom).*

We expect the ratio of $\hat{\nu}_{n,k}((r,\infty])$ and $r^{-\hat{\alpha}_{n,k}}$, called the scaling ratio, to be approximately 1, at least for values of r in a neighborhood of 1, if we have made the correct choice of k. In the Stărică plot, we plot the above scaling ratio against the scaling constant r, and choose k so that the graph hugs the horizontal line of height 1. The interesting point to be noted is the fact that the rate of the transmission has a much heavier tail than the length of transmission. So a model with a random transmission rate with heavy tails seems appropriate.

When two random variables are heavy tailed, a modest assumption on the joint distribution $F(\boldsymbol{x})$ of the pair is that the distribution be in the domain of attraction of an extreme value distribution. When both marginal distributions of $F(\boldsymbol{x})$ are the same, the extreme value distribution will have an exponent measure $\nu_*(\cdot)$ in standard form [43, 45, 46, 139]. However, when the two marginal distributions are not the same (which is always the case except in simulations), we must first rescale each component of the bivariate vector by its quantile function (or its estimate, the k-th largest marginal order statistic) and then adjust for the difference in tail exponents by powering each component of the random vector by the tail exponent (or its estimate). The transformation to polar coordinates

$$T : \boldsymbol{x} \mapsto (r, \theta)$$

induces a product form on ν_*,

$$\nu_* \circ T^{-1}(dr, d\theta) = cr^{-1}S(d\theta),$$

for some constant $c > 0$ and $S(\cdot)$ a probability measure. When the bivariate vector $\boldsymbol{X}$ concentrates on $\mathbb{R}_+^2$, $S(\cdot)$ concentrates on $[0, \pi/2]$. Asymptotic independence corresponds to S concentrating on $\{0\}$ and $\{\pi/2\}$. If the bivariate observations are $\boldsymbol{X} = \{(X_{1,i}, X_{2,i}), 1 \le i \le n\}$, then we transform

$$\boldsymbol{X} \to \left\{\left(\left(\frac{X_{i1}}{X_{1,(k)}}\right)^{\alpha_1}, \left(\frac{X_{i2}}{X_{2,(k)}}\right)^{\alpha_2}\right); i = 1, \ldots, n\right\} \overset{T}{\mapsto} \left\{\left(R_i^{(n)}, \Theta_i^{(n)}\right),\ 1 \le i \le n\right\},$$

and we need to estimate S from

$$\left\{\Theta_i^{(n)} : R_i^{(n)} > 1\right\}.$$

If we make a density plot of these Θs, we should be able to get an impression of the tendency toward asymptotic dependence. Modes at 0 and $\pi/2$ indicate a tendency toward asymptotic independence while a mode at $\pi/4$ indicates a tendency towards asymptotic dependence.

For the (buL,buF) data, that is, the length of download against the file size downloaded, the estimated angular measure density, given in Figure 6.32, shows two modes at zero and $\pi/2$ indicating a tendency toward independence. The angular measure estimation was made using $k = 100$. This is based on the Stărică plot given in Figure 6.33, which shows the superiority of the choice $k = 100$.

Interestingly, the analysis of (buL,buR), the download length vs. the inferred transmission rate, shows much less tendency toward asymptotic independence [108] and the whole issue of the dependence between download time, download rate, and downloaded file size deserves more careful study. See [20] for more comments on this point.

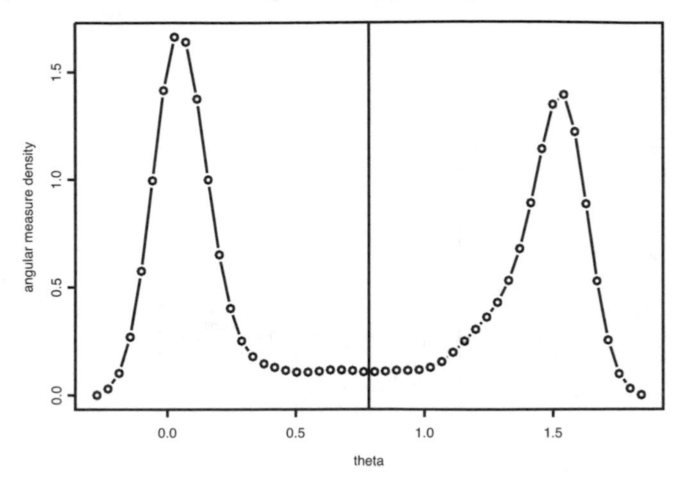

Figure 6.32 *The data (buL,buF): density estimate for the angular measure.*

6.8.2.1 *A model for asymptotic independence*

An obvious generalization of the infinite source Poisson model is to retain Poisson connection times, but generalize the transmission mechanism. We assume that there are i.i.d. pairs $\{(L_k, R_k), k \geq 1\}$ giving length of a download and rate of transmission so that when a download is initiated, a random rate is chosen for the transmission. Empirically it is often the case that each marginal distribution of L_k and R_k is heavy tailed. The file transmitted will, of course, be $L_k R_k$ and analyzing the infinite source Poisson model will require knowledge of how the product $L_k R_k$ behaves. The classical extreme value notion of asymptotic independence is not strong enough to generate useful information about the product and a strengthening is required. We briefly discuss the approach assuming that L has the heavier tail; a minor modification is necessary if R has the heavier tail.

Define the measure ν_α on $(0, \infty]$ by

$$\nu_\alpha(dx) = \alpha x^{-\alpha-1} dx, \quad x > 0, \alpha > 0.$$

We assume the following.

(1) Asymptotic independence. There exists a probability distribution G on $\mathbb{R}_+$ and a sequence $b_n \to \infty$ such that

$$nP\left[\left(\frac{L}{b_n}, R\right) \in \cdot\right] \xrightarrow{v} \nu_\alpha \times G,$$

vaguely on $\mathbb{E} := (0, \infty] \times [0, \infty]$. This implies that L has a regularly varying tail with parameter α and that for large n, R and large values of L tend to be independent.

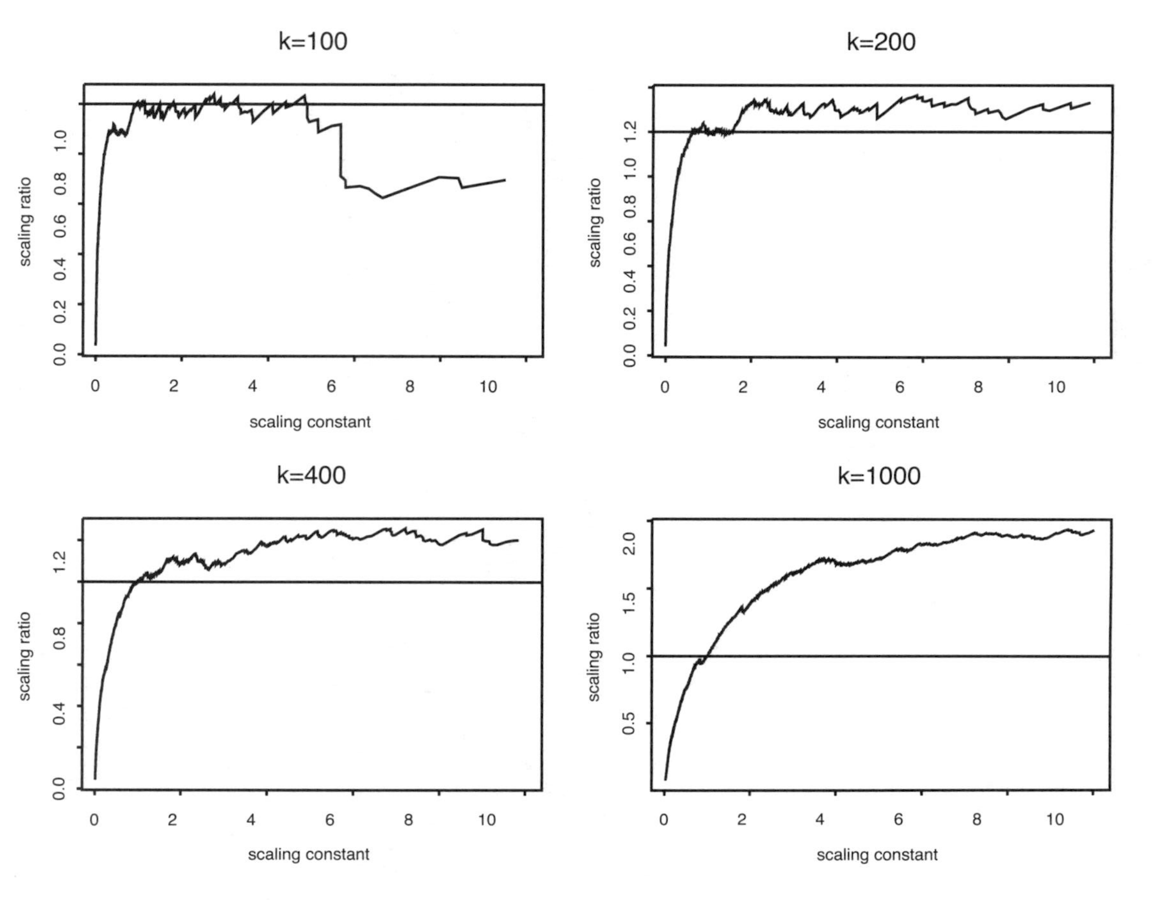

Figure 6.33 *The data (buL,buF): Stărică plot showing preference for choice* $k = 100$.

(2) Tightness. For some $\eta > 0$,

$$\lim_{\epsilon \downarrow 0} \limsup_{n\to\infty} nE\left[\left(\frac{L}{b_n}R\right)^{\alpha+\eta} 1_{[L \le b_n \epsilon]}\right] = 0.$$

Then provided

$$c := \int_0^\infty u^\alpha G(du) < \infty$$

we have

$$P[LR > x] \sim cP[L > x], \quad x \to \infty.$$

It turns out that (1) holds if multivariate regular variation [8, 9, 139] holds and suggests that (L, R) is multivariate regularly varying but in the wrong coordinate system:

$$nP\left[b_n^{-1}(LR, R) \in \cdot\right] \xrightarrow{v} \mu$$

on $\mathbb{E}$. It is not hard to see that μ and $\nu_\alpha \times G$ are related.

The implication for the modified infinite source Poisson model is that cumulative traffic at large time scales is still approximated by a stable Lévy motion. It is undoubtedly true, in addition, that under the right asymptotic regime, FBM limits are possible. This is either good or bad news depending on whether you believe cumulative traffic is or is not either FBM or stable Lévy motion. And keep in mind that in the data example given, it was (buL,buF) that seemed to be asymptotically independent (in the classical sense) and not the pair (buL,buR). When L and R both have regularly varying tails with parameters α_L and α_R in the range $(1, 2)$ but L and R are not asymptotically independent, $F = LR$ has tail index $\alpha_F < 1$ [108].

6.9 A model with a control

One of the obvious problems with the previous models is that they do not incorporate controls and protocols that have been designed to make the real internet robust and scalable. One reason why it may be unlikely to observe stable Lévy motion as an approximation to cumulative traffic is that controls prevent the big buildup of traffic that a heavy tailed model would predict. Protocols, such as TCP (transport control protocol), are designed to keep congestion at the end user when packets are being lost by the simple mechanism that the sending machine is signaled to slow transmission. Roughly, how it works is as follows [92]: files are broken into packets and the first packet is sent. If an acknowledgement from the destination is received within a specified time window, then the sender sends two packets. As long as acknowledgements are received from the destination machine, transmission continues to increase linearly. However, if no acknowledgement is received, the packet is presumed lost and resent but the transmission rate is halved. The rate continues to be halved until an acknowledgement is received, at which point linear build-up of the rate resumes.

It is a curious fact that the transmission rate is halved when packets are lost. Why not divide the rate by 3? We sought a model that would help us study this question as well as the delay the protocol imposes on the sender. What price in terms of delay does a sender pay for the socially or globally optimal behavior of allowing the transmission rate to be slowed?

Because it is very difficult to model the complex interactions in the internet in a realistic way, we have studied these questions in a restricted, simplified context of a single channel on/off source destination pair with a buffer in between [17].

The model consists of i.i.d. jobs or files of size $\{L_n\}$, which have common distribution F_L as well as off periods denoted by the i.i.d. random variables $\{Y_n\}$, which have common distribution F_Y and which are independent of the file sizes. There is a buffer between the source and destination and the buffer empties at constant rate r, assuming the buffer is nonempty. We assume $r \in (0, 1)$ to ensure that buffer content can exit from zero. In place of the system of acknowledgements, we assume that there is a critical buffer level $\gamma > 0$. During a transmission in this single channel model, if

- buffer content $<\gamma$, then the transmission rate increases linearly like $1 + at$; and if
- buffer content $\geq\gamma$, then the transmission rate decreases geometrically like e^{-dt}.

We suppose X_n is the buffer content at time of the beginning of the n-th transmission. The process $\{X_n\}$ will have a Markov chain structure. Here is a more formal outline of the construction of the process.

6.9.1 Construction

Step 1. Cumulative input function. Begin by defining the cumulative input function $I_x(t)$, which is the amount of work injected into the system in $[0, t]$ starting from initial content x, assuming the job size is ∞. There are two cases: $x < \gamma$ and $x \geq \gamma$.

Case $x \geq \gamma$. Define

$$t_0(x) = \inf\left\{u > 0 : x + \int_0^u e^{-dw} dw - ru = \gamma\right\},$$

which is the time necessary for buffer content to move from level x down to γ. Provided $t \in (0, t_0(x))$, we have

$$\frac{d}{dt} I_x(t) = e^{-dt}.$$

At time $t_0(x)$, we reset the input rate to 1, and then let it decrease exponentially again. Repeat this at times $t_0(x) + nt_0(\gamma)$, $n \geq 1$.

Case $x < \gamma$. Define

$$t(x) = \inf\left\{u > 0 : x + \int_0^u (1 + as)ds - ru = \gamma\right\}$$

as the time required to move from x up to level γ. At time $t(x)$, the input rate is $1 + at(x)$ and the input rate begins to decrease exponentially until time $t_a(x) + t(x)$, where

$$t_a(x) = \inf\left\{u : \gamma + \int_0^u (1 + at(x))e^{-dw} dw - ru = \gamma,\right.$$

at which time, buffer content is again γ. Then at $t(x) + t_a(x)$,

- reset the input rate to 1, and let the rate decrease exponentially until the content level is again γ. Then

- reset the input rate to 1, let the rate decrease exponentially until content is again γ. Then
- $\cdots$.

Step 2. The transmission time $\tau(x, l)$. We have defined $I_x(t)$ assuming there was an infinite file to send. What if the file is of size l? The time necessary for transmitting a file of size l when content is initially x is

$$\tau(x, l) = I_x^{\leftarrow}(l) = \inf\{t > 0 : I_x(t) = l\}.$$

Step 3. The embedded MARKOV chain. Define the embedded Markov chain $\{X_n\}$. With $\{L_n\}, \{Y_n\}$ representing i.i.d. file sizes and off periods, define

$$X_n = (X_{n-1} + L_n - r\tau(X_{n-1}, L_n) - rY_n)_+$$

to be the buffer content at the beginning of transmissions of files. This says the buffer content when a transmission begins is the content when the previous file was sent, plus the file sent, minus what left the buffer during the transmission of the file, minus what left the buffer during an off period.

Step 4. A time scale. If X_n is the buffer content when the n-th transmission commences, let S_n be the time, and we define

$$S_0 = 0, \quad S_n = S_{n-1} + \tau(X_{n-1}, L_n) + Y_n.$$

We think of $[S_n, S_{n+1})$ as the nth transmission cycle, and it consists of an on period of length $\tau(X_{n-1}, L_n)$ plus an off period. Then we can define a continuous time process between the times $\{S_n\}$: for $t \in [S_{n-1}, S_{n-1} + \tau(X_{n-1}, L_n))$

$$X(t) = X_{n-1} + I_{X_{n-1}}(t - S_{n-1}) - r(t - S_{n-1})$$

and for $t \in [S_{n-1} + \tau(X_{n-1}, S_n)$,

$$X(t) = (X(S_{n-1} + \tau(X_{n-1}, L_n)) - r(t - [S_{n-1} + \tau(X_{n-1}, L_n)])_+.$$

6.9.2 Results

Here is an outline of features of this model.

1. Controlled growth. The Markov chain can increase only in a controlled way due to the effect of the control. If $X_n \geq \gamma$ then

$$X_{n+1} - X_n \leq \sup_{t \geq 0} \left(\int_0^t e^{-dw} dw - rt \right) = \frac{1 - r + r \log r}{d} =: \delta.$$

2. Thin tails. Since the service rate satisfies $r < 1$, if L is too small and $Y = 0$, the content would always increase. To prevent circumstances like this, it is desirable to assume

$$P\left[L > \frac{1}{d}\right] > 0.$$

In this case, $\{X_n\}$ possesses a stationary distribution π and

$$\pi(x, \infty) \leq c_1 e^{-c_2 x},$$

for sufficiently large x. Thus, even if we assume heavy tailed file sizes, the control keeps buffer content moderate and the stationary distribution tail is exponentially bounded. Since buffer content is thin tailed, we have a right to worry about the effect on user transmission times.

3. Ergodicity. Under mild additional conditions, $\{0\}$ is an atom and $\{X_n\}$ is geometrically ergodic, strongly mixing, and Harris recurrent. There exists $0 < \rho < 1$ such that

$$\rho^{-n} \| P_x[X_n \in \cdot] - \pi(\cdot) \|_{\mathrm{TV}} \to 0,$$

where $\| \cdot \|_{\mathrm{TV}}$ denotes the total variation metric.

4. Stationary transmission times. For any $y \geq 0$

$$P_y[\tau(X_{n-1}, L_n) \in A] \to P_\pi[\tau(X, L) \in A],$$

where X and L are independent random variables with $L \stackrel{d}{=} L_1$, and X has the stationary distribution π.

5. Transmission time tail behavior. If L has a subexponential distribution,

$$P_\pi[\tau(X, L) > x] \sim \bar{F}_L(rx).$$

Thus, the transmission tail is asymptotically equivalent to the file size tail, modulo the factor r for the work rate. Note this expression is independent of d, the exponential rate parameter. So the exponential decrease parameter does not affect the asymptotic form of the tail. (Whether you halve the rate or divide by 17 does not influence the asymptotic tail of the transmission time distribution.)

Where is the effect of d felt? Based on simulations, it appears to influence $E(\tau(X, L))$ mildly, but we only have bounds on $E(\tau(X, L))$, and the influence of d is difficult to quantify.

6.10 Conclusion

Data networks are fascinating to the curious and broadly trained investigator. Their thorough study benefits from techniques taken from probability, statistics, applied mathematics, computer science and electrical and computer engineering. Many studies today have isolated one small part of the network for study and ignored many interactions and dependencies. The challenge for the future is to increase model realism by accounting for interaction among users competing for network resources. Most of the current attempts to realistically take into account competition have been empirical or simulation based.

Heavy tail and extreme value analyses are fraught with imprecisions. Modeling assumptions based on asymptotic properties are inherently controversial. The assumption to model with a heavy tail or to incorporate long range dependence into the model may be viewed in the same light as conviction about religion: it works for the user, but there may be other models that work for other users. There are usually several models consistent with collected data measurements. Imagine the difficulty of trying to statistically distinguish between a Pareto distribution and a truncated Pareto. It is useful to maintain a focus that suggests the key question is not, does a particular model fit the data?, but rather, what is the class of models consistent with the data?

Acknowledgments

Sidney Resnick's research was partially supported by NSF grant DMS-97-04982 at Cornell University.

Bibliography

[1] P. Abry. *Ondelettes et Turbulences—Multirésolutions, Algorithmes de Décompositions, Invariance d'Echelle et Signaux de Pression*. Diderot, Editeur des Sciences et des Arts, Paris, 1997.

[2] P. Abry, P. Flandrin, M.S. Taqqu, and D. Veitch. Wavelets for the analysis, estimation and synthesis of scaling data. In *Self-Similar Network Traffic and Performance Evaluation*, pages 39–88. Wiley, New York, 2000.

[3] P. Abry and D. Veitch. Wavelet analysis of long-range-dependent traffic. *IEEE Trans. Inform. Theory*, 44(1):2–15, 1998.

[4] R.J. Adler. *An Introduction to Continuity, Extrema and Related Topics for General Gaussian Processes*, volume 12 of *IMS Lectures Notes*. Institute of Mathematical Statistics, Hayward, 1990.

[5] R.J. Adler, R.E. Feldman, and M.S. Taqqu (editors). *A Practical Guide to Heavy Tails. Statistical Techniques and Applications.* Birkhauser, Boston, 1998. xvi, 533 pages.

[6] E. Altman, K. Avratchenkov, and C. Barakat. A stochastic model of TCP/IP with stationary random losses. In *Proceedings of the ACM SIGCOMM*, pages 231–242, 2000.

[7] M. Arlitt and C. Williamson. Web server workload characterization: the search for invariants. Master's thesis, University of Saskatchewan, 1996.

[8] B. Basrak. *The sample autocorrelation function of nonlinear time series*. PhD thesis, Rijksuniversiteit Groningen, Groningen, Netherlands, 2000.

[9] B. Basrak, R. Davis, and T. Mikosch. A characterization of multivariate regular variation. Preprint, 2000.

[10] J. Beirlant, P. Vynckier, and J. Teugels. Tail index estimation, Pareto quantile plots, and regression diagnostics. *J. Amer. Statist. Assoc.*, 91(436):1659–1667, 1996.

[11] J. Beran. Statistical methods for data with long-range dependence. *Statistical Science*, 7(4):404–416, 1992. With discussions and rejoinder, pages 404–427.

[12] J. Beran. *Statistical Methods for Long Memory Processes*. Chapman and Hall, London, 1994.

[13] R.N. Bhattacharya, V.K. Gupta, and E. Waymire. The Hurst effect under trends. *J. Applied Probability*, 20:649–662, 1983.

[14] N.H. Bingham, C.M. Goldie, and J.L. Teugels. *Regular Variation*. Cambridge University Press, 1987.

[15] D.C. Boes. Schemes exhibiting Hurst behavior. In J. Srivastava, editor, *Essays in Honor of Franklin Graybill*, pages 21–42, Amsterdam, North Holland. 1988.

[16] D.C. Boes and J.D. Salas-La Cruz. On the expected range and expected adjusted range of partial sums of exchangeable random variables. *J. Applied Probability*, 10:671–677, 1973.

[17] M. Borkovec, A. Dasgupta, S. Resnick, and G. Samorodnitsky. A single channel on/off model with tcp-like control. web available at http://critical.orie.cornell.edu/trlist/trlist.html; to appear: *J. Applied Probability*, 2000.

[18] P.J. Brockwell and R.A. Davis. *Time Series: Theory and Methods*. Springer-Verlag, New York, second edition, 1991.

[19] P.J. Brockwell, N. Pacheco-Santiago, and S. Resnick. Weak convergence and range analysis for dams with Markovian input rate. *J. Applied Probability*, 19:272–289, 1982.

[20] F.H. Campos, J.S. Marron, S. Resnick, and K. Jaffay. Extremal dependence: Internet traffic applications. Web available at http://www.orie.cornell.edu/~sid, 2002.

[21] J. Cao, W. Cleveland, D. Lin, and Don X. Sun. The effect of statistical multiplexing on internet packet traffic: Theory and empirical study. http://cm.bell-labs.com/cm/ms/departments /sia/InternetTraffic/webpapers.html, 2001.

[22] J. Cao, W. Cleveland, D. Lin, and Don X. Sun. On the nonstationarity of internet traffic. *Proceedings of the ACM SIGMETRICS '01*, pages 102–112, 2001. http://cm.bell-labs.com/cm/ms/departments/sia/InternetTraffic/webpapers.html.

[23] P. Carlsson and M. Fiedler. Multifractal products of stochastic processes: fluid flow analysis. In *Proceedings of 15th Nordic Teletraffic Seminar NTS-15, August 22–24, 2000; Lund, Sweden*, pages 173–184, 2000.

[24] E. Castillo. *Extreme Value Theory in Engineering*. Academic Press, San Diego, 1988.

[25] J. Cohen, S. Resnick, and G. Samorodnitsky. Sample correlations of infinite variance time series models: an empirical and theoretical study. *J. Appl. Math. Stochastic Anal.*, 11(3):255–282, 1998.

[26] A.G. Constantine and P. Hall. Characterizing surface smoothness via estimation of effective fractal dimension. *J.R. Statist. Soc. B*, 56:97–113, 1994.

[27] D.R. Cox. Long-range dependence: a review. In H.A. David and H.T. David, editors, *Statistics: An Appraisal*, pages 55–74. Iowa State University Press, 1984.

[28] M. Crovella and A. Bestavros. Explaining world wide web traffic self-similarity. Preprint available as TR-95-015 from {crovella,best}@cs.bu.edu, 1995.

[29] M. Crovella and A. Bestavros. Self-similarity in world wide web traffic: evidence and possible causes. In *Proceedings of the ACM SIGMETRICS '96*

International Conference on Measurement and Modeling of Computer Systems, 24:160–169, 1996.

[30] M. Crovella and A. Bestavros. Self-similarity in world wide web traffic: evidence and possible causes. *Performance Evaluation Review*, 24:160–169, 1996.

[31] M. Crovella and A. Bestavros. Self-similarity in world wide web traffic: evidence and possible causes. *IEEE/ACM Transactions on Networking*, 5(6):835–846, 1997.

[32] M. Crovella, A. Bestavros, and M.S. Taqqu. Heavy-tailed probability distributions in the world wide web. In M.S. Taqqu, R. Adler, R. Feldman, editor, *A Practical Guide to Heavy Tails: Statistical Techniques for Analysing Heavy Tailed Distributions*. Birkhäuser, Boston, 1999.

[33] M. Crovella, G. Kim, and K. Park. On the relationship between file sizes, transport protocols, and self-similar network traffic. In *Proceedings of the Fourth International Conference on Network Protocols (ICNP'96)*, pages 171–180, 1996.

[34] S. Csörgő, P. Deheuvels, and D. Mason. Kernel estimates for the tail index of a distribution. *Ann. Statist.*, 13:1050–1077, 1985.

[35] S. Csörgő and D. Mason. Central limit theorems for sums of extreme values. *Math. Proc. Camb. Phil. Soc.*, 98:547–558, 1985.

[36] C. Cunha, A. Bestavros, and M. Crovella. Characteristics of www client-based traces. Preprint available as BU-CS-95-010 from {crovella,best}@cs.bu.edu, 1995.

[37] R.A. Davis and T. Mikosch. The sample autocorrelations of heavy-tailed processes with applications to ARCH. *Ann. Statistics.*, 26(5):2049–2080, 1998.

[38] R.A. Davis and S.I. Resnick. Limit theory for moving averages of random variables with regularly varying tail probabilities. *Ann. Probability*, 13(1):179–195, 1985.

[39] R.A. Davis and S.I. Resnick. More limit theory for the sample correlation function of moving averages. *Stochastic Processes and their Applications*, 20:257–279, 1985.

[40] R.A. Davis and S.I. Resnick. Limit theory for the sample covariance and correlation functions of moving averages. *Ann. Statistics*, 14(2):533–558, 1986.

[41] R.A. Davis and S.I. Resnick. Limit theory for bilinear processes with heavy tailed noise. *Ann. Applied Probability*, 6:1191–1210, 1996.

[42] L. de Haan. *On Regular Variation and Its Application to the Weak Convergence of Sample Extremes*. Mathematisch Centrum Amsterdam, 1970.

[43] L. de Haan and J. de Ronde. Sea and wind: multivariate extremes at work. *Extremes*, 1(1):7–46, 1998.

[44] L. de Haan and L. Peng. Comparison of tail index estimators. *Statist. Neerlandica*, 52(1):60–70, 1998.

[45] L. de Haan and S.I. Resnick. Limit theory for multivariate sample extremes. *Z. Wahrscheinlichkeitstheorie*, 40:317–337, 1977.

[46] L. de Haan and S.I. Resnick. Estimating the limit distribution of multivariate extremes. *Stochastic Models*, 9(2):275–309, 1993.

[47] L. de Haan and S.I. Resnick. On asymptotic normality of the Hill estimator. *Stochastic Models*, 14:849–867, 1998.

[48] L. de Haan and H. Rootzén. On the estimation of high quantiles. *J. Statist. Plann. Inference*, 35(1):1–13, 1993.

[49] A.L.M. Dekkers and L. de Haan. On the estimation of the extreme-value index and large quantile estimation. *Ann. Statist.*, 17:1795–1832, 1989.

[50] A.L.M. Dekkers and L. de Haan. Optimal choice of sample fraction in extreme-value estimation. *Journal of Mult. Anal.*, 1993.

[51] A.L.M. Dekkers, J.H.J. Einmahl, and L. de Haan. A moment estimator for the index of an extreme-value distribution. *Ann. Stat.*, 17:1833–1855, 1989.

[52] H. Drees, L. de Haan, and S. Resnick. How to make a Hill plot. *Ann. Statistics*, 28(1):254–274, 2000.

[53] N.G. Duffield, J.T. Lewis, N. O'Connell, R. Russell, and F. Toomey. Statistical issues raised by the bellcore data. In *Proceedings of 11th IEE UK Teletraffic Symposium, Cambridge UK*, pages 23–25, London, 1994. IEE.

[54] D.E. Duffy, A.A. McIntosh, M. Rosenstein, and W. Willinger. Analyzing telecommunications traffic data from working common channel signaling subnetworks. In M.E. Tarter and M.D. Lock, editors, *Computing Science and Statistics Interface, Proceedings of the 25th Symposium on the Interface*, 25:156–165, San Diego, 1993.

[55] R. Durrett and S.I. Resnick. Weak convergence with random indices. *Stochastic Processes and Their Applications*, 5:213–220, 1977.

[56] P. Embrechts, C. Kluppelberg, and T. Mikosch. *Modelling Extreme Events for Insurance and Finance*. Springer-Verlag, Berlin, 1997.

[57] K. Falconer. *Fractal Geometry: Mathematical Foundations and Applications*. John Wiley & Sons Ltd., Chichester, 1990.

[58] P. Feigin and S. Resnick. Linear programming estimators and bootstrapping for heavy tailed phenomena. *Advances in Appl. Probability*, 29:759–805, 1997.

[59] P. Feigin and S.I. Resnick. Estimation for autoregressive processes with positive innovations. *Stochastic Models*, 8:479–498, 1992.

[60] P. Feigin and S.I. Resnick. Limit distributions for linear programming time series estimators. *Stochastic Processes and their Applications*, 51:135–165, 1994.

[61] P. Feigin and S.I. Resnick. Pitfalls of fitting autoregressive models for heavy-tailed time series. *Extremes*, 1:391–422, 1999.

[62] A. Feldmann, A.C. Gilbert, and W. Willinger. Data networks as cascades: Investigating the multifractal nature of Internet WAN traffic. In *Proceedings of the ACM Sigcomm '98*, pages 25–38, Vancouver, B.C., 1998.

[63] W. Feller. The asymptotic distribution of the range of sums of independent random variables. *Ann. Math. Statistics*, 22:427–432, 1951.

[64] W. Feller. *An Introduction to Probability Theory and Its Applications*, volume 2. Wiley, New York, 2nd edition, 1971.

[65] M.W. Garrett and W. Willinger. Analysis, modeling and generation of self similar vbr video traffic. In *Proceedings of the ACM SigComm*, London, 1994.

[66] J. Geluk, L. de Haan, S. Resnick, and C. Stărică. Second-order regular variation, convolution and the central limit theorem. *Stochastic Processes and their Applications*, 69(2):139–159, 1997.

[67] A. Gilbert, Y. Joo, and N. McKeown. Congestion control and periodic behavior. Preprint: AT&T Labs–Research, 180 Park Ave, Florham Park, NJ 07932-0971, 2000.

[68] A.C. Gilbert, W. Willinger, and A. Feldmann. Scaling analysis of conservative cascades, with applications to network traffic. *IEEE Transactions on Information Theory*, 45(3):971–991, 1999.

[69] L. Giraitis, P. Kokoszka, and R. Leipus. Testing for long memory in the presence of a general trend. *J. Applied Probability*, 38:1033–1054, 2001.

[70] L. Giraitis, P. Kokoszka, R. Leipus, and G. Teyssiere. Semiparametric estimation of the intensity of long memory in conditional heteroskedasticity. *Statistical Inference for Stochastic Processes*, 3:113–128, 2000.

[71] E.G. Gladyshev. A new limit theorem for processes with Gaussian increments. *Theory Probab. Appl.*, 6:52–61, 1961.

[72] C.A. Guerin, H. Nyberg, O. Perrin, S. Resnick, H. Rootzen, and C. Stărică. Empirical testing of the infinite source poisson data traffic model. *Stochastic Models*, 19:151–200, 2003.

[73] P. Hall. On some simple estimates of an exponent of regular variation. *J. Roy. Stat. Assoc.*, 44:37–42, 1982. Series B.

[74] D. Heath, S. Resnick, and G. Samorodnitsky. Patterns of buffer overflow in a class of queues with long memory in the input stream. *Ann. Applied Probability*, 7(4):1021–1057, 1997.

[75] D. Heath, S. Resnick, and G. Samorodnitsky. Heavy tails and long range dependence in on/off processes and associated fluid models. *Math. Oper. Res.*, 23(1):145–165, 1998.

[76] D. Heath, S. Resnick, and G. Samorodnitsky. How system performance is affected by the interplay of averages in a fluid queue with long range dependence induced by heavy tails. *Ann. Appl. Probab.*, 9:352–375, 1999.

[77] D. Heyman and T.V. Lakshman. What are the implications of long-range dependence for vbr-video traffic engineering? *IEEE-ACM Transactions on Networking*, 4(3):301–317, 1996.

[78] B.M. Hill. A simple general approach to inference about the tail of a distribution. *Ann. Statist.*, 3:1163–1174, 1975.

[79] T. Hsing. On tail estimation using dependent data. *Ann. Statist.*, 19:1547–1569, 1991.

[80] H.E. Hurst. Long-term storage capacity of reservoirs. *Transactions of the American Society of Civil Engineers*, 116:770–808, 1951.

[81] H.E. Hurst. Methods of using long-term storage in reservoirs. *Proceedings of the Institution of Civil Engineers, Part I*, pages 519–577, 1955.

[82] I.A. Ibragimov and Y.A. Rozanov. *Gaussian Random Processes*. Springer-Verlag, New York, 1978. Translated from the Russian by A.B. Aries.

[83] J. Istas. Wavelet coefficients of a Gaussian process and applications. *Ann. Inst. H. Poincaré Probab. Statist.*, 28(4):537–556, 1992.

[84] J. Istas and G. Lang. Variations quadratiques et estimation de l'exposant de Hölder local d'un processus gaussien. *C. R. Acad. Sci. Paris Sér. I Math.*, 319(2):201–206, 1994.

[85] J. Istas and G. Lang. Quadratic variations and estimation of the local Hölder index of a Gaussian process. *Ann. Inst. H. Poincaré Probab. Statist.*, 33(4):407–436, 1997.

[86] J. Istas and C. Laredo. Estimation de la fonction de covariance d'un processus gaussien stationnaire par méthodes d'échelles. *C. R. Acad. Sci. Paris Sér. I Math.*, 316(5):495–498, 1993.

[87] S. Jaffard. The multifractal nature of Lévy processes. *Probab. Theory Related Fields*, 114(2):207–227, 1999.

[88] P. Jelenković and A. Lazar. Subexponential asymptotics of a Markov-modulated random walk with queueing applications. *J. Applied Probability*, 35(2):325–347, 1998.

[89] P. Jelenković and A. Lazar. Asymptotic results for multiplexing subexponential on-off processes. *Advances in Applied Probability*, 31:394–421, 1999.

[90] P. Jelenković and A.A. Lazar. A network multiplexer with multiple time scale and subexponential arrivals. In Glasserman P., K. Sigman, and D.D. Yao, editors, *Stochastic Networks*, volume 117 of *Lecture Notes in Statistics*, pages 215–235. Springer, New York, 1996.

[91] J. Kent and A.T.A. Wood. Estimating the fractal dimension of a locally self-similar gaussian process by using increments. *J.R. Statist. Soc. B*, 59(3):679–699, 1997.

[92] S. Keshav. *An Engineering Approach to Computer Networking; ATM Networks, the Internet, and the Telephone Network*. Addison-Wesley, Reading, MA, 1997.

[93] T. Konstantopoulos and S.J. Lin. Macroscopic models for long-range dependent network traffic. *Queueing Systems Theory Appl.*, 28(1-3):215–243, 1998.

[94] M. Kratz and S. Resnick. The qq-estimator and heavy tails. *Stochastic Models*, 12:699–724, 1996.

[95] H. Künsch. Discrimination between monotonic trends and long range dependence. *J. Applied Probability*, 23:1025–1030, 1986.

[96] J.W. Lamperti. Semi-stable stochastic processes. *Transaction of the American Mathematical Society*, 104:62–78, 1962.

[97] J.W. Lamperti. Semi-stable Markov processes i. *Z. Wahrscheinlichkeitstheorie verw. Geb.*, 22:205–225, 1972.

[98] M.R. Leadbetter, G. Lindgren, and H. Rootzén. *Extremes and Related Properties of Random Sequences and Processes.* Springer-Verlag, New York, 1983.

[99] W.E. Leland, M.S. Taqqu, W. Willinger, and D.V. Wilson. On the self-similar nature of Ethernet traffic. *ACM/SIGCOMM Computer Communications Review*, pages 183–193, 1993.

[100] W.E. Leland, M.S. Taqqu, W. Willinger, and D.V. Wilson. Statistical analysis of high time-resolution ethernet Lan traffic measurements. In *Proceedings of the 25th Symposium on the Interface between Statistics and Computer Science*, pages 146–155, 1993.

[101] W.E. Leland, M.S. Taqqu, W. Willinger, and D.V. Wilson. On the self-similar nature of Ethernet traffic (extended version). *IEEE/ACM Transactions on Networking*, 2:1–15, 1994.

[102] J. Levy and M. Taqqu. Renewal reward processes with heavy-tailed interrenewal times and heavy-tailed rewards. *Bernoulli*, 6:23–44, 2000.

[103] J. Levy and M.S. Taqqu. On renewal processes having stable inter-renewal intervals and stable rewards. *Les Annales des Sciences Mathematiques du Quebec*, 11:95–110, 1987.

[104] B.B. Mandelbrot and M.S. Taqqu. Robust R/S analysis of long-run serial correlation. In *Proceedings of the 42nd Session of the International Statistical Institute*, Manila, 1979. Bulletin of the I.S.I. Vol. 48, Book 2, pp. 69–104.

[105] P. Mannersalo, I. Norros, and R. Riedi. Multifractal products of stochastic processes: a preview. Technical Document COST257TD(99)31, September, 1999; available at http://www.vtt.fi/tte/staff2/petteri/research.html, 1999.

[106] D. Mason. Laws of large numbers for sums of extreme values. *Ann. Probability*, 10:754–764, 1982.

[107] K. Maulik and S. Resnick. Small and large time scale analysis of a network traffic model. Queuing Systems, 43:221–250, 2003.

[108] K. Maulik, S. Resnick, and H. Rootzén. A network traffic model with random transmission rate. *J. Applied Probability*, 39:671–699, 2002.

[109] T. Mikosch, S. Resnick, H. Rootzén, and A.W. Stegeman. Is network traffic approximated by stable Lévy motion or fractional Brownian motion? *Ann. Applied Probability*, 12(1):23–68, 2002.

[110] T. Mikosch and A. Stegeman. The interplay between heavy tails and rates in self-similar network traffic. Technical Report. University of Groningen, Department of Mathematics, Available via www.cs.rug.nl/eke/iwi/preprints/, 1999.

[111] T. Mikosch and C. Stărică. Change of structure in financial data. Available at: http://www.math.ku.dk/mikosch/preprint.html, 1999.

[112] T. Mikosch and C. Stărică. Long range dependence effects and ARCH modeling. Available at: http://www.math.ku.dk/mikosch/preprint.html, 2000.

[113] P.A.P. Moran. On the range of cumulative sums. *Ann. Inst. Stat. Math.*, 16:109–112, 1964.

[114] G.L. O'Brien and W. Vervaat. Marginal distributions of self-similar processes with stationary increments. *Z. Wahrscheinlichkeitstheorie verw. Gebiete*, 64:129–138, 1983.

[115] G.L. O'Brien and W. Vervaat. Self-similar processes with stationary increments generated by point processes. *Ann. Probility*, 13:28–52, 1985.

[116] T.J. Ott, J.H.B. Kemperman, and M. Mathis. The stationary behavior of ideal tcp congestion avoidance. In *Proceedings of IEEE INFOCOM'99*, New York, 1999.

[117] T.J. Ott and A. Misra. The window distribution of idealized tcp congestion avoidance with variable packet loss. In *INFOCOM (3)*, pages 1564–1572, 2000.

[118] J. Padhye, V. Firoiu, D. Towsley, and J. Kurose. Modeling TCP throughput: a simple model and its empirical validation. In *ACM SIGCOMM '98 Conference on Applications, Technologies, Achitectures, and Protocols for Computer Communication*, pages 303–314, Vancouver, Canada, 1998. citeseer.nj.nec.com/padhye98modeling.html.

[119] K. Park and W. Willinger. Self-similar network traffic: an overview. In K. Park and W. Willinger, editors, *Self-Similar Network Traffic and Performance Evaluation*. Wiley-Interscience, New York, 2000.

[120] M. Parulekar and A.M. Makowski. Tail probabilities for a multiplexer with a self-similar traffic. *Proceedings of the 15th Annual IEEE INFOCOM*, pages 1452–1459, 1996.

[121] M. Parulekar and A.M. Makowski. Tail probabilities for M/G/∞ input process (i): Preliminary asymptotics. University of Maryland Institute for Systems Research Technical Report TR 96-41, 1996.

[122] L. Peng. *Second Order Condition and Extreme Value Theory*. PhD thesis, Tinbergen Institute, Erasmus University, Rotterdam, 1998.

[123] V. Pipiras and M. Taqqu. The limit of a renewal-reward process with heavy-tailed rewards is not a linear fractional stable motion. *Bernoulli*, 6:607–614, 2000.

[124] R.-D. Reiss and M. Thomas. *Statistical Analysis of Extreme Values*. Birkhäuser Verlag, Basel, second edition, 2001.

[125] S. Resnick. Heavy tail modeling and teletraffic data. *Ann. Statist.*, 25:1805–1869, 1997.

[126] S. Resnick. Why non-linearities can ruin the heavy tailed modeler's day. In M.S. Taqqu, R. Adler, R. Feldman, editors, *A Practical Guide to Heavy Tails: Statistical Techniques for Analysing Heavy Tailed Distributions*, pages 219–240. Birkhäuser, Boston, 1998.

[127] S. Resnick and P. Greenwood. A bivariate stable characterization and domains of attraction. *Journal of Multivariate Analysis*, 9:206–221, 1979.

[128] S. Resnick and G. Samorodnitsky. A heavy traffic approximation for workload processes with heavy tailed service requirements. *Management Science*, 46:1236–1248, 2000.

[129] S. Resnick and G. Samorodnitsky. Limits of on/off hierarchical product models for data transmission. Technical Report 1281 available at www.orie.cornell.edu/trlist/trlist.html, to appear in *Annals of Applied Probability*.

[130] S. Resnick, G. Samorodnitsky, and F. Xue. How misleading can sample acf's of stable ma's be? (Very!). *Ann. Applied Probability*, 9(3):797–817, 1999.

[131] S. Resnick, G. Samorodnitsky, and F. Xue. Growth rates of sample covariances of stationary symmetric α-stable processes associated with null recurrent markov chains. *Stochastic Processes and Their Applications*, 85:321–339, 2000.

[132] S. Resnick and C. Stărică. Consistency of Hill's estimator for dependent data. *J. Applied Probability*, 32(1):139–167, 1995.

[133] S. Resnick and C. Stărică. Asymptotic behavior of Hill's estimator for autoregressive data. *Stochastic Models*, 13:703–723, 1997.

[134] S. Resnick and C. Stărică. Smoothing the Hill estimator. *Adv. Applied Probability*, 29:271–293, 1997.

[135] S. Resnick and E. van den Berg. Sample correlation behavior for the heavy tailed general bilinear process. *Stochastic Models*, 16(2):233–258, 2000.

[136] S. Resnick and E. van den Berg. A test for nonlinearity of time series with infinite variance. *Extremes*, 3(4):145–172, 2000.

[137] S. Resnick and E. van den Berg. Weak convergence of high-speed network traffic models. *J. Applied Probability*, 37(2):575–597, 2000.

[138] S.I. Resnick. Point processes, regular variation and weak convergence. *Adv. Applied Probability*, 18:66–138, 1986.

[139] S.I. Resnick. *Extreme Values, Regular Variation and Point Processes*. Springer-Verlag, New York, 1987.

[140] S.I. Resnick. *Adventures in Stochastic Processes*. Birkhäuser, Boston, 1992.

[141] S.I. Resnick. *A Probability Path*. Birkhäuser, Boston, 1998.

[142] S.I. Resnick and H. Rootzén. Self-similar communication models and very heavy tails. *Ann. Applied Probability*, 10:753–778, 2000.

[143] R.H. Riedi. Multifractal processes. Technical Report, ECE Dept., Rice University, TR 99-06, 1999.

[144] R.H. Riedi and J. Levy-Vehel. TCP traffic is multifractal: a numerical study. Preprint, 1997.

[145] R.H. Riedi and W. Willinger. Toward an improved understanding of network traffic dynamics. In *Self-Similar Network Traffic and Performance Evaluation*. Wiley, New York, 2000.

[146] H. Rootzén, M.R. Leadbetter, and L. de Haan. Tail and quantile estimation for strongly mixing stationary sequences. Technical Report 292, Center for Stochastic Processes, Department of Statistics, University of North Carolina, Chapel Hill, 1990.

[147] H. Rootzén, M.R. Leadbetter, and L. de Haan. On the distribution of tail array sums for strongly mixing stationary sequences. *Ann. Applied Probability*, 8(3):868–885, 1998.

[148] J.D. Salas and D. Boes. Expected range and adjusted range of hydrologic sequences. *Water Resoures Research*, 10(3):457–463, 1974.

[149] J.D. Salas and D. Boes. Nonstationarity of mean and Hurst phenomenon. *Water Resources Research*, 14((1)):135–143, 1978.

[150] J.D. Salas, D. Boes, and V. Yevjevich. Hurst phenomenon as a preasymptotic behavior. *J. Hydrology*, 44((1-2)):1–15, 1979.

[151] G. Samorodnitsky and M. Taqqu. *Stable NonGaussian Random Processes: Stochastic Models with Infinite Variance*. Stochastic Modeling. Chapman & Hall, New York, xviii, 632, 1994.

[152] R. Srikant. Control of communication networks. In Tariq Samad, editor, *Perspectives in Control Engineering: Technologies, Applications, New Directions*, pages 462–488. IEEE Press, 2000.

[153] A. Stegeman. Modeling traffic in high-speed networks by on/off models. Master's thesis, University of Groningen, Department of Mathematics, The Netherlands, 1998.

[154] A. Stegeman. Nonstationarity versus long-range dependence in computer network traffic measurements. Technical report, University of Groningen, Department of Mathematics, The Netherlands, 2002. available at http://www.math.rug.nl/stegeman/.

[155] C. Stărică. Multivariate extremes for models with constant conditional correlations. *J. Empirical Finance*, 6:515–553, 1999.

[156] M.S. Taqqu and J. Levy. Using renewal processes to generate long-range dependence and high variability. In E. Eberlein and M.S. Taqqu, editors, *Dependence in Probability and Statistics*, pages 73–89, Birkhäuser, Boston, 1986.

[157] M.S. Taqqu, W. Willinger, and R. Sherman. Proof of a fundamental result in self-similar traffic modeling. *Computer Communications Review*, 27:5–23, 1997.

[158] D. Veitch and P. Abry. Estimation conjointe en ondelettes des paramètres du phénomène de dépendance longue. In *Proc. 16ième Colloque GRETSI, Grenoble, France*, pages 1451–1454, 1997.

[159] D. Veitch and P. Abry. A wavelet based joint estimator for the parameters of LRD. *IEEE Trans. Info. Th.*, 1998. special issue "Multiscale Statistical Signal Analysis and Its Application."

[160] W. Vervaat. Sample paths of self-similar processes with stationary increments. *Ann. Probabability*, 13:1–27, 1985.

[161] W. Vervaat. Properties of general self-similar processes. *Bulletin of the International Statistical Institute*, 52(Book 4):199–216, 1987.

[162] W. Whitt. *Stochastic Processs Limits: An Introduction to Stochastic-Process Limits and their Application to Queues*. Springer-Verlag, New York, 2002.

[163] W. Willinger and V. Paxson. Where mathematics meets the Internet. *Notices of the American Mathematical Society*, 45(8):961–970, 1998.

[164] W. Willinger, M.S. Taqqu, M. Leland, and D. Wilson. Self-similarity through high variability: statistical analysis of ethernet lan traffic at the source level. *Computer Communications Review*, 25:100–113, 1995. Proceedings of the ACM/SIGCOMM'95, Cambridge, MA.

[165] W. Willinger, M.S. Taqqu, M. Leland, and D. Wilson. Self-similarity through high variability: statistical analysis of ethernet lan traffic at the source level (extended version). *IEEE/ACM Transactions on Networking*, 5(1):71–96, 1997.

CHAPTER 7

Multivariate Extremes

Anne-Laure Fougères
INSA de Toulouse–Université Paul Sabatier

Contents

7.1 Introduction

A wide variety of situations concerned with extreme events has an inherent multivariate character, as pointed out by Coles and Tawn (1991). Let us consider, for example, the oceanographic context, and focus on the sea-level process. Such a variable can be divided into several physical components like mean-level, tide, surge, and wave, which are driven by different physical phenomena (see, for example, Tawn (1992) for details). Moreover, extreme sea conditions leading to damages are usually a consequence of extreme values jointly in several components. The joint structure of the processes has therefore to be studied. Another type of dependence that can be of great interest is the temporal one: high sea levels can be all the more dangerous when they last for a long period of time. Therefore, a given variable observed at successive times is likely to contain crucial information. Other examples of applications have been listed recently by Kotz and Nadarajah (2000), concerning, among others, pollutant concentrations (Joe, Smith, and Weissman, 1992), reservoir safety (Anderson and Nadarajah, 1993), or Dutch sea dikes safety (Bruun and Tawn, 1998; de Haan and de Ronde, 1998).

Historically, the first direction that had been explored concerning multivariate extreme events was the modeling of the asymptotic behavior of componentwise

1-58488-411-8/04/$0.00+$.50

maxima[1] of independent and identically distributed (i.i.d.) observations. Key early contributions to this domain of research are, among others, the papers of Tiago de Oliveira (1958), Sibuya (1960), de Haan and Resnick (1977), Deheuvels (1978), and Pickands (1981). The general structure of the multivariate extreme value distributions has been explored by de Haan and Resnick (1977). Useful representations in terms of max-stable distributions, regular variation functions, or point processes, have been established. The next section is devoted to the asymptotic model for componentwise maxima. The main results are sketched, after a brief summary of the univariate extreme value context. Statistical inference developed in this setup will also be summarized. The limitations of this way of modeling multivariate extreme events will then be envisaged in Section 7.3, where we focus on asymptotically independent events. Recent alternatives introduced by Ledford and Tawn (1996, 1997) will be presented. Finally, the problem of how to measure extremal dependence is tackled, and some tools are reviewed.

7.2 Classical results on componentwise maxima

7.2.1 Univariate extreme events: Summary

The problem of how to model the tails of a univariate distribution has been widely studied, and presents a myriad of applications, as recently listed by Kotz and Nadarajah (2000, Section 1.1, Section 1.9, and Section 2.8). The key assumption that underlies all the methods of modeling is the existence of a domain of attraction for the maxima, that is: if $X_1, \ldots, X_n$ are i.i.d. observations of a random phenomenon with distribution function (d.f.) F, then there exist two sequences $(a_n)_n$ and $(b_n)_n$, where $a_n > 0$, $b_n \in \mathbb{R}$, and a nondegenerate d.f. G such that

$$\lim_{n\to\infty} P\left\{\frac{\max X_i - b_n}{a_n} \leq x\right\} = \lim_{n\to\infty} F^n(a_n x + b_n) = G(x). \tag{7.1}$$

The set of d.f. G such that (7.1) holds is now referred as the generalized extreme value (GEV) family, introduced by von Mises (1954) and Jenkinson (1955). The d.f. G has the following parametric form:

$$G(x) = \exp\left\{-\left(1+\xi\,\frac{x-\mu}{\sigma}\right)_+^{-1/\xi}\right\},$$

where the notation a_+ stands for a if $a > 0$ and zero otherwise, and where $\sigma > 0$ and $\mu, \xi \in \mathbb{R}$. This distribution function was originally called Fréchet, Weibull, or Gumbel distribution, depending on whether the shape parameter ξ is positive, negative, or zero (as a limiting case). Fisher and Tippett (1928) exhibited these three types of distributions for G, see also Gnedenko (1943).

Under the fundamental hypothesis of the existence of a domain of attraction for the maxima, two main ways of modeling extreme events have emerged. The first concerns

[1] Note that results on minima can immediately be deduced using the property that for any variable X, $\min X = -\max(-X)$. All that follows will just be written for maxima.

models for block maxima, based on the representation (7.1), where the asymptotic distribution of the (renormalized) maxima is considered as an approximation for the distribution of the maxima over a fixed (large enough) number of observations. Statistical inference in the GEV parametric family has been largely studied, see, for example, a review in Kotz and Nadarajah (2000), Section 2.2 to Section 2.6. Secondly, threshold methods have been considered, based on the asymptotic form of the distribution of excesses over a given threshold. More precisely, two avenues have been exploited. The first one, referred to as the peaks over threshold (POT) method, is based on the generalized Pareto approximation. If the d.f. of a random variable (r.v.) X is in the domain of attraction of a GEV distribution with parameters (μ, σ, ξ), then the conditional distribution function of exceedances has the following property

$$P(X > u + x \mid X > u) \sim \left(1 + \xi \, \frac{x}{\widetilde{\sigma}}\right)_+^{-1/\xi}, \qquad u \to \infty$$

where $\widetilde{\sigma} = \sigma + \xi(u - \mu)$. This approach is due to Balkema and de Haan (1974) and Pickands (1975), and has been widely studied (see Leadbetter, Lindgren, and Rootzén (1983), for example). Davison and Smith (1990) make a review of the statistical properties of this method, and focus also on the problem of the choice of the threshold. The second way of using threshold models is to approximate the point process associated with observations greater than u by a nonhomogeneous Poisson process with intensity measure of (x, ∞) given by $(1 + \xi \; (x - \mu)/\sigma \,)_+^{-1/\xi}$. This has been studied by Pickands (1971) and Smith (1989). Adapted methods have also been developed when extremes of dependent sequences are of interest, which is actually the usual case when considering notably environmental data. See, for example, Leadbetter, Lindgren, and Rootzén (1983), Smith (1989), and Davison and Smith (1990). In addition, the nonstationary frame has been explored by Leadbetter, Lindgren, and Rootzén (1983) and Hüsler (1986), among others. We refer to Coles (2001) for a review of practical methodologies when dealing with such data. For an introduction to univariate extreme value theory with applications to insurance and finance, see Embrechts, Klüppelberg, and Mikosch (1997).

7.2.2 Multivariate extreme value distributions

As mentioned in the introduction, exploration of how to model multivariate extreme events began with the study of the limiting behavior of componentwise maxima. All the theory developed is based, as in the univariate case, on the existence of a domain of attraction. Denote in bold-face elements $\mathbf{x} = (x_1, \ldots x_d)$ of $\mathbb{R}^d$. If $\mathbf{X}_i = (X_{i,1}, \ldots X_{i,d}), i = 1, \ldots, n$, are i.i.d. random vectors of dimension d with d.f. F, one assumes that there exist $\mathbb{R}^d$-sequences $(\mathbf{a}_n)_n$ and $(\mathbf{b}_n)_n$, where $a_{n,j} > 0$ and $b_{n,j} \in \mathbb{R}$ for all $j = 1, \ldots, d$, and a d.f. G with nondegenerate margins such that

$$P\big\{\big(\max_{i=1,\ldots,n} \mathbf{X}_i - \mathbf{b}_n\big)/\mathbf{a}_n \leq \mathbf{x}\big\} = F^n(\mathbf{a}_n\,\mathbf{x} + \mathbf{b}_n) \to G(\mathbf{x}), \tag{7.2}$$

when $n \to \infty$. The d.f. G is then called a multivariate EV distribution function, and one says that F is in the (multivariate) domain of attraction of G (for the maxima). Note in particular that the univariate margins of G are EV distributions.

Example 1 (i) Consider the multivariate normal d.f. $F_{\mathcal{N}}$, with all univariate margins equal to $\mathcal{N}(0, 1)$, and with all its correlations less than 1 ($\mathbb{E}X_i X_j < 1$, for all $i, j = 1, \ldots, d$). Such a distribution is in the domain of attraction of the independence with univariate Gumbel margins (Sibuya, 1960). Indeed, one has that

$$F_{\mathcal{N}}^n(a_n\mathbf{x} + \mathbf{b}_n) \to G(\mathbf{x}) = \prod_{j=1}^{d} \exp\{-e^{-x_j}\}.$$

The norming constants are respectively equal to $a_n = (2\log n)^{-1/2}$ and $\mathbf{b}_n = b_n\mathbf{1}$, where $b_n = (2\log n)^{1/2} - 1/2(\log\log n + \log 4\pi)/(2\log n)^{1/2}$, and $\mathbf{1} = (1, \ldots, 1)$ (see, for example, Resnick (1987), Example 2).
(ii) Next consider the Archimedean d.f. F_ϕ, with all univariate margins uniformly distributed on $[0, 1]$, introduced by Genest & MacKay (1986a, 1986b), and defined by

$$F_\phi(\mathbf{x}) = \phi^{-1}\left\{\sum_{j=1}^{d} \phi(x_j)\right\},$$

where ϕ is a function defined on $(0, 1]$ such that $\phi(1) = 0$ and $(-1)^j \mathrm{d}^j \phi^{-1}(t)/\mathrm{d}t^j \geq 0$, for all $j = 1, \ldots, d$. Moreover, assume that $\phi(1 - 1/t)$ is a regularly varying function at infinity with index $-m$, for some $m \geq 1$. Recall that a function $\psi : (0, \infty) \to (0, \infty)$ is said to be regularly varying at infinity with index ρ, denoted $\psi \in RV_\rho$, if and only if $\lim_{t\to\infty} \psi(st)/\psi(t) = s^\rho$ for all $s > 0$ (e.g., Bingham, Goldie, and Teugels, 1989). These distributions are in the domain of attraction of the logistic EV distribution (see further in Example 2), namely:

$$F_\phi^n(\mathbf{x}/n + \mathbf{1}) \to \exp\left[-\left\{\sum_{j=1}^{d}(-x_j)^m\right\}^{1/m}\right],$$

for all $\mathbf{x} < \mathbf{0}$, where $\mathbf{0} = (0, \ldots, 0)$. This last result is due to Genest and Rivest (1989).

Even if the parametric character of the univariate EV family of distributions is now lost in the multivariate context, as a subset of the max-infinite divisible distributions, a specific structure still remains. The results sketched here are essentially due to de Haan and Resnick (1977), and are, for example, presented in Galambos (1987), Resnick (1987, Chapter 5) and Kotz and Nadarajah (2000, Chapter 3).

Let us first assume for convenience that the univariate extreme value margins follow unit Fréchet distributions (with d.f. defined for all $y > 0$ by $\phi_1(y) = e^{-1/y}$). This standardization leads to a separation of the marginal behavior and the dependence part of the distribution. There is no loss of generality in assuming specific margins, as stated in Proposition 5.10 by Resnick (1987). Note that in the case of unit Fréchet margins, normalization sequences $(a_{n,j})$ and $(b_{n,j})$ can be shown to be respectively equal to $a_{n,j} = n$ and $b_{n,j} = 0$, for all $j = 1, \ldots, d$.

The following characterizations of the multivariate EV distributions can then be obtained (see, for example, Proposition 5.11 of Resnick, 1987). The set E denotes here $E = [0, \infty]^d \setminus \{\mathbf{0}\}$. The symbol $\bigvee$ is used for supremum. The function $\mathbb{1}$ is defined by $\mathbb{1}_{z\in C} = 1$ if $z \in C$, and 0 otherwise, and $\|\cdot\|$ denotes any norm on $\mathbb{R}^d$.

Theorem 1 *The following assertions are equivalent:*

[C1] *G is a multivariate EV distribution with unit Fréchet margins.*

[C2] *There exists a finite measure S on $\mathcal{B} = \{\mathbf{y} \in E : ||\mathbf{y}|| = 1\}$ such that for each $\mathbf{x} = (x_1, \ldots, x_d) \in E$, one has that*

$$G(\mathbf{x}) = \exp\left\{ - \int_{\mathcal{B}} \bigvee_{j=1}^{d} \frac{w_j}{x_j} \, dS(\mathbf{w}) \right\},$$

with

$$\int_{\mathcal{B}} w_j \, dS(\mathbf{w}) = 1, \quad \text{for all } j = 1, \ldots, d. \tag{7.3}$$

[C3] *There exists a nonhomogeneous Poisson process $\sum_k \mathbf{1}_{(t_k, \mathbf{j}_k) \in \cdot}$ on $[0, \infty) \times E$ with intensity measure Λ defined, for $t > 0$ and $B \subset E$, by $\Lambda([0, t] \times B) = t\mu^*(B)$, where for all $A \subset \mathcal{B}$ and $r > 0$,*

$$\mu^* \left\{ \mathbf{y} \in E \; : \; ||\mathbf{y}|| > r \; ; \; \frac{\mathbf{y}}{||\mathbf{y}||} \in A \right\} = \frac{S(A)}{r}, \tag{7.4}$$

and S is a finite measure such that (7.3) *holds and*

$$G(\mathbf{x}) = P\left(\bigvee_{t_k \leq 1} \mathbf{j}_k \leq \mathbf{x} \right) = \exp\left(-\mu^*\{(\mathbf{0}, \mathbf{x}]^c\}\right).$$

Remark 1 Conditions (7.3) secure that the margins are all unit Fréchet distributed. Equation (7.4) shows that the measure μ^* composed with the application T^{-1} defined from $T : E \to (0, \infty] \times \mathcal{B}, \; \mathbf{y} \mapsto (||\mathbf{y}||, \mathbf{y}/||\mathbf{y}||)$, is a product measure of a simple function of the radial component and a measure S of the angular component. More precisely, one has $\mu^* \circ T^{-1}\{(r, \infty) \times A\} = S(A)/r$ for all $A \subset \mathcal{B}$, $r > 0$, so that

$$\mu^*\{(\mathbf{0}, \mathbf{x}]^c\} \;=\; \mu^* \circ T^{-1}(T\{(\mathbf{0}, \mathbf{x}]^c\}) \;=\; \int_{T\{(\mathbf{0},\mathbf{x}]^c\}} \frac{1}{r^2} \, dS(\mathbf{w}) dr. \tag{7.5}$$

Moreover, writing $T((\mathbf{0}, \mathbf{x}]^c) = \{(r, w) \in (0, \infty) \times \mathcal{B} \; : \; rw \in (\mathbf{0}, \mathbf{x}]^c\}$

$$= \{(r, w) \in (0, \infty) \times \mathcal{B} \; : r > \bigwedge_{j=1}^{d} \frac{x_j}{w_j}\},$$

and using this last expression in (7.5) leads to

$$\mu^*\{(\mathbf{0}, \mathbf{x}]^c\} = \int_{\mathcal{B}} \bigvee_{j=1}^{d} \frac{w_j}{x_j} \, dS(\mathbf{w}).$$

The measure S is often called spectral measure, and μ^* is the exponent measure. Finally, without going into the proof, note that the key result leading to Theorem 1 is that the EV distributions coincide with the max-stable distributions. Assuming unit Fréchet margins, these distributions are of the form $G(\mathbf{x}) = \exp(-\mu^*\{(\mathbf{0}, \mathbf{x}]^c\})$, for all

$\mathbf{x} \in E$, where μ^* is a measure satisfying the homogeneity property $t\mu^*(tB) = \mu^*(B)$, for all $t > 0$ and B a Borel set of E.

Example 2 (i) A particular and important case is the case of independence. It corresponds in the representation [C2] to a measure S, which is concentrated on $\{e_i, i = 1, \ldots, d\}$, where $e_i = (0, \ldots, 0, 1, 0, \ldots, 0)$ are the vectors of the canonical basis of $\mathbb{R}^d$ (see, for example, Corollary 5.25 of Resnick, 1987).

(ii) Several parametric families of bivariate and multivariate EV distributions have been proposed by Tawn (1988), Coles and Tawn (1991), Joe (1990), and Tawn (1990), among others. See Kotz and Nadarajah (2000) for a recent review of these existing parametric models. One of the most classical is the so-called logistic model, proposed by Gumbel (1960), and defined for $\mathbf{x} > \mathbf{0}$ by

$$G_\alpha(\mathbf{x}) = \exp\left\{-\left(\sum_{j=1}^{d} x_j^{-\alpha}\right)^{1/\alpha}\right\}, \tag{7.6}$$

for some parameter $\alpha \geq 1$. The limit case $\alpha = 1$ corresponds to the independence between the variables. Different asymmetric generalizations of this family have been proposed; see, for example, the nested logistic model (Coles and Tawn, 1991).

In the bivariate case, the family of EV distributions can be represented in a different way. Indeed, Pickands (1981) has shown that a bivariate d.f. G is an EV d.f. with unit Fréchet margins if and only if

$$G(\mathbf{x}) = \exp\left\{-\left(\frac{1}{x_1} + \frac{1}{x_2}\right) A\left(\frac{x_2}{x_1 + x_2}\right)\right\}, \tag{7.7}$$

where A is a convex function, $A : [0, 1] \to [1/2, 1]$, such that $\max(t, 1-t) \leq A(t) \leq 1$ for all $0 \leq t \leq 1$. See also Sibuya (1960) and Tiago de Oliveira (1975, 1980) for other bivariate representations. The measure S defined in representation [C2] is thus related to the function A by the following

$$A(t) = \int_B \max\{tw_1, (1-t)w_2\}\, dS(\mathbf{w}).$$

Hence, except for the margins, the d.f. G is characterized by a one-dimensional function A, referred to as the dependence function. Particular examples for A are $A(t) = 1$, for all $0 \leq t \leq 1$, which corresponds to the independence for G, or $A(t) = \max(t, 1-t)$ corresponding to total positive dependence. The logistic model defined in (7.6) corresponds to $A_\alpha(t) = \exp\{t^\alpha + (1-t)^\alpha\}^{1/\alpha}$, for $0 \leq t \leq 1$ and $\alpha \geq 1$.

At this stage, it is of practical importance to examine what the different characterizations yield when formulated from a d.f. F belonging to a specific domain of attraction. As before, one considers i.i.d. observations $\mathbf{X}_i = (X_{i,1}, \ldots X_{i,d})$, $i = 1, \ldots, n$, which are assumed to have unit Fréchet margins. In practice, when the margins have unknown distributions, one may, for example, transform the observations $X_{i,j}$, ($i = 1, \ldots, n$, and $j = 1, \ldots, d$) into the pseudo-observations $Z_{i,j} = 1/\log\{n/(R_{i,j} - 1/2)\}$, where

$R_{i,j}$ is the rank of $X_{i,j}$ among $X_{1,j}, \ldots, X_{n,j}$. Such a transformation, suggested in Joe, Smith, and Weissman (1992), ensures the $Z_{i,j}$ to be in the univariate domain of attraction of a unit Fréchet distribution. Refer also to Mason and Huang's work for the estimation of the dependence function using ranks only (see Huang, 1992 or Drees and Huang, 1998). We use the same notations as in Theorem 1.

Theorem 2 *The following statements are equivalent:*

[D1] *The d.f. F of the $\mathbf{X}_i$s $(i = 1, \ldots, n)$ is in the domain of attraction of a multivariate EV distribution G with unit Fréchet margins.*

[D2] $$\lim_{t\to\infty} \frac{-\log F(t\mathbf{x})}{-\log F(t\mathbf{1})} = \lim_{t\to\infty} \frac{1 - F(t\mathbf{x})}{1 - F(t\mathbf{1})} = \frac{-\log G(\mathbf{x})}{-\log G(\mathbf{1})} = \frac{\mu^*([\mathbf{0}, \mathbf{x}]^c)}{\mu^*([\mathbf{0}, \mathbf{1}]^c)},$$
where the measure μ^* is defined in Theorem 1.

[D3] $$\lim_{t\to\infty} t\, P\left\{||\mathbf{X}_i|| > t\ ;\ \frac{\mathbf{X}_i}{||\mathbf{X}_i||} \in A\right\} = S(A),$$ *for each A Borel set of $\mathcal{B}$ and $i = 1, \ldots, n$, where S is defined in Theorem 1.*

[D4] *The point process associated with $\{\mathbf{X}_1/n, \ldots, \mathbf{X}_n/n\}$ converges weakly to a nonhomogeneous Poisson process on E with intensity measure μ^*.*

Remark 2 Note that expression [D3] is also equivalent to

$$[D3'] \qquad \lim_{t\to\infty} P\{\mathbf{X}_i \,/\, ||\mathbf{X}_i|| \in A \mid ||\mathbf{X}_i|| > t\} = S(A)/S(\mathcal{B}).$$

Formulation [D3] clearly suggests a simple nonparametric way to estimate the measure S associated with an EV d.f. G from observations that are in the domain of attraction of G. If a sample $(\mathbf{X}_1, \ldots, \mathbf{X}_n)$ is available from F, then, provided that t is a well chosen function of n, which ensures convergence, a natural candidate to estimate S is deduced from the empirical measure of $(||\mathbf{X}_i||/t, \mathbf{X}_i/||\mathbf{X}_i||)$, that is $1/n \sum_{i=1}^{n} \mathbf{1}\{(||\mathbf{X}_i||/t, \mathbf{X}_i/||\mathbf{X}_i||) \in \cdot\ \}$. More precisely, such a convergence is achieved as soon as $t = n/k_n$, where $(k_n)_n$ is a sequence of integers such that $k_n \to \infty$ and $n/k_n \to \infty$ when $n \to \infty$ (Resnick, 1986, Proposition 5.3). An estimator of S can therefore be obtained using the observations $\mathbf{X}_i$ such that $||\mathbf{X}_i|| > n/k_n$. From a practical point of view, it is usually more convenient to replace the condition $\{||\mathbf{X}|| > n/k_n\}$ by $\{||\mathbf{X}|| > ||\mathbf{X}||_{[k_n]}\}$, where $||\mathbf{X}||_{[k_n]}$ denotes the $(n - k_n + 1)$th order statistic of $||\mathbf{X}||_i$. Both conditions are asymptotically equivalent (see, for example, Appendix 3 of Capéraà and Fougères, 2000); the second one offers the advantage of keeping a fixed number of observations for estimating S. This finally leads to the estimator S_n of S, defined for any Borel set A of $\mathcal{B}$, by:

$$S_n(A) = \frac{1}{k_n} \sum_{i=1}^{n} \mathbf{1}\left\{||\mathbf{X}_i|| > ||\mathbf{X}||_{[k_n]},\ \frac{\mathbf{X}_i}{||\mathbf{X}_i||} \in A\right\}. \tag{7.8}$$

A variety of nonparametric estimation techniques have been developed from the useful representation [D3], referred as multivariate threshold methods. The estimators proposed essentially differ from each other in terms of the choice of the norm and of a function closely related to the mapping T defined in Remark 1. The developments have been mostly formulated in the bivariate case. See, for example, de Haan (1985a), Joe, Smith, and Weissman (1992), de Haan and Resnick (1993), Einmahl, de Haan, and Huang (1993), Einmahl, de Haan, and Sinha (1997), as well as Capéraà and Fougères (2000) for a small-sample study comparing several methods from the previously cited works. Further interesting references are de Haan and de Ronde (1998), de Haan and Sinha (1999), Abdous, Ghoudi, and Khoudraji (1999), and Einmahl, de Haan, and Piterbarg (2001). Note that no theoretical result has been obtained yet concerning an optimal choice for the threshold k_n in the multivariate setup. In practice, choosing k_n is a tricky problem, and usually several values are considered, for which a relative stability of the estimations is expected. Improving this empirical choice, Abdous and Ghoudi (2002) proposed recently an interesting and convenient procedure for an optimal threshold selection, based on a double kernel technique (Devroye, 1989). Abdous and Ghoudi also suggested a unifying approach that includes most of the estimates previously mentioned. Parametric approaches have been proposed by Coles and Tawn (1991, 1994) and Joe, Smith, and Weissman (1992), among others. These approaches make use of the point process representation [D4] and of parametric families of multivariate EV distributions. Finally, some parametric models based on the representation [D2] have also been introduced by Ledford and Tawn (1996) and Smith, Tawn, and Coles (1997). Even if traditionally parametric and nonparametric schools seem to confront each other, they present complementary advantages, and nonparametric estimations can notably be used as a starting point for inference, on which flexible parametric models can be built.

Remark 3 In the particular case where observations from an EV d.f. G are directly available, specific techniques that differ from the threshold methods have been developed in the bivariate case by Pickands (1981), Tawn (1988), Tiago de Oliveira (1989), Smith, Tawn, and Yuen (1990), Deheuvels (1991), Coles and Tawn (1991), Capéraà, Fougères, and Genest (1997), Hall and Tajvidi (2000), among others. In such a situation where no selection is needed above a sufficiently high threshold, estimation techniques are, of course, much more accurate.

Below we summarize for a simple case the practical estimation of the probability of an extreme event via multivariate EV models. Given a sample of random vectors $\mathbf{X}_i, i = 1, \ldots, n$, with d.f. F, consider the problem of estimating the probability $P(\mathbf{X} \in A)$, where A is an exceptional set in which no data have been observed. We assume that F is in the domain of attraction of a multivariate EV d.f. G, and again, for simplicity we deal with known margins and assume that they are unit Fréchet distributed. If A is the complement of a rectangle, $A = (\mathbf{0}, n\mathbf{u}]^c$, one may write, for example, using [D2]:

$$P(\mathbf{X} \in A) = 1 - F(n\mathbf{u}) \approx -\frac{1}{n} \log G(\mathbf{u}) = \frac{1}{n}\mu^*\{(\mathbf{0}, \mathbf{u}]^c\} = \frac{1}{n} \int_{\mathcal{B}} \bigvee_{j=1}^{d} \frac{w_j}{u_j} dS(\mathbf{w}).$$

Making use of the empirical measure S_n defined in (7.8), an estimator of $P(\mathbf{X} \in A)$ is then given by

$$\frac{1}{nk_n} \sum_{i=1}^{n} \left(\bigvee_{j=1}^{d} \frac{X_{i,j}}{u_j ||X_{i,j}||} \right) \mathbb{1}\{||\mathbf{X}_i|| > ||\mathbf{X}||_{[k_n]}\}.$$

This is one possible nonparametric way to make use of the multivariate EV model. The choice of the proportion of data used for the estimation of S is a delicate point in practice. Dealing with any form of extreme event A is of course not so straightforward, and needs care. We refer, for example, to de Haan and de Ronde (1998), or Bruun and Tawn (1998), for a complete application and evaluation of failure probabilities. Note that, even if from a theoretical point of view the methods based on multivariate EV models that have been developed in the literature are available in a d-dimensional context, the complexity linked to the solution of problems in practice increases rapidly with the dimension d. Most of the work done in the multivariate context concerns examples where $d = 2$ or 3.

An important point is that both parametric and nonparametric threshold estimation techniques present some problems in the particular case where the data are asymptotically independent, i.e., when their distribution is in the domain of attraction of the independence. This limit situation corresponds in the representation [C2] to the case where S is singular, concentrated on some boundary points of $\mathcal{B}$ (see Example 2). Hence parametric methods, as maximum likelihood estimation, face a problem of non-regularity, and nonparametric methods also present less satisfying results, as shown via simulations by Capéraà and Fougères (2000) in the bivariate case. Moreover, EV models come from componentwise maxima $(\max_{i=1,\ldots,n} X_{i,1}, \ldots, \max_{i=1,\ldots,n} X_{i,d})$, which in practice typically do not correspond to any observation $\mathbf{X}_k$. In case of asymptotic dependence, the componentwise maxima, however, tend to occur jointly, so EV models are useful in such a situation. This is not the case anymore in case of asymptotic independence. Asymptotic independence seems, however, to be an important case in practice, as pointed out by Marshall and Olkin (1983) or de Haan and de Ronde (1998). Indeed, it actually corresponds to most of the classical families of distributions, as listed in Marshall and Olkin (1983) or Capéraà, Fougères, and Genest (2000), among others. For example, see the wind and rain data considered by Anderson and Nadarajah (1993) and Ledford and Tawn (1996). Some alternatives and refined models have been proposed in this particular case by Ledford and Tawn (1996, 1997), and will be presented in the following section.

7.3 An alternative modeling approach

According to Sibuya (1960), a bivariate pair of r.v. (X_1, X_2) with common marginal d.f. F_1 is said to have asymptotically independent components if and only if

$$\lim_{u \to x_{F_1}} P(X_1 > u \mid X_2 > u) = 0,$$

where $x_{F_1} = \sup\{x \in \mathbb{R} : F_1(x) < 1\}$. For the distribution of (X_1, X_2) this property is equivalent to being in the domain of attraction of the independence. This follows from the next result, which also states how multivariate asymptotic independence in general actually reduces to the bivariate case (Berman, 1961; see, for example, Resnick, 1987, Proposition 5.27).

Theorem 3 *Let $\{\mathbf{X}_n, n \geq 1\}$ be a sequence of i.i.d. random vectors in $\mathbb{R}^d$ $(d \geq 2)$ with d.f. F. Assume for simplicity that all the univariate margins are the same, with* *common d.f. F_1 belonging to the univariate domain of attraction of G_1. So one has, for some $a_n > 0$, $b_n \in \mathbb{R}$, the convergence $F_1^n(a_n x + b_n) \to G_1(x)$, as $n \to \infty$. The following assertions are then equivalent:*

(i) *The d.f. F is in the domain of attraction of the independence:*

$$F^n(a_n\mathbf{x} + b_n\mathbf{1}) = P\left(\bigvee_{i=1}^{n} \mathbf{X}_i \le a_n\mathbf{x} + b_n\mathbf{1}\right) \to \prod_{j=1}^{d} G_1(x_j).$$

(ii) *For all $1 \le k < \ell \le d$,*

$$P\left(\bigvee_{i=1}^{n} X_{i,k} \le a_n x_k + b_n \ , \ \bigvee_{i=1}^{n} X_{i,\ell} \le a_n x_\ell + b_n\right) \to G_1(x_k)G_1(x_\ell).$$

(iii) *For all $1 \le k < \ell \le d$, and x_k, x_ℓ such that $G_1(x_k), G_1(x_\ell) > 0$,*

$$\lim_{n\to\infty} n\, P(X_{1,k} > a_n x_k + b_n \ , \ X_{1,\ell} > a_n x_\ell + b_n) = 0.$$

(iv) *For all $1 \le k < \ell \le d$,*

$$\lim_{t\to x_{F_1}} P(\, X_{1,k} > t \mid X_{1,\ell} > t\,) = 0.$$

For simplicity consider the bivariate case of asymptotic independence. Note that in this case, the probability mass of joint tails, that is, of sets of the form $\{(X_1 - b_{n,1})/a_{n,1} > x_1,\ (X_2 - b_{n,2})/a_{n,2} > x_2\}$, is of lower order than that for sets like $\{(X_1 - b_{n,1})/a_{n,1} > x_1$ **or** $(X_2 - b_{n,2})/a_{n,2} > x_2\}$. Therefore, models based on bivariate extreme value distributions do not provide any satisfying way to estimate such joint tails.

In order to fill this gap in the bivariate setup, Ledford and Tawn (1996, 1997) proposed joint tail models adapted to the asymptotic independence case. When stated with unit Fréchet margins, these models are essentially based on the model

$$P(Z_1 > z_1,\ Z_2 > z_2) \sim \frac{\mathcal{L}(z_1, z_2)}{z_1^{c_1} z_2^{c_2}}, \tag{7.9}$$

when $z_1, z_2 \to \infty$, where $c_1, c_2 > 0$ are such that $c_1 + c_2 \ge 1$, and $\mathcal{L}$ is a bivariate slowly varying function. Recall that a measurable function $\mathcal{L} : \mathbf{x} \in \mathbb{R}^d \mapsto \mathcal{L}(\mathbf{x}) > 0$ is called a multivariate slowly varying function if there exists a positive function λ satisfying $\lambda(t\mathbf{x}) = \lambda(\mathbf{x})$ for all $\mathbf{x} \in \mathbb{R}^d$, $t > 0$, and such that

$$\lim_{t\to\infty} \frac{\mathcal{L}(t\mathbf{x})}{\mathcal{L}(t\mathbf{1})} = \lambda(\mathbf{x}),$$

for all $\mathbf{x} \in \mathbb{R}^d$. See, for example, de Haan (1985b), Basrak, Davis, and Mikosch (2000) or Mikosch (2001) for further properties. A measure of extremal dependence is then provided by the so-called coefficient of tail dependence $\eta = 1/(c_1 + c_2) \in (0, 1]$. Asymptotic dependence corresponds to $\eta = 1$ and $\mathcal{L}(r, r) \to 0$ as $r \to \infty$, whereas $\eta < 1$ implies asymptotic independence. Since because of (7.9) one has

$$P(T > u + t \mid T > u) \sim \frac{\mathcal{L}(u + t, u + t)\,(u + t)^{-1/\eta}}{\mathcal{L}(u, u)\,u^{-1/\eta}} \approx \left(1 + \frac{t}{u}\right)^{-1/\eta},$$

Ledford and Tawn (1996) suggest to estimate η as the shape parameter of the GPD for $T = \min(Z_1, Z_2)$. Peng (1999) proposed another estimator of η, for which he obtained the asymptotic normality.

Note that (7.9) is ensured as soon as a second order condition is assumed for (Z_1, Z_2), which strengthens the multivariate domain of attraction condition (7.2). More precisely, one assumes the existence of a positive function ψ and a finite and nonzero function h such that

$$\lim_{t\to\infty} \frac{1}{\psi(t)} \{t\, P(Z_1 > tz_1 \text{ or } Z_2 > tz_2) + \log G(z_1, z_2)\} = h(z_1, z_2).$$

As mentioned by de Haan and de Ronde (1998), the function ψ is then necessarily a regularly varying function with nonpositive exponent ρ (that is, $\psi(tu)/\psi(t) \to u^\rho$ as $t \to \infty$, for all $u > 0$). The relation between ρ and η is $\rho = 1 - 1/\eta$.

Ledford and Tawn (1997) and Bruun and Tawn (1998) implemented different submodels from (7.9), specifying further conditions concerning the form of $\mathcal{L}$. For example, Bruun and Tawn modelled the joint tails assuming that

$$P(Z_1 > z_1,\ Z_2 > z_2) = \frac{a_0 + \mathcal{L}_*(z_1, z_2, V)}{(z_1 z_2)^{1/2\eta}}, \tag{7.10}$$

for z_1, z_2 large enough, in terms of $\eta \in (0, 1]$, and $a_0 \in \mathbb{R}^+$. The function $\mathcal{L}_*$ is defined by

$$\mathcal{L}_*(z_1, z_2, V) = \left(\frac{z_1}{z_2}\right)^{1/2} + \left(\frac{z_2}{z_1}\right)^{1/2} - (z_1 z_2)^{1/2} V(z_1, z_2),$$

where V is an exponent measure as in Theorem 1, $V(z_1, z_2) = \mu^*\{(\mathbf{0}, \mathbf{z}]^c\}$. Bruun and Tawn used more specifically a parametric model for V. They compared, through Monte-Carlo simulations and for different degrees of dependence, the behavior of three models, namely: the model (7.10), the bivariate extreme value model with d.f. $F(z_1, z_2) = \exp\{-V(z_1, z_2)\}$, and a model based on univariate extreme value models for the margins. In case of asymptotic independence, the model (7.10) really outperforms the bivariate extreme value model in terms of relative percentage error when estimating a specific 10^{-3} quantile (see Bruun and Tawn (1998), Section 4, for details). The bivariate extreme value model has actually the tendency to overestimate the probability of failure when asymptotic independence occurs.

7.4 Measuring extremal dependence

One of the main topics strongly connected with multivariate extreme events modelling is the problem of how to measure the dependence in the extreme observations. A first way to look at this problem is to use multivariate EV models and to measure the strength of the dependence in the limiting distribution. Representation [C2] clearly exhibits that all the dependence is contained in the measure S, but summaries of this information can be useful. Note in passing that the correlation coefficient is useless in the EV context, as on the one hand it is not invariant under transformations of the margins, and on the other hand several cases of univariate EV d.f. may not have a finite variance, as the Fréchet d.f., for example. Focusing on the EV model, several measures have been proposed, mainly in the bivariate case. Consider a pair (X_1, X_2) from an EV distribution, with common marginal d.f. F. Tiago de Oliveira (see de Haan, 1985a) proposed as an index of extremal dependence $\theta \in [1, 2]$ defined by

$$P(\max(X_1, X_2) \leq x) = F^{\theta}(x). \tag{7.11}$$

Using Pickands' representation (7.7) gives another expression of θ, that is $\theta = 2A(1/2)$. Such a measure has been considered by Tawn (1988) for testing independence in the bivariate EV context (see also Capéraà, Fougères, and Genest, 1997).

Alternative measures for bivariate EV distributions are Kendall's τ and Spearman's ρ. They are nonparametric measures of dependence, which have the following closed forms in terms of A (see, e.g., Ghoudi, Khoudraji, and Rivest, 1998):

$$\tau = \int_0^1 \frac{t(1-t)}{A(t)}\, dA'(t), \qquad \rho = 12 \int_0^1 \{A(t)+1\}^{-2}\, dt - 3.$$

The definition (7.11) can clearly be extended by considering $\theta \in [1, d]$ such that

$$P(\max(X_1, \ldots, X_d) \leq x) = F^{\theta}(x).$$

In a more general context than EV distributions, Buishand (1984) considered, for all r.v. X_1, X_2 with common d.f. F and joint d.f. H, the function θ_H defined by

$$P(\max(X_1, X_2) \leq x) = F^{\theta_H(x)}(x). \tag{7.12}$$

In the particular case of the EV distribution, such a function is constant, $\theta_H(x) = \theta$.

Another way to introduce the measure θ can be deduced from Joe (1993). Indeed, in the bivariate case he defines the upper tail dependence parameter λ for a d.f. H with univariate marginal d.f. F as

$$\lambda = \lim_{x \to x_F} \frac{\bar{H}(x, x)}{1 - F(x)} = \lim_{x \to x_F} P(X_1 > x | X_2 > x), \tag{7.13}$$

where (X_1, X_2) is a random pair from H, and $\bar{H}$ denotes the survival function defined by $\bar{H}(x_1, x_2) = P(X_1 > x_1, X_2 > x_2)$. Asymptotic independence is therefore equivalent to $\lambda = 0$. The tail dependence parameter of a distribution is linked to the domain of attraction of this distribution. More precisely, if H is a d.f. belonging to the domain of attraction of an EV d.f. L, and if λ is the tail dependence parameter of H, then λ is also the tail dependence parameter of L (see Joe, 1997, Theorem 6.8). Moreover, as

$$\frac{\bar{H}(x, x)}{1 - F(x)} \sim 2 - \frac{\log H(x, x)}{\log F(x)}$$

in the neighborhood of x_F, it follows from (7.12) and (7.13) that

$$\lambda = 2 - \lim_{x \to x_F} \frac{\log H(x, x)}{\log F(x)} = 2 - \lim_{x \to x_F} \theta_H(x).$$

Consider a bivariate d.f. H, with upper tail dependence parameter λ. Assume that H belongs to the domain of attraction of an EV d.f. L, with univariate margins G, and with index θ in (7.11). Then Joe's result above yields for θ_L constant:

$$\lambda = 2 - \lim_{x \to x_F} \theta_L(x) = 2 - \theta.$$

One should note that the upper tail dependence parameter just depends on the dependence structure, so it can be expressed in terms of the copula, as did Joe (1993). Recall that a copula is a multivariate d.f. with all its univariate margins uniformly distributed on [0, 1] (Sklar, 1959). Given a multivariate d.f. H with continuous univariate margins $F_1, \ldots, F_d$, there exists a unique function C_H associated to H such that

$$H(\mathbf{x}) = C_H(F_1(x_1), \ldots, F_d(x_d)).$$

This function C_H, called the copula of H, does not depend on the margins of H, and contains therefore all the information relative to the dependence between the different components $X_1, \ldots, X_d$. It is the d.f. of the vector $\mathbf{U} = (U_1, \ldots, U_d) = (F_1(X_1), \ldots, F_d(X_d))$. See, for example, Kimeldorf and Sampson (1975), Schweizer and Sklar (1983), or Nelsen (1998), for further details. The upper tail dependence parameter defined by (7.13) is equivalently defined by:

$$\lambda = \lim_{u \to 1} \frac{\bar{C}_H(u, u)}{1 - u} = 2 - \lim_{u \to 1} \frac{\log C_H(u, u)}{\log u}. \tag{7.14}$$

In case of asymptotic independence ($\lambda = 0$), Ledford and Tawn (1996) proposed to measure the tail dependence via the coefficient $\eta = 1/(c_1 + c_2)$ defined from (7.9). Coles, Heffernan, and Tawn (2000) make use of both measures λ and η (denoted χ and $(1 + \bar{\chi})/2$, respectively) for diagnostics of asymptotic independence on different sets of data.

7.5 Conclusion

The aim of this chapter was to present a review of the results obtained in the area of multivariate extreme events. The models developed in the literature are essentially based on the limiting behavior of renormalized componentwise maxima. The structure of the family of limiting distributions is actually quite rich, and can be studied in terms of max-stable distributions, as well as via point process representations. Statistical inference methods deduced from this family of EV distributions have also been reviewed. The limits of the multivariate EV models have been pointed out, especially in the case of asymptotic independence. Alternative models have been summarized. Finally, some measures of extremal dependence have been presented. One should note that this domain of research is currently very active, and promising alternatives have recently been proposed, for example, Heffernan and Tawn (2001).

References

ABDOUS, B. and GHOUDI, K. (2002). Nonparametric estimates of multivariate extreme dependence functions. Personal communication.

ABDOUS, B., GHOUDI, K., and KHOUDRAJI, A. (1999). Nonparametric estimation of the limit dependence function of multivariate extremes. *Extremes*, 2, 243–265.

ANDERSON, C. W. and NADARAJAH, S. (1993). Environmental factors affecting reservoir safety. In *Statistics for the Environment.* Barnett, V. and Turkman, K. F., Editors, 163–182, New York: Wiley.

BALKEMA, A. A. and DE HAAN, L. (1974). Residual life time at great age. *Ann. Probab.* 2, 792–804.

BASRAK, B., DAVIS, R. A., and MIKOSCH, T. (2000). A characterization of multivariate regular variation. *Annals of Applied Probability* 12, 908–920.

BERMAN, S. M. (1961). Convergence to bivariate limiting extreme value distributions. *Ann. Inst. Statist. Math.* 13, 217–223.

BINGHAM, N. H., GOLDIE, C. M., and TEUGELS, J. L. (1989). *Regular variation.* Cambridge University Press.

BRUUN, J. T. and TAWN, J. A. (1998). Comparison of approaches for estimating the probability of coastal flooding. *Appl. Statist.* 47, 405–423.

BUISHAND, T. A. (1984). Bivariate extreme-value data and the station-year method. *J. Hydrol.* 69, 77–95.

CAPÉRAÀ, P. and FOUGÈRES, A.-L. (2000). Estimation of a bivariate extreme value distribution. *Extremes* 3, 311–329.

CAPÉRAÀ, P., FOUGÈRES, A.-L., and GENEST, C. (1997). A nonparametric estimation procedure for bivariate extreme value copulas. *Biometrika* 84, 567–577.

CAPÉRAÀ, P., FOUGÈRES, A.-L., and GENEST, C. (2000). Bivariate distributions with given extreme value attractor. *J. Multivariate Anal.* 72, 30–49.

COLES, S. G. (2001). *An Introduction to Statistical Modeling of Extreme Values.* London: Springer.

COLES, S. G., HEFFERNAN, J. E., and TAWN, J. A. (2000). Dependence measures for extreme value analyses. *Extremes* 2, 339–365.

COLES, S. G. and TAWN, J. A. (1991). Modelling multivariate extreme events. *J. R. Statist. Soc.* B 53, 377–392.

COLES, S. G. and TAWN, J. A. (1994). Statistical methods for multivariate extremes: an application to structural design (with discussion). *Appl. Statist.* 43, 1–48.

DAVISON, A. C. and SMITH, R. L. (1990). Models for exceedances over high thresholds. *J. R. Statist. Soc.* B 52, 393–442.

DEHEUVELS, P. (1978). Caractérisation complète des lois extrêmes multivariées et de la convergence aux types extrêmes. *Publ. Inst. Statist. Univ. Paris* 23, 1–36.

DEHEUVELS, P. (1991). On the limiting behavior of the Pickands estimator for bivariate extreme value distributions. *Statist. Probab. Lett.* 12, 429–439.

DEVROYE, L. (1989). The double kernel method in density estimation. *Annales de l'Institut Poincaré* 25, 533–580.

DREES, H. and HUANG, X. (1998). Best attainable rates of convergence for estimators of the stable dependence function. *J. Multivariate Anal.* 64, 25–47.

EINMAHL, J., DE HAAN, L., and HUANG, X. (1993). Estimating a multidimensional extreme value distribution. *J. Multivariate Anal.* 47, 35–47.

EINMAHL, J., DE HAAN, L., and SINHA, A. K. (1997). Estimating the spectral measure of an extreme value distribution. *Stoch. Proc. Appl.* 70, 143–171.

EINMAHL, J., DE HAAN, L., and PITERBARG, V. (2001). Nonparametric estimation of the spectral measure of an extreme value distribution. *Ann. Statist.*, 29, 1401–1423.

EMBRECHTS, P., KLÜPPELBERG, C., and MIKOSCH, T. (1997). *Modelling Extremal Events for Insurance and Finance*. Berlin: Springer.

FISHER, R.A. and TIPPETT, L. H. C. (1928). Limiting forms of the frequency distributions of largest or smallest member of a sample. *Proc. Cambridge Philos. Soc.* 24, 180–190.

GALAMBOS, J. (1987). *The Asymptotic Theory of Extreme Order Statistics*, 2nd edition. Malabar, FL: Kreiger.

GENEST, C. and MACKAY, R. J. (1986a). Copules archimédiennes et familles de lois bidimensionnelles dont les marges sont données, *Canad. J. Statist.* 14, 145–159.

GENEST, C. and MACKAY, R. J. (1986b). The joy of copulas: Bivariate distributions with uniform marginals, *Amer. Statist.* 40, 280–283.

GENEST, C. and RIVEST, L.-P. (1989). A characterization of Gumbel's family of extreme value distributions, *Statist. Probab. Lett.* 8, 207–211.

GHOUDI, K., KHOUDRAJI, A., and RIVEST, L.-P. (1998). Propriétés statistiques des copules de valeurs extrêmes bidimensionnelles. *Canad. J. Statist.* 26, 187–197.

GNEDENKO, B.V. (1943). Sur la distribution limite du terme maximum d'une série aléatoire. *Ann. Math.* 44, 423–453.

GUMBEL, E. J. (1960). Distributions des valeurs extrêmes en plusieurs dimensions. *Publ. Inst. Statist. Univ. Paris* 9, 171–173.

DE HAAN, L. (1985a). Extremes in higher dimensions: The model and some statistics. In *Proc. 45th Session Inter. Statist. Inst.*, Amsterdam, paper 26.3. Amsterdam: International Statistical Institute.

DE HAAN, L. (1985b). Multivariate regular variation and applications in probability theory. In *Multivariate analysis, VI*, P. R. Krishnaiah ed., Amsterdam: North-Holland, 281–288.

DE HAAN, L. and RESNICK, S. I. (1977). Limit theory for multidimensional sample extremes. *Z. Wahr. verw. Geb.* 40, 317–337.

DE HAAN, L. and RESNICK, S. I. (1993). Estimating the limit distribution of multivariate extremes. *Comm. Statist. Stoch. Models* 9, 275–309.

DE HAAN, L. and DE RONDE, J. (1998). Sea and wind: Multivariate extremes at work. *Extremes* 1, 7–45.

DE HAAN, L. and SINHA, A. K. (1999). Estimating the probability of a rare event. *Ann. Statist.* 27, 732–759.

HALL, P. and TAJVIDI, N. (2000). Distribution and dependence-function estimation for bivariate extreme-value distributions. *Bernoulli* 6(5), 835-844.

HEFFERNAN, J. E. and TAWN, J. A. (2001). A conditional approach for multivariate extreme values. *Submitted.*

HUANG, X. (1992). *Statistics of bivariate extreme values*. Thesis, Erasmus University Rotterdam, Tinbergen Institute Research Series 22.

HÜSLER, J. (1986). Extreme values of nonstationary random sequences. *J. Appl. Probab.* 23 (4), 937–950.

JENKINSON, A. F. (1955). The frequency distribution of the annual maximum (or minimum) values of meteorological elements. *Q. J. Roy. Meteorol. Soc.* 87, 158–171.

JOE, H. (1990). Families of min-stable multivariate exponential and multivariate extreme value distributions. *Statist. Probab. Lett.* 9, 75–81.

JOE, H. (1993). Parametric family of multivariate distributions with given margins. *J. Multivariate Anal.* 46, 262–282.

JOE, H. (1997). *Multivariate Models and Dependence Concepts*. London: Chapman and Hall.

JOE, H., SMITH, R. L., and WEISSMAN, I. (1992). Bivariate threshold models for extremes. *J. R. Statist. Soc.* B 54, 171–183.

KIMELDORF, G. and SAMPSON, A. R. (1975). Uniform representations of bivariate distributions. *Comm. Statist.* 4, 617–627.

KOTZ, S. and NADARAJAH, S. (2000). *Extreme value distributions. Theory and applications*. Imperial College Press.

LEADBETTER, M. R., LINDGREN, G. and ROOTZÉN, H. (1983). *Extremes and Related Properties of Random Sequences and Series*, New York: Springer-Verlag.

LEDFORD, A. W. and TAWN, J. A. (1996). Statistics for near independence in multivariate extreme values. *Biometrika* 83, 169–187.

LEDFORD, A. W. and TAWN, J. A. (1997). Modelling dependence within joint tail regions. *J. R. Statist. Soc.* B 59, 475–499.

MARSHALL, A. W. and OLKIN, I. (1983). Domains of attraction of multivariate extreme value distributions. *Ann. Probab.* 11, 168–177.

MIKOSCH, T. (2001). Modeling dependence and tails of financial time series. In: *Extreme Values in Finance, Telecommunications, and the Environment*, Finkenstädt, B. and Rootzén, H., Eds., CRC/Chapman & Hall, Boca Raton, Ch. 5.

MISES, R., VON (1954). La distribution de la plus grande de n valeurs. *Selected papers*, Vol. II, p. 271–294. Providence, R. I.: Amer. Math. Soc.

NELSEN, R. B. (1998). *An introduction to copulas*. New York: Springer-Verlag.

PENG, L. (1999). Estimation of the coefficient of tail dependence in bivariate extremes. *Statist. Probab. Lett.* 43, 399–409.

PICKANDS, J. (1971). The two-dimensional Poisson process and extremal processes. *J. Appl. Probab.* 8, 745–756.

PICKANDS, J. (1975). Statistical inference using extreme order statistics. *Ann. Statist.* 3, 119–131.

PICKANDS, J. (1981). Multivariate extreme value distributions. *Bull. Int. Statist. Inst.* 859–878.

RESNICK, S. I. (1986). Point processes, regular variation, and weak convergence. *Adv. Appl. Probab.* 18, 66–138.

RESNICK, S. I. (1987). *Extreme Values, Regular Variation, and Point Processes*. New York: Springer.

SCHWEIZER, B. and SKLAR, A. (1983). *Probabilistic Metric Spaces*. New York: North-Holland.

SIBUYA, M. (1960). Bivariate extreme statistics. *Ann. Inst. Math. Statist.* 11, 195–210.

SKLAR, A. (1959). Fonctions de répartition à n dimensions et leurs marges, *Publ. Inst. Statist. Univ. Paris* 8, 229–231.

SMITH, R. L. (1989). Extreme value analysis of environmental time series: an application to trend detection in ground-level ozone, *Statist. Sc.* 4, 367–393.

SMITH, R. L., TAWN, J. A., and YUEN, H. K. (1990). Statistics of multivariate extremes. *Inter. Statist. Rev.* 58, 47–58.

SMITH, R. L., TAWN, J. A., and COLES, S. G. (1997). Markov chain models for threshold exceedances. *Biometrika* 84, 249–268.

TAWN, J. A. (1988). Bivariate extreme value theory: models and estimation. *Biometrika* 75, 397–415.

TAWN, J. A. (1990). Modelling multivariate extreme value distributions. *Biometrika* 77, 245–253.

TAWN, J. A. (1992). Estimating probabilities of extreme sea-levels. *Appl. Statist.* 41, 77–93.

TIAGO DE OLIVEIRA, J. (1958). Extremal distributions. *Rev. Fac. Sci. Lisboa, Ser. A* 7, 215–227.

TIAGO DE OLIVEIRA, J. (1975). Bivariate extremes; extensions. In *Proceedings of the 40th Session of the International Statistical Institute. Bull. Inst. Internat. Statist.* 46, no. 2, 241–254.

TIAGO DE OLIVEIRA, J. (1980). Bivariate extremes: foundations and statistics. In *Multivariate analysis, V, Proceedings of the 5th International Symposium, Univ. Pittsburgh*. Amsterdam-New York: North-Holland, 349–366.

TIAGO DE OLIVEIRA, J. (1989). Intrinsic estimation of the dependence structure for bivariate extremes. *Statist. Prob. Letters* 8, 213–218.

Index

C

D

E

N